Lecture Notes in Computer Science 4736

Commenced Publication in 1973
Founding and Former Series Editors:
Gerhard Goos, Juris Hartmanis, and Jan van Leeuwen

Stephan Winter   Matt Duckham   Lars Kulik
Ben Kuipers (Eds.)

Spatial
Information Theory

8th International Conference, COSIT 2007
Melbourne, Australia, September 19-23, 2007
Proceedings

 Springer

Volume Editors

Stephan Winter
The University of Melbourne
Australia
E-mail: winter@unimelb.edu.au

Matt Duckham
The University of Melbourne
Australia
E-mail: mduckham@unimelb.edu.au

Lars Kulik
The University of Melbourne
Australia
E-mail: lars@csse.unimelb.edu.au

Ben Kuipers
The University of Texas
USA
E-mail: kuipers@cs.utexas.edu

Library of Congress Control Number:2007933826

CR Subject Classification (1998): E.1, I.2, F.1, H.2.8, H.1, J.2

LNCS Sublibrary: SL 1 – Theoretical Computer Science and General Issues

ISSN   0302-9743
ISBN-10   3-540-74786-9 Springer Berlin Heidelberg New York
ISBN-13   978-3-540-74786-4 Springer Berlin Heidelberg New York

Springer is a part of Springer Science+Business Media

springer.com

© Springer-Verlag Berlin Heidelberg 2007

Typesetting: Camera-ready by author, data conversion by Scientific Publishing Services, Chennai, India
Printed on acid-free paper   SPIN: 12119318   06/3180   5 4 3 2 1 0

Preface

Spatial information theory explores the foundations of space and time. It searches to model perceptions and cognition of space and time, their efficient representation, reasoning about these representations, and the communication of knowledge about space and time. The Conference on Spatial Information Theory, COSIT, focuses especially on the geographic scale, the scale beyond immediate vista. Even within the limits to geographical scale, spatial information theory covers interests of multiple disciplines, such as philosophy, cognitive psychology, linguistics, mathematics, artificial intelligence, and geography. This list is not exhaustive, but it shows the need and desire to talk across disciplinary boundaries. COSIT is the place for this exchange and search for common foundations. To foster the encounter, COSIT is a single-track conference with a limited number of papers, and is traditionally held at secluded locations.

For COSIT 2007, a record number of 102 submissions were received and carefully reviewed by the international Program Committee and many additional reviewers. In the end, 27 papers were selected for presentation, which corresponds to an acceptance rate of 26%. To preserve the single-track format of COSIT, the number of accepted papers was kept within the range of previous COSITs. Consequently, COSIT 2007 was the most competitive COSIT yet, with a lower acceptance rate than any previous COSIT. At COSIT 2007, 25 peer-reviewed posters were also presented, with three keynotes given by Deborah McGuinness (Knowledge Systems, Stanford University, USA), Mandyam Srinivasan (Queensland Brain Institute, Australia) and Kim Marriott (Constraint Solving and Optimization, Monash University, Australia).

The day before the conference, participants had the difficult choice between four international workshops: the Workshop on Spatial Cognition in Architectural Design chaired by Thomas Barkowsky (Germany), Zafer Bilda (Australia), Christoph Hölscher (Germany) and Georg Vrachliotis (Switzerland); the Workshop on Semantic Similarity Measurement and Geospatial Applications chaired by Krzysztof Janowicz (Germany), Angela Schwering (Germany), Martin Raubal (USA) and Werner Kuhn (Germany); the Joint Workshop on Distributed Geoinformatics and Sensing, Ubiquity, and Mobility organized by Patrick Laube (Australia) and co-chaired by Matt Duckham (Australia), Arie Croitoru (Canada), Peggy Agouris (USA), Lars Kulik and Egemen Tanin (both Australia); and the Workshop on Social Space and Geographic Space chaired by Stephan Winter and Garry Robins (both Australia). All four workshops were full-paper peer-reviewed, and the proceedings were published separately.

The last conference day was reserved for the doctoral colloquium, a highlight of every COSIT. Doctoral students had the opportunity to present their research in a supportive environment to an international audience of researchers, industry and fellow students. Participants received valuable feedback on issues such as how to identify and refine research questions, how to present research orally, and how to publish research, as well as insights into how to complete a PhD successfully in an interdisciplinary field.

COSIT 2007 was held at the Melbourne Business School in Mount Eliza, Victoria: a quiet place on top of the cliffs of Port Phillip Bay, and some distance from buzzing Melbourne. This location was chosen in the spirit of previous conference locations. COSITs were held in Pisa (Italy), Elba (Italy), Semmering (Austria), Laurel Highlands (Pennsylvania, USA), Stade (Germany), Morro Bay (California, UDA), Ittingen (Switzerland) and Ellicottville (New York, USA). This list shows that COSIT 2007 was already the eighth conference in this series (Pisa counts traditionally as "COSIT 0").

It also shows that COSIT 2007 was the first COSIT taking place in the Asia-Pacific region. The declared goal of this decision was to create better opportunities for researchers from Asia and Oceania to join the ongoing scientific discussion, and to promote in this region spatial information theory as one of the central foundations of an information society, considering that almost all decisions have a spatial component. This goal came out evenly: five papers and 11 posters had authors or co-authors from this region, not to mention the doctoral colloquium. But a place that is easier to access for some, is more difficult to access for others. The majority of participants had made a special effort to come to COSIT 2007, and this is recognized and appreciated by the organizers.

Any conference needs many contributors to make it a success, and COSIT 2007 was no exception. In scientific terms, success can only be linked to the encounter with new ideas, the discussions with new and old colleagues, and the feedback, encouragement, inspiration or plans for collaborations taken home from Mount Eliza. In this respect, all thanks go to the participants.

In financial terms, success was largely supported by our generous sponsors. ESRI was the COSIT 2007 primary sponsor. Ordnance Survey and Multimedia Victoria were COSIT 2007 key sponsors. Shell was the COSIT 2007 poster session sponsor. To support the participation of doctoral students, special thanks go to Google Inc. as the COSIT 2007 doctoral colloquium sponsor and PSMA Australia Ltd. as COSIT 2007 student scholarship sponsor. The Cooperative Research Center for Spatial Information (CRCSI) supported some of their doctoral students to participate. The Spatial Sciences Institute credited participating Australian professionals with points in its continuing professional development program. The Department of Computer Science and Software Engineering and the Department of Geomatics of the University of Melbourne helped with pre-financing the deposits. All this support made COSIT 2007 possible and accessible, and we would like to express our thanks to all our sponsors.

In terms of organization, the Chairs were actively supported by the COSIT Steering Committee. Thanks also go to the COSIT Program Committee and the additional reviewers, who had to deal with the largest number of submissions of any COSIT so far. Tessa Fitzpatrick from Melbourne Conference Management took the burden of managing the conference registrations, and Shauna Houston from the Melbourne Business School in Mount Eliza was the local organizer. Students from the University of Melbourne gave a helping hand. And last, but not least, thanks to Alexander Klippel from Penn State University, who organized the COSIT 2007 doctoral colloquium.

September 2007

Stephan Winter
Benjamin Kuipers
Matt Duckham
Lars Kulik

Organization

General Chairs

Stephan Winter, The University of Melbourne, Australia
Benjamin Kuipers, University of Texas at Austin, USA

Program Chairs

Matt Duckham, The University of Melbourne, Australia
Lars Kulik, The University of Melbourne, Australia

Steering Committee

Anthony Cohn, University of Leeds, UK
Michel Denis, LIMSI-CNRS, Paris, France
Max Egenhofer, University of Maine, Orono, USA
Andrew Frank, Technical University Vienna, Austria
Christian Freksa, University of Bremen, Germany
Stephen Hirtle, University of Pittsburgh, USA
Werner Kuhn, University of Münster, Germany
Benjamin Kuipers, University of Texas, Austin, USA
David Mark, State University New York, Buffalo, USA
Dan Montello, University of California, Santa Barbara, USA
Barry Smith, State University New York, Buffalo, USA
Sabine Timpf, University of Würzburg, Germany
Barbara Tversky, Stanford University, USA
Michael Worboys , University of Maine, Orono, USA

Program Committee

Pragya Agarwal, UK
Thomas Barkowsky, Germany
John Bateman, Germany
Brandon Bennett, UK
Michela Bertolotto, Ireland
Thomas Bittner, USA
Mark Blades, UK
Gilberto Camara, Brazil
Roberto Casati, France
Christophe Claramunt, France

Eliseo Clementini, Italy
Anthony G. Cohn, UK
Helen Couclelis, USA
Matteo Cristani, Italy
Leila de Floriani, Italy
Michel Denis, France
Maureen Donnelly, USA
Geoffrey Edwards, Canada
Max Egenhofer, USA
Carola Eschenbach, Germany

Sara Fabrikant, Switzerland
Andrew Frank, Austria
Christian Freksa, Germany
Mark Gahegan, USA
Antony Galton, UK
Christopher Gold, UK
Reg Golledge, USA
Mary Hegarty, USA
Stephen Hirtle, USA
Hartwig H. Hochmair, USA
Kathleen Hornsby, USA
Christopher B. Jones, UK
Marinos Kavouras, Greece
Alexander Klippel, USA
Barry J. Kronenfeld, USA
Markus Knauff, Germany
Werner Kuhn, Germany
Michael Lutz, Germany
David M. Mark, USA
Harvey Miller, USA
Daniel Montello, USA
Bernard Moulin, Canada
Reinhard Moratz, Germany

Bernhard Nebel, Germany
Dimitris Papadias, Hong Kong, China
Juval Portugali, Israel
Jonathan Raper, UK
Martin Raubal, USA
Jochen Renz, Australia
Thomas Roefer, Germany
Andrea Rodriguez, Chile
Christoph Schlieder, Germany
Michel Scholl, France
Barry Smith, USA
John Stell, UK
Holly Taylor, USA
Sabine Timpf, Switzerland
Andrew Turk, Australia
Barbara Tversky, USA
David Uttal, USA
Laure Vieu, France
Robert Weibel, Switzerland
Michael Worboys, USA
Wai-Kiang Yeap, New Zealand
May Yuan, USA

Additional Reviewers

A. Reyyan Bilge
Stefano Borgo
Tad Brunye
David Caduff
Lutz Frommberger
Björn Gottfried
Kai Hamburger
Krzysztof Janowicz
Peter Kiefer
Yohei Kurata
Patrick Laube
Paola Magillo
Claudio Masolo
Sebastian Matyas
Patrick McCrae

Mohammed Mostefa Mesmoudi
Nicole Ostländer
Laura Papaleo
Marco Ragni
Kai-Florian Richter
Urs-Jakob Rüetschi
Jochen Schmidt
Lutz Schröder
Klaus Stein
Martin Tomko
Allan Third
Jan Oliver Wallgrün
Jochen Willneff
Chunyan Yao

Sponsoring Institutions

Primary Sponsor: ESRI

Key Sponsor:
Ordnance Survey (OSGB), UK

Key Sponsor:
Multimedia Victoria, Australia

Doctoral Colloquium Sponsor: Google Inc.

Poster Session Sponsor:
Shell

Student Scholarship Sponsor:
PSMA Australia Ltd

In addition, the COSIT 2007 organizers gratefully acknowledge the support of the Cooperative Research Center for Spatial Information (CRCSI); the Spatial Sciences Institute (SSI); and the Department of Computer Science and Software Engineering and the Department of Geomatics, University of Melbourne.

Table of Contents

Perception and Cognition

Reasoning and Algorithms

Navigation and Landmarks

Uncertainty and Imperfection

Progress on Yindjibarndi Ethnophysiography

David M. Mark[1], Andrew G. Turk[2], and David Stea[3]

[1] Department of Geography, National Center for Geographic Information and Analysis
University at Buffalo, Buffalo, NY 14261, USA
dmark@buffalo.edu
[2] School of Information Technology, Murdoch University,
Perth, Western Australia 6150, Australia
a.turk@murdoch.edu.au
[3] Department of Geography, Texas State University
San Marcos, Texas 78666, USA
ds34@txstate.edu

Abstract. This paper reviews progress on the Ethnophysiography study of the Yindjibarndi language from the Pilbara region of Western Australia. Concentrating on terms for water-related features, it concludes that there are significant differences to the way such features are conceptualized and spoken of in English. Brief comments regarding a similar project with the Diné (Navajo) people of Southwestern USA are provided, together with conclusions regarding Ethnophysiography.

1 Introduction

Granö [8] divided the human perceptual environment into two zones: the proximity and the landscape. People interact directly with objects in the proximity, while the landscape remains at a distance, perceived mainly through vision. Spatial cognition research at landscape scales has principally examined imagery, navigation and wayfinding, whereas research on conceptualization and categorization has mainly dealt with objects in the proximity. To begin filling this gap, Mark and Turk [12,13] coined the term "Ethnophysiography" to cover a new research field that examines the similarities and differences in conceptualizations of landscape held by different language and/or cultural groups. Ethnophysiography also examines emotional and spiritual bonds to place and landscape, and the role of landscape features in traditional knowledge systems.

In this paper, we present a case study of such issues among the Yindjibarndi people of northwestern Australia. After a brief review of Ethnophysiography, we present results on how water-related features are conceptualized and expressed in the Yindjibarndi language. We then review some findings regarding traditional spiritual connections to landscape, before providing a summary and pointers to future research.

S. Winter et al. (Eds.): COSIT 2007, LNCS 4736, pp. 1–19, 2007.

1.1   What Is Ethnophysiography?

The core of Ethnophysiography is the investigation (for any particular language) of categories of landscape features, especially those denoted by common words (usually nouns or noun-phrases). Those terms and their definitions form a research topic of considerable importance in their own right. But an understanding of the landscape vocabulary also provides foundations for understanding other important dimensions of Ethnophysiography, including the study of knowledge systems, beliefs and customs of a people concerning landforms and landscapes. Thus, Ethnophysiography is related to the study of 'place' and 'place attachment', termed *Topophilia* by Tuan [26], and examines how these significances are tied into the traditional beliefs, often embedded in creation stories, which help to make sense of the world, of its physiographic entities and their relationship to everyday activities, including traditional cultural practices. As broadly constituted, Ethnophysiography also includes study of the nature of place names (toponyms) and their relationship to generic landscape terms.

Mark and Turk [12,13,14] initiated the Ethnophysiography topic in the ongoing case study with the Australian Indigenous language Yindjibarndi. To some extent it was a natural extension of earlier work on geographical ontology and especially on geographic categories in some European languages [21]. Since 2004, the authors have expanded their Ethnophysiography research to include a case study of landscape terminology and conceptualization employed by the Diné (Navajo) people of South Western USA. This paper, however, reports mainly on results of the Yindjibarndi project.

The authors have also been collaborating with researchers at the Max Planck Institute for Psycholinguistics (MPI) in Nijmegen. Researchers in the MPI's Language and Cognition Group have recently concluded a set of case studies of landscape terms (and some place names) in ten languages in a wide variety of geographic locations (although all in tropical regions) [6]. This work has extended very significantly the range of Ethnophysiography case studies and strengthened its linguistic basis. In the introduction to this collection of studies Burenhult and Levinson ([6], p. 1) discuss the theoretical basis of the work and its relationship to Ethnophysiography. They review the results of the case studies and state that *"The data point to considerable variation within and across languages in how systems of landscape terms and place names are ontologised. This has important implications for practical applications from international law to modern navigation systems."*

Ethnophysiography addresses several fundamental research issues in cognitive linguistics and category formation. Ethnophysiography also has important implications for the design of Indigenous mapping systems and GIS. Also, the results of ethnophysiographic studies can be of value to communities where cultural and language preservation is important, by providing contributions to Indigenous information systems and pictorial dictionaries of landscape features.

1.2 Landscape Terms and Categories

Of all the countless possible ways of dividing entities of the world into categories, why do members of a culture use some groupings and not use others? What is it about the nature of the human mind and the way that it

*interacts with the nature of the world that gives rise to the categories that
are used?* [11], p. 85

Barbara Malt went on from those questions to present an excellent review of research
issues regarding categories. The issues raised by Malt frame the terminological
aspects of Ethnophysiography. For geographic entities and categories, the problem of
defining the sets of instances is even more complicated than that for domains
commonly studied, since to some extent instances of landforms and many other
geographic features are themselves contingent on the definitions of the available
types. Thus while a bird might be a robin, but otherwise is still a bird but of some
other type, a height of land might be a hill in one speech community "but not a
coherent entity at all under different landform type definitions. Even within a single
language different words may attach to different landscape entities, depending upon
frames/scales of reference, e.g. a landform near where one of the authors spent his
childhood was called "Thunder Mountain" by locals, but "Thunder Hill" by people
from areas of more spectacular mountains. The existence and classification of
landforms and other landscape elements are not facts of a mind-independent reality,
either as objects or as categories [22]. Thus they make an excellent domain for the
investigation of whether categorization of natural entities is universal, or whether it
differs significantly across languages and cultures.

1.3 Cultural and Spiritual Significance of Places

The preceding section concentrated implicitly on 'realist' aspects of topography and
of cognitive responses. However, for Yindjibarndi speakers (and many other
Indigenous peoples, as well as perhaps non-Indigenous people) conceptualization of
landscape elements also incorporates what could be termed beliefs, cultural
considerations, or spirituality – aspects of what makes 'space' a 'place'. A review of
some literature on how lived experience (engagement with social and physical
worlds) influences conceptions of place and topophilia is provided in [15].

In [13], it was noted that for *"Indigenous Australians, including the Yindjibarndi
people, spirituality and topography are inseparable (e.g. all yinda have warlu)"* – for
Yindjibarndi people, each permanent water hole (**yinda**) has a spirit (**warlu**) which
determines how one should act at that place. Yu [29] has described similar beliefs and
associated practices for another indigenous group living several hundred kilometers
northeast of the Yindjibarndi. Malpas ([10], p. 95) emphasizes the critical importance
of consideration of the observer (agent) in the process: *"Understanding an agent,
understanding oneself, as engaged in some activity is a matter both of understanding
the agent as standing in certain causal and spatial relations to objects and of
grasping the agent as having certain attitudes - notably certain relevant beliefs and
desires - about the objects concerned"*.

Toussaint *et al.* [25] discuss the relationship of Indigenous Australians to landscape
features, especially water sources. They *"stress the interpretive significance of
human/habitat interactions by emphasizing their connectivity rather than their
independence"* (p. 61). Their exploration of these issues concentrates on the beliefs
and practices of Aboriginal groups living in and around the valley of the Fitzroy River
in the Kimberly region of the far North of Western Australia. They discuss creation
stories, rainmaking ceremonies and other community activities relating to *"both river*

systems and a range of permanent, seasonal or intermittent water sources", and suggest that these indicate that *"land and water were and are regarded not only as physical but also as cosmological and integral to the past in the present"* (p. 63). They go on to quote local Indigenous narratives that explain *"the contingent relationship between human knowledge and activity, maintenance of water and land as place, the actual realization of myth"* (p. 64). The cultural responsibility for the water and land are expressed through a range of common practices, including:

- Teaching children about creation traditions, social relationships, religious associations, care of the environment, hunting and gathering techniques and age and gender restrictions regarding use of resources. It is worth noting that traditional myths and tales are often used as a way of teaching children good environmental practice;
- Practicing 'water etiquette', such as the songs or chants to use when approaching a site and the proper way to introduce a stranger to the place;
- Carrying out ceremonies for the *"ritually induced increase of natural phenomena"* (p. 66), including particular species of plants and animals;
- Preserving and communicating knowledge about key places in the landscape (such as water sources) and their cultural significance.

One of the authors (Turk) has observed (or participated in) each of these activities with Yindjibarndi speakers at their places of cultural significance. These matters are not just some disappearing folk tradition, but rather are part of the everyday life of Yindjibarndi speakers. Hence, they have a crucial impact on the way they conceptualize landscape. This is verified by an extract from interview transcripts provided in a later section of this paper.

In their COSIT03 paper, Mark and Turk [13] noted the difficulty of integrating such aspects of Indigenous knowledge with Western Philosophy (in the 'realist' tradition) and indicated that further attempts would be made to undertake this task. Initial efforts, such as the COADs approach [13], have not proved particularly useful, so one of the authors (Turk) has commenced a major study of the relationship between these aspects of Ethnophysiography and Phenomenology [20].

2 Yindjibarndi Ethnophysiography: A Progress Report

2.1 The Yindjibarndi People, Language and Country

The traditional 'country' of the Yindjibarndi people is mostly along the middle part of the valley of the Fortescue River in the Pilbara region of northwestern Australia, and on adjacent uplands [23]. As a result of the European colonization process (from the 1860s), most of the Yindjibarndi speakers now live in and around the small town of Roebourne, in what traditionally was Ngarluma country. The Yindjibarndi language belongs to the Coastal Ngayarda language group, in the South-West group of Pama-Nyungan languages [7]. According to Thieberger [24], there are about 1,000 remaining Yindjibarndi speakers. The Roebourne community is mostly Indigenous and people use their own languages and English to differing degrees, depending on the context, sometimes mixing words from different languages in the same utterance.

Several linguists have studied the Yindjibarndi language and a number of partial draft dictionaries have been used in this research - as detailed in [13].

Mark and Turk ([13], p. 31 summarized the nature of Yindjibarndi country as follows:

> There are no permanent or even seasonal rivers or creeks in Yindjibarndi country. Larger watercourses have running water only after major precipitation events, usually associated with cyclones (hurricanes). Between such major rain events, rivers continue to 'run', however, the water is underground, beneath the (usually sandy) surface. Permanent pools occur where the lie of the land and the geology cause the water table to break the surface of the ground. Permanent sources of water include permanent pools along the channels of the Fortescue and other larger rivers, as well as some permanent small springs, and soaks where water can be obtained by digging. Unlike many areas of inland Australia, there are no significant intermittent or seasonal lakes in Yindjibarndi country. Local relief (elevation differences) within most of the traditional country of the Yindjibarndi is relatively low, with rolling hills and extensive flats.

2.2 Research Method

Mark and Turk have been carrying out Ethnophysiographic fieldwork with Yindjibarndi people since 2002 [12,13,14]. Together with the third author, they have refined methods of eliciting landscape terms and developed a large set of photographs of landscape features in the Roebourne area.

For the section of the Yindjibarndi case study reported here, the method utilized was interviews with (mainly) groups of speakers of the language, conducted in June 2006 in the small town of Roebourne, and surrounding areas, in the Pilbara region of Western Australia. All three authors were present and took part in all interviews. The participants were requested to discuss the landscape features displayed in a set of photos, with special reference to the Yindjibarndi terms that were appropriate. The interviews were facilitated by one of the researchers (who has visited the community many times), while the others researchers took notes and made occasional contributions to the dialogue. The sessions were audio taped and the researchers took notes. The procedures were carried out with the informed consent of all participants, under the provisions of University at Buffalo and Murdoch University Ethics Committee approvals, and the participants were given small payments to compensate for their time.

The materials used during the interviews were a fixed set of 40 numbered color photo prints (approximately 27 x 20 cm; 11 x 8 inches). Each photo showed a landscape scene, and they were chosen (and ordered) to display a good cross-section of landscape features, without any recognizable sequence in feature type or location. Sixteen of the photos were from traditional Yindjibarndi 'country' on the Tablelands inland from Roebourne or along the Fortescue River; fifteen other photos were from Ngarluma 'country', towards the coast, where Roebourne is located and where many Yindjibarndi people have lived for almost a century. The remaining nine photographs were taken in other areas of Australia, including five from Banjima/Gurrama 'country' inland from the Yindjibarndi homeland (in the Karajini National Park);

three from areas near the coast to the south-west of Ngarluma country, and one from the Northern Territory, near Alice Springs.

There were four separate interview sessions, as follows:

#1: Held in a small room in the Roebourne Telecentre on 13th June. Participants were a husband and wife (TS and AS) of late middle age, with mixed Yindjibarndi / Ngarluma heritage. The female (AS) is a teacher of language. The session lasted 1 hr and 45 min.

#2: Held by the banks of Miaree Pool, Maitland River, south of Roebourne on 16th June. The main participants were three female elders (DS, NW and PM), with occasional contributions from the woman who participated in Session 1 (AS). The session lasted 38 min.

#3: Held at Jindawurru (Millstream National Park) outside the visitor centre on 18th June. The participant was a very senior male Yindjibarndi elder (NC). The session lasted 1 hr and 10 min.

#4: Held in a large room at the Juluwarlu cultural centre. The main participant was a senior female elder (CC, wife of participant in Session 3) with contributions also made by her two daughters and a cultural worker from Juluwarlu (LC). The session lasted 1 hr and 25 min.

Detailed transcriptions of three of the audio recordings were undertaken and reviewed by two researchers, with reference to their notes and existing Yindjibarndi dictionaries. We are awaiting a copy of video footage of the 4th session. All transcripts will be reviewed by linguists at the culture/language organizations Juluwarlu and/or Wangka Maya and copies of the final versions (and the audio recordings) provided to the participants.

Detailed analysis of the landscape terms used by the participants has allowed the researchers to revise their previous understanding of Yindjibarndi landscape concepts. A subset of these terms is reported on here, and subsequent publications will deal with the rest. The researchers will also prepare a pictorial dictionary of Yindjibarndi landscape terms for publication (in collaboration with Juluwarlu and Wangka Maya) and copies will be provided to the Roebourne community and to relevant linguists.

Alternative research methods, including recordings of participants' comments during extended vehicle trips through their 'country', have been employed in the Diné (Navajo) case study. These are discussed in [15].

2.3 Yindjibarndi Terms and Categories Related to Water in the Landscape

Given the limitations in length of papers for this conference, we have decided to present a review of the main Yindjibarndi terms for one landscape domain, namely water in the landscape and its effects. Thus the review will include Yindjibarndi terms for water bodies, watercourses, water sources, and water events at landscape scales.

Myers ([16], pp. 26-27) has provided a clear account of the role and critical importance of a typology of water sources for the Pintupi, a group who live in the Western Desert of Australia:

For hunters and gatherers, the unavailability of water supplies poses the fundamental subsistence challenge. It is important to understand the nature of this resource. Although there are no permanent surface waters in the area, the Pintupi have found it possible to exploit other types of water supply. They have used large, shallow, and transient pools (pinangu) formed by heavy rain, claypans (maluri) and rock reservoirs (walu) in the hills that might be filled from lighter rainfalls, soakage wells (tjurnu) in sandy creek beds, and "wells" (yinta) in the sand or in the rock between the sand ridges. ... Water is, then, a geographically specific resource, and Pintupi subsistence technology depends on knowledge of the location of water sources and the conditions under which they are likely to be usable as well as on movement to use them."

Recall that Yindjibarndi country and much of the rest of Western Australia is also very dry, and that surface runoff is generally a rare event, most often associated with precipitation from tropical cyclones. Thus, except for a few small permanent surface springs, water is either restricted to permanent pools (**yinda**), or is a transient or ephemeral condition of concavities in the landscape.

2.3.1 Watercourses

Watercourses are channels in which water sometimes or always flows. In much of Europe, larger watercourses have permanently flowing water, for example, those features referred to as "rivers" in English. As the English language has been extended to refer to geographic features in arid or semiarid areas, words such as "river" or "creek" often are still used, even if water is seldom present at the surface, and English speakers normally indicate frequency of flow by adding adjectives such as 'seasonal', 'intermittent', or 'ephemeral' to nouns that canonically refer to flowing water features. In the American southwest, English speakers also borrowed the words 'arroyo' and "canyon" from Spanish, and extended the word 'wash' to refer to dry sandy stream beds.

Longitudinal depressions in the landscape are referred to as valleys, canyons, gorges, gullies, etc., in English. Normally, English seems to make a clear distinction between such concave topographic features and the watercourses that often run along them. However, in arid and semiarid regions, the distinction between topographic and hydrological features quite literally dries up.

Wundu

Fig. 1. Photos 15 (left) and 35 (right) are among the many that illustrate **wundu**

The Yindjibarndi language has at least two words that refer to longitudinal depressions in the landscape: **wundu** and **garga**.[1] The meanings of these terms, and the distinction between them, are discussed in the following subsections.

Wundu is one of the most frequent terms in the transcripts. Apparently, **wundu** is used to refer to stream beds and stream channels, that is, watercourses with flat cross-profiles and relatively low-gradient long profiles. Figure 1 shows two of the photos used in the elicitation study, both of which were called **wundu** by all three groups. In the Yindjibarndi dictionary, Wordick [28] and Anderson [1,2] listed the definition of wundu as "river(bed), gorge". Von Brandenstein [5] spells this term "wurndu" and lists the meaning simply as "river". We would argue, however, that since normally the English word "river" refers to a stream of water, **wundu** definitely is not the semantic equivalent of river. Furthermore, there appear to be no distinct Yindjibarndi terms for different sizes of **wundu**, as there are for streams in English (river, creek, brook, etc.).

Garga

The Yindjibarndi word **garga** refers to a longitudinal concavity in a hillside. To be a **garga**, it appears that the depression must have a significant longitudinal gradient, and to lack a flat alluvial floor. Thus it seems to be a *topographic* term, but the hydrologic function of a **garga** is widely recognized by the Yindjibarndi speakers we talked to. The dictionaries by Wordick and by Anderson list the meaning of the term **garga** as "wash, arroyo", terms in American English that do not appear to be good semantic matches to **garga** as we understand it. Von Brandenstein [5] lists the meaning of **garga** as "gully", which appears to be a better English fit to the features called **garga** by the Yindjibarndi participants.

Fig. 2. The central feature in 0050hoto 7 (left) was referred to as a **garga** by all three groups, and the feature in Photo 21 (right) was referred to as **garga** by two of the three

The distinction between Wundu and Garga

The differences between the meanings of the Yindjibarndi terms **wundu** and **garga** is illustrated by the following excerpt from the transcripts.

---

[1] One additional term, **yarndirr**, was used by one of our interview groups to refer to a gorge, but since other groups did not use it, we will reserve discussion of this term until its meaning and use are confirmed.

From the transcript for Photo #15 (see Figure 1, above), session 1:
[0:33:52]

AT: So what makes it a **wundu** rather than a **garga**?

AS: (correcting) graga[2]

TS: Oh, it's not a **garga**! Its where the water's been running, and makes it a **wundu**. But at the moment it's dry. What makes the difference between a **garga** and a **wundu**, this is bit wider, quite wider than a **garga**, **garga** is a small thing – which begins small then creates the bigger …

AT: So is it that it's got a flat bottom?

AS: Mmm. [indicating agreement]

TS: Yes. This one here. This is flat. Whereas a gully …

AT: So once it gets a flat bottom, because there is a lot of water running through it, it becomes a **wundu**.

TS: It becomes a **wundu**, whether it's running or dry.

From the above discussion and from references in other transcripts, it appears clear that **garga** refers to a small to moderately sized longitudinal depression in a hillside that carries runoff during heavy precipitation. For the word **garga** to apply, a feature apparently requires a significant longitudinal gradient.

2.3.2 Flowing and Still Water

Recall that there are no permanently flowing rivers or creeks in Yindjibarndi country. Thus flowing water appears to be more of an event than a topographic feature. We know of four words that are used in the Yindjibarndi language to refer to temporary water in the landscape. Briefly, a strong, fast, deep flow of water is referred to as **manggurdu**, and often is translated to the English word "flood" by the participants. A more gentle trickle of water is referred to as **yijirdi**, while temporary pools of still water in a channel are referred to simply as **bawa**, meaning "water". Lastly, small pools of water that remain as water in a riverbed is drying up may be referred to as **thula**, a word that also means "eye". An additional term, "**jinbi**", appears to refer to water flowing from small permanent springs. Differences in the meanings of these terms are illustrated below using excerpts from the transcripts.

---

[2] The appropriate order of the letters "ar" or "ra" in this word (and others) is still under investigation.

From the transcript for Photo #6, session 4:

LC: When there's **bawa** like this, they want to know if you got a **yini** for it. [11:17]

....

CC: Yes, **yijirdi**.
It's running downstream isn't it ... **gubija** stream ...

[**bawa** = water; **yini** = word; **gubija** = small]

This appears to confirm that **yijirdi** is the Yindjibarndi term for a small or shallow flow of water. The participants in the other two sessions also used the term **yijirdi** for the stream in this photo

From the transcript for Photo #14, session 1:

[0:29:29]

AT: What is the difference between a **jinbi** and a **manggurdu?**

TS: This is rushing water. **Jinbi** is only just a little water, like a spring, you know?

DM: And then there was that other word, **yijirdi**?

TS: That's a **yijirdi**.

AS: You can have a **yijirdi** or **manggurdu**.

TS: Depends on how you describe it.

TS: This can be a **yijirdi**

AS: **Manggurdu** is a lot of rushing water, when it is smaller, and smaller, it becomes **yijirdi**.

TS: This is a **manggurdu**.

AT: So a **yijirdi** and a **jinbi** are the same.

TS: Yeah – **yijirdi** and **jinbi**. But a **yijirdi** runs a bit faster than a **jinbi**. Like a trickle.

AS: **Jinbi** just be layered ...... like a spring water.

TS: **Jinbi** is like a spring that's always there.

AT: **Yijirdi** is a bit more water?

TS: Yeah.

AT: And a **manggurdu** is a lot of water?

AS: A lot of water running. Flood.

TS: This one is a flood.

If (as claimed above) **wundu** is not equivalent to "river", neither is **manggurdu**. A **manggurdu** is not a permanent feature of the landscape, but is an event in or a condition of a channel. It is interesting that none of the three groups used the term **wundu** to refer to picture #20, and all three used the term **manggurdu**.

Fig. 3. Photo 20. The Ashburton River at Nanutarra Roadhouse, January 2006

As noted above, there are two main Yindjibarndi terms for still (non-flowing) water in the landscape. Temporary still water is referred to simply as **bawa** (water), whereas permanent pools, **yinda**, are places of great cultural and spiritual significance, as discussed further below.

Photo 18

From the transcript for Photo #18, session 4:
FV: That's a **wundu**. It's not a permanent pool and its not a little stream.
CC: It's a **wundu**, yeah.
FV: Be there, then it'll dry up.
[33:56]
CC: Yeah.
FV: Say that's a **thurla**?
CC: **Thurla**, yeah. There's a little … with a **thurla**, **bawa**, there's only a little bit of water there now, we call them **thurla**, little bit, …

gubija thurla bawa, that's that one, and there's a **wundu**, isn't it.
That'll dry up, that'll dry up later on.

Fig. 4. Photographs 14 (left) and 28 (right) depict "waterfalls" in Karijini National Park

From the transcript for Photo #14, session 4:
LC: You got a **yini** for waterfall?
CC: No, no, we haven't one.
     Yijirdi [some other Yindjibarndi talk]. **Yijirdi**.
FV: Little stream, stream water,
     A **yinda**, which is a permanent pool.
[27:57]
CC: Running down there.
FV: And then you got a **wundu**, you know, that flows, but is not a **yinda**, not a,
     its smaller than a **yinda**, and its bigger than a ... **yijirdi**, you know, little
     stream. [28:14] **Wundu**.
CC: But big one like this one, permanent pool, we call them **yinda**. Cause they
     gotta be there all the time.

From the transcript for Photo #28, session 1:
 TS: **Bawa. Yalimbirr**, where the waterfall is, that's a **yalimbirr**. It's a drop.
 AS: **Bawa yindi**
 TS: We borrowed that word. Don't know if we've *got* any name for a waterfall.
 TS: This is a **yinda**
 AS: **Bawa yindi**? **Bawa yindi marnda**.
 TS: **Bawa yindi**. Sort of like a waterfall
 AS: letting this one out.
[1:01:05]

As illustrated above, all three groups used the term **yinda** for the permanent pool,
and **yijirdi** for the water trickling down the rocks into it. In this context, the
participants said there was no Yindjibarndi word for waterfall. However, in the
discussions of Photo 28, as illustrated below, two of three groups used the phrase
"**bawa yindi**" meaning "water descends and remains in sight". Perhaps the details of
the fall, trickling over the rocks versus falling vertically, influence the word choice;
more research is needed on this point.

2.4 Spiritual and Metaphysical Aspects of Yindjibarndi Connections to Landscape

The cultural and spiritual aspects of relationships to landscape were already introduced above. For Yindjibarndi people, these include strong spiritual relationships with permanent water features, as well as other significant landscape elements. This point was emphasized in the course of the fieldwork at Roebourne in June 2006. During an audio recording session at Juluwarlu (the Yindjibarndi culture maintenance group) examining photos of landscape features, with the authors asking for landscape terms, the informants spontaneously referred to spiritual aspects of the features. The following example (edited) from the transcripts illustrates this:

Photo 31 (very large pool at Millstream, discussed in Session 4)

[57:15]
ED: Must be, look like Millstream...
      [some Yindjibarndi talk]
LC: This is a **wundu**.
CC: (correcting) This is a **yinda**. This is Deep Reach.
ED: **Yinda**.
......
CC: I think we will have bin explained to you what this is (referring to previous discussions with AT)
ED: It means a lot to us.
AT: yes
CC: That's where the two men been taken, you know?
ED: And buried - very frightening for us to go and stop there, you know ....
CC: You see the hole on top of the bank in Millstream (addressing DM)?
DM: I have not seen it.
ED: You ought to see it. Maybe more bigger than this (showing size with her hands). Snake got up, got the two men, and come back - this wasn't like this, it was a dry riverbed - 'til that. Camping ground for the old people as well, long ago.
CC: In the **wundu**, yes.
ED: In the Dreamtime, yes.
......

CC: They bin crying, crying, and that **warlu** got wild, sent the big flood come down.

FV: Made them permanent pools.

Here the informants were talking about how the **warlu** (mythic snake) had come up out of the ocean at Onslow and traveled up the route of the Fortescue River chasing two boys who broke the Law, until he got to Jindawurru (Millstream). One account of that event says:

> *He finally got up at Nunganunna [Deep Reach Pool, at Millstream] and lifted the law breakers up in the sky in a willy willy ... when they fell, he swallowed them through his thumbu (anus) and drowned the whole tribe in the biggest flood of water. Today Barrimindi lives deep down in the pool he made at Nunganunna.* [9], p. 1

This is part of an explanation of what Yindjibarndi call the 'learning times', 'when the world was soft'. For many this is known as a 'dreamtime' story. The 'dreamtime' is a crude translation of Indigenous explanations of the formation of the world into its current landscape. However, *"The "dreaming" is not a set of beliefs which is being lost because it is no longer valid, it is rather a way of talking, of seeing, of knowing, and a set of practices, which is as obtuse, as mysterious and as beautiful as any poetry"* ([4], p. 12).

Articles by Bednarik and Palmer review then (1970s) current and historical (mid 19th Century) interviews with Yindjibarndi speakers regarding landscape terms and place names. Bednarik [3] provides support for the 'lifestyle, utility and affordance' aspects of traditional Indigenous conceptions of ranges, hills and associated valleys and hydrologic features. He discusses the absence of camp sites near a particular deep and narrow canyon with a permanent waterhole. He suggests that: *"The setting would be an outstanding hunting trap. Quite likely camps were not made near this well concealed gorge less [sic] hunting be jeopardized. However, sacredness of the well would be an equally convincing explanation ..."* (p. 57). This discussion highlights the link between landscape affordances and important human activities, but also introduces the possibility of sacred associations with landscape being of equal or even greater importance.

Palmer [17] discusses evidence of links between landscape features (e.g. rocks; mounds; hills) and sacred sites (e.g. *dalu* - 'increase' sites). He also links these spiritual aspects of culture (and landscape sites) to petroglyphs in the Pilbara area. Palmer extends this discussion, with particular emphasis on links between landscape and important cultural 'story lines' [18]. The importance and meaning of sacred sites is also discussed in [19], where extremely significant cultural traditions are linked to specific features (places) in the landscape of Yindjibarndi country. He says that *"To discuss any of these sites in isolation is wrong because the river constitutes a unity over and through which the great mythic ancestors were believed to have traveled. Their function as creators, instigators and founders of traditional ritual practice is inextricable from the natural features that proclaim their validity"* (p. 232).

3 Discussion and Future Research

The results reported above just scratch the surface of landscape terminology in Yindjibarndi. Further analysis of the transcripts from June 2006, in comparison with published dictionaries and our previous fieldwork, should confirm the semantics of a core of landscape terms. The field experiences and the audio recordings and transcripts also suggest several research issues that were not apparent at the onset of the project.

3.1 The Importance of Perspective or Viewpoint

In our investigation of Yindjibarndi words for landscape features it has become clear that speakers of this language will use different words for the same physical feature depending on their 'point-of-view'. In a physical sense, this can relate to the position of the observer relative to the landscape feature – for example for the cliff feature in photo 33 (Figure 5) the informants in Session 4 said: [1:04:52] LC: "If you sort of, looking upwards ..." FV: "**Gankalangga**"[3]. LC: "And if you're standing at the top?" ED: "Looking down, that's **gunkurr**." The edge at the top of the cliff in Photo 34 was talked about in similar fashion. It is not clear whether these terms are alternative terms for the same feature, depending on perspective, or whether they are denoting the perspective itself. However, preliminary results from the Navajo Ethnophysiography study suggests a similar distinction for Navajo speakers, where the term for a rock canyon might be different when on the canyon floor than it would be when standing on the rim above. Clearly this possibility requires further research. Perspective-dependent terminology would provide a challenge for GIS implementation.

Fig. 5. Photos 33 (left) and 34 (right)

3.2 Importance of Country, Territory

In most of the photograph discussions with Yindjibarndi participants, and in most discussions with Navajo subjects as well, participants either stated where they thought the photo was taken, or asked about the location. There was often a strong concern for knowing the real-world location before beginning the description of features. In many

---

[3] The dictionaries by Anderson have: **gunkurr** = "downwards"; **gankala** = "up, high, above, at the top"; **-ngga** = at (locative suffix).

cases, the participants then brought in their background knowledge of the places, including features or properties not visible in the photographs. The concern for place and locality may run deeper than just background knowledge: at several points in the discussions of the photographs from Karijini National Park, Yindjibarndi participants expressed some reluctance to give terms for features in the traditional country of another group. The observed responses may related to Tuan's ideas of topophilia and the different perceptions of landscape by 'natives' and 'visitors', or to ideas of territoriality and respect for others, or both. Follow-up discussions will be conducted to try to assess the reasons for this added complexity of the elicitation of terminology for landscape.

3.3 Update on the Navajo Study

We have conducted considerably more fieldwork and landscape photo response sessions with Navajo speakers than we have with Yindjibarndi speakers, but transcription and analysis lag behind. We have conducted 10 'field interviews' in Navajo country, recording almost 28 hours of audio. We also have shown 39 photographs of Navajo country to three groups of elders and one other language expert, recording more than 9 hours of audio. Conceptual systems for many landscape domains appear to be organized differently in the Navajo language than they are in Yindjibarndi or English, but details will have to await completion of the transcription and translation phases.

4 Conclusions

Consideration of the definitions and semantics of Yindjibarndi terms for water-related features indicates several important differences between the Yindjibarndi and English languages regarding the conceptualizations, the ontology, of water in the landscape. First of all, permanence of water features is a matter of categorical predication in the Yindjibarndi ontology, that is, otherwise similar entities that differ only in being permanent or temporary apparently are considered to be different kinds of things and are referred to using different terms. In contrast, the English language makes permanence a matter of accidental predication and denotes it by using adjectives. In English, water features are first classified by flowing or not, and then by size, such that size is often of categorical importance (e.g., lake vs. pond). We have found no terms for water-related features in Yindjibarndi that are differentiated only by size, and size, where specified, is indicated by generic modifiers for large or small.

Our own case studies, and those by MPI, provide strong (perhaps conclusive) support for the basic Ethnophysiography hypothesis – i.e. *that people from different language groups/cultures have different ways of conceptualizing landscape, as evidenced by different terminology and ways of talking about and naming landscape features.*

The case studies also provide some indications of the sort of factors that could be causal agents in producing such differences – what could be termed 'Ethnophysiography dimensions'. These include:

a. The **topography** of the region occupied by the language group – whether mountainous, hilly or flat; the presence or absence of particular landscape features, such as, volcanic cinder cones; sand dunes; coral reefs, etc.;
b. The **climate** of the region – the strength of seasons (e.g. does it snow in winter); its variability (e.g. does rainfall come only from cyclones); etc.;
c. The **vegetation** in the region – its variability in space and time; its density; its uses (e.g. for shelter, food and medicine); its affordance re travel; etc.;
d. The **lifestyle** and **traditional economy** of the people – whether they pursue hunter/gather activities, are cultivators, etc.;
e. Religious **beliefs** or spiritual concerns linked to landscape - e.g. creation beliefs, presence of spirits in landscape features, etc. and cultural practices (e.g. ceremonies; taboos) which accompany such beliefs;
f. **Historical factors** – such as: movement of the people into their current region; colonization of the people by outsiders with a significantly different language; major changes in lifestyle and economy; etc.;
g. The **grammar** of the language – the roles played by nouns, verbs, adjectives, compound words, noun-phrases, etc.;
h. The structure of **place names** – if descriptive; if include generic landscape terms; how they arise and are constructed.

It is assumed that these factors are unlikely to be independent of each other, rather they will frequently interact – e.g. climate partly determines vegetation, which influences lifestyle and traditional economy. By conducting an increasing number of Ethnophysiography case studies in regions providing similarities and comparisons between these factors it is hoped that it will be possible to develop a descriptive model of how these factors (and perhaps others) work together to determine conceptualizations of landscape, and hence terminology and place names for particular language groups. Clearly this objective requires considerable trans-disciplinary collaboration by researchers in different parts of the world.

Use of this framework should lead to an improved understanding of differences (and similarities) in conceptualizations of landscape by people from different language/cultural groups. It is hoped that this improved understanding will facilitate efforts in design of effective generic (cross-cultural) aspects of GIS, aid in efforts to produce culturally-specific GIS (especially for Indigenous users) and also the production of pictorial dictionaries to aid in the preservation of traditional culture.

Acknowledgments. Many members of the Roebourne community, the cultural organization Juluwarlu and the Pilbara Aboriginal Language Centre (Wangka Maya) provided invaluable assistance regarding the Yindjibarndi language. Special thanks go to: Trevor Soloman, Allery Sandy, Dora Soloman, Nelly Wally, Pansy Manda, Cherry Cheedy, Ned Cheedy, Lorraine Coppin, Michael Woodley, Beth Smith and Sue Hanson. We also wish to thank the Australian Institute of Aboriginal and Torres Strait Islander Studies for providing material from the Aboriginal Studies Electronic Data Archive (ASEDA). Funding support by the US National Science Foundation (grants BCS-0423075 and BCS-0423023) and from Murdoch University is gratefully acknowledged. Work with the Diné (Navajo) people has been facilitated by Carmelita Topaha and conducted under permit C0513-E from the Historic Preservation Office of the Navajo Nation.

References

1. Anderson, B.: Yindjibarndi dictionary. Photocopy (1986)
2. Anderson, B., Thieberger, N.: Yindjibarndi dictionary. Document 0297 of the Aboriginal Studies Electronic Data Archive (ASEDA) Australian Institute of Aboriginal and Torres Strait Islander Studies, GPO Box 553, Canberra, ACT 2601, Australia (No date)
3. Bednarik, R.G.: A survey of Prehistoric Sites in the Tom Price Region, North Western Australia. Oceania 12(1), 51–76 (1977)
4. Benterrak, K., Muecke, S., Roe, P.: Reading the Country. Fremantle Arts Centre Press, Fremantle, Western Australia (1984)
5. Von Brandenstein, C.G.: Wordlist from Narratives from the north-west of Western Australia in the Ngarluma and Jindjiparndi languages. Canberra, ASEDA Document #0428 (1992)
6. Burenhult, N., Levinson, S., (eds.): Language and landscape: A cross-linguistic perspective. Language Science (in press)
7. Gordon Jr., R.G. (ed.): Ethnologue: Languages of the World. Fifteenth edition. Dallas, Tex.: SIL International (2005), Online version, http://www.ethnologue.com/show_lang_family.asp?code=YIJ
8. Granö, J.G.: Pure Geography. Baltimore, MD: Johns Hopkins University Press. Book originally published in German in 1929 and in Finnish in 1930 (1997)
9. King, W.: Ngurra - Homelands. In: Rijavec, F. (ed.) Know the Song Know the Country, Ieramugadu Group Inc. Roebourne, Western Australia (1994)
10. Malpas, J.E.: Place and Experience: A Philosophical Topography. Cambridge University Press, Cambridge (1999)
11. Malt, B.C.: Category Coherence In Cross-Cultural-Perspective. Cognitive Psychology 29(2), 85–148 (1995)
12. Mark, D.M., Turk, A.G.: Ethnophysiography. In: Pre-COSIT Workshop (2003)
13. Mark, D.M., Turk, A.G.: Landscape Categories in Yindjibarndi: Ontology, Environment, and Language. In: Kuhn, W., Worboys, M.F., Timpf, S. (eds.) COSIT 2003. LNCS, vol. 2825, pp. 31–49. Springer, Heidelberg (2003)
14. Mark, D.M., Turk, A.G.: Ethnophysiography and the Ontology of the Landscape. In: Egenhofer, M.J., Freksa, C., Miller, H.J. (eds.) GIScience 2004. LNCS, vol. 3234, pp. 152–155. Springer, Heidelberg (2004)
15. Mark, D.M., Turk, A.G., Stea, D.: Ethnophysiography of Arid Lands: Categories for Landscape Features. In: Hunn, E., Meilleur, B., Johnson, L.M., (eds.) Landscape Ethnoecology, Concepts of Physical and Biotic Space (in press)
16. Myers, F.R.: Pintupi Country, Pintupi Self: Sentiment, Place, and Politics among Western Desert Aborigines. University of California Press, Berkeley (1986)
17. Palmer, K.: Petroglyphs and Associated Aboriginal Sites in the North West of Western Australia. Oceania 10(2), 152–160 (1975)
18. Palmer, K.: Myth, Ritual and Rock Art. Oceania 12(1), 38–50 (1977)
19. Palmer, K.: Aboriginal Sites and the Fortescue River, North West of Western Australia. Oceania 12(3), 226–233 (1977)
20. Patocka, J.: An Introduction to Husserl's Phenomenology. Translated by Kohak, E. Open Court Publishing, Chicago, IL (1996)
21. Smith, B., Mark, D.M.: Geographic categories: An ontological investigation. International Journal of Geographical Information Science 15(7), 591–612 (2001)
22. Smith, B., Mark, D.M.: Do Mountains Exist? Towards an Ontology of Landforms. Environment and Planning B 30(3), 411–427 (2003)

23. Tindale, N.B.: Aboriginal Tribes of Australia. Their Terrain, Environmental Controls, Distribution, Limits, and Proper Names. Australian National University Press, Canberra (1974)
24. Thieberger, N.: compiler: Handbook of Western Australian languages south of the Kimberley region. Pacific Linguistics, Australian National University (1996, accessed August 2003), Web version, http://coombs.anu.edu.au/WWWVLPages/AborigPages/LANG/WA/contents.htm
25. Toussaint, S., Sullivan, P., Yu, S.: Water Ways in Aboriginal Australia: An Interconnected Analysis. Anthropological Forum 15(1), 61–74 (2005)
26. Tuan, Y.-F.: Topophilia: A Study Of Environmental Perception, Attitudes, and Values. Prentice-Hall, Englewood Cliffs, NJ (1974)
27. Turk, A.G., Mark, D.M.: Talking in COADs: A New Ontological Framework for Discussing Interoperability of Spatial Information Systems. In: Kuhn, W., Worboys, M.F., Timpf, S. (eds.) COSIT 2003. LNCS, vol. 2825, Springer, Heidelberg (2003)
28. Wordick, F.J.F.: The Yindjibarndi language. Pacific linguistics. Series C., no. 71. Canberra: Dept. of Linguistics, Research School of Pacific Studies, Australian University (1982)
29. Yu, S.: Ngapa Kunangkul (Living Water): An Indigenous View of Groundwater. In: Gaynor, A., Trinca, M., Haebich, A. (eds.) Country: Visions of Land and People in Western Australia, Western Australian Museum, Perth, Western Australia (2002)

Study of Cultural Impacts on Location Judgments in Eastern China

Danqing Xiao[1] and Yu Liu[2]

[1] Department of Spatial Information Science and Engineering,
University of Maine Orono, Maine 04469, USA
danqing.xiao@maine.edu
[2] Institute of Remote Sensing & GIS, Peking University, Beijing 100871, China
liuyu@urban.pku.edu.cn

Abstract. This paper examined cultural impacts on absolute and relative location estimates of 12 Eastern Chinese cities, based on questionnaires of each city's latitude and distances between city pairs. Linear regression analysis of the latitude estimates revealed that the estimated latitude of a city is statistically significantly related to its actual latitude. MDS analysis of the distance estimates revealed the gap that divided Eastern China into two regions and cultural-related causation of the gap was explained in detail. In particular, the Chinese language and its impact on spatial cognition were addressed. Results were compared with North America; some important features of spatial cognition were similar: the categorical storage of spatial information and the absolute-relative location reasoning process.

1 Introduction

1.1 Cultural Characteristics of China

Culture is defined as a body of knowledge and beliefs that is more or less shared between individuals within a group and transmitted across generations [1]. Cultural differences and the impact on geographical regions is a traditional topic in Geographical Information Science. It reflects language differences and their impact on spatial cognition [1,2,3]. There are three important cultural characteristics of China that influence the geographical cognition process:

1. China is divided mainly into two dialect regions: Northern (including Mandarin, the official language of China.) and Southern. All three Northern dialects have one same generation thus widely understandable, while southern dialects have diverse generations and are only understood locally. These language differences may reflect in the result of regional divisions.

2. Traditional geography divides Eastern China into three regions: Northeastern, North Hua and South Hua. The border between North and South Hua is "Qinling Mountain-Huai River" (Q-H border). It is also the border of North and South Climate region, on which the mean temperature of January is 32°F. Q-H border plays an important administrative role that

S. Winter et al. (Eds.): COSIT 2007, LNCS 4736, pp. 20–31, 2007.

regions south of Q-H border are not allowed to install central heating systems. Thus people experience significantly different winter climate in North Hua and South Hua, although the natural climate doesn't change dramatically. For example city Nanyang (33°N) and city Nanjing (32°N) have the same outdoor temperature all year long, while the indoor temperature during winter is 40°F in Nanjing and 66°F in Nanyang because of the central heating availability in Nanyang.

3. Chinese use absolute frames such as cardinal directions to express location at multiple scales [4], while the egocentric frames (left and right) is weak and seldom used in spatial description. For example the linguistic description of routing is mostly of the time expressed as "turn east" rather than "turn right". Further proof in traditional Chinese is that the character for "left" also stands for east because the ancient Chinese egocentric direction system is always facing the south [5].

1.2 Cultural Impacts on Regionalization

Regionalization is one of the most distinctive characteristics when individuals make geographical estimates due to its direct relation to national regions [6]. The traditional claim says that geography is the "study of regions", in which the taxonomy of geographic regions and the formation processes reflects some of the universal characteristics of human thoughts [7], especially spatial cognition. There are four major types of geographic regions: administrative, thematic, functional and cognitive [8]. Cognitive region, formed by people's informal perceptions, will be emphasized in this paper because it best reflects the cultural differences. Administrative region, formed by legal or political action, is also studied because administrative boundary is precise compared to other three types [9] and it is peculiarly combined with climate in China.

  Culture impacts regional cognition when individuals categorize space and form biases in judgments [4]. Location estimate tasks, for example, allow researchers to examine the biases quantitatively and related with geographical regions. Scholars have studied the biases for latitude estimates of European cities [10] and North American cities [11] under global scale and found three common rules of location judgments [12, 13]:

1. People use plausible reasoning during their estimate tasks based on knowledge available and the representation shows the trend of regionalization.
2. Both absolute and relative location estimates are based on the same geographic representation.
3. The participants tend to stretch cities to global landmarks during the estimating tasks. Study of North American cities reveals participants stretched North America southward toward the equator.

1.3 The Role of Culture in Spatial Cognition

Culture and its role in geography have a long history of being discussed. The most frequently discussed one is the language influences in GIS as a by-product of cultural differences [3,4]. Importance of languages has been stressed; however, the trend was to emphasize on cross-cultural differences rather than admitting cross-cultural generalities. Later research focused on cultural influences on spatial cognition found that items below are the same across cultures: differential treatment of spatial information in memory, reasoning, language, as a function of scale; categorical and hierarchical organization of regions [1,14]. As influential as the previous conclusion became, the study of cultural differences in spatial cognition became less noticed.

Yet the role of culture in spatial cognition is far from determined that previous studies are not complete in two aspects: first, researchers studied spatial cognition under such small scale (e.g., cross–culture study of cartography of villages [15]) that cultural differences may not be significant at all. Second, previous scholars seldom select oriental cultures for cross-culture comparison. It is likely that all dated back to ancient Greek-Rome [16], western cultures, especially geographical related ones, tend to be similar that the differences is too weak to appear.

There are still two important research topics in cultural impacts on spatial cognition: cultural differences within one region to study its genesis and impacts; and cross-cultural comparison between oriental and western civilizations to study its generality. Previous researchers compared Japanese and American city guide books, which leads to the findings of the influence of the distinctive oriental city pattern and address system on people's use of maps. Result shows that "the contents of linguistic information were entirely influenced by socio-cultural factors rather than environmental conditions such as the street pattern regularity" [17]. This study showed the influence of cultural differences on spatial languages, especially the pictorial information and linguistic description of maps.

This paper focuses on location judgments of Chinese cities under national scale and it is based on the same experiment (the stimuli are changed to Chinese cities) by European and North American scholars to eliminate the difference caused by experiment design. Both latitude and distance estimates of 12 selected cities are investigated and the result is analyzed using Multidimensional Scaling method. It is expected this modified experiment will reveal the characteristics of Chinese cognitive pattern, which should be different from previous North American and European characteristics considering cultural influences. Still universal cultural generalities are expected to be found, especially those related to categorical organization of regions.

2 Objectives

The first objective is to study the influences of culture, especially languages on spatial cognition. According to the Sapir-Whorf Hypothesis [18] that a cultural group's language determines the way members perceive and think about the world, significant differences are expected to be found in China because the language is based on hieroglyphic characters whereas Latin is based on pronunciation-determined alphabet. Although Whorf hypothesis is being criticized these days, the influence of Chinese language is still not ignorable [19]. As a socialistic and developing country, it is also expected that influences caused by ideological and economic differences exist.

The second objective is to find cultural universals by comparing our study with previous research [10,11,12] based on the same experiment. It is agreed that many important aspects of spatial cognitive structures and processes are universally shared by humans everywhere [1]. In China despite its distinctive culture, universals are still expected to be found as the nature of mankind. Moreover, what aspects are still universal and how they influence spatial cognition are of high importance.

3   Method

The study of geographical judgments is based on a survey of latitude and distance estimates of 12 cities selected from Eastern China. Volunteers answer a series of questions requiring their geographical knowledge of the cities. Plausible reasoning process is also needed to answer the question in order to provide data for advanced analysis of spatial cognition [20].

3.1 Participants and Design

Participants of the experiment are 60 volunteers from three sources:

  a)   21 undergraduate students from the class: "Introduction to Cartography";
  b)   10 undergraduate students majoring in Geographical Information System;
  c)   29 graduate students majoring in physics (Number=6) and earth sciences (Number=23).

Of the 60, 43 provided valid questionnaires; the rest questionnaires were eliminated due to data integrity (i.e., 13 participants didn't finish the questionnaires because participants claimed to have limited geographical knowledge; 4 participants provided invalid data by providing numbers in sequence). For the 43 who accomplished the 30-minute questionnaire (36 males, 7 females), there's no extra credit gained.

All the participants are students of Peking University and had lived in Beijing at least for one year and a half. Up to 85% of the participants are from Eastern China. The mean age of participants was 22.1, ranging from 18 to 24. 90% of the participants had solid background in geography.

3.2 Stimuli

Twelve cities are selected from Eastern China. Each city is the capital city of a province (except Nanyang) to make sure participants have abundant knowledge so they can make reasonable estimates to provide valid data. The cities' longitude ranged from 112.3°E to 126.3°E to reduce the distances cause by longitude differences therefore better connect latitude estimates with distance estimates.

City locations and their dialect regions are shown in Fig. 1:

Harbin: 46°N Changchun: 44°N Shenyang: 42°N Dalian: 39°N;
Beijing: 40°N Jinan: 36.4°N   Zhengzhou: 34.5°N Nanyang: 33°N;
Nanjing: 32°N Ningbo: 30°N Fuzhou: 26°N Kaosiung: 22.4°N.

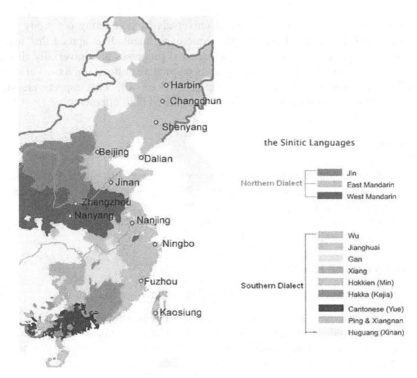

Fig. 1. Selected stimuli cities and their location

3.3 Procedure

Participants are asked to fill out questionnaires which included three parts: personal information, latitude estimates with knowledge rating and distance estimates. Participant may withdraw from the study at any time and sign the agreement on the cover page which guarantee the researchers to use the information collected for scientific purposes. After signing, participants provide their personal information (i.e., sex, age and major).

For latitude estimates, participants are asked to rate their familiarity of each city on a scale from 0 (none) to 6 (quite familiar). Each city's latitude is asked to be reasonably estimated due to one's related knowledge. For distance estimates participants are asked to estimate distances of 66 pairs of cities. All the tasks are asked to be done individually and instinctually. To avoid giving estimates in sequences, the order of city pairs was randomized. To make sure latitude task and distance task are done individually, on 30 type "A" questionnaires the distance task came first, on the remaining 30 type "B" the latitude task came first. The possibility of each type a participant will have is 50%.

3.4 Data Analysis

The spatial structure of the distance estimates was analyzed using Multidimensional Scaling (MDS). MDS is a set of data analysis techniques that display the structure of distance-like data as a geometrical picture [21]. Psychologists used MDS to study the category of a set of objects by their distances, similarities or proximity rankings. It is then introduced in cartography to study the spatial structure of mental maps. Here classical MDS is used for the reason that according to the mathematical definition of Classical MDS, if the original distances matrix is entered, the result will represents the actual location relations.

The data are entered into latitude estimates and distance estimates datasheets. Data of distances between cities are entered into a 12*12 matrix, where the element $A[m,n]$ is the distance estimate between city m and city n. The MDS software is introduced to analyze the distance matrix and generate the structure map of city distances using Classical MDS [22]. The software selected is SPSS 11.0 (Statistical Program for Social Sciences 11.0).

To make a comparison, the actual distances matrix is also analyzed by MDS following the same procedure. The actual distances data are from http://www. geobytes.com /CityDistanceTool.htm by calculating the distance of two points on the sphere according to each point's longitude and latitude.

4 Results

4.1 Latitude Estimates

The latitude estimates were examined for two features: (1) how cities are sorted by region; (2) whether gaps among regions are discernable [11]. To better evaluate the biases in latitude estimates, Fig. 2 is made by displaying each city's actual latitude and estimate latitude according to its location.

As shown in Fig. 2, Cities located north of 35° N are northward estimated (represent in the figure that the estimated value is above actual value). However, regionalization is not typical in Fig. 2.

It is found estimate error $\Delta Lati$ is significantly related with that city's actual latitude. Estimate error of a city is defined as

$$\Delta Lati = Lati_{est} - Lati_{actual} \cdot \tag{1}$$

where $Lati_{est}$ is the estimated latitude, and $Lati_{actual}$ the actual latitude of the city.

The linear regression of $\Delta Lati$ and $Lati_{actual}$ results in:

$$\Delta Lati = 0.16377 * Lati_{actual} - 4.65053 . \tag{2}$$

The correlation coefficient R of $\Delta Lati$ and $Lati_{actual}$ is 0.778. According to the significance R-test at the probability level p =0.01 (the lowers the p-level, the more significant the relationship) and degrees of freedom 10, r=0.7079<0.778, which means that the relationship is significant.

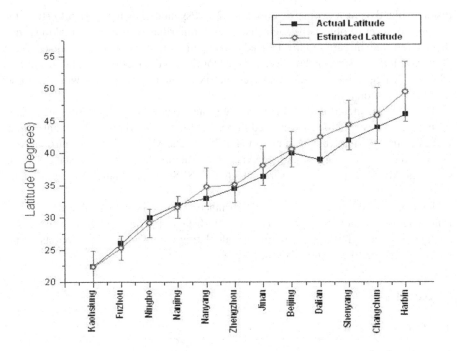

Fig. 2. Latitude estimates result and comparison

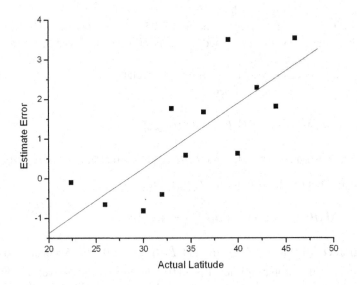

Fig. 3. Linear regression of actual latitude and estimate error

A city's estimated latitude can be deduced from its actual latitude based on equation 2:

$$Lati_{est} = 1.16377*Lati_{actual} - 4.65053 \qquad (3)$$

Equation 3 shows a strong categorization of cities into two regions: northward estimated cities and southward estimated cities. Generally the farther north the city is located, the more north-stretched it is to be estimated. Cities south of certain latitude (in the studied case it is 28.4°N) will be southward estimated that its estimated latitude is below the actual latitude.

4.2 Distance Estimates

The output of both actual and estimated MDS is re-scaled and rotated (the relative location of each points not changed) to be projected to the administrative map of Eastern China. Fig. 4 shows the comparison of actual MDS (on the left) and estimated MDS (on the right). Actual MDS and its projection on the maps is the same as locations of the cities in real world, while estimated MDS could be considered the cognition map of those cities. Moreover, to assure the comparison, estimated MDS has the same re-scale and rotating angle as actual MDS.

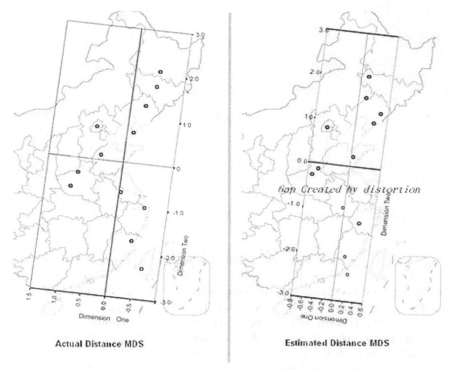

Fig. 4. Actual MDS and estimated MDS

Estimated distance MDS shows that a gap forms from 32° N to 33° N, which divides cities largely into two groups: northern cities and southern cities. The north-south discrimination of distance estimates is also accordant with the latitude estimates. This phenomenon is best explained by regulation-related climate factor because the gap is where the Q-H border located. More proofs are that only cities near Q-H border are significantly stretched away because their experienced-climates are more influenced by the Q-H border. Especially the city Nanyang, although it is on the Q-H border and belongs to the South Hua geographically, it is also in a typical North Hua province thus has heating for winter. It is concluded that the gap between two cognitive regions of Eastern China is mainly influenced by regulation-related climate difference.

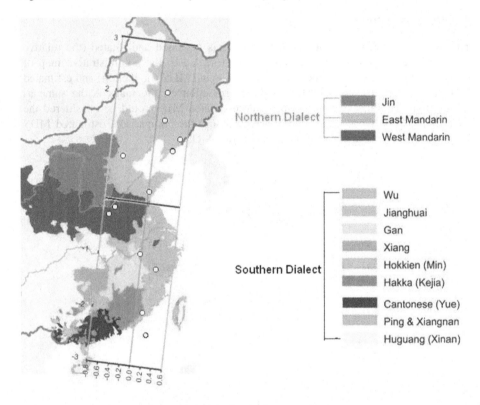

Fig. 5. Estimated MDS and its projection on dialect map

Projecting estimated MDS onto the dialect map, we will find that cities within the same dialect region are estimated to be closer. The north-south gap also corresponds with northern dialect and southern dialect division. It shows the influence of culture because people speaking the same dialect are more likely to share the same culture [18], therefore more likely to consider themselves are within the same geographical region. However, the dialect border does not correspondent with the gap which indicates that the regionalization of eastern China is less influenced by dialects.

Moreover, the gap created by estimates does not correspondent with the landmark used for the task, where the estimate error is zero according to the linear regression equation (equation 2). Although Q-H border is the cause of the gap of the latitude estimates, it is not the landmark because the landmark, if there is one, should be located around 28.4°N.

4.3 Correlations Between Latitude and Distance Estimates

If the coordinates of the MDS model points are projected to a Euclidean metric scale (distances under global-scale should be described as spherical metrics. However, assuming that participants' spatial knowledge is not precise enough to distinguish among these spherical and Euclidean metrics, distance estimates in Euclidean metric are acceptable [11]), a set of latitude values will be generated as the projection into latitude domain called indirect latitude estimates (i.e. estimates generated from latitude estimates part of the questionnaires are called direct estimates).

The Pearson correlation coefficient of direct and indirect latitude estimates is 0.998, showing a strong linear relationship. It is concluded that both latitude estimates (absolute location judgments) and distance estimates (relative location judgments) are based on a common representation at global scale.

5 Conclusion

There are four characteristics of Eastern China's location estimates: (1) participants divided Eastern China into two non-overlapping psychological regions: the north and the south; (2) there is a large gap between the north and the south; (3) a city's estimated latitude is strongly related to its location. The farther north the city is located, the more north-stretched it is to be estimated. Cities south of certain latitude will be southwardly estimated; (4) both latitude estimates and distance estimates are based on a common representation at global scale.

Similarly, previous researches show four characteristics of North America's location estimates: "(1) participants divided the continents into non-overlapping psychological regions that could be independently influenced by new information; (2) the regions typically had large "boundary zones" between them; (3) there was relatively little north-south discrimination among the locations of cities within most regions; (4) for both the Old and the New World, the estimates became more biased as the cities being estimated were actually located farther south; (5) Latitude estimates and distance estimates are generated from the same representation of geographical knowledge" [11].

By comparing those characteristics, three cultural universal aspects are found:

1. Categorical organization of regions.
2. The existence of relation between estimated latitudes and actual location.
3. The existence of common representation for absolute location judgments and relative location judgments [23].

Both the Chinese and the North Americans have the most important features of spatial cognition in common---the categorical storage of spatial information and the reasoning process. Despite the emphasis on absolute frames to express location in Chinese language, it has little influence on Chinese people's general cognitive behavior.

However, landmarks used by the subjects during estimation are quite different between North America and Eastern China. Subjects in China did not show the clear preference of any geographical features as the landmark, while that European and North American subjects use the Equator as the landmark in their experiment [11, 12, 13]. The lack of global landmarks during the estimates tasks in Eastern China is due to the cities chosen as the stimuli in the task. While most cities using in previous study located from Canada to Mexico, in our study the cities are all within the small region of Eastern China.

6 Discussion

People tend to exaggerate the difference between cultures. As the process of globalization and the prevalence of English language, cultural differences become minor yet still worth paying attention to. The results of latitude estimates in Eastern China approve universalities across cultures shared by human beings when facing location estimate tasks. Thus it is not surprising to see the triumph of English-based GIS software such as ArcGIS in Chinese market.

Although there are differences between Chinese and North American while making the location estimates, the differences are more caused by local knowledge such as the unique regulation of one region. These are parts of the culture, but very weak parts because it has not been transmitted across generations. It is likely that in other regions of the world, the differences in geographical cognition are mainly caused by non-cultural or weak cultural issues such as the climate. What cultural and non-cultural issues are related with spatial cognition? What roles do they play? These are all interesting questions for the future researchers to answer.

While admitting generalities of geographical estimates at global scale, there are still important disparities at smaller scale caused by cultural differences that may interfere with daily geographical behaviors (e.g. navigation, [17]). The different spatial frames used by different cultures are also important when people study the path-finding or route memorizing behavior across cultures [24].

Acknowledgments. The authors are indebted to our sincere gratitude to Dr. George Taylor and Dr Chaowei Yang for their valuable advices during the preparation of the manuscript, and Dr. Daniel Montello for his constructive comments on data interpretation.

References

1. Montello, D.R.: How significant are cultural differences in spatial cognition? In: Kuhn, W., Frank, A.U. (eds.) COSIT 1995. LNCS, vol. 988, pp. 485–500. Springer, Heidelberg (1995)
2. Frank, A.U., Mark, D.M.: Language issues for GIS. In: Maguire, D.J., Goodchild, M.F., Rhind, D.W. (eds.) Geographical Information Systems: Principles and Applications, vol. 1, pp. 147–163. Longmans, London (1991)
3. Mark, D.M., Gould, M.D., Nunes, J.: Spatial language and geographic information systems: cross-linguistic issues. In: Proceedings, 2nd Latin American Conference on Applications of Geographic Information Systems, Merida, Venezuela, pp. 105–130 (1989)

4. Huttenlocher, J., Hedges, L.V., Duncan, S.: Categories and particulars: Prototype effects in estimating spatial location. Psychological Review 98, 352–376 (1991)
5. Advanced Chinese dictionary. People's Education Press, Beijing, China (2005)
6. Saarinen, A.O.: Improving Information Systems Development Success under Different Organizational Conditions. In: Proceedings Urban and Regional Information Systems Association 1987 Annual Conference, Washington, DC: URISA, vol. 4, pp. 1–12 (1987)
7. Montello, D.R.: Spatial cognition. In: Smelser, N.J., Baltes, P.B. (eds.) International Encyclopedia of the Social & Behavioral Sciences, vol. 7, pp. 14771–14775. Pergamon, Oxford (2001)
8. Montello, D.R.: Regions in geography: Process and content. In: Duckham, M., Goodchild, M.F., Worboys, M.F. (eds.) Foundations of Geographic Information Science, pp. 173–189. Taylor & Francis, London (2003)
9. Couclelis, H.: Location, place, region, and space. In: Abler, R.F., Marcus, M.G., Olson, J.M. (eds.) Geography's Inner Worlds, pp. 215–233. Rutgers University Press, New Brunswick, NJ (1992)
10. Friedman, A., Brown, N., McGaffey, A.: A basis for bias in geographical judgments. Psychonomic Bulletin & Review 9, 151–159 (2002)
11. Friedman, A., Montello, D.R.: Global-scale location and distance estimates: Common representations and strategies in absolute and relative judgments. Journal of Experimental Psychology: Learning, Memory, and Cognition 32, 333–346 (2006)
12. Friedman, A., Kerkman, D., Brown, N.R., Stea, D., Cappello, H.: Cross-cultural similarities and differences in North Americans' geographic location judgments. Psychonomic Bulletin & Review 12(6), 1054–1060 (2005)
13. Friedman, A., Kerkman, D.D., Brown, N.: Spatial location judgments: A cross-Psychonomic national comparison of estimation bias in subjective North American geography. Bulletin & Review 9, 615–623 (2002)
14. Levinson, S.: Language and space. Annual Review of Anthropology 25, 353–382 (1996)
15. Stea, D., Blaut, J.M., Stephens, J.: Cognitive mapping and culture: Mapping as a cultural universal. In: Portugali, J. (ed.) The construction of cognitive maps, Springer, Berlin (1996)
16. von Humboldt, W.: On language: On the Diversity of Human Language Construction and its Influence on the Mental Development of Human Species. Cambridge University Press, Cambridge, UK (1999)
17. Suzuki, K., Wakabayashi, Y.: Cultural Differences of Spatial Descriptions in Tourist Guide books. In: Spatial Cognition IV: Reasoning, Action, and Interaction, Springer, Berlin (2005)
18. Sapir, E., Mandelbaum, D.G., Hymes, D.H., Sapir, E.: Selected Writings of Edward Sapir in Language, Culture and Personality. University of California Press, Berkeley, CA (1986)
19. Talmy, L.: How language structures space. In: Pick, H., Acredolo, L. (eds.) Spatial Orientation: Theory, Research and Application, Plenum Press (1983)
20. Collins, A.M., Michalski, R.: The logic of plausible reasoning: A core theory. Cognitive Science 13, 1–49 (1989)
21. Young, F.W.: Multidimensional Scaling. In: Kotz, J. (ed.) Encyclopedia of Statistical Sciences, pp. 649–658. Wiley, Chichester (1985)
22. Young, F.W., Hamer, R.M.: Multidimensional Scaling: History, Theory and Applications. Erlbaum, New York (1987)
23. Friedman, A., Brown, N.: Reasoning about geography. Journal of Experimental Psychology: General 129, 193–219 (2000)
24. Davies, C., Pederson, E.: Grid patterns and cultural expectations in urban wayfinding. In: Montello, D. (ed.) Spatial information theory: Foundations of geographic information science, pp. 400–414. Springer, Heidelberg (2001)

Cross-Cultural Similarities in Topological Reasoning

Marco Ragni[1], Bolormaa Tseden[1], and Markus Knauff[2]

[1] University of Freiburg, Department of Computer Science
{ragni,tseden}@informatik.uni-freiburg.de
[2] Justus-Liebig-Universität Gießen, Department of Psychology
markus.knauff@psychol.uni-giessen.de

Abstract. How do we reason about topological relations? Do people with different cultural backgrounds differ in how they reason about such relations? We conducted two topological reasoning experiments, one in Germany and one in Mongolia to analyze such questions. Topological relations such as "A overlaps B", "B lies within C" were presented to the participants as premises and they had to find a conclusion that was consistent with the premises ("What is the relation between A and C?"). The problem description allowed multiple possible "conclusions". Our results, however, indicate that the participants had strong preferences: They consistently preferred one of the possible conclusions and neglected other conclusions, although they were also consistent with the premises. The preferred and neglected conclusions were quite similar in Germany and Mongolia.

1 Introduction

Imagine the following conversation between two friends:

A says: "My bag is *in* your car."
B says: "My car is *in* my garage."

It is easily inferable from these two statements that A's bag must be in B's garage. Now imagine the following conversation:

A says: "The green car is *between* the red car and the blue car."
B says: "The green car is *beside* the red car."

What do you infer from these two statements? In fact, the cars can be in two different arrangements.

RED   GREEN   BLUE   or   BLUE   GREEN   RED

Now imagine that the two friends are talking about colored regions on a thematic map.

A says: "The white region is *inside* the black region."
B says: "The gray region *overlaps* the black region."

S. Winter et al. (Eds.): COSIT 2007, LNCS 4736, pp. 32–46, 2007.

What can be inferred from these two statements? There are, from a topological perspective (discerning boundaries as well), five possibilities how the three areas can be located on the map (cf. Fig. 1).

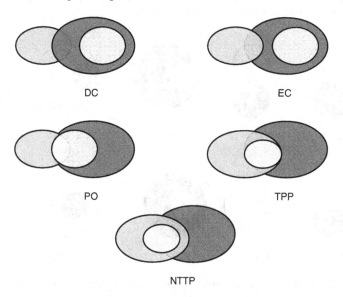

Fig. 1. All possible models for the premises "The white region is inside the black region" and "The gray region overlaps the black region"

All three problems are typical examples of everyday experiences. We frequently use spatial expressions that describe the topological relations between regions or solid objects, but we do not communicate metrical information such as the distances between regions or even the ordering of objects. We use the term "beside" but do not specify whether we want to indicate that the object is to the left or to the right of the other object.

Although reasoning about topological relations is an important issue in AI-research, only very little cognitive research has been done on *human topological reasoning* (e.g., [20]). The aim of this paper is to strengthen this link between AI-research and experimental cognitive research on topological reasoning. As the formal basis of our research we use the RCC-8 calculus, which was developed by Randell, Cui, and Cohn [31]. By using a set-theoretic approach Egenhofer and Franzosa [9] could derive the same relations within their 4-Intersection Calculus. In RCC8, two regions can be in eight different topological relations (see Fig. 2). They can be *disconnected* (DC) - both regions have no common point - two regions can be *externally connected* (EC) - they have common points on the border - or two regions can *partially overlap* (PO) - both regions share a proper subregion. Other possibilites are that the first region can be part of the second and the two borders touch each other we get *tangentially proper part* (TPP) or the first region can contain the second region (TPPI). If one region contains the other and the borders do not touch, we get *non-tangentially proper part* (NTTP) and for the inverse relation (NTTPI). Finally, two regions can be *equal* (EQ). It is easy to show that by taking two regions in a two- or three-dimensional Euclidean space, the relation between these two regions can be expressed exactly by one of these eight base relations.

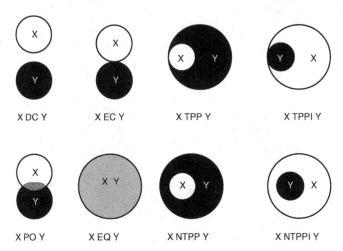

Fig. 2. Two-dimensional examples for the eight basic relations of RCC-8

In the two experiments discussed here we used the RCC relations to construct a type of reasoning tasks that is referred to as *three-term-series-problems* (3ts-problems) in cognitive research (e.g., [13,16]). In these tasks always two statements are used as premises and the task of the participants is to generate a statement that is consistent with the premises – the conclusion. Take for instance,

A overlaps B.
B overlaps C.
Which relation can hold between A and C?

The 3ts-problems can be formally described by the composition of two base relations and the question for a (all) satisfiable relations. The set of all possible relations that has X r1 Y, Y r2 Z as its premise are denoted by the composition $c(r1, r2)$. This is normally presented as a composition table (cf. Table 1).

For the above example $c(PO, PO)$ contains the following four relations: DC, PO, EQ, EC. Since RCC-8 consists of eight base relations, there are 64 possible compositions of two base relations. In other words, exactly 64 different three-term-series problems exist. If we omit all one-relation cases (cells in Table 1), it results in 37 multiple relation cases (the shaded cells in Table 1) out of the 64 possible compositions. The participants of our studies were confronted with these 37 problems and had to infer a relation ("conclusion") about the first and the third object. Since RCC-8 allows one to model arbitrary regions, we decided to limit the discourse universe to two-dimensional disks.

Our research was motivated by two main research questions: First, we were interested in how individuals deal with problems for which the composition allows more than one relation. Knauff and colleagues were able to show in another domain (reasoning with ordering information such as in the Allen-interval calculus) that whenever a composition has multiple relations, human reasoners consistently prefer to infer only one of the relations and ignore others [20,21,27,28,32]. We sought to explore whether such preferences also hold in reasoning about topological relations. Therefore, in the experiment, we only used the compositions with more than one relation and also dropped the (trivial) compositions that contained "equal" relations.

Table 1. The formal composition table. White cells contain a single relation only. * stands for the union of all relations. Taken from [30].

	DC	EC	PO	EQ	TPP	TPPI	NTPP	NTPPI
DC	*	DC, EC, PO, TPP, NTPP	DC, EC, PO, TPP, NTPP	DC	DC, EC, PO, TPP, NTPP	DC	DC, EC, PO, TPP, NTPP	DC
EC	DC, EC, PO, TPPI, NTPPI	DC, EC, PO, TPP, TPPI, EQ	DC, EC, PO, TPP, NTPP	EC	EC, PO, TPP, NTPP	DC, EC	PO, TPP, NTPP	DC
PO	DC, EC, PO, TPPI, NTPPI	DC, EC, PO, TPPI, NTPPI	*	PO	PO, TPP, NTPP	DC, EC, PO, TPPI, NTPPI	PO, TPP, NTPP	DC, EC, PO, TPPI, NTPPI
EQ	DC	EC	PO	EQ	TPP	TPPI	NTPP	NTPPI
TPP	DC	DC, EC	DC, EC, PO, TPP, NTPP	TPP	TPP, NTPP	DC, EC, PO, TPP, TPPI, EQ	NTPP	DC, EC, PO, TPPI, NTPPI
TPPI	DC, EC, PO, TPPI, NTPPI	EC, PO, TPPI, NTPPI	PO, TPPI, NTPPI	TPPI	PO, EQ, TPP, TPPI	TPPI, NTPPI	PO, TPP, NTPP	NTPPI
NTPP	DC	DC	DC, EC, PO, TPP, NTPP	NTPP	NTPP	DC, EC, PO, TPP, NTPP	NTPP	*
NTPPI	DC, EC, PO, TPPI, NTPPI	PO, TPPI, NTPPI	PO, TPPI, NTPPI	NTPPI	PO, TPPI, NTPPI	NTPPI	PO, TPPI, TPP, EQ, NTPP, NTPPI	NTPPI

The second motivation for our study was the intuition that the way humans reason about topological relations might depend on their cultural background. Many researchers have claimed that thinking in general is highly culturally determined [14]. One reason might be that different geographical, cultural, and social environments foster different thinking styles [2,3,4]. Another view is that thinking is inextricably linked to "language" and that people with different languages would also think differently [22,23].

2 Experiment 1 – Germany

2.1 Participants

We tested 20 undergraduate students from the University of Freiburg. They received course credits for their participation. Half of them were female and half were male; the youngest was 20 the oldest 35 years old. They were all native speakers of German.

2.2 Methods, Materials, and Procedure

The experiment used the set of RCC-relations presented in Table 1. An unequivocal verbal description was developed for all these relations. The verbal translations were based on our experiences in earlier studies [20]. However, to make sure that the obtained data could be clearly related to the involved inference processes, a detailed "definition phase" had to be conducted by all participants prior to the main part of the experiment. In this definition phase the participants saw pictorial examples of the spatial relations and also received a detailed verbal description of the "semantic" of each of the eight relations.

In the main part of the experiment all participants had to solve the same set of 37 3ts-problems. These were presented to each participant in a randomized order. Here is an example-problem:

The blue circle is within the green circle.
The green circle overlaps the red circle.
Which relation does hold between the blue and the red circle?

From the 8X8 possible models we only used the 37 problems for which the composition consists of more than one relation and in which the equal relation does not appear. The experiment was conducted as a group experiment, in which all participants sat in a lecture room and received the experimental materials on an experimental block. Each problem was presented on two pages. On the first page, the two premises were presented as centered sentences. On the second page, the participants were asked for the conclusion. They had to write into a gap between the two phrases "The blue circle ... the red circle". During the entire experiment a second paper was available on which they could check the exact wording of the verbal relations. The participants were not allowed to turn back the pages and they were also asked not to draw sketches or to use anything else that could help to externally solve the problem. The complete inference had to be performed mentally.

2.3 Results and Discussion

Overall, 93 percent of the problems were correctly solved. The results regarding the preference effects are reported in Figure 3 and Table 2. As shown in Table 2, out of the given 37 problems exactly 27 problems (73%) were solved with a clear preference for one relation. However, it is remarkable that several relations could have been chosen as a possible conclusion, but, in fact, the participants chose just one of them and their preferences also often corresponded. Table 2 shows the preferences for each of the compositions separately. In each cell, the first relation (printed in bold) is the preferred relation. The percentages in the line below show the relative frequency of this relation, i.e. how often it was chosen by the participants. The first value is the mean of the two experiments reported here (Germany and Mongolia), the second refers to the present experiment (in German), and the third value refers to the second experiment in Mongolia (reported below). An inspection of the percentages for the present experiment shows how strong the *preference effect* is. The most impressive result can be observed for the DC-EC-problem, where 18 of the participants (90%) chose the DC relation as the conclusion, while only one participant used the relations EC and PO, respectively (5% each), and no one chose the NTTP relation (0%). Most of the results in the table are so pronounced that we do not need to report the statistical tests. It is clearly detectable that the probability to obtain these preferences purely by chance is between 0.34 (for cells with three relations) and 0.14 (for cells with all seven relations).

The reasoners chose sometimes a relation that is inconsistent with the premises. These "errors" are the shaded values in Table 2. However, there were small differences between the tasks. For instance, if participants had to deal with inside, which is TPP, NTPP or the converse, then errors appeared. Conversely, if participants had to deal with the relations DC, EC or PO, no errors appeared. Our results indicate, first that humans perform well in solving topological reasoning problems based on the RCC-relations. The second finding is that there are strong preferences.

There is an unequivocal cognitive preference for DC, EC and PO. From 27 tasks DC had been generated 17 times as an answer, PO was inferred five times, NTPP and NTPPI two times and EQ once. There is obviously a clear difference between the relation DC and PO, which even holds in those cases where these relations are not logically valid, e.g., the composition EC and TPP. While some relations are inferred by most participants, other relations are generated very rarely or even completely neglected, like the relations TPP and NTTP. By analyzing (cf. Table 3) the chosen relations with respect to all consistent relations, the pattern can be described in the following way: If the relation DC is consistent, then it is chosen first, if not, then the relation EC. If there are compositions of TPP and TPPI or NTPP and NTPPI then participants choose EQ.

How can these findings be explained? The preferred relations of the experiment reveal that for humans it is easier to represent and reason with objects 'distinct' or 'identical' than objects 'overlapping' or 'contained in'. The reason might be that if two objects are disjoint (or identical), the number of regions is smaller than when two regions share common subregions. For example, if two objects A and B overlap then there are three regions to represent: a region A, a region B, and the 'subregion' C, where A and B overlap. In this case, several regions have to be represented mentally, so the load on working memory is increased. But why is DC then preferred over EQ?

The reason might lie in the fact, that participants follow the unique names assumption, i.e., different names indicate different objects. Accordingly, a reasoner has to represent in case of DC that two objects are distant, whereas in the case of EQ that two different objects lie identical. In other words, if two regions are represented separately then it already follows that these two cannot be identical. Conversely, if two regions share the same place then it does not follow logically that both regions are different. Therefore, in the second case additional information has to be processed. So the representation with DC is less complex than the representation with EQ.

3   Experiment 2 — Mongolia

This experiment was conducted in Mongolia. 20 students (9 female and 11 male) of the National University of Mongolia (Ulaanbaatar) participated in the experiment. The youngest was 18 years the oldest 28. All were logically naïve reasoners. They were all native speakers of the Mongolian language.

3.1   Methods, Materials, and Procedure

The methods and procedures were identical to those in Experiment 1. The only difference was that the material had to be translated into Mongolian language. This was done by this paper's second author and double-checked by a second Native speaker of Mongolian.

Mongolian is the primary language, i.e. spoken by the majority of Mongolian residents, and officially written in the Cyrillic alphabet. It is also spoken in some of the surrounding areas in China and Russia. Structurally, it has a rich number of morphemes enabling the speaker to construct rather complex words from a 'simple' root. A detailed description of the morphology, lexicon, and syntax can be found in [15].

3.2   Results and Discussion

Overall, the performance of the Mongolian participants was poorer than the ones in Germany. Of 740 answers, 109 were incorrect and 631 correct. The mean number of correct conclusions was 85.2%. Again there were slight differences between the tasks. The participants made most mistakes with the relation DC (45 times), followed by EC (18 times) and EQ (15 times).

Again there were strong preferences. Out of the 37 problems exactly 23 problems (63%) were solved with a clear preference for DC. Here the most impressive result is that for the DC-EC-problem, where 18 of the participants (90%) chose the DC relation as conclusion, while only 2 chose another relation. The preferences for each composition separately are given in Table 2. Again, the results are so clear that we do not need to report statistical values.

EQ is always chosen if a relation and its converse appeared. This concurs with the unification principle described by Rauh and colleagues [32]. Here is as well a clear preference for DC, EC, and PO. Unlike, the first experiment, participants in the second experiment chose EC instead of EQ.

Table 2. The "cognitive" composition table. The bold relation is the preferred one, where the grey shaded are incorrect ones but also chosen by the participants. The tuple gives first the mean over both experiments, the second the percentage of the first experiment, and the third gives the percentage of the second experiment. The entries are sorted w.r.t. the mean values.

	DC	EC	PO	TPP	TPPI	NTPP	NTPPI
DC	**DC** (75%,80%,70%) EC (7,5%,10%,5%) EQ (7,5%,5%,10%) PO (7,5%,5%,10%) NTTPI (2,5%,-,5%)	**DC** (87,5%,90%,85%) EC (5%,5%,5%) PO (5%,5%,5%) NTTP (2,5%,-,5%)	**DC** (85%,80%,90%) PO (10%,15%,5%) EC (5%,5%,5%)	**DC** (80%,75%,85%) PO (12,5%,15%,10%) EC (7,5%,10%,5%)		**DC** (65%,60%,70%) PO (17,5%,25%,10%) EC (12,5%,15%,10%) NTPP (2,5%,-,5%) TPPI (2,5%,-,5%)	
EC	**DC** (85%,80%,90%) EC (7,5%,10%,5%) PO (5%,5%,5%) NTTPI (2,5%,5%,-)	**DC** (70%,70%,70%) EC (22,5%,20%,25%) EQ (5%,10%,-) NTTP (2,5%,-,5%)	DC (80%,85%,75%) PO (12,5%5%,20%) EC (5%,10%,-) NTPP (2,5%,-,5%)	EC (52,5%,30%,75%) NTPP (17,5%,30%,5%) TPP (12,5%,15%,10%) PO (2,5%,5%,-) DC (7,5%,5%,10%) NTPPI (2,5%,5%,-)	DC (62,5%,70%,55%) EC (32,5%,25%,40%) EQ (2,5%,-,5%,-) TPP (2,5%,-,5%)	PO (60%,60%,60%) NTPP (12,5%,20%,5%) EC (12,5%,5%,20%) TPP (2,5%,-,5%) EC (12,5%,5%,20%) DC (12,5%,10%,15%)	
PO	DC (70%,65%,80%) EC (10%,10%,10%) PO (5%,10%,-) TPP (5%,-,10%)	DC (77,5%90%,65%) PO (12,5%,5%,20%) EC (5%,5%,5%) NTPP (2,5%,-,5%) TPP (2,5%,-,5%)	PO (70%,65%,75%) PO (17,5%,25%,10%) EC (7,5%,-,15%) EQ (5%,10%,-)	PO (40%,35%,45%) NTPP (30%,50%,10%) TPP (2,5%,-,5%) DC (17,5%,5%,30%) EC (7,5%,5%,10%) NTPPI (2,5%,5%,-)	DC (30%,30%,30%) PO (30%,35%,25%) EC (25%,25%,25%) NTPPI (5%,5%,5%) TPPI (2,5%,-,5%) EQ (7,5%,5%,10%)	PO (45%,40%,50%) NTPP (30%,35%,25%) TPP (5%,10%,-) DC (10%,5%,15%) NTTPI (2,5%,5%,-) TPPI (2,5%,5%,-) EC (2,5%,-,5%) EQ (2,5%,-,5%)	DC (40%,55%,25%) EC (30%,25%,35%) PO (27,5%,15%,40%) TPPI (2,5%,5%,-)
TPP		DC (67,5%,75%,60%) EC (30%,20%,40%) PO (2,5%,5%,-)	DC (35%,50%,20%) PO (32,5%,25%,40%) NTPP (10%,-,20%) EC (12,5%,10%,15%) TPP (2,5%,5%,-) NTPPI (12,5%,10%,5%)	NTPP (45%,55%,35%) TPP (25%,25%,25%) DC (17,5%,10%,25%) NTPPI (10%,10%,10%) EC (2,5%,-,5%)	EQ (40%,35%,45%) PO (20%,30%) DC (17,5%,30%,5%) EC (7,5%,10%,5%) TPPI (2,5%,-,5%) NTPPI (7,5%,5%,10%)		PO (45%,60%,30%) DC (22,5%,20%,25%) NTPPI (10%,10%,10%) EC (2,5%,-,5%) EQ (10%,5%,15%) TPP (7,5%,5%,10%) NTPP (2,5%,-,5%)
TPPI	DC (80%,75%,85%) PO (7,5%,10%,5%) NTTPI (5%,5%,5%) EC (5%,5%,5%) TPPI (2,5%,5%,-)	EC (47,5%,35%,60%) NTPPI (20%,30%,10%) PO (20%,25%,15%) TPPI (5%,5%,5%) DC (7,5%,5%,10%)	PO (52,5%,30%,75%) PO (10%,15%,5%) NTPP (5%,5%,5%) NTPPI (27,5%,45%,10%) EC (5%,5%,5%)	EQ (47,5%,60%,35%) PO (20%,20%,20%) TPPI (2,5%,-,5%) TPP (5%,5%,5%) DC (12,5%,5%,20%) NTPPI (7,5%,5%,10%) EC (2,5%,-,5%) NTPP (2,5%,5%,-)	NTPPI (37,5%,55%,20%) TPPI (15%,10%,20%) TPPI (5%,5%,5%) EQ (5%,10%,-) PO (5%,10%,-) EC (15%,-,30%) DC (20%,15%,25%)	PO (47,5%,55%,40%) NTPP (15%,30%,-) EQ (15%,-,30%) DC (10%,5%,15%) EC (5%,-,10%) NTPPI (5%,5%,5%) TPPI (2,5%,5%,-)	
NTPP			DC (30%,20%,40%) EC (30%,40%,20%) PO (22,5%,15%,30%) NTPP (7,5%,10%,5%) TPP (2,5%,5%,-) EQ (5%,-,5%) NTPPI (2,5%,5%,-)		PO (55%,60%,50%) DC (20%,15%,25%) NTPP (7,5%,10%,5%) EC (5%,-,10%) NTPPI (7,5%,10%,5%) EQ (5%,5%,5%)		PO (30%,40%,20%) EQ (27,5%,15%,40%) DC (22,5%,25%,20%) NTPPI (12,5%,10%,15%) NTPP (7,5%,10%,5%)
NTPPI	DC (57,5%,50%,65%) EC (15%,10%,20%) NTPPI (12,5%,25%,-) PO (10%,15%,5%) TPPI (5%,-,10%)	PO (52,5%,55%,50%) NTTPI (20%,25%,15%) DC (12,5%,10%,15%) EC (10%,10%,10%) TPP (5%,-,10%)	PO (37,5%,25%,50%) TPPI (7,5%,10%,5%) DC (10%,5%,15%) TPP (5%,5%,5%) EQ (2,5%,-,5%) NTPPI (32,5%,55%,10%)	PO (45%,40%,30%) NTPP (40%,45%,35%) EQ (5%,10%,-) DC (10%,5%,15%) EC (5%,-,10%) NTPP (2,5%,-,5%) TPP (2,5%,-,5%)		EQ (42,5%,40%,45%) NTPPI (27,5%,30%,25%) PO (5%,10%,-) NTPP (10%,10%,10%) TPP (2,5%,-,5%) DC (12,5%,10%,15%)	

3.3 Overall Analysis and Cross-Cultural Comparison

The two experiments differed mainly in the groups of participants having different cultural backgrounds. Nevertheless, in both experiments we found clear preferences. We did not treat the experiments as a single study with the two different groups of participants as a between-subjects factor, because the experimental set-ups were not absolutely identical, and a conjoint analysis would result in statistical problems (e.g. inhomogeneity of variance). However, a direct comparison between the two groups can provide additional evidence in support of our preference account and, in particular, is directly related to the hypothesis that the way human reason with topological relations might depend on the cultural background of the individuals. In the following, we, therefore report an overall analysis of the two experiments. All together out of 740 (20x37=740) answers, 55 were incorrect, and 685 were correct (error rate of about 7%). The relations which were given most frequently as an answer were: DC, PO, NTPPI and NTTP. Even in those tasks in which DC had not been expected it was given nevertheless as an answer. If one compares the analyses of both experiments, it can be seen that the wrong answers were given within the same tasks. Most subjects had no difficulties with relations like "disconnected", "touched" but with relations like TPP, NTPP, NTPPI. In both experiments we have the relations DC, EC and PO. A difference is that Mongolians chose more often relations like EC (see Table 2), whereas German participants did not use this relation often. Likewise, in the second experiment the relation NTPP was chosen only 44 times, while in the first experiment the relation NTTP was chosen 82 times.

4 General Discussion

We reported two major findings. First, we found that human reasoners do not draw all inferences that are possible from a formal point of view. They only choose a subset of compositions from the formal composition table. These preferences seem not to be affected by cultural aspects, since there was no significant difference in the preferences of the different relations by the Mongolian and German participants. The second finding is that humans prefer those relations which have the smallest overlapping complexity. In the following we discuss both findings.

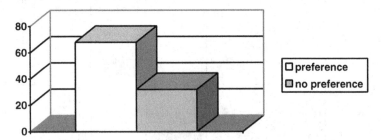

Fig. 3. The overall preference effect for both experiments. The figure shows that in about 68 percent of the cases participants had a clear preference for a relation.

There is an impressive similarity between the preferences in Germany and Mongolia. This finding has two theoretical consequences. The first is that our results present serious problems for the idea that the way humans reason with topological relations might depend on the cultural background of the individuals. The German and Mongolian languages are quite dissimilar and the theory of linguistic relativity would have predicted strong differences between the German and Mongolian participants. The same would have been the prediction of accounts that claim that different geographical, cultural, and social environments foster different thinking styles. We did not find such differences. Instead, we found strong universalities. Our interpretation is that the preferences are an immediate result of the structure of our neuro-cognitive system [11]. They rely on very elementary functions of working memory and reasoning and these are almost encapsulated processes that are quite unaffected by cultural aspects (but see in specific cases [6,14,34]).

This is related to the second corollary from our study. If the reasoners draw the same inferences they seem to use the same cognitive strategy to solve the problems. We believe that they all construct, inspect, and validate preferred mental models. The preferred models account has been developed in our group in the last years [20,27,28]. The main assumption is that the major strategy employed in spatial reasoning is the successive construction of a simulation, or model of the "state of affairs", which contains all the information given in the premises in an integrated representation. New information, such as a reasoning problem's conclusion, is generated or evaluated by inspecting and varying the possible models [19]. In the domain of reasoning with spatial relations, the theory of mental models has received much empirical support. For example, the process of integrating the second premise into the mental model has been directly demonstrated by an increased demand on processing resources during this stage of problem presentation. The *preferred models* theory describes the reasoning process in three distinct phases [19]. In the construction phase, reasoners construct a mental model that reflects the information from the premises. If new information is encountered during the reading of the premises it is immediately used in the construction of the model. During the inspection phase, this model is inspected to find new information that is not explicitly given in the premises. Finally, in the variation phase alternative models are searched that refute this putative conclusion. The main difference from the other versions of the mental model theory is that the empirical data do not support the notion that people construct several models. Instead, people only construct one model and then vary this model by a process of local transformation. This model, which is called the preferred mental model (PMM) is easier to construct and to maintain in working memory compared to all other possible models [28]. In the model variation phase this PMM is varied to find alternative interpretations of the premises (e.g., [32]). In the present context that means that our participants only constructed the preferred model for the problems they were confronted with. As there was no need to find alternative interpretations of the premises – we asked for a *possible* relation, not a logically *necessary* one – they stopped as soon as they had constructed the first model that was consistent with the premises. This preferred model is the computationally the cheapest as it is less demanding in terms of processing and in terms of representational costs (see [27,28]). From this model they could "read off" just one of the formally possible relations and this resulted in the preference for that specific relation or conclusion, respectively. The preferred model leads to a specific relation and this in turn results in the preferred conclusion. In the following we explain why some models (relations) could be preferred over others.

4.1 An Approach to Explain the Preferences

The starting point of our account to explain the preferences in reasoning with the RCC relations is to set the preferred relations in Table 2 in relation with all consistent relations (Table 3).

Whenever the relation DC is consistent, participants (both Mongolian and Germans) have preferred this relation (cf. Table 3). Otherwise, they chose the relation EC or EQ. But the latter is chosen only if a relation and its converse is composed (cf. Table 2). Then a cultural difference appears: Mongolians prefer the relation PO over NTPPI (Fig. 5) and Germans NTPPI over PO (Fig. 6). The sequence of preferred relations (DC, EC) avoid the overlapping of regions (cf. Fig. 5).

Table 3. The chosen relations with respect to the set of consistent relations for the first and the second experiment. (N)TPP is an abbreviation for the disjunction NTTP and TPP.

	MONGOLIA								GERMANY							
	Relation chosen								Relation chosen							
	DC	EC	PO	TPP	TPPI	NTTP	NTPPI	EQ	DC	EC	PO	TPP	TPPI	NTTP	NTPPI	EQ
ALL	33 55,0%	4 6,7%	8 13,3%	0 0%	0 0%	1 1,7%	4 6,7%	10 16,7%	34 56,7%	2 3,3%	14 23,3%	0 0%	0 0%	2 3,3%	2 3,3%	6 10%
DC EC	23 57,5%	16 40,0%	0 0%	1 2,5%	0 0%	0 0%	0 0%	0 0%	29 72,5%	9 22,5%	1 2,5%	0 0%	0 0%	0 0%	0 0%	1 2,5%
DC EC PO (N)TPP	92 58,2%	18 11,4%	34 21,5%	0 0%	1 0,6%	9 5,7%	2 1,3%	2 1,3%	99 61,9%	15 9,4%	33 20,6%	2 1,3%	0 0%	4 2,5%	5 3,1%	2 1,3%
DC EC PO TPP TPPI EQ	15 37,5%	6 15%	6 15%	0 0%	1 2,5%	1 2,5%	2 5,0%	9 22,5%	20 50%	6 15%	4 10%	0 0%	0 0%	0 0%	1 2,5%	9 22,5%
DC EC PO (N)TPPI	94 58,8%	20 12,5%	26 16,3%	5 3,1%	3 1,9%	1 0,6%	6 3,8%	5 3,1%	96 60%	18 11,3%	31 19,4%	1 0,6%	2 1,3%	0 0%	10 6,3%	2 1,3%
EC PO (N)TPP	2 10%	15 75%	0 0%	2 10%	0 0%	1 5,0%	0 0%	0 0%	1 5,0%	6 30%	3 15%	3 15%	0 0%	6 30%	1 5%	0 0%
EC PO (N)TPPI	2 10%	12 60%	3 15%	0 0%	1 5,0%	0 0%	2 10%	0 0%	1 5,0%	7 35%	5 25%	0 0%	1 5,0%	0 0%	6 30%	0 0%
PO EQ TPP TPPI	4 20%	1 5%	4 20%	1 5%	1 5%	0 0%	2 10%	7 35%	1 5%	0 0%	4 20%	1 5%	0 0%	1 5,0%	1 5,0%	12 60,0%
PO TPP NTPP	15 18,8%	9 11,3%	39 48,8%	1 1,3%	0 0%	8 10%	1 1,3%	7 8,8%	5 6,3%	2 2,5%	38 47,5%	3 3,8%	2 2,5%	27 33,8%	3 3,8%	0 0%
PO (N)TPPI	9 11,3%	5 6,3%	41 51,3%	4 5,0%	2 2,5%	5 5,0%	14 17,5%	1 1,3%	4 5,0%	3 3,8%	30 37,5%	1 1,3%	5 6,3%	1 1,3%	34 42,5%	2 2,5%
PO EQ (N)TPPI (N)TPP	3 15,0%	0 0%	0 0%	0 0%	1 5,0%	2 10%	5 25%	9 45%	2 10%	0 0%	2 10%	0 0%	0 0%	2 10%	6 30%	8 40%
TPP NTPP	5 25%	1 5%	0 0%	5 25%	0 0%	7 35%	2 10%	0 0%	2 10%	0 0%	0 0%	5 25%	0 0%	11 55%	2 10%	0 0%
TPPI NTPPI	5 25%	6 30%	0 0%	1 5,0%	4 20%	0 0%	4 20%	0 0%	3 15%	0 0%	1 5%	1 5%	2 10%	2 10%	11 55%	0 0%
Total	302 40,9%	113 15,3%	161 21,8%	20 2,7%	14 1,9%	34 4,6%	44 6,0%	50 6,8%	297 40,1%	68 9,2%	166 22,4%	17 2,3%	12 1,6%	56 7,6%	82 11,1%	42 5,7%

It is remarkable that the sequence of relations in both cultures (Figure 5 and Figure 6) adhere to a principle of "overlapping avoidance". Vögele, Schlieder, and Visser [35] proposed this concept for predicting preferences in RCC-5. Since their main assumption was "A configuration with singularities ... never acts as preferred model" (p. 245) their concept could not predict that EC is even preferred over PO or NTPP. Modifying and extending their concept could be a help in explaining then the preferences of the German experiment (but not the preference of the Mongolian experiment), since NTPPI would be computationally cheaper than PO. The reason why EQ is chosen (cf. Table 2) in the case where (N)TPP and (N)TPPI are composed can be easily explained by the so-called *unification principle* [32]. This principle claims that if participants compose a relation and its converse then they generally tend to choose the equal relation. Since TPP and NTTP are very similar this unification principle is applied for both relations together.

It seems to be the case that representations of separated regions are preferred over those representations where regions share a common subregion. The reason might be that conceptual chunking is easier in the first case, since only two separate regions have to be stored and not three kinds of regions, both original regions and a common subregion.

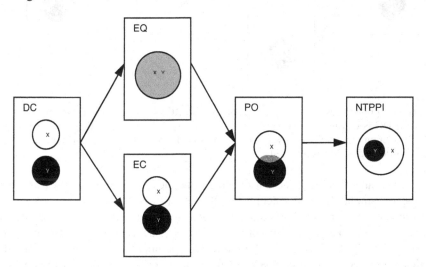

Fig. 5. Human preferences (Mongolia) in reasoning with RCC-8 relations for Experiment 2 (cf. Table 3). If DC is consistent, then participants prefer this relation, if not, they choose EC or EQ. Both relations are preferred over PO and NTPPI. There are 5 cases not consistent with this preferences namely the cells corresponding to EC;NTPP, and (N)TPP;(N)TPPI.

Since we have now identified a principle that can account for the empirical findings, namely the principle of overlapping avoidance, we can compare this with the other principles that were found for a very similar calculus. A study by Knauff, Rauh & Schlieder [20] analyzed the conceptual and inferential adequacy of Allen's interval calculus in the spatial domain, and it could have shown similar results: that participants do really construct preferred mental models in multiple-model cases and that all participants consistently prefer the same solution [20]. Rauh and colleagues [32] could explain the results with the help of three main preference strategies: The principles of linearization, regularization, and unification. Since the principle of linearization depends on an ordering, and since we have pure topological objects (without any ordering) this principle cannot be applied to our results. Neither does the principle of regularization hold for our results: The relation EC (comparable to the relation *meets* and *is met by* in Allen's Calculus) is the second most preferred relation in our findings. The only principle that can be transferred is the regularization principle as was already outlined.

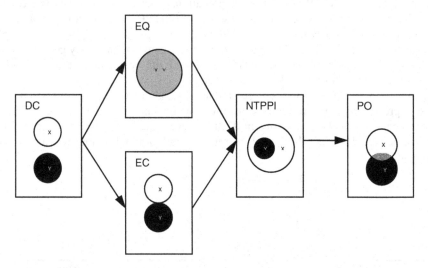

Fig. 6. Human preferences (Germany) in reasoning with RCC-8 relations for Experiment 2 (cf. Table 3). If DC is consistent, then participants prefer this relation, if not, they choose EC or EQ. Both relations are preferred over NTPPI and PO. In contrast to Fig. 5 (Mongolia) the sequence of PO and NTPPI is reversed.

It is remarkable that only one out of three principles can be transferred from the topologically similar Interval calculus to RCC-8. It would, however, be of great interest to conduct a cross-cultural study with Allen's Calculus, too. Especially since the Interval Calculus implies an ordering, cultural dependencies might play a substantial role in its expression. Furthermore, an empirical as wells as formal investigation is necessary to compare the found preferences for RCC8 with similar findings of Mark and Egenhofer [26] based on the analysis of sketches of line region relations reflecting natural language phrases in English and Spanish.

Acknowledgments. This research was supported by grants to MK from the DFG (German National Research Foundation) in the Transregional Collaborative Research Center, SFB/TR 8 project. The authors would like to thank Bernhard Nebel, Stefan Wölfl, and Andreas Bittner for comments, suggestions, and proof-reading of an earlier version of this paper. We would also like to thank three anonymous reviewers for their helpful comments.

References

1. Allen, J.F.: Maintaining knowledge about temporal intervals. Communications of the ACM 26, 832–843 (1983)
2. Berry, J.W.: Human ecology and cognitive style: Comparative studies in cultural and psychological adaptation. Sage/Halstead/Wiley, New York (1976)
3. Berry, J.W.: An ecological approach to understanding cognition across cultures. In: Altarriba, J. (ed.) Cognition and culture: A cross-cultural approach to cognitive psychology, pp. 361–375. North-Holland, Amsterdam (1993)
4. Berry, J.W., Saraswathi, T.S., Dasen, P.R. (eds.): Handbook of cross-cultural Psychology, Allyn & Bacon, Boston, MA. Basic processes and human development, vol. II (1997)
5. Byrne, R.M.J., Johnson-Laird, P.N.: Spatial reasoning. Journal of Memory and Language 28, 564–575 (1989)
6. Chan, T.T., Bergen, B.: Writing direction influences spatial cognition. In: Bara, B., Barsalou, L.W., Bucciarelli, M. (eds.) Proceedings of the 27th Annual Conference of the Cognitive Science Society, pp. 412–417. Erlbaum, Mahwah, NJ (2005)
7. Cohn, A.G.: Qualitative spatial representation and reasoning techniques. In: Brewka, G., Habel, C., Nebel, B. (eds.) KI-97: Advances in Artificial Intelligence, pp. 1–30. Springer, Heidelberg (1997)
8. Egenhofer, M.J.: Reasoning about binary topological relations. In: Günther, O., Schek, H.J. (eds.) Proceedings of the Second Symposium on Large Scaled Spatial Databases, pp. 143–160. Springer, Berlin (1991)
9. Egenhofer, M.J., Franzosa, R.: Point-set topological spatial relations. International Journal of Geographical Information Systems 2, 133–152 (1991)
10. Egenhofer, M.J., Clementini, E., Di Felice, P.: Topological relations between regions with holes. International Journal of Geographical Information Systems 2, 129–144 (1994)
11. Fangmeier, T., Knauff, M., Ruff, C.C., Sloutsky, V., FMRI,: evidence for a three-stage model of deductive reasoning. Journal of Cognitive Neuroscience 18, 320–334 (2006)
12. Freksa, C.: Temporal reasoning based on semi-intervals. Artificial Intelligence 54, 199–227 (1992)
13. Hunter, I.M.L.: The solving of three-term series problems. British Journal of Psychology 48, 286–298 (1957)
14. Jahn, G., Knauff, M., Johnson-Laird, P.N.: Preferred Mental Models in Reasoning about Spatial Relations. Memory & Cognition (to appear)
15. Janhunen, J. (ed.): The Mongolic languages. Routledge, London (2003)
16. Johnson-Laird, P.N.: The three-term series problem. Cognition 1, 57–82 (1972)
17. Johnson-Laird, P.N.: Mental models. Towards a cognitive science of language, inference, and consciousness. Harvard University Press, Cambridge, MA (1983)
18. Johnson-Laird, P.N.: Mental models and deduction. Trends in Cognitive Sciences 5, 434–442 (2001)
19. Johnson-Laird, P.N., Byrne, R.M.J.: Deduction. Erlbaum, Hove, UK (1991)
20. Knauff, M., Rauh, R., Schlieder, C.: Preferred mental models in qualitative spatial reasoning: A cognitive assessment of Allen's calculus. In: Moore, J.D., Lehman, J.F. (eds.) Proceedings of the Seventeenth Annual Conference of the Cognitive Science Society, pp. 200–205. Lawrence Erlbaum Associates, Mahwah, NJ (1995)
21. Knauff, M.: The cognitive adequacy of Allen's interval calculus for qualitative spatial representation and reasoning. Spatial Cognition and Computation 1, 261–290 (1999)
22. Levinson, S., Kita, S., Haun, D., Rasch, B.: Returning the tables: Language affects spatial reasoning. Cognition 84, 155–188 (2002)

23. Levinson, S.C., Meira, S.: 'Natural concepts' in the spatial topological domain. Language 79(3), 485–516 (2003)
24. Manktelow, K.I.: Reasoning and Thinking. Psychology Press, Hove, UK (1999)
25. Mark, D., Egenhofer, M.: Modeling spatial relations between lines and regions: combining formal mathematical models and human subjects testing. Cartography and Geographic Information Systems 21, 195–212 (1994)
26. Mark, D., Comas, D., Egenhofer, M., Freundschuh, S., Gould, J., Nunes, J.: Evaluating and refining computational models of spatial relations through cross-linguistic human-subjects testing. In: Frank, A., Kuhn, W. (eds.) Spatial Information Theory: A theoretical basis for GIS, pp. 553–568. Springer, Berlin (1995)
27. Ragni, M., Fangmeier, T., Webber, L., Knauff, M.: Complexity in Spatial Reasoning. In: Proceedings of the 28th Annual Cognitive Science Conference, Lawrence Erlbaum Associates, Mahwah, NJ (2006)
28. Ragni, M., Knauff, M., Nebel, B.: A Computational Model for Spatial Reasoning with Mental Models. In: Bara, B.G., Barsalou, L., Bucciarelli, M. (eds.) Proceedings of the 27th Annual Cognitive Science Conference, pp. 1064–1070. Lawrence Erlbaum, Mahwah, NJ (2005)
29. Ragni, M., Wölfl, S.: On Generalized Neighborhood Graphs. In: Furbach, U. (ed.) KI 2005: Advances in Artificial Intelligence, 28th Annual German Conference on AI, Springer, Berlin (2005)
30. Randell, D.A., Cohn, A.G., Cui, Z.: Computing transitivity tables: A challenge for automated theory provers. In: Proceedings of the 11th CADE, Springer, Berlin (1992)
31. Randell, D.A., Cui, Z., Cohn, A.G.: A spatial logic based and regions and connection. In: Nebel, B., Swarthout, W., Rich, C. (eds.) Proceedings of the third Conference on Principles of Knowledge Representation and Reasoning, pp. 165–176. Morgan Kaufmann, Cambridge, MA (1992)
32. Rauh, R., Hagen, C., Knauff, M., Kuβ, T., Schlieder, C., Strube, G.: Preferred and alternative mental models in spatial reasoning. Spatial Cognition and Computation (2005)
33. Schlieder, C., Berendt, B.: Mental model construction in spatial reasoning: A comparison of two computational theories. In: Schmid, U., Krems, J.F., Wysotzki, F. (eds.) Mind modelling: A cognitive science approach to reasoning, pp. 133–162. Pabst Science Publishers, Lengerich (1998)
34. Spalek, T.M., Hammad, S.: The left-to-right bias in inhibition of return is due to the direction of reading. Psychological Science 16, 15–18 (2005)
35. Vögele, T., Schlieder, C., Visser, U.: Intuitive Modelling of Place Name Regions for Spatial Information Retrieval. In Kuhn, W., Worboys, M.F. Timpf, S., eds.: Spatial Information Theory. Foundations of Geographic Information Science, International Conference, Proceedings of the COSIT 2003. Springer, Berlin (2003)

Thalassographeïn: Representing Maritime Spaces in Ancient Greece

Jean-Marie Kowalski[1], Christophe Claramunt[1], and Arnaud Zucker[2]

[1] Naval Academy Research Institute, Lanvéoc-Poulmic, BP 600, 29240 Brest Naval, France
{kowalski,claramunt}@ecole-navale.fr
[2] CEPAM - UMR 6130, CNRS, UNS. 98 Bd. Edouard Herriot, BP 3209, 06204
Nice Cedex 3
zucker@unice.fr

Abstract. Ancient Greek literature gives us many relevant traces of the way people represented maritime spaces. Greek narratives endow these spaces with meaning and make them legible. Indeed, these testimonies are a significant contribution to our knowledge of the "odological" space at sea, a space of description including numerous cultural features rather than a space described. This space is based on experiential data: Greek sailors did not have at their disposal scientific means of calculating a fix or using dead reckoning navigation techniques. In these conditions, their representation of maritime spaces integrates both qualitative and quantitative features, that somehow renew Aristotle's *Categories* of "quality" and "quantity". These features are mainly based on chronological grounds and on an experiential vision of spatial relations, that match the organization of space with the organization of human maritime activities and modify the representation of boundaries.

1 Introduction

Geographers "form [their] ideas of shape and size and also other characteristics, qualitative and quantitative, precisely as the mind forms its ideas from sense impressions – for our senses report the shape, colour and size of an apple, and also its smell, feel, and flavour; and from all this the mind forms the concept of an apple. So, too, even in the case of large figures, while the senses perceive only the parts, the mind forms a concept of the whole from what the senses have perceived." These words might reflect a contemporary point of view of a geographer perceiving and representing a geographical space from both qualitative and quantitative aspects. However, they were actually written by the Greek geographer Strabo (Geography, 2.5.11 [21]), more than two thousand years ago.

Strabo's knowledge of the geographic space is partly based on and derived from the description of maritime travels reported in books, such as *Harbours* or *Coasting Voyages* (*Geography*, 1.1.21 [21]), with the reserve that he does not completely rely on their authors, as "they have failed to add all the mathematical and astronomical information which properly belonged in their books." As denoted in Strabo's *Geography* (2.5.11 [21]), he derived from these written experiences and reported a personal global perception of space, i.e. his "autopsy" and concrete knowledge of the

S. Winter et al. (Eds.): COSIT 2007, LNCS 4736, pp. 47–60, 2007.

places by travelling, but a great part of his material is based on other geographers' testimonies. According to him, even if sight is a major sense, hearing is nevertheless superior. Strabo's claim is quite surprising as hearing is usually associated with the narration of past events while "autopsy" is a means of personal enquiry about geography, the environment or whatever is visible (see Herodotus, 2.99.1 [8]). The scientist's task is therefore to make a synthesis of the information provided by man's senses. The nature and quality of these data is a major concern whenever Ancient Greek literature deals with the description of maritime spaces.

There is no specific literature dedicated to the representation and description of maritime spaces in Ancient Greece, meant to be used similarly as our contemporary pilot charts. One must acknowledge that literature containing geographic data is scattered through various literary genres, from Homer's *Odyssey* to Ptolemy's *Geography* and different kinds of navigation narrative. No sailor would dare sail around the Mediterranean with, for instance, the Pseudo-Skylax's description of this sea, as this text is undoubtedly a good synthesis of the knowledge of the Mediterranean coastline in the first half of the 4th Century B.C., but it is fragmentary and absolutely useless as far as navigation is concerned because of the lack of information considered crucial by contemporary sailors.

Before being set down on paper through texts or maps, geographic data on maritime spaces were orally transmitted and mainly based on the experience of individuals. The main consequence of this statement is that Greek representation of maritime spaces is basically a representation of what Janni called the "odological space" [11], a space that differs from the cartographical one because of the introduction of human cultural features that influence the perception of natural elements. At sea, this representation depends on the conditions of this particular environment, which is situated – in the Greek conception of space – outside the inhabited world. This inhabited world the Ancient Greeks called *oikouménè* could more easily and differently be measured by man.

Strabo's attempt to lay down in a text a description of the inhabited world based on a global synthesis of odological representations is a truly challenging task which is bound to confront the difficulty of giving a steady representation of data coming from moving, dated and biased points of view of the various observers. This challenge echoes Levi-Strauss' statement (*The savage mind*, [12]) that works of art and scale models are not mere passive counterparts of the object they represent, but an experience of this object, as far as they are "man made" and "hand made".

Moreover, any attempt to find in these texts such methods of representation of maritime spaces as those in use in modern times would be bound to fail. Indeed, Pytheas – who supposedly discovered in the late 4th century B.C. the island of Thule, north of Great Britain - is not Columbus: the lack of modern navigation techniques compelled the Ancient Greek sailors to imagine different ways to represent their trip over the seas. Therefore, they could not determine the coordinates of their position precisely. They could not as well use dead reckoning navigation techniques. In such conditions, determining one's position was not based on the calculus of fixed coordinates, but on a dynamic reference to the surrounding environment. That is why Ancient geographers lay the emphasis on qualitative aspects rather than on quantitative aspects of the organization of maritime spaces. Furthermore, qualitative aspects often underlie the quantitative ones.

Therefore, the purpose of this paper is neither to say if Strabo and other Ancient Greek geographers are right or wrong, nor to identify precisely the places they show in their representations, but to highlight the main concepts and features of the way people represented maritime spaces in Greek Antiquity. Indeed, geographic literature gives us relevant traces of the Greek perception of these "odological" spaces, based on the experiential models [15] of individuals and potentially revealing some principles of Naïve Geography [7], rather than the early beginnings of a method of estimating one's position in a geographical space.

2 Ontologies in Maritime Spaces

The study of ontologies in maritime spaces in such sailors' narrative as the Pseudo-Skylax [16,17] makes it clear that their definition is based on descriptive characteristics, that depend on the narrator's point of view along his trip at sea, rather than on a mentally reconstructed bird's eye view. This narrator selects and organizes what he sees, and his personal point of view endows these elements with meaning, i.e. "legibility" [14] inside the global context of his navigation around the Mediterranean and the Black Sea.

2.1 The Sea: A Space of Itinerary Descriptions Rather Than a Space Described

Even if the sea is a constantly moving and uneven surface, modern navigation charts give much information about it: its minimum depth, the nature of the bottom, the rocky places, the areas subject to dangerous waves, and the main streams. Any 21st century reader feels quite disappointed when he comes to realize that the sea itself is never directly described, even if it is often mentioned in Greek descriptions of maritime spaces. It looks as if oceans and seas were to be considered as the space of description rather than the space described, a space in which elements are separated by a deep void rather than by the deep liquid element.

Indeed, there are numerous occurences of the terms "*thalassa*" and "*pelagos*", that refer to the sea in Greek, but the narrator barely says that the sea is rough or calm, shallow or deep, wide or narrow. For instance, the Pseudo-Skylax underlines only once that the sea is rough (Pseudo-Skylax, 1), near the Pillars of Hercules (Gibraltar).

It is also a quite challenging task to propose a definition of the limit between sea and shore. Things are not as clear as the conventional limit of high tide on contemporary nautical charts. Nevertheless, as far as the sea seems to be considered as the space of navigation, its limits are these of the boats' draught. That's why the island of Kerne is the southern limit of the description made by the Pseudo-Skylax (Pseudo-Skylax, 112). Boats cannot sail beyond this point because the sea is said to be "shallow and muddy, and filled with sea-weed." Therefore, the limit of the space described is fixed by the draught of the boat.

This space - which looks more like a void than the liquid element - runs alongside littoral regions that a modern reader, used to at least two-dimensional representations of the inhabited world, can hardly delimit. Indeed, those regions are usually referred to by the name of the people living in the area. Their definition is more ethnic than physical. *Bona fide* boundaries [19] such as rivers or mountains are often mentioned, but they are not strictly required. For instance, what is referred to as Mount Orion is the physical boundary between Iapygia and the Samnites, but no other boundary is clearly mentioned between Iapygia and the neighbouring territories (Pseudo Skylax, 14-15).

Furthermore, these regions look one-dimensional: their length is measured, usually in terms of time, but no information is given about their width, and the relief is hardly mentioned. Mount Orion is referred to as a boundary, but neither its height nor its extent are even estimated. That is why there is no descriptive model of the littoral territories. They just look like the one-dimensional coastline the sailor can sail along, but they are not strictly delimited and cannot be fully considered as well-defined geographic entities.

2.2   Experiential Data of Navigation and Coastline Interpretation

Since the Pseudo-Skylax, like the other ancient Greek geographers and sailors, does not give much information about the general aspect of the territories described, it is sometimes very difficult for the modern reader to identify these places with a sufficient level of certainty: Greek geographers and sailors usually does not give a comprehensive description of maritime spaces, but they pick up unsystematic elements from the surrounding environment. The general aspect of the coastline is not described, but circumnavigation narratives from the Hispanic side of Gibraltar to the African side let us imagine a circle-like trip around the Mediterranean sea and the Black sea. This kind of information is far from complete and is not sufficient to build a global map of the Greeks' sea in the 4th century BC, but a closer examination of the Pseudo-Skylax description reveals the existence of descriptive models of the "odological space", the basic features of which are based on the experiential data of sailing rather than on the information given by the sailor's senses. Therefore, the representation of the coastline apparently relies only on what has been *seen*, but it highly depends on the route followed by the ship, i.e. what has been *done*.

For instance, whenever the sailor follows a winding route along the shore, the navigation is said to be *"kolpôdès"*. The substantive *"kolpos"* from which this adjective derives can hardly be translated by a unique word in our modern languages since it refers to various geographic entities such as bays, gulfs and even seas. The Great Syrtis, today's Gulf of Sidra, the Minor Syrtis, today's Gulf of Gabès (Pseudo-Skylax, 110, 27), are called *"kolpoï"*. Both of them can easily be identified as gulfs or bays, with qualitative information given about their size. The trouble for the modern reader is that the Adriatic Sea is also said to be a *"kolpos"* (Pseudo-Skylax, 14), although it does not look at all like what we would call a "gulf" or a "bay".

Therefore, the substantive *"kolpos"* seems to be a generic functional term, scale-independent, which refers to an odological concept and means any kind of geographic entity which compels the sailors to follow a winding route along the shore. One would say that this definition also includes such geographic entities as capes but a closer reading of the description of *"kolpoï"* shows that they have common specific features: a mouth (*stoma*) which is generally delimited and measured, and a *"mukhos"*, the "innermost part", i.e. supposedly the most distant place from the mouth. This distance is also generally measured.

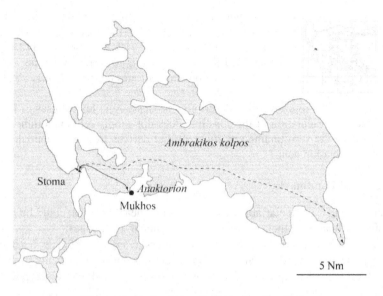

Fig. 1. The Ambrakikos Gulf (Ambracian Gulf)

Figure 1 is based on today's cartographical representation of the Ambrakikos Gulf. According to the Pseudo-Skylax (31.7-8), the mouth of the "Anaktorikos kolpos" is said to be 4 stadia wide: the *stoma* therefore seems to be the inner part of the entrance of this *kolpos*, but the author does not even mention the limits of the mouth, while they are mentioned for the Ionian Sea (Pseudo-Skylax, 27.3) which is delimited in the West by Cape Rizzuto and in the east by the Ceraunian Mountains – today's Kanalit Mountains, in the south of Albania. The estimated width of this mouth is "about 500 stadia".

The *"mukhos"* of the "Anaktorikos kolpos" is 120 stadia distant from the mouth, according to the Pseudo-Skylax. The distance must here be considered as an equivalent of a time-based estimation of the trip: 120 stadia correspond to a one-hour diurnal navigation, while 500 stadia correspond to half a 24-hour navigation, i.e. (very) approximately the distance sailed during a one-day diurnal navigation ([2] p. 79).

The dashed line on figure 1 shows the route between the *stoma* and the furthest place in today's Ambracian Gulf. The length of this route makes it highly improbable that the *mukhos* should be identified as the furthest place in the Ambracian Gulf as it is approximately 23 Nm away from the *stoma*. Furthermore, the Pseudo-Skylax (Pseudo-Skylax, 34) mentions the harbour of Anaktorion. This harbour is only 5 Nm away from the *stoma*, a distance ancient ships could more probably sail in a one-hour diurnal navigation.

From this, we can draw the conclusion that the Pseudo-Skylax does not describe today's entire Ambracian Gulf, but only its first part, as the *mukhos* is obviously not the furthest geographical place from the *stoma*, but the furthest one as maritime routes are concerned. This does not mean that ancient sailors ignored the existence of the whole Ambracian Gulf, but the description of the main maritime routes did not imply that it should have been mentioned.

While *"kolpoï"* draw figures in which sea penetrates the land, *"aktaï"* draw figures in which land penetrates the sea. Both the region of Lucania – south of Italy - (Pseudo-Skylax, 12.3) and Libya – today's Africa – (Pseudo-Skylax, 112.71) are supposed to be *"aktaï"*. Lucania and Libya have nothing in common, neither size nor descriptive features, but in both cases sailors can sail around them. Unlike *"kolpoï"*, *"aktaï"* are not measured.

On the base of the representative descriptive processes and lexical uses of Pseudo-Skylax, Greek representation of the coastline does not simply rely on common sense representation of the surrounding environment, but derives from experiential models rebuilt from navigation data.

2.3 Structural Landmarks

Greek sailors paid a particular attention to capes and headlands that could be referred to as useful landmarks on their way between harbours or anchorages.

"Akrôtèrion" is a generic substantive which refers to any kind of cape, headland or promontory. These very different geographic entities make it difficult to build a typology based on descriptive features, while Sorrows' and Hirtle's typology of landmarks [20] seems to be more relevant in this case. Indeed, they distinguish between visual, cognitive and structural landmarks.

A close examination of the Pseudo-Skylax's description of Sicily (Pseudo-Skylax, 13) makes it clear that landmarks are not cultural, but are obviously visual and mainly structural. Indeed, Sicily is said to be "in front of Reggio". Then, cape Peloro (North-East of Sicily) is mentioned as an *"akrôtèrion"* of Sicily. This cape is the first one that can be seen from Calabria, inasmuch as it is the closest one from the continent. Therefore, the first landmark mentioned is the first one a sailor can see on his trip from the continent to Sicily. The second landmark is Cape Pakhynos (South-East of Sicily). This cape is not described, and a quick glance at today's nautical charts makes it quite clear that the ancient Greek sailors supposedly knew the existence of the small rocky islands surrounding it. Visual information seems to be fragmentary, but this landmark, like the first one, is a structural element of the trip. Indeed, this cape is the southernmost one of Sicily. What is more, Bernard's [4] artistic view (Fig. 2) of this cape reveals that the small rocky islands could not be seen easily from the sea. In other words, the apparent lack of information is faithful to what sailors could actually see.

Cape Lilybaeum is the westernmost limit of Sicily; that is why it is considered as an important landmark even if it is not the most prominent one on the western coast of Sicily. Here again, the Pseudo-Skylax does not mention the existence of small rocky islands around this cape, but Bernard's view of the western end of Sicily (Fig. 3) leads one to think that they could merge into the surrounding environment for a sailor coming from the high sea.

Sicily is not explicitly described as a triangle, but the Pseudo-Skylax mentions its three vertices (Fig. 4). These are not the most prominent capes around this island, but ancient Greek sailors knew that they were the basic structural features of their routes around Sicily.

L'apparence de *Cap de Paffaro* en *Sicile*, quand en navigant on y paffe : & le coing eft un rivage de blancq fablon·
Tenant le coing il y a trois ou quatre petites Ifles.

Fig. 2. Cape Pakhynos. "What Cape Passaro looks like in Sicily, when sailing along. The corner is a white sand shore. In this corner, there are three or four small islands." [4]

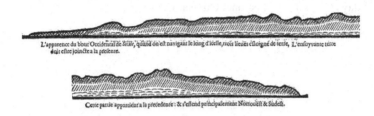

L'apparence du bout Occidental de *Sicile*, quand on eft navigant le long d'icelle, trois lieuës éloigné de terre, L'enfuyvante terre
doit eftre joinéte a la prefente.

Cette partie appartient a la precedente : & s'eftend principalement Nortoüeft & Sudeft.

Fig. 3. Western end of Sicily [4]. "What the western end of Sicily looks like, when sailing along three leagues away from the shore. The next picture must be connected to this one. This picture is connected to the previous one. This land mainly lies from the South-East to the North-West."

Fig. 4. The three "vertices" of Sicily

Visual information is very fragmentary, but the reader knows for instance that he will see an altar on the cape situated west of Gibraltar on the African side (Pseudo-Skylax, 112.27), or that the city of Phalasarna is situated near the cape of Crete that sailors coming from Sparta can see first (Pseudo-Skylax, 47.4). Lastly, the description provides scarcely any information about the orientation of these landmarks, even if Cape Granos (Creta) is said to lay "towards the sunrise" (Pseudo-Skylax, 47.31).

Therefore, what makes a landmark for a Greek sailor is neither its general aspect, nor clearly identified descriptive features, but its structural function in the organization of maritime space, even if the very fragmentary visual information given generally fits what sailors could truly see from their ships. That is why Greek descriptions generally focus the reader's attention on some landmarks such as capes which are not actually the most prominent and visible ones. This does not mean at all that the Greek geographers overestimated the dimensions of some geographic entities, but they undoubtedly insisted on them in order to put the emphasis on their structural function in the organization of space.

For example, the identification of the cape referred to in Pytheas' (Pytheas, F6a = Strabo, *Geography*, 1.4.5) description of Brittany (France) has been much discussed ([5], p. 126–133). Pytheas wrote during the 4th century B.C. that Europe's prominence "lies over against Iberia" and that there are "capes, among which the cape of the Ostimnioï, called Kabaion, and the islands situated in front of it." He added that "it takes three days to sail to the last one, Ouxisamé."

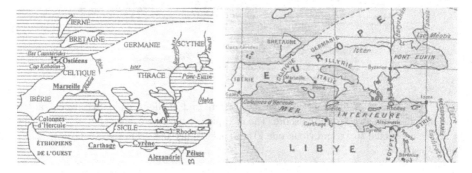

Fig. 5. Artists' views of the north-west part of the inhabited world according to Eratosthene [18] and Strabo [3]

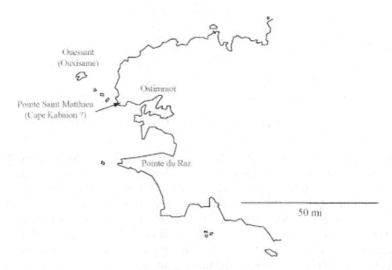

Fig. 6. Brittany (France)

Strabo (*Geography*, 1.4.3) blames Eratosthenes for relying too much on Pytheas' narrative and for saying subsequently that the Armorican peninsula (Brittany, France) is 3000 stadia (ca. 325 miles) prominent in the sea (Fig. 5). The geographer thinks that Eratosthenes overestimates the dimensions of this prominence. Obviously, Strabo knew the existence of maritime routes from Iberia to Britain, but he did not have at his disposal experiential data given by sailors who sailed along Brittany and could use Cape Kabaion as an essential landmark on their way north to Britain.

Certainly, any sailor coming from the south would first see the "Pointe du Raz", which seems to be the most prominent cape, but Pytheas clearly mentions several islands, while there is only one west of the Pointe du Raz. On the contrary, the Pointe Saint Matthieu is not as prominent as the Pointe du Raz, but its structural function in the organization of space makes it an essential landmark: this is the last safe place for boats sailing north, before Ouessant and the British islands. Moreover, the people of the Ostimnioï live in the East of the Pointe Saint Matthieu, and it is possible from there to head east and sail to their land. That is why Cape Kabaion should probably be identified with the Pointe Saint Matthieu rather than with the Pointe du Raz.

2.4 Harbours and Man-Made Buildings

The quality of anchorages and harbours, of port installations, the language spoken by the people living in the area and their behaviour towards foreigners are a major concern even for contemporary sailors. Greek texts give at least two levels of information.

First, they generally give qualitative information about the opportunity of finding port facilities in a delimited area. These harbours are not systematically mentioned, but the sailor knows that he will find them on his way between two places. For example, the Pseudo-Skylax (Pseudo-Skylax, 4) says that it takes four days and four nights to sail from the estuary of the Rhone to the city of Antion (nowadays Anzio). Along this trip, the only harbour mentioned is Massalia (nowadays Marseille), but the author lets us know that this coast gives good port facilities: the coast is said to be "*euliménos*", which means "*with good harbours*". The description also mentions a series of harbours in Acarnania, a region situated in the west of continental Greece (Pseudo-Skylax, 34), which is said to be "*euliménos*". This means that the list is not meant to be complete and that it is quite easy to call at a port in this region which takes two days to sail along. Therefore, these two examples show that the existence of harbours seems to be a qualitative feature determining the nature of the coast and the safety of navigation.

A second level of information explicitly mentions the existence of harbours in the cities listed. The generic substantive "*limèn*" refers to harbours in general, but there is a great difference between a strong and well protected military harbour and a simple place where boats can go ashore safely. That is why descriptions often distinguish between "*emporia*" where ships can unload their goods and "*limenes kleïstoï*" (closed harbours), which were military harbours, but also harbours closed with sea walls.

The cities around these harbours are never described: no information is given about their extent, their organization or their architecture. Greek descriptions usually just give cultural information about the people living there, with a clear distinction between Greek and Barbarian people. The only buildings mentioned are generally those that can be considered as useful landmarks. For example, the Pseudo-Skylax refers to a temple dedicated to Apollo, placed on the Cape Sounion (Pseudo-Skylax, 57.6-7), but no information is given about its visual aspect.

3 Aristotelian Categories and the Organization of the "Odological Space"

Strabo's scientific purpose, as it is exposed in the *Geography* (*Geography*, 2.5.11), is to appreciate "shape and size and also other characteristics, qualitative and quantitative", of geographical objects. At first glance, notions of "quality" and "quantity" respectively refer to Aristotelian categories of *"poion"* and *"poson"*; but Strabo introduces a significant evolution in the meaning of these categories. Greek descriptions of maritime spaces not only refer to the "quantitative" and "qualitative" categories but also to the one called "relation".

3.1 Quantity Matters

Things are said to be "quantified" whenever they are measurable. For instance (*Categories*, 5a) a line in a geometrical plane is something measurable and continuous, as any of its parts is in contact with another part [1]. These parts can also be located in a geometrical plane and they have a position that can be clearly determined (*Categories*, 5a15). The fact is that time is also measurable (as parts of time are in contact together), according to Aristotle (*Categories*, 5a25-30), but its parts do not have a position like points in a plane: parts of time have a rank (*taxis*) rather than a position.

Inside the "odological space", what makes the relation between two cities or two regions is not the course and the distance between them, but the route followed by the sailor who places them in a chronological order. A quick overview of the Greek vocabulary of spatial relations clearly shows that these relations are fully chronological. "After" Megara is Corinth, "after" Corinth is Sikyôn (Pseudo-Skylax, 40-42). The preposition *"meta"* massively predominates, even if it is sometimes replaced by the adverb "then" (*eïta*).

The chronological organization of the "odological space" makes it a seemingly impossible task to determine the position of islands. In these conditions, the sailor does not have at his disposal indications of longitude and latitude. It seems to be rather difficult to place them in the general arrangement of the description, which is mainly focused on the continental coastline. Whenever islands are concerned, the organization of space is no longer merely chronological. Islands are not "after" another place, but "in front of" (*kata*) another one. This preposition undoubtedly belongs in Greek to the semantic field of spatial relations, but it is actually impossible to draw the conclusion from this statement that a position can be determined, just like in a geometrical plane. The gap between geometrical positioning and the kind of representation involved by temporal measurement of space still remains. Sicily is indeed supposed to be "in front of" Reggio. The trouble is that no indication is given about the direction or the heading of the island. We could here imagine that "in front of" is to be considered from the place on the shore where the distance is the shortest between the island and the continent. Unfortunately, the shortest crossing between the continent and Sicily is not from Reggio, but this harbour is the most important one to cross over to Sicily. Therefore, maritime routes rather than geometry define what "in front of" means, just like Corsica is "in front of" Marseille for a French sailor, while it is "in front of" Genoa for an Italian one.

That is why islands have a particular status in Greek descriptions of maritime spaces. For instance, whenever the Pseudo-Skylax mentions an island, the author comes back to the continent at the point from which he had left it: after his description of Sicily, he continues the description of the coast, starting from Reggio. Islands consequently do not properly have a "rank" in the chronological organization of the description, but they introduce chronological digressions inside it.

3.2 Quality and Relation Refine

Because of the lack of reliable scientific means to measure maritime spaces, geographers were often compelled to use experiential grounds to estimate distances. Strabo's *Geography* (2.5.24) gives a vivid example of this statement: "The sea-passage from Rhodes to Alexandria is, with the north wind, approximately four thousand stadia, while the coasting-voyage is double that distance. Eratosthenes says that this is merely the assumption made by navigators in regard to the length of the sea-passage, some saying it is four thousand stadia, others not hesitating to say it is even five thousand stadia, but that he himself, by means of the shadow-catching sun-dial, has discovered it to be three thousand seven hundred and fifty stadia." This passage is an important contribution to our knowledge of the representation of maritime spaces. We can here perceive the implicit asymmetry of distances [7]: distances are tightly related to the orientation of the wind. Strabo makes it clear that the distance is much shorter wind aft than wind ahead, especially with ancient ships. The distance is not the same from Rhodes to Alexandria as from Alexandria to Rhodes. Moreover, this distance depends on the route followed by the ship, as the coastal-voyage is double the distance through high sea.

The last information is not the least: thanks to Eratosthenes' scientific calculation of the latitude of both places, Strabo could indicate a much more accurate measure of this distance. In spite of this, and in spite of the discrepancy of the data gathered among sailors, he still relies on the most consensual estimation of the distance, 4000 stadia, which corresponded to a navigation of four days and four nights ([2], p. 69).

What emerges from this passage is that "quantity" in Greek representations of maritime spaces relies on experiential data that imply the asymmetry of distances and the tight subordination of distance to the route followed.

The importance of experience is even more obvious as the categories of quality and relation are concerned. The Aristotelian conception of quality is quite different from our own, as far as this category does not only apply to spatial and temporal reasoning. Indeed, Aristotle distinguishes between four kinds of qualities. The first one concerns "habitus" and "dispositions" such as health and illness, cold and hot (*Categories*, 8b25). The second one concerns abilities (*Categories*, 9a15-25). The third one concerns affective qualities or affections (*Categories*, 9a25). The last one (*Categories*, 10a10) is the most important for us, as it concerns figures (*skhèmata*) and shapes (*morphaï*). This is obviously the kind of quality Strabo refers to. His description of the Nile gives an example of how the geographer could liken geographical spaces to geometrical figures (*Geography*, 17.1.4): "being 'split at the head', as Plato says, the Nile makes this place as it were the vertex of a triangle, the sides of the triangle being formed by the streams that split in either direction and extend to the sea [...] An island, therefore, has been formed by the sea and the two streams of the river; and it is called the Delta on account of the similarity of its shape."

Strabo also says that our senses let us know the "size" of objects. Size is first a matter of quantity, as size is measurable, but it is also a matter of relation. Indeed, "big" is not the correlative of "small" like "a lot" is not the correlative of "few" out of any context (*Categories*, 5b15-20). Nothing is big or long itself, but as it looks so when it is compared to something else. A six-foot man is tall, but a six-foot mountain is very small: relation is highly based on experiential data and is meaningless out of context.

Quality and relation consequently introduce a second dimension in a seemingly one-dimensional chronological representation of space, but this dimension highly depends on the sailor's experience.

3.3 Boundaries: Contiguity, Separation and Inclusion

Notions such as "territorial waters" or "exclusive economic zone" are meaningless for the Ancient Greeks and are useless to understand the sharing of maritime spaces and of the coastline. Indeed, this coastline is delimited by physical *"bona fide"* but also cultural *"fiat"* boundaries [19], some crisp and others vague.

According to Aristotle, something is measurable whenever any of its parts is in contact with another one. Contiguity is indeed a major feature of the regions mentioned. For instance, the main term used in Pseudo-Skylax to designate the kind of connection between two regions is the verb *"ékhesthai"* whose adaptation in modern languages is usually "after". The trouble is that "after" does not put into relief the original meaning of *"ékhestaï"*: "come next to, follow closely" (LSJ, s.v. *ékhô*). So, the region inhabited by the Latins is said to be contiguous to Tyrrhenia (Pseudo-Skylax, 8.1). In this case, as in many other situations, physical boundaries are not mentioned and ethnography is the first distinctive feature of coastal regions.

In these conditions, boundaries are often explicitly vague. According to the Pseudo-Skylax, the Ligurians do not come immediately after the Iberians along the Mediterranean coast, but these two people are mixed between the Iberians' territory and the Rhone.

Nevertheless, coastal regions, but also bays, gulfs and seas are usually separated by physical boundaries. Ancient geographers indeed generally consider [9] that Europe is separated from Asia by the river Tanais (nowadays the Don), and that Libya, i.e. Africa, is separated from Asia by the Canopic mouth of the Nile (Pseudo-Skylax, 70.107). Inside these great areas, the inhabited regions and maritime spaces are often delimited by physical boundaries. For example, the Ligurians live between the Rhone River and the city of Antion, and the Ceraunian Mountains (today's Kanalit Mountains) are considered as the eastern limit of the Ionian Sea. It looks as if geographic entities considered as physical boundaries became one-dimensional. The length of the course of the Nile through Egypt does not matter: the Canopic mouth is to be considered as the point where Asia and Africa meet together. No matter what the extent of the city of Antion is: it shares the same status as the Rhone River as a contact point between the Ligurians and their neighbours. In the same way, the Kanalit Mountains lie in Albania from south-east to northwest, but what really matters is the point where the Ionian Sea is in contact with the Mediterranean. Coastal regions and maritime spaces seem to be generally contiguous. Physical boundaries are sometimes mentioned, but they have no extent and should be considered as points of contact rather than as true separations.

Such wide spaces as these coastal regions include smaller elements which cannot be precisely located. This is often the case with cities inside coastal regions. For example, in his description of the Asiatic side of the Bosphorus Strait, the Pseudo-Skylax (Pseudo-Skylax, 73) mentions a series of cities which are said to be Greek in the region described. These cities are neither described nor located. The author does not even say if there are other Greek cities there, but one knows that they are included in the region inhabited by the people of the Sindoï.

4 Conclusion

Helen Couclelis [6] had already shown the genius of Aristotle's *Physics* that "laid the foundations of the mechanistic model of space and time some 2,000 years before Newton", but also "struggled with the necessity to go beyond that model." The link made by Strabo between Aristotelian *Categories* and the geographical representation of the world shows how surprisingly modern they are and how both quantitative and qualitative reasoning were a major concern for ancient Greek geographers.

The examples mentioned in this paper put into relief the main features of the Ancient Greeks' vision of maritime spaces, which was predominately based on experiential data. Both qualitative and quantitative aspects tend to draw the routes of navigation on chronological grounds rather than a textual representation of the maritime spaces themselves. The sea and the coastline are therefore considered more as the space of representation than the space represented. In spite of the constant improvement of the scientific knowledge Greek sailors and geographers had of their surrounding environment, Strabo's example shows that they would always chose approximate indications based on the accumulation of experiences rather than a more accurate quantification of distances that would certainly be closer to truth, but useless for the sailor. Further studies should put the emphasis on permanent features and evolutions of this vision from Homer to Ptolemy, in the particular Greek cultural space, but also with a cross-cultural point of view.

References

1. Cooke, H.-P., Tredennick, H.: Aristotle: Categories, on interpretation, prior analytics. Harvard University Press, Cambridge, MA (1938, reprinted 1996)
2. Arnaud, P.: Les routes de la navigation antique, itinéraires en Méditerranée. Editions Errance, Paris (2005)
3. Aujac, G.: Strabon et la science de son temps, Les Belles Lettres, coll. d'études anciennes, Paris (1966)
4. Bernard, G.: Description de la Méditerranée, Amsterdam (1607)
5. Bianchetti, S.: Pitea di Massalia, L'Oceano. Introduzione, testo, traduzione e commento, Istituti Editoriali e Poligrafici Internazionali, Pisa-Roma (1998)

6. Couclelis, H.: Aristotelian Spatial Dynamics in the Age of Geographic Information Systems. In: Egenhofer, M.J., Golledge, R.G. (eds.) Spatial and Temporal Reasoning in Geographic Information Systems, pp. 109–118. Oxford University Press, Oxford (1998)
7. Egenhofer, M.J., Mark, D.: Naïve Geography. In: Kuhn, W., Frank, A.U. (eds.) COSIT 1995. LNCS, vol. 988, pp. 1–15. Springer, Heidelberg (1995)
8. Herodotus: The Persian wars. T.I, translated by Godley, A.D. Harvard University Press, Cambridge, (Massachusets), London (England) (1920)
9. Jacob, C.: Inscrire la terre habitée sur une tablette. In: Détienne, M.: Les savoirs de l'écriture en Grèce ancienne. Presses Universitaires de Lille, Lille, France, pp. 273–304 (1988)
10. Jacob, C.: L'Empire des cartes. Albin Michel, Paris (1992)
11. Janni, P.: La mappa e il periplo. Cartografia antica e spazio odologico. Bretschneider, Roma (1984)
12. Levi-Strauss, C.: The savage mind. University of Chicago Press, Chicago (1968 ed. fr. 1962)
13. Liddell, H.G., Scott, R.: Greek-English lexicon, 9th edn. with a revised supplement The Clarendon Press, Oxford, UK (1996)
14. Lynch, K.: The image of the city. The MIT Press, Cambridge, MA (1960)
15. Mark, D.M., Frank, A.U.: Experiential and Formal Models of Geographic Space. Environment and Planning B 23, 3–24 (1996)
16. Pseudo-Skylax In: Müller, K.: Geographi Graeci minores, vol. I, Firmin Didot, Paris (1855)
17. Pseudo-Skylax: Le Périple du Pont-Euxin, (ed.) transl. comm. Counillon P., de Boccard, Bordeaux (2004)
18. Pseudo-Skymnos: Les géographes grecs. T.I Introduction générale. Texte établi et traduit par D. Marcotte. Collection des Universités de France, Paris (2000)
19. Smith, B., Varzi, A.C.: Fiat and Bona Fide Boundaries: Towards an Ontology of Spatially Extended Objects. In: Frank, A.U. (ed.) COSIT 1997. LNCS, vol. 1329, pp. 103–119. Springer, Heidelberg (1997)
20. Sorrows, M.E., Hirtle, S.C.: The Nature of Landmarks for Real and Electronic Spaces. In: Freksa, C., Mark, D.M. (eds.) COSIT 1999. LNCS, vol. 1661, pp. 37–50. Springer, Heidelberg (1999)
21. Strabo: Geography, Books 1–2. English translation by Jones, H.L. Harvard University Press, Cambridge, MA (1917, reprinted 2005)
22. Strabo: Geography, vol. VIII. English translation by Jones, H. L. Harvard University Press, Cambridge, MA (1932)

From Top-Level to Domain Ontologies: Ecosystem Classifications as a Case Study

Thomas Bittner

Department of Philosophy, Department of Geography
New York State Center of Excellence in Bioinformatics and Life Sciences
National Center for Geographic Information and Analysis
State University of New York at Buffalo

Abstract. This paper shows how to use a top-level ontology to create robust and logically coherent domain ontology in a way that facilitates computational implementation and interoperability. It uses a domain ontology of ecosystem classification and delineation outlined informally Bailey's paper on 'Delineation of Ecoregions' as a running example. Baily's (from an ontological perspective) rather imprecise and ambiguous definitions are made more logically rigorous and precise by (a) restating the informal definitions formally using the top-level terms whose semantics was specified rigorously in a logic-based top-level ontology and (b) by enforcing the clear distinction of types of relations as specified at the top-level and specific relations of a given type as they occur in the ecosystem domain. In this way it becomes possible to formally distinguish a number of relations which logical interrelations are important but which have been confused and been taken to be a single relation before.

1 Introduction

Ontologies are tools for specifying the semantics of terminology systems in a well defined and unambiguous manner [1]. *Domain ontologies* are ontologies that provide the semantics for the terminology used to describe phenomena in a specific discipline or a specific domain. In this paper the domain of ecosystem classification and delineation is used as an example. Other domains include hydrology and environmental science, as well as medicine, biology, and politics.

In contrast to domain ontologies, *top-level ontologies* specify the semantics for very general terms (called here top-level terms) which play important foundational roles in the terminology used in nearly every domain and discipline. Top-level terms that are relevant to this paper are listed in Table 1.

Building a domain ontology is an expensive and complex process [3]. Research has shown that robust domain ontologies must be [1,4]: (i) based on a well designed top-level ontology; (ii) developed rigorously using formal logic. This means that the semantics of the domain vocabulary is specified within a logic-based framework using top-level terms with an already well established semantics. One advantage of this approach is that top-level ontologies need to be developed only once and then can be

S. Winter et al. (Eds.): COSIT 2007, LNCS 4736, pp. 61–77, 2007.

Table 1. Types of top-level relations, their signatures, and their abbreviated top-level terms. (Adopted from [2]).

first arg.	second arg.	relational top-level terms	Symbolic representation
individual	individual	individual-part-of	*IP*
individual	universal	instance-of	*Inst*
individual	collection	member-of	ϵ
universal	universal	sub-universal-of (is a)	$\sqsubseteq$
universal	universal	(up/down) universal-part-of	*uUP, dUP*
collection	universal	extension-of	*Ext*
collection	collection	sub-collection-of	$\leq$
collection	collection	(up/down) partonomically-included-in	*uPI, dPI*
collection	individual	sums-up-to	*Sum*
collection	individual	partition-of	*Pt*

reused in many different domain ontologies. Another advantage is that a top-level ontology provides semantic links between the domain ontologies which are based on it.

A *logic-based* ontology is a logical theory [5]. The terms of the terminology, whose semantics is to be specified, appear as names, predicate and relation symbols of the formal language. Logical axioms and definitions are then added to express relationships between the entities, classes, and relations denoted by those symbols. Through the axioms and definitions the semantics of the terminology is specified by admitting or rejecting certain interpretations. In [2] a logic-based ontology for the top-level terms listed in Table 1 was presented.

Disciplines in which logic-based domain ontologies are quite common include medicine, biomedicine, and microbiology. Examples of logic-based medical domain ontologies are GALEN [6], SNOMED CT [7], and the NCI Thesaurus [8]. An example of a domain ontology for biomedicine and microbiology is the description logic based version of the Gene Ontology [9]. Currently efforts are being made to create a single suite of interoperabele biomedical ontologies with a common top-level ontology as unifying ontological and formal basis [10,11].

There are still only preliminary attempts to provide logic-based domain ontologies within the geo-spatial domains [12,13]. Examples are [14,15] for general ontologies of geographic categories, [16] for a domain ontology for hydrology, and [17,18] for domain ontologies for ecosystems. The latter provide a starting point for this paper.

This paper uses the example of ecosystem classification and delineation to demonstrate how top-level ontologies can help to enhance the degrees to which information processing tools can be used in the retrieval, management and integration of data by improving the robustness and logical rigor of domain ontologies the are used to structure the data to be processed. In Section 2 a simplified version of a logic-based ontology for the top-level terms in Table 1. Section 3 discusses important distinctions in the use of top-level terms in top-level and domain ontologies domain ontologies. Section 4 demonstrates how to build a robust logic-based domain ontology of ecosystem classification and ecoregion delineation by using the top-level ontology of Section 2 and the informal definitions of domain specific terms presented in [19].

2   A Simple Top-Level Ontology

Following [2] three disjoint sorts of entities are distinguished: (i) individual endurants
(New York City, New York State, Planet Earth); (ii) endurant universals (*human being,
heart, human settlement, socio-economic unit*); and (iii) collections of individual endurants (the collection of grocery items in my shopping bag at this moment in time,
the collection of all human beings existing at a given time). In the logical theory this
dichotomy of individuals, universals, and collections is reflected by distinguishing different sorts of variables – one sort for each category.

The theory is presented in a sorted first-order predicate logic with identity and use
the letters $w, x, x_1, y, z, \ldots$ as variables ranging over (endurant) individuals; c, d, e, g
as variables ranging over universals; and $p, q, r, p_1, \ldots$ as variables ranging over collections. The logical connectors $\neg, =, \wedge, \vee, \rightarrow, \leftrightarrow$ have their usual meanings (not,
identical-to, and, or, if ... then, and if and only if (iff), respectively). The symbol $\equiv$
is used for definitions. (x) symbolizes universal quantification (for all x ...) and $(\exists x)$
symbolizes existential quantification (there is at least one x ...). All quantification is
restricted to a single sort. Restrictions on quantification will be understood by conventions on variable usage. Leading universal quantifiers are omitted. Labels for axioms
begin with 'A' and labels for definitions begin with 'D'.

Please note that the aim of this section is to give a self-contained and simplified
axiomatic theory which is sufficient to demonstrate how to use a top-level ontology to
build an atemporal domain ontology of ecosystem classification and delineation. For a
more fully developed ontology see [2]. For additional discussions of universals see for
example [20].

2.1   Mereology of Individuals

Individual-part-of relations hold between individual endurants. For example, my heart
is an individual part of my body, the Niagara Falls are individual parts of the Niagara
River, Nebraska is an individual part of the United States of America. *IP* xy signifies
that individual x is part of individual y.

The individual x *overlaps* the individual y if and only if there exists an individual z
such that z is a part of x and z is a part of y (D_O).

$$D_O \quad O\,xy \equiv (\exists z)(IP\,zx \wedge IP\,zy)$$

For example, Yellowstone National Park overlaps Wyoming, Montana, and Idaho.

The standard axioms requiring that individual parthood is reflexive (AM1), antisymmetric (AM2), transitive (AM3) are included in the theory. In addition it is required that
if every z that overlaps x also overlaps y then x is part of y (AM4).

AM1 *IP* xx	AM3 *IP* $xy \wedge IP\,yz \rightarrow IP\,xz$
AM2 *IP* $xy \wedge IP\,yx \rightarrow x = y$	AM4 $(z)(O\,zx \rightarrow O\,zy) \rightarrow IP\,xy$

2.2   Collections, Sums, Partitions, and Partonomic Inclusion

Collections are like (finite) sets of individuals with at least one member. Examples of
collections include: the collection of Hispanic people in Buffalo's West Side as specified

in the 2000 census records, the collection of federal states of the USA, the collection of postal districts in the USA, etc.

The symbol ' ϵ ' stands for the member-of relation between individuals and collections. The notation $\{x_1, \ldots, x_n\}$ is used to refer to a finite collection having $x_1, \ldots, x_n$ as members. A minimal set of axioms requires: collections comprehend in every case at least one individual (AC1) and that two collections are identical if and only if they have the same members (AC2); for every x there is a collection having x as its only member (AC3); the union of two collection always exists (AC4).

$$AC1\ (\exists x)(x \in p) \qquad\qquad AC3\ (\exists p)(p = \{x\})$$
$$AC2\ p = q \leftrightarrow (x)(x \in p \leftrightarrow x \in q) \qquad AC4\ (\exists r)(x)(x \in r \leftrightarrow x \in p \vee x \in q)$$

The following definitions are included: collection p is a *sub-collection* of the collection q ($p \leq q$) if and only if every member of p is also a member of q ($D_\leq$); Collection p is *discrete*, $D\ p$, if and only if the members p do not overlap (D_D); The individual y is the *sum of the members of the collection p* if and only if every individual w overlaps y if and only if y overlaps some member of p D_{Sum}; Collection p *partitions* the individual y if and only if y is the sum of p and p is discrete (D_{Pt}).

$$
\begin{array}{ll}
D_\leq & p \leq q \equiv (x)(x \in p \rightarrow x \in q) \\
D_D & D\ p \equiv (x)(y)(x \in p \wedge y \in p \wedge O\ xy \rightarrow x = y) \\
D_{Sum} & Sum\ py \equiv (x)(O\ xy \leftrightarrow (\exists z)(z \in p \wedge O\ xz)) \\
D_{Pt} & Pt\ pyt \equiv Sum\ pyt \wedge D\ pt
\end{array}
$$

For example, the collection which has the federal states of the USA and the District of Columbia as its only members is discrete. The USA is the sum of this collection. Moreover, this collection partitions the USA.

Collection p is *upwards partionomically included* in collection q if and only if every member of p is an individual part of some member of q (*uPI*). Collection p is *downwads partionomically included* in collection q if and only if every member of q has some member of p as an individual part (*dPI*).

$$
\begin{array}{ll}
D_{uPI} & uPI\ pq \equiv (x)(x \in p \rightarrow (\exists y)(y \in q \wedge IP\ xy)) \\
D_{dPI} & dPI\ pq \equiv (y)(y \in q \rightarrow (\exists x)(x \in p \wedge IP\ xy))
\end{array}
$$

For example, let USC be the collection which has all the counties[1] of the USA as its members and let USF be the collection that has all the federal states of the USA as its members. Then USC is up- and downwards partionomically included in USF: every member of USC (a county) is part of some member of USF (a federal state) and every member of USF (a federal state) has some member of USC (a county) as its part.

2.3   Universals, Instantiation, and Universal Parthood

The variables c, d, e, g are used for universals (classes, types) like (*human being, federal state, mountain, forest, tree, plant,* and so forth). The relation of instantiation holds

---

[1] To keep matters simple I ignore the fact that in Louisiana counties are called 'parish' and in Alaska counties are called 'borough'.

between individuals and universals. For example New York City is an instance of the universal *city*. '*Inst xc*' signifies that the individual x instantiates the universal c. In terms of *Inst* one can define: c is a sub-universal-of d if and only if the instances of c are also instances of d $(D_{\sqsubseteq})$; c is a proper sub-universal-of d if and only if c is a sub-universal of d and d is not a sub-universal-of c $(D_{\sqsubset})$; collection p is the extension of universal c if and only if for all x, x is a member of p if and only if x instantiates c (D_{Ext}).

$$D_{\sqsubseteq} \quad c \sqsubseteq d \equiv (x)(Inst\ xc \rightarrow Inst\ xd)$$
$$D_{\sqsubset} \quad c \sqsubset d \equiv c \sqsubseteq d \wedge d \not\sqsubseteq c$$
$$D_{Ext} \quad Ext\ pc \equiv (x)(x \in p \leftrightarrow Inst\ xc)$$

For example, the universal *federal state* is a sub-universal-of the universal *socio-economic unit*. Therefore every instance of *federal state* (e.g., New York State) is also an instance of *socio-economic unit*. The extension of the universal *federal state* is the collection of all federal states. This collection has as members the federal states of the USA, the federal states of Germany, etc.

The following axioms are included: every universal has an instance (AU1); there is maximal universal (AU2); two universals are identical if and only if they have the same instances (AU3);[2] if two universals share a common instance then one is a sub-universal of the other (AU4); and if c is a proper sub-universal of d then there is a proper sub-universal e of d such that c and d have no instance in common (A5).

$$AU1\ (\exists x)Inst\ xc$$
$$AU2\ (\exists x)(y)(y \sqsubseteq x)$$

$$AU3\ (x)(Inst\ xc \leftrightarrow Inst\ xd) \leftrightarrow c = d$$
$$AU4\ (\exists x)(Inst\ xc \wedge Inst\ xd) \rightarrow c \sqsubseteq d \vee d \sqsubseteq c$$
$$AU5\ c \sqsubset d \rightarrow (\exists e)(e \sqsubset d \wedge \neg(\exists x)(Inst\ xc \wedge Inst\ xe)$$

From these axioms it follows that universals form tree-like hierarchies ordered by the sub-universal relation. In the scientific realm such tree-like structures most closely resemble classification hierarchies established using the Aristotelean method of classification. Using this method classification trees (intended to resemble hierarchies of universals) are built by defining a universal lower down in the hierarchy by specifying the parent universal together with the relevant differentia, which specify what marks out instances of the defined universal or species within the wider parent universal or genus, as in: *human =$_{df}$ rational animal* where 'rational' is the differentia [21,18]. Differentia need to be such that the immediate sub-universals of a given universal are jointly exhaustive and pair-wise disjoint. Thus besides rational animals there are non-rational animals and all animals are either rational or non-rational.

Corresponding to the partonomic inclusion relations *uPI* and *dPI* between collections the relations of upward and downward universal parthood, *uUP* and *dUP*, between universals are introduced: c is an *upward-universal-part-of* universal d if and only if every instance of c is an individual part of some instance of d (D_{uUP}); c is a *downward-universal-part-of* universal d if and only if every instance of d has some instance of c as an individual part (D_{dUP}).

---

[2] This certainly is an oversimplification but will not cause problems for the limited scope of this paper.

$$D_{uUP} \quad uUP\ cd \equiv (x)(Inst\ xc \to (\exists y)(Inst\ yd \wedge IP\ xy))$$

$$D_{dUP} \quad dUP\ cd \equiv (y)(Inst\ yc \to (\exists x)(Inst\ xc \wedge IP\ xy))$$

For example, the universal *waterfall* is an upwards-universal-part of the universal *river*, since every instance of *waterfall* is individual-part-of some instance of *river*.

The formal theory presented in this section is called TLO. A computational representation of this theory can be found at http://www.buffalo.edu/~bittner3/Theories/Papers/Cosit2007Theory.html. A more sophisticated version which constitutes the basis of Basic Formal Ontology (BFO) can be accessed via http://www.ifomis.org/bfo/fol.

3   From Top-Level Ontologies to Domain Ontologies

The terms of a top-level ontology refer to *classes* of relations in certain intended domains of interpretation which satisfy the relevant axioms of the top-level ontology. In the reminder I will refer to the relations that satisfy the axioms associated to the top-level term T of TLO in the intended domains of interpretation *relations of type T*. For example, relations (in the intended domains of interpretation) which satisfy the axioms (AU1-5, $D_{\sqsubseteq}$, and $D_{\sqsubset}$) associated with the term 'sub-universal-of' (abbreviated by $\sqsubseteq$) are called *relations of type sub-universal-of*, or sub-universal-of relations for short. Similarly, relations (in the intended domains of interpretation) which satisfy the axioms (AM1-4 and D_O) associated with the term 'individual-part-of' (abbreviated by *IP*) are called *relations of type individual-part-of*, or individual-part-of relations for short.

In domain ontologies top-level terms often refer *specific relations* in a particular domain. For example, in a domain ontology of socio-economic units, the term 'sub-universal-of' may refer to a specific relation which holds between socio-economic units, and which satisfies the axioms associated with the term 'sub-universal-of' in TLO. Thus in some sense, a domain ontology is a formal representation of one specific *model* (in the model-theoretic sense) of the underlying top-level ontology.

The distinction between types of relations as they are specified in top-level ontologies and particular relations of a given type in a given domain becomes even more important in domains where there is more than one relation of a given type. In the ecosystem example in Section 4 there will be three distinct relations of type sub-universal-of and three distinct relations of type universal-part-of.

The remainder of this section addresses more formally the distinction between a specific relation on a particular domain and types of relations as specified in a top-level ontology.

3.1   Specific Binary Relations

A (specific) *binary relation R with domain of discourse $\mathcal{D}(R)$* is a set of ordered pairs of members of the set $\mathcal{D}(R)$, i.e., $R \subseteq \mathcal{D}(R) \times \mathcal{D}(R)$.[3] If R is a binary relation with domain $\mathcal{D}(R)$ then I will also say that R is a binary relation on $\mathcal{D}(R)$. I write $R(x, y)$

---

[3] In the meta-language the language of set theory is used to talk about specific relations and their properties in a general but precise way.

to say that R holds between $x, y \in \mathcal{D}(R)$, i.e., $R(x, y)$ if and only if $(x, y) \in R$. If R is a binary relation on $\mathcal{D}(R)$ then one can define the relations uR and dR on the powerset (the set of all subsets) of $\mathcal{D}(R)$, i.e., $\mathcal{D}(uR) = \mathcal{D}(dR) = \mathcal{P}(\mathcal{D}(R))$, as follows:

$$uR(X, Y) =_{df} \forall x \in X : \exists y \in Y : R(x, y)$$
$$dR(X, Y) =_{df} \forall y \in Y : \exists x \in X : R(x, y)$$
(1)

R is called an *individual-level* relation on $\mathcal{D}(R)$ while uR and dR are *class-level* relations on $\mathcal{P}(\mathcal{D}(R))$ [11,22]. For example, if R is the individual-part-of relation between human body parts (e.g., my left arm is an individual-part-of my body) then the class-level relation uR is the relation upwards-universal-part-of between body part universals (the universal *left arm* is a upwards-(and downwards)-universal-part-of the universal *human body*.)

For any binary (individual-level or class-level) relation R, one can define the *immediate-R-relation*, R_i on $\mathcal{D}(R)$ in terms of R: The immediate-R-relation, R_i, holds between x and y if and only if the relation R holds between x and y and there is not member z of the domain of R such that $R(x, z)$ and $R(z, y)$.[4]

$$R_i(x, y) =_{df} R(x, y) \text{ and } \neg(\exists z)(z \in \mathcal{D}(R) \text{ and } R(x, z) \text{ and } R(z, y))$$
(2)

Consider Fig. 1 which depicts the graph of the relation immediate-sub-universal-of between ecosystem universals such that the set of nodes of the graph is the domain of this relation. The relation immediate-sub-universal-of is a specific example of an R_i-relation in the sense of Definition 2. The corresponding R-relation is the sub-universal-of relation between the ecosystem universals represented by the nodes of the graph.

In general, let $\Gamma = (N, E)$ be the graph of the relation R_i. Then the set of nodes of Γ is the domain of R_i, i.e., $N = \mathcal{D}(R_i) = \mathcal{D}(R)$, and $(x, y) \in R_i$ if and only if there is an edge from node x to node y in the graph, i.e., $V = \{(x, y) | R_i(x, y)\}$. In the remainder of this paper it is always assumed that the set of nodes, N, is finite, and it will always be the case that if Γ is the graph of the relation R_i then R is the reflexive and transitive closure of R_i. R_i-relations will be used in Section 4 to draw graphs of specific binary relations between individual ecoregions and ecosystem classes. These graphs serve as graphic representations of a 'formal' domain ontology. Under the assumptions made here one can always obtain R from the graph of R_i and vice versa.

Any given binary (individual-level or class-level) relation either has or lacks each of the logical properties listed in Table 2 (and of course others). The properties of reflexivity, antisymmetry, transitivity, as well as the root property are standard and do not need further discussion. Relation R has the *no-single-immediate-predecessor* property (NSIP) if and only if for all $x, y \in \mathcal{D}(R)$, if $R_i(x, y)$ then there is a $z \in \mathcal{D}(R)$ such that $R_i(z, y)$ and $x \neq z$. For example, the relation immediate-sub-universal-of in Fig. 1, has the NSIP property: every non-leaf node of the graph has at least two children. The relation which graph is depicted in Fig. 2(b) lacks the NSIP property.

Relation R has the *single-immediate-successor* property (SIS) if and only if some members of $\mathcal{D}(R)$ stand in the R_i relation and no $x \in \mathcal{D}(R)$ stands in the R_i relation

---

[4] There may be situations where R_i is the empty relation – See [23] for details. For most non-empty R_i, R_i is intransitive – See [23] and Table 2.

Table 2. Properties which a binary relation R either has or lacks. [23].

property	description
reflexive	$\forall x \in \mathcal{D}(R) : R(x,x)$
antisymmetric	$\forall x,y \in \mathcal{D}(R)$: if $R(x,y)$ and $R(y,x)$ then $x = y$
transitive	$\forall x,y,z \in \mathcal{D}(R)$: if $R(x,y)$ and $R(y,z)$ then $R(x,z)$
intransitive	$\forall x,y \in \mathcal{D}(R)$: if $R(x,y)$ and $R(y,z)$ then not $R(x,z)$
root	$\exists x \in \mathcal{D}(R)$: $[\exists y \in \mathcal{D}(R): R(y,x)$ and $\forall y \in \mathcal{D}(R): R(y,x)$ or $y = x]$
NSIP	$\forall x,y \in \mathcal{D}(R)$: (if $R_i(x,y)$ then $\exists z \in \mathcal{D}(R)$: $(R_i(z,y)$ and $x \neq z))$
SIS	$\exists x,y \in \mathcal{D}(R) : R_i(x,y)$ and
	$\forall x,y,z \in \mathcal{D}(R)$: (if $R_i(x,y)$ and $R_i(x,z)$ then $y = z)$

to distinct members of $\mathcal{D}(R)$. For example, the relation immediate-sub-universal-of in Fig. 1, has the SIS property: every node has at most one parent. Graphs of relations that lack the SIS property form lattice structures, in which there may be nodes with more than one parent, rather than trees. (See [23] for extended discussions.)

3.2 Properties and Types of Relations

One can classify binary relations according to their logical properties. Relations of type *partial ordering* are relations which are reflexive, antisymmetric, and transitive. Relations of type *individual-part-of* are relations that hold between individual entities and that are reflexive, antisymmetric, transitive, and in addition satisfy axiom AM4. Hence relations of type *individual-part-of* are also of type *partial ordering*. Relations of type *rooted partial ordering* are relations of type *partial ordering* that also have the root property. Since relations satisfy the axioms of the top-level ontology in virtue of the properties they possess or lack, the types of relations defined in terms of satisfaction of axioms of the top-level ontology corresponds to the typing of relations according to their logical properties.

Relations of type *finite list-or-tree forming* are relations of type *rooted partial ordering* which are such that if R is a relation of type *rooted partial ordering* then R_i (the immediate R-relation in the sense of Def. 2) has the SIS property. All the relations whose R_i-graphs are depicted in this paper are of type *finite list-or-tree forming*. Intended interpretations of the top-level term universal-part-of are relations of type *finite tree forming* that hold between between universals and which are related to the individual-part-of relation between their instantiating individuals in the ways specified in D_{uUP} and D_{dUP}.

Relations of type *finite tree forming* are relations of type *rooted partial ordering* which are such that if R is a relation of type *rooted partial ordering* then R_i has the NSIP and the SIS properties. The relations whose R_i-graphs are depicted in Figures 1, 2(a), 3, 4, 5(a) and 5(b) all are of type *finite tree forming*. Intended interpretations of the top-level term sub-universal-of of TLO are relations of type *finite tree forming* that hold between between universals and are related in the appropriate ways to the instance-of relation between individuals and universals, i.e., satisfy the axioms AU1-5, $D_{\sqsubseteq}$, and $D_{\sqsubset}$.

In the remainder of this paper I use SMALL CAPITAL LETTERS to signify top-level terms referring to types of entities such as INDIVIDUAL and UNIVERSAL, and for top-level terms referring to types of relations such as INDIVIDUAL-PART-OF, UNIVERSAL-PART-OF, SUB-UNIVERSAL-OF, etc. These terms correspond to the symbols used in the formal theory, TLO, as summarized in Table 1. I use `typewriter font` and superscripts to distinguish specific relations in the domain ontology from relation-types denoted by top-level terms in the top-level ontology. I.e., I write `sub-universal-of`[1] and `sub-universal-of`[2] to refer to distinct specific relation among ecosystem universals, both of which satisfy the axioms of the SUB-UNIVERSAL-OF relation in the top-level ontology in virtue of their logical properties.

4   Ecosystem Classification and Delineation

In this section a domain ontology of ecosystem classification and delineation is developed. The domain ontology os based on Bailey's influential paper "Deliniation of ecosystem regions" [19]. Baily's specification of his domain ontology is from a formal-ontological perspective rather imprecise and ambiguous. It will be made more rigorous and precise by (a) using the semantically well-defined top-level terms listed in Table 1 and (b) by enforcing the clear distinction of types of relations as specified in the top-level ontology and specific relations of a type as they occur in the ecosystem domain. This will make it possible to explicitly distinguish a number of relations which have been confused and been taken to be a single relation before. Graphic rather than symbolic representations for those relations will be used in this section. The symbolic representations can be obtained from the graphs as described in Section 3.1.

4.1   Classification of Geographic Ecosystems with Respect to Broad Climatic Similarity, and Definite Vegetational Affinities

Consider the following sentence:

(BL1) "Ecoregions are large ecosystems of regional extent that contain a number of smaller ecosystems. They are geographical zones that represent geographical groups of similarly functioning ecosystems" [19, p. 365]

Using the terms of the top-level ontology one can rephrase (BL1) as follows: Ecoregions are INDIVIDUALS that are INDIVIDUAL-PARTS-OF the biosphere on the surface of the Earth (an INDIVIDUAL). Ecosystems are UNIVERSALS and are INSTANTIATED BY ecoregions[5] with similar functional characteristics. Geographic ecosystems are ecosystem UNIVERSALS which instantiating ecoregions are of geographic scale or larger. Every ecoregion has smaller ecoregions as INDIVIDUAL-PARTS all of which are INSTANCES-OF the UNIVERSAL *ecosystem*.

According to this interpretation of (BL1), 'are' (in BL1) is intended to mean INSTANCE-OF, 'contains' is intended to mean HAS-INDIVIDUAL-PART, 'represent' is

---

[5] Ecoregions here are individual ecosystem, i.e., 'ecoregion' and 'ecosystem individual' treated as synonyms. Following Bailey no distinction between an individual ecosystem and the region it occupies at a given point in time is made.

intended to mean INSTANCE-OF, 'geographical groups of similarly functioning ecosystems' is intended to mean UNIVERSALs that are INSTANTIATED BY similarly functioning ecoregions (INDIVIDUALS).

(BL2) "Regional boundaries may be delineated ... by analysis of the environmental factors that most probably acted as selective forces in creating variation in ecosystems" [19, p. 366]

(BL2) indicates that the *differentia* used for distinguishing ecosystem UNIVERSALS are environmental factors that create the variations between the ecoregions that INSTANTIATE distinct ecosystem UNIVERSALS. Environmental factors used as differentia fall into the two major groups of climate and vegetation [19]. The climate categorization is based on the annual and monthly averages of temperature and precipitation [24].

Thus, geographic ecosystems are UNIVERSALS which INSTANCES, ecoregions, are characterized by broad climatic similarity, definite vegetational affinities, etc. The resulting classification hierarchy is depicted in Fig. 1 where the nodes of the depicted graph are ecosystem UNIVERSALS that form the domain of the relation sub-universal-of[1] and the directed edges represent the relation sub-universal-of$_i^1$ (the immediate sub-universal-of relation in the sense of equation 2 which domain corresponds to the nodes of the graph). The root of the tree is the ecosystem UNIVERSAL *geographic ecosystem*. The relevant differentia which indicate what marks out INSTANCES of the immediate SUB-UNIVERSALS of the ROOT are roughly Koeppen's climate groups [19][6]. For example,

Humid Temperate Ecosystem $=_{df}$ *Geographic Ecosystem* with humid temperate climate.

The relevant differentia which tell us what marks out INSTANCES of the immediate SUB-UNIVERSALS of the ecosystem UNIVERSALS that are differentiated by climate groups, are roughly Koeppen's climate types [19]. For example,

Prairie Ecosystem $=_{df}$ *Humid Temperate Ecosystem* with prairie climate.

The relevant differentia which tell us what marks out INSTANCES of the immediate SUB-UNIVERSALS of the ecosystem UNIVERSALS that are differentiated by climate groups *and* climate types are climax plant formations [19]. For example,

Prairie Bushland Ecosystem $=_{df}$ *Prairie Ecosystem* with climax vegetation type Bushland.

It follows that all ecoregions that INSTANTIATE the UNIVERSAL *Humid Temperate Ecosystem* are characterized by the humid temperate climate group. Similarly, ecoregions that INSTANTIATE the UNIVERSAL *Prairie Bushland Ecosystem* are characterized by the humid temperate climate group, by the prairie climate type, and by bushland vegetation. In other words: *Prairie Bushland Ecosystem* is a sub-universal-of[1] *Prairie Ecosystem* which in turn is a sub-universal-of[1] *Humid temperate domain* and thus, according to definition $(D_{\sqsubseteq})$, every INSTANCE-OF *Prairie Bushland*

---

[6] Bailey collapses Koeppen's subtropical and temperate climate groups into 'Humid temperate'.

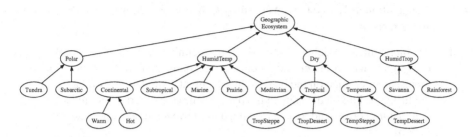

Fig. 1. Graphs of the relations `sub-universal-of`$_i^1$ and `upwards-universal`
`-part-of`$_i^1$. Notice that in Bailey's classification all the leaf nodes in the depicted graph have further sub-universals which are omitted here. See [19, p. 369] for a more complete tree.

Ecosystem is an INSTANCE-OF *Prairie Ecosystem* and is also an INSTANCE-OF *Humid Temperate Ecosystem.*

4.2 Universal Parthood Relations Between Ecosystem Universals

One can make (BL1) even more precise by using the top-level relation UNIVERSAL-PART-OF to specify more precisely what is meant by "Ecoregions are large ecosystems of regional extent that contain a number of smaller ecosystems". This portion of (BL1) indicates that, corresponding to the relation `sub-universal-of`^{1} between ecosystem UNIVERSALS, there is a hierarchical order in which smaller ecosystem UNIVERSALS are parts of larger ecosystem UNIVERSALS.

Let ecosystem UNIVERSAL E_1 be a `sub-universal-of`^{1} ecosystem UNIVERSAL E_2 as depicted in Fig. 1. One can verify that every INSTANCE-OF UNIVERSAL E_1 is an INDIVIDUAL-PART-OF some INSTANCE-OF UNIVERSAL E_2. Thus, in addition to the relation `sub-universal-of`^{1}, the relation `upwards-universal-part-of`^{1} holds between the ecosystem UNIVERSALS depicted in Fig. 1. For example, every ecoregion which is an INSTANCE-OF the UNIVERSAL *Prairie Bushland Ecosystem* is INDIVIDUAL-PART-OF an ecoregion that is an INSTANCE-OF the UNIVERSAL *Prairie Ecosystem.* Similarly, every ecoregion that is an INSTANCE-OF the UNIVERSAL *Prairie Ecosystem* is in turn INDIVIDUAL-PART-OF some ecoregion that is an INSTANCE-OF the UNIVERSAL *Humid Temperate Ecosystem.*

One can see that there is a correspondence between the relation `sub-universal-of`^{1} and `upwards-universal-part-of`^{1} in the sense that for all nodes E_1 and E_2 in the graph of Fig. 1: E_1 is a `sub-universal-of`^{1} E_2 if and only if E_1 is a `upwards-universal-part-of`^{1} E_2. However, from the top-level ontology it is clear that these two relations are very different: `upwards-universal-part-of`^{1} is a class-level relation corresponding to an individual-level relation of type INDIVIDUAL-PART-OF (Definitions (1) and (D_{uUP})). By contrast, the relation `sub-universal-of`^{1} is NOT a class-level version of an individual-level relation.[7]

---

[7] This indicates some serious limitations of the extensional conception of relations as introduced in Section 3.1, which identifies the relations `sub-universal-of`^{1} and `upwards-universal-part-of`^{1}.

4.3 Classification of Ecosystems According to Kinds of Climatic and Vegetation Characteristics

(BL3) "A hierarchical order is established by defining successively smaller ecosystems within larger ecosystems ... subcontinental areas, termed domains, are identified on the basis of broad climatic similarity ... domains ... are further subdivided, again on the basis of climatic criteria, into divisions ... divisions correspond to areas having definite vegetational affinities" [19, p. 366]

(BL3) indicates that, in addition to the classification of ecosystems according to *particular* climatic and vegetational affinities among the INSTANTIATING ecoregions (Fig. 1), Bailey also classifies ecosystems according to the *kinds* of climatic and vegetation characteristics that characterize the INSTANTIATING ecoregions. Bailey [19] distinguishes the following (additional) geographic ecosystem UNIVERSALS: *domains* are ecosystem UNIVERSALS, which INSTANTIATING ecoregions are characterized *only* by the climatic group; *divisions*, which INSTANTIATING ecoregions are characterized by climatic group *and* climatic type but *not* by plant formations; and *provinces*, which INSTANTIATING ecoregions are characterized by climatic group *and* climatic type *and* the climax plant formations [19, Fig. 1, p. 367].[8] The graph of the resulting relation sub-universal-of$_i^2$ is depicted in Fig. 2(a), which shows ecosystem UNIVERSALS as nodes and the relation sub-universal-of$_i^2$ as directed edges.

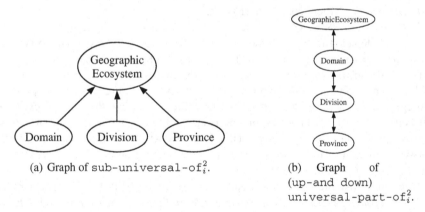

(a) Graph of sub-universal-of$_i^2$.

(b) Graph of (up-and down) universal-part-of$_i^2$.

Fig. 2. Relations among ecosystem universals whose instances are characterized by the same kinds of climatic and vegetation characteristics

Under the assumption that ecoregions that are homogeneous in climate group, type, etc. are also spatially maximal[9], there is an interrelationship between the size of an ecoregion and the kinds of climatic, vegetation, etc. characteristics that characterize that ecoregion in the sense described in the previous paragraph. As pointed out in (BL3)

---

[8] *Sections* are omitted in this paper.

[9] Bailey seems to mean maximal but not necessarily singly-connected parts of a given continent.

the universal *domain* is INSTANTIATED BY ecoregions of subcontinental scale. Ecoregions that INSTANTIATE the universals *division* and *province* are of successively smaller scales.

The hierarchical 'nesting' of provinces into divisions into domains is captured by relations of type UNIVERSAL-PART-OF. Let $\texttt{upwards-universal-part-of}^2$ be the relation which holds between the universals *domain, division, province* and *Geographic ecosystem* such that every INSTANCE-OF the UNIVERSAL *division* is INDIVIDUAL-PART-OF some INSTANCE-OF the UNIVERSAL *domain* and similarly for *province* and *division*, and *domain* and *Geographic Ecosystem*. The graph of $\texttt{upwards-}$ $\texttt{universal-part-of}_i^2$ is represented by the upwards pointing arrows in Fig. 2(b).

Between the universals *domain, division*, and *province* in addition the relation $\texttt{downwards-universal-part-of}^2$ holds: every INSTANCE-OF the UNIVERSAL *domain* has some INSTANCE-OF the UNIVERSAL *division* as an INDIVIDUAL-PART and every INSTANCE-OF the UNIVERSAL *division* has some INSTANCE-OF the UNIVERSAL *province* as an INDIVIDUAL-PART. Notice, that NOT every INSTANCE-OF *Geographic Ecosystem* has some INSTANCE-OF *domain* as an INDIVIDUAL PART. (The Subtropical Division is an INSTANCE OF *geographic ecosystem* but there is no INSTANCE OF *domain* that is as an INDIVIDUAL PART OF the Subtropical Division.) The graph of the relation $\texttt{downwards-universal-part-of}_i^2$ is represented by the downwards arrows in Fig. 2(b).

One can see that, in contrast to the graphs of $\texttt{sub-universal-of}_i^1$ and $\texttt{upwards-universal-part-of}_i^1$, the graphs of $\texttt{sub-universal-of}_i^2$ and $\texttt{upwards-universal-part-of}_i^2$ are quite different. This is another reason why it is important to distinguish between relations of type SUB-UNIVERSAL-OF and relations of type (UPWARDS-)UNIVERSAL-PART-OF.

4.4 'Intersecting' Both Classifications

There are ecoregions that are INSTANCES-OF both, the UNIVERSAL *domain* and the UNIVERSAL *Humid Temperate Ecosystem*. Bailey [19] calls the UNIVERSAL that has as INSTANCES all ecoregions that are INSTANCES-OF both, *domain* and *Humid Temperate Ecosystem, Humid Temperate Domain. Dry domain* is the UNIVERSAL that has as INSTANCES all ecoregions that INSTANTIATE *Dry ecosystem* and *domain*. Similarly for *Polar domain, Tundra division*, etc. (See also [17] for a similar approach or 'intersecting' classification trees.)

Let $\texttt{sub-universal-of}^3$ be the SUB-UNIVERSAL-OF relation which has as its domain the set which has as its members UNIVERSALS that are constructed in the way described in the previous paragraph and, in addition, the UNIVERSAL *Geographic ecosystem*. An important feature of the relation $\texttt{sub-universal-of}^3$ is, that *Geographic ecosystem* is the only UNIVERSAL that a has proper SUB-UNIVERSAL. Thus, the graph of $\texttt{sub-universal-of}_i^3$ is a flat but rather broad tree as indicated in Fig. 3.

The graph of the relation $\texttt{sub-universal-of}_i^3$ is a *refinement* of the graph of the relation $\texttt{sub-universal-of}_i^2$ in Fig. 2(a) in the sense that each of the nodes *domain, division*, and *province* in the graph of the relation $\texttt{sub-universal-of}_i^2$ is replaced by a set of jointly exhaustive and pairwise disjoint UNIVERSALS. For example,

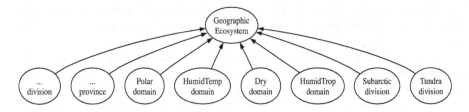

Fig. 3. Graph of the relation `sub-universal-of`$_i^3$

the node *Domain* in Fig. 2(a) is replaced by the nodes *PolarDomain*, *Humid Temperate Domain*, *Dry domain*, and *Humid Tropical domain* in Fig. 3. Similarly for the other nodes in Fig. 2(a).

Let `upwards-universal-part-of`^{3} be the UPWARDS-UNIVERSAL-PART-OF relation on the domain of `sub-universal-of`^{3}. The graph of `upwards-universal-part-of`$_i^3$ (Fig. 4) is a refinement of the graph of `upwards-universal-part-of`$_i^2$ (Fig. 2(b)) in the sense that the nodes *domain*, *division*, and *province* of the graph of `upwards-universal-part-of`$_i^2$ correspond to layers or *cuts* [18] in the tree formed by the graph of `sub-universal-of`$_i^3$.

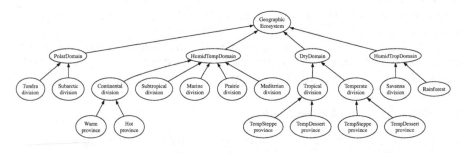

Fig. 4. Graph of the relation `universal-part-of`$_i^3$

Notice, that there is a graph-isomorphism between the graphs of the relations `sub-universal-of`$_i^1$, `upwards-universal-part-of`$_i^1$, and `upwards-universal-part-of`$_i^3$, i.e., the graphs of the relations are structurally identical. Notice, however, that the *domains* of the relations `upwards-universal-part-of`^{1} and `upwards-universal-part-of`^{3} are quite different, since the SUB-UNIVERSAL-OF relations, `sub-universal-of`$_i^1$ and `sub-universal-of`$_i^3$ are very different as easily recognizable in Figures 1 and 3.

4.5   Ecosystem Delineation

The discussion so far has focused on ecosystem classifications (SUB-UNIVERSAL-OF relations) and on the hierarchical spatial nestings that are induced by these classifications through the corresponding UNIVERSAL-PART-OF relations. However Bailey [19] also emphasizes the delineation of ecoregions. Delineation here refers to the establishing of fiat boundaries [25][10] that separate ecoregions which INSTANTIATE ecosystem UNIVERSALS that are *differentiated* in the SUB-UNIVERSAL hierarchy. That is, *delineation* at the level of ecoregions (INDIVIDUALS) corresponds to establishing *differentia* between ecosystem UNIVERSALS.

Consider the classification of geographic ecosystem universals into *domain, division, province,* etc. (formally represented by the relation sub-universal-of^2 as depicted via sub-universal-of$_i^2$ in Fig. 2(a)). This classification is such that the EXTENSION of the UNIVERSAL *domain* (i.e., the COLLECTION of ecoregions that IN-STANTIATE the UNIVERSAL *domain*) PARTITIONS the (biosphere on) the surface of the Earth in the sense that that (i) no distinct MEMBER-OF the EXTENSION-OF *domain* have a common INDIVIDUAL-PART and (ii) jointly the MEMBERS-OF the EXTENSION-OF *domain* SUM-UP-TO the INDIVIDUAL '(Biosphere on) the surface of Earth'. That is, the ecoregions that are MEMBERS OF the EXTENSION OF *domain* are jointly ex-haustive and pair-wise disjoint. Similarly, the EXTENSIONs of *division* and *province* all PARTITION the INDIVIDUAL *Surface of Earth.* (Fig. 5(a).)

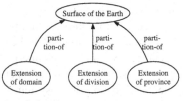

(a) The extensions of the univer-sals *domain, division,* and *province* partition the surface of Earth. (partition-of^E)

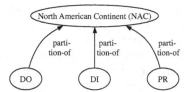

(b) Collections of geographic ecoregions that partition *NAC* . partition-of^{NAC}

Fig. 5. Partitions formed by collections of ecoregions

Consider the UNIVERSAL *Dry Domain.* Obviously, not all ecoregions that are IN-STANCES OF this UNIVERSAL are INDIVIDUAL-PARTS-OF the North American conti-nent (*NAC*). In fact, there is only a single INSTANCE of *Dry Domain* that is INDIVIDUAL-PART-OF *NAC*. The other INSTANCES-OF *Dry Domain* are INDIVIDUAL-PARTS-OF other continents.

For many practical purposes it is useful to refer not to all INSTANCES of UNI-VERSALS like *domain, division,* and *province* but only to those INSTANCES that are INDIVIDUAL-PART-OF the North American Continent (*NAC*). For this purpose the no-tion of COLLECTION is used. Let *DO* be the COLLECTION of ecoregions that are INSTANCES-OF the UNIVERSAL *domain* and that are INDIVIDUAL-PART-OF *NAC*; let *DI* be the COLLECTION of ecoregions that are INSTANCES-OF the UNIVERSAL *division* and that are INDIVIDUAL-PART-OF *NAC*; and let *PR* be the COLLECTION of ecore-gions that are INSTANCES-OF the universal *province* and that are INDIVIDUAL-PART-OF *NAC*. The COLLECTION *DO* is a PARTITION-OF the INDIVIDUAL North American Continent (*NAC*) in the sense that (i) no distinct MEMBERS-OF *DO* have a common

---

[10] Of course, those boundaries are subject to vagueness, as Bailey himself points out.

INDIVIDUAL-PART and (ii) jointly the MEMBERS of DO SUM-UP-TO NAC. Similarly, the COLLECTIONS DI and PR are PARTITIONS-OF NAC. The graph of the relation partition-of^{NAC} is depicted in Fig. 5(b).

The COLLECTIONS DO, DI, and PR not only PARTITION NAC they are also (up- and downward) PARTONOMICALLY-INCLUDED in one another in the sense that they are hierarchically structured such that every MEMBER-OF PR is an INDIVIDUAL-PART-OF some MEMBER-OF DI and every MEMBER-OF DI has some MEMBER-OF PR as INDIVIDUAL-PART. Similarly for DI and DO and for PR and DO. Of course this mirrors the UNIVERSAL-PART-OF relations upwards and downwards-universal-part-of[2] between the universals *province*, *division*, and *domain* (Fig. 2(b)).

5   Conclusions

In this paper a logic-based top-level ontology was used to create a domain ontology of ecosystem classification and delineation. The aim was to express the domain ontology underlying Bailey's paper 'Delineation of Ecoregions' [19] in a logically rigorous form that is accessible not only to human domain specialists but also to computers.

Notice that the claim of the paper is not that Bailey's definitions must be interpreted in the ways suggested here. In fact it is a weakness of Bailey's definitions that they are imprecise (at least from a logical and computational perspective) and can be interpreted in different ways, and thus leave (at least from the perspective of a non-domain specialist) the exact nature of the relationships between the classification of ecosystems into different kinds and the spatial nesting of ecoregions that instantiate those kinds implicit.

It was the aim of this paper to make one possible interpretation of Bailey's work as precise as possible by using notions like INDIVIDUAL, UNIVERSAL, INSTANCE-OF, PART-OF, UNIVERSAL-PART-OF, PARTITION-OF, etc., which exact meaning was specified using an axiomatic theory. This makes it easier for other researchers to understand and to criticize this particular interpretations.

The analysis of this paper also showed is that it is important to distinguish types of relations as specified at the top-level from specific relations of a given type as they occur in a specific domain. By taken this distinction into account it became possible to explicitly distinguish a number of relations among ecosystems and ecoregions which have been confused and been taken to be a single relation before.

References

1. Guarino, N.: Formal ontology and information systems. In: Guarino, N. (ed.) Formal Ontology and Information Systems (FOIS'98), pp. 3–15. IOS Press, Amsterdam (1998)
2. Bittner, T., Donnelly, M., Smith, B.: Individuals, universals, collections: On the foundational relations of ontology. In: Varzi, A., Vieu, L. (eds.) Proceedings of the third International Conference on Formal Ontology in Information Systems, FOIS04, pp. 37–48. IOS Press, Amsterdam (2004)
3. Rector, R.: Modularization of domain ontologies implemented in description logics and related formalisms including OWL. In: Proceedings of the international conference on Knowledge capture, pp. 121–128 (2003)

4. Gangemi, A., Guarino, N., Masolo, C., Oltramari, A., Schneider, L.: Sweetening ontologies with DOLCE. AI Magazine 23(3), 13–24 (2003)
5. Copi, I.: Symbolic Logic. Prentice Hall, Upper Saddle River, NJ (1979)
6. Rector, A., Rogers, J.: Ontological issues in using a description logic to represent medical concepts: Experience from GALEN: Part 1 - principles. Methods of Information in Medicine (2002)
7. Spackman, K., Campbell, K., Cote, R.: SNOMED RT: A reference terminology for health care. In: Proceedings of the AMIA Annual Fall Symposium, pp. 640–644 (1997)
8. Sioutos, N., de Coronado, S., Haber, M.W., Hartel, F.W., Shaiu, W.L., Wright, L.W.: Nci thesaurus: a semantic model integrating cancer-related clinical and molecular information. J. Biomed. Inform. 40(1), 30–43 (2007)
9. The Gene Ontology Consortium: Creating the gene ontology resource: Design and implementation. Genome Res. 11 (2001) 1425–1433
10. OBO: Open biomedical ontologies (2006), http://obofoundry.org
11. Smith, B., Ceusters, W., Klagges, B., Köhler, J., Kumar, A., Lomax, J., Mungall, C., Neuhaus, F., Rector, A., Rosse, C.: Relations in biomedical ontologies. Genome Biology 6(5), r46 (2005)
12. Abdelmoty, A.I., Smart, P.D., Jones, C.B., Fu, G., Finch, D.: A critical evaluation of ontology languages for geographic information retrieval on the internet. Journal of Visual Languages & Computing 16(4), 331–358 (2005)
13. Agarwal, P.: Ontological considerations in giscience. International Journal of Geographical Information Science 19(5), 501–536 (2005)
14. Grenon, P., Smith, B.: SNAP and SPAN: Towards dynamic spatial ontology. Spatial Cognition and Computation 4(1), 69–103 (2004)
15. Mark, D., Smith, B., Tversky, B.: Ontology and geographic objects: An empirical study of cognitive categorization. In: Freksa, C., Mark, D.M. (eds.) COSIT 1999. LNCS, vol. 1661, pp. 283–298. Springer, Heidelberg (1999)
16. Feng, C.C., Bittner, T., Flewelling, D.: Modeling surface hydrology concepts with endurance and perdurance. In: Egenhofer, M.J., Freksa, C., Miller, H.J. (eds.) GIScience 2004. LNCS, vol. 3234, pp. 67–80. Springer, Heidelberg (2004)
17. Sorokine, A., Bittner, T.: Understanding taxonomies of ecosystems: a case study. In: Fisher, P. (ed.) Developments in Spatial Data Handling, pp. 559–572. Springer, Berlin (2005)
18. Sorokine, A., Bittner, T., Renscher, C.: Ontological investigation of ecosystem hierarchies and formal theory for multiscale ecosystem classifications. geoinformatica 10(3), 313–335 (2006)
19. Bailey, R.G.: Delineation of ecosystem regions. Environmental Management 7, 365–373 (1983)
20. Lowe, E.J.: Recent advances in metaphysics (keynote). In: Welty, C., Smith, B. (eds.) International Conference on Formal Ontology in Information Systems (2001)
21. Smith, B., Koehler, J., Kumar, A.: On the application or formal principles to life science data: a case study in the gene ontology. In: Rahm, E. (ed.) DILS 2004. LNCS (LNBI), vol. 2994, pp. 79–94. Springer, Heidelberg (2004)
22. Donnelly, M., Bittner, T., Rosse, C.: A formal theory for spatial representation and reasoning in bio-medical ontologies. Artificial Intelligence in Medicine 36(1), 1–27 (2006)
23. Bittner, T., Donnelly, M.: Logical properties of foundational relations in bio-ontologies. Artificial Intelligence in Medicine 39, 197–216 (2007)
24. Koeppen, W.: Grundriss der Klimakunde. W. de Gruyter, Berlin (1931)
25. Smith, B.: Fiat objects. Topoi 20(2), 131–148 (2001)

Semantic Categories Underlying the Meaning of 'Place'

Brandon Bennett[1] and Pragya Agarwal[2]

[1] School of Computing, University of Leeds
brandon@comp.leeds.ac.uk
[2] Department of Geomatic Engineering, University College London
pagarwal@ge.ucl.ac.uk

Abstract. This paper analyses the semantics of natural language expressions that are associated with the intuitive notion of 'place'. We note that the nature of such terms is highly contested, and suggest that this arises from two main considerations: 1) there are a number of logically distinct categories of place expression, which are not always clearly distinguished in discourse about 'place'; 2) the many *non-substantive* place count nouns (such as 'place', 'region', 'area', etc.) employed in natural language are highly ambiguous. With respect to consideration 1), we propose that place-related expressions should be classified into the following distinct logical types: a) 'place-like' count nouns (further subdivided into abstract, spatial and substantive varieties), b) proper names of 'place-like' objects, c) locative property phrases, and d) definite descriptions of 'place-like' objects. We outline possible formal representations for each of these. To address consideration 2), we examine meanings, connotations and ambiguities of the English vocabulary of abstract and generic place count nouns, and identify underlying elements of meaning, which explain both similarities and differences in the sense and usage of the various terms.

1   Introduction

'Place' is a basic notion in everyday communication. It is a fundamental concept in geography and plays a key role in almost every field of human enquiry [8,20,12,13]. Despite this ubiquity and importance, the semantics of 'place' is poorly understood and controversial [16,19]. Massey ([14], p22) claimed that places 'do not necessarily mean the same thing to everybody' and that 'there is an increasing uncertainty about what we mean by place and how we relate to place'.

A clear semantic model of its basic concepts is a critical condition for establishing an adequate ontology for a domain. In GIScience and related geographic discourse, the notion of place is a basic category that is employed to individuate meaningful portions of space and to describe spatial locations of physical objects. 'Places' are the conceptual entities that enable cognitive structuring of the spatial aspects of reality.

Previous work [1,2,3] has demonstrated the significance of 'place' in the development of an integrated geographic ontology and elaborated on the need to resolve the contested nature of place, which is caused primarily by the lack of a clear semantic account. Related work has also shown that disagreement in establishing the meanings or rules associated with place terminology arises primarily because the relationships

S. Winter et al. (Eds.): COSIT 2007, LNCS 4736, pp. 78–95, 2007.

and distinctions between place and other spatial concepts such as 'neighbourhood', 'region', 'area' and 'location' are not clearly defined [1]. The lack of consensus on the meaning of such terms is compounded by the presence of vagueness and ambiguity in place-related vocabulary [4,6,22].

The current paper promotes the view that the notion of 'place' can only be adequately understood by examining the logical role that place-related concepts play in natural language and developing a semantic theory of place which formalises this logic. In fact, we shall see that place enters into language in a number of distinct but closely related ways, and these have to be distinguished and separately analysed before a more comprehensive theory of place can be articulated.

In addition to the various logical roles of place expressions, natural language also employs a large vocabulary of abstract and non-substantive terms for place-like entities — for instance: 'region', 'neighbourhood', 'district', 'location', 'area' etc., as well as 'place' itself. Understanding the differences between these closely related terms is critical to defining an unambiguous semantic framework that can support a general ontology of place.

The structure of the paper is as follows. In the next section we identify fundamental properties of the concept of place, which will underpin our semantic analysis. In Section 3 we consider the grammatical categories of different kinds of place terminology used in natural language, and indicate how these correspond to different logical categories of place expression. Section 4 examines semantic attributes that are relevant to the interpretation and differentiation of place-related terminology. In Section 5 the category of 'place-like' count nouns is analysed in detail. These are sub-categorised into three levels of abstraction: substantive place count nouns (such as 'church' or 'town'), primarily spatial count nouns (such as 'region' or 'neighbourhood') and abstract or generic count nouns (such as 'place' itself). This analysis also identifies typical modes of use and connotations of specific natural language terms. Finally, Section 6 concludes by considering how far the present work takes us in understanding the semantics of place and identifying directions for further work.

2   The Nature of 'Place'

2.1   Vagueness and Ambiguity in the Meaning of 'Place'

The pervasive vagueness and ambiguity of place-related terminology might lead one to the view that 'place' itself is a vague and ambiguous concept that is not amenable to semantic analysis. In this paper we take a somewhat different view. Though we readily acknowledge that the terminology used to describe place is often vague and ambiguous, we do *not* consider this to be intrinsic to the underlying semantic relationships associated with space and place. Rather it stems from the following two considerations: 1) generic place terms ('place', 'region', 'area' etc.) are typically ambiguous in that their meaning is compounded from a number of distinct though closely related senses; 2) concepts of place are in most cases dependent on other concepts, such as geographic feature types, which are themselves vague.

Many authors have discussed the particular character of our notion of 'place' and have come to a variety of different views about what special feature is distinctive of place [16,14,11,9]. We believe that the lack of consensus is partly due to insufficient generality in the views that have been adopted and also due to a lack of attention to the ambiguities in many concepts associated with place.

Our view is that a clear account of the nature of 'place' can be elicited from a detailed analysis of the semantic content of natural language terminology. Following the methodology for analysing vague and ambiguous concepts outlined by Bennett [7], our approach will be to identify semantic categories and logical principles underlying the usage of place terminology, and incorporate these within a semi-formal theory encompassing the most significant semantic attributes of place-related concepts and articulating their modes of expression. Because of the ambiguity of natural language, there will not be a definite mapping between natural language terms and elements of this semantic theory. Rather, we shall find that each natural language term has a number of distinct senses, corresponding to different ways in which it can be interpreted within our semantic framework.

2.2   Similarity, Continuity and Integrity

A fundamental ingredient of our awareness of the world is our recognition of correlations between *continuity* and *similarity*. Consider our visual perception of a scene. We may conceive of this in very general terms as a distribution of colour over space. Colours are seen as more or less similar.[1] Space is manifest as having a certain continuity — we can traverse space along a line or identify connected *subspaces* of the global space. Given these fundamental aspects of perception, the recognition of correlations between continuity and similarity is the basis for our division of the world into entities. Thus (at a rather primitive level of perception) a visual scene may be divided into parts, each of which is spatially-contiguous (i.e. self-connected) and whose points are all similar in colour.

Similarity is not, of course, an absolute relationship, but is manifest to varying degrees: in some cases we find relatively sharp boundaries separating different types or qualities of matter that occupy space; but in other cases the quality of matter varies in graduated way. For instance, the colour of (the surface of) a material object may vary continuously over its extension. This gives rise to an intrinsic vagueness in partitions of space that are dependent on similarity. Nevertheless, even where boundaries are indeterminate, similarity and continuity still impose a structure on space, such that we can, for example, distinguish dark patches on a surface whose colour varies continuously.

While similarity and continuity determine the most basic partition of space into integral units, the individuation of objects in general depends on more complex principles of integrity. Many kinds of physical object are not uniformly constituted, but rather consist of complex arrangements of heterogeneous parts. So, the objects that we consider to exist in the world are individuated according to sophisticated principles of integrity

---

[1] In fact colour itself can be regarded as constituting a kind of space. This would mean that correlations between space and colour can be seen as arising from juxtaposition of two distinct kinds of space.

that relate to complex combinations of more basic units. Moreover, we also apprehend and individuate spatial objects in terms of their sociological or political status. Hence, 'places' may be distinguished on the basis of ownership, control or jurisdiction, which is only indirectly related to physical reality.

2.3   Locating, Hosting and Anchoring

One of the most important functions of places is to designate a *location* for other objects. This function can be achieved *via* a variety of different types of spatial constraint that can be associated with a place. Perhaps the most typical and easiest to model semantically is what we describe as *hosting*. This occurs where a place is associated with a *subspace* of a larger embedding space; and a hosting relation obtains when the spatial extension of an object is contained within this subspace.

We shall see later that there are various different ways in which a hosting region can be associated with objects of a place-like character. The simplest is where the hosting region is simply the extension of a physical object. However, this is not a typical case, since, if a physical object or person is located within a building, for example, they do not normally interpenetrate the physical material of the building, but rather their extension is within a region of free space that is circumscribed by the building. Thus, the hosting region associated with a place object may be associated with the region corresponding to the concavities in that object.[2] More generally, the means by which a place serves to locate an object may be more complex than hosting within a subspace. For example, 'on the top of the mountain' or 'on the other side of the wall' involve more subtle spatial constraints.

A further consideration, relevant to the locating aspect of places is that they are typically *anchored* in relation to objects that are more or less permanently fixed relative to what is taken as a global reference frame (typically the surface of the Earth).

3   Linguistic and Logical Categories of Place Terminology

Place is a notion that is manifest in a variety of different grammatical roles in natural language. For instance, we have names of places (e.g. London, Europe etc.), common nouns expressing types of place (e.g. town, forest, country etc.) and we also have spatial properties (e.g. under the table, on a train etc.). To provide a semantics of place that accounts for the natural language usage of place terminology, it is essential to distinguish these different grammatical categories. Moreover, the diverse modes of linguistic functionalty of different grammatical categories suggests that a formal theory of place should include a corresponding number of different logical categories.

Before examining them in detail, we first list what we believe to be the most important categories of linguistic expression relating to place:

---

[2] More precisely, using an extension operator **ext** and a *convex-hull* operator **conv** one might specify that the hosting region associated with an object x is the region $\mathsf{conv}(\mathsf{ext}(x)) - \mathsf{ext}(x)$. Although, in some cases, topological containment rather then inclusion the convex-hull would be more appropriate.

- *Count Nouns.*[3] Many count nouns are used to categorise things that we regard as places. These may refer to classes of physical object (such as town, city, room, building, forest), or to types of primarily spatial entity (such as 'area' or 'neighbourhood'). We must also consider more abstract count nouns relating to place (such as 'location' or 'place' itself).
- *Locative Property Phrases.* Phrases such as 'in London', 'on the hill', 'by the sea-side' are predicative expressions that characterise the location of an object. They generally contain a preposition referring to a spatial relationship and also a reference to one or more objects which act as an *anchor* for the relation.
- *Place-Names.* Nominal terms, such as 'London', 'England', 'the Black Forest' refer to objects that are normally considered to be places.[4]
- *Definite Descriptions.* These are phrases such as 'the library', 'the shed at the end of the garden', which function as complex nominal expressions referring to places.

3.1 Place-Like Count Nouns

By count nouns we mean those expressions of natural language that characterise *types* of object. Many count nouns characterise types of object that we may consider to be places; for instance: 'room', 'town', 'forest', 'country', etc.. We shall call these *place-like* count nouns. In accordance with ordinary usages of the word 'place', such count nouns are not themselves places; rather they refer to classes of objects of a place-like nature.

A key attribute of a place-like count noun is that its instances are capable of *locating* other objects, either by *hosting* or some more complex mode of spatial constraint. Another consideration in determining whether a count noun is place-like is that its instances are objects that are fixed in space (relative to the position of the Earth). For example, a count noun such as 'tree' may also function in a place-like way. Thus, the distinction between place-like and non-place-like count nouns is not absolute. Nouns such as 'town' or 'country' are perhaps the most prototypical cases, whereas 'tree' or 'bag' have some but not all of the typical semantic attributes of a place-like count noun.

Despite this lack of precision in what constitutes a place-like count noun, we believe it is a useful sub-categorisation of count nouns and is correlated with certain objective linguistic phenomena which support this distinction and may sharpen our intuitions. One could make the classification more precise by distinguishing more refined categories of locating, hosting, anchoring (etc) count nouns, but this is beyond the scope of the present work.

A good indication of a count noun being place-like is that it is typically associated with a 'default' spatial preposition that is used much more commonly than any other to form a locative property phrase from the count noun. Typically this will be 'in' or 'at'

---

[3] These are often called 'common nouns' but we prefer the term 'count noun' as it gives an indication of the semantic character of this kind of word, since one can test if a term is a count noun by considering whether phrases of the form 'one κ', 'two κs' *etc.* make sense.

[4] It is clear that the structure of composite place names such as 'the Black Forest' has some semantic significance. However, in the present work we shall treat all place-names as atomic symbols.

(or possibly both of these, since they are often interchangeable without affecting the meaning of the locative phrase). One test that we can apply to identify place-like count nouns is to consider whether associated definite descriptions can be given as answers to 'Where?' questions. For instance 'library' is place-like, since the question 'Where is John', might be answered simply by 'the library'. (Of course the usage of place-like count nouns in this way may be regarded as an abbreviated or degenerate form of a more explicit place expression such as 'in the library'.)

3.2 Locative Property Phrases

Phrases such as 'in London', 'on the hill', 'by the sea-side', 'between the church and the oak tree', are predicative expressions which characterise the location of an object. We shall refer to such expressions as *locative property phrases*. A very wide variety of such phrases can be found in natural language, and in our investigations we have spent considerable effort classifying and formalising the logical structure of such phrases. But, because of their complexity and diversity, a comprehensive account of these is beyond the scope of the present work. Here, we shall consider only the most typical form of locative property phrase.

A typical locative property phrase contains a preposition referring to a (localising) spatial relation, R, and also a reference to one or more objects which act as an *anchor* for the relation. Whether a particular object x, whose spatial extension is $\mathsf{ext}(x)$, satisfies the locative property is determined according to whether the relation

$$R(\mathsf{ext}(x), s_1, \ldots, s_n)$$

is true.[5] Each anchor region s_i will typically be determined by an anchor object α_i, such that $\mathsf{ext}(\alpha_i) = s_i$.

For example, in the sentence "The villa is in Barcelona" a locative property is expressed by the phrase "in Barcelona". In this phrase, the name Barcelona is associated with a spatial extent s. The fact that Barcelona is the name of an entity of the category "city", determines the way that its spatial extent is defined (primarily in terms of the geometrical configuration of its urban fabric). Vagueness in the notion of "city" (and possibly also in the conditions of appellation of the name "Barcelona") will result in an associated vagueness in the extent s.

The place description also includes the preposition 'in', which determines the spatial relation, R. In this case we could interpret the relation as the *parthood* relation $\mathsf{P}(r, s)$, where r and s are 2-dimensional spatial regions corresponding to the *footprints* of (the spatial extensions of) the villa and Barcelona. Thus, the meaning of the sentence can be formally represented by

$$\mathsf{P}(\mathsf{footprint}(\mathsf{ext}(\mathsf{the\_villa})), \mathsf{footprint}(\mathsf{ext}(\mathsf{barcelona}))) \; .$$

Suppose we have a place description involving a more subtle spatial relationship — for example, "The villa is on the edge of Barcelona". Here, the place description also

---

[5] In most cases there will only be a single anchor region, so the description will take the form $R(\mathsf{ext}(x), s)$.

includes the phrase "on the edge of", which determines the spatial relation, R. In this case there is some ambiguity and vagueness in the spatial relation being referred to. To give a precise semantics, we would need to identify a well-defined spatial relationship. (For instance we might associate every region with a "thick border", whose width is proportional to the size of the region; and being on the edge of a region could be defined as lying within this border.) However, this vagueness is not essentially due to the fact that this is a locative property; but arises from the particular relation used in this case.

Natural languages contain many spatial prepositions and these often correspond to different spatial constraints according to the context in which they are employed [21]. Hence a comprehensive theory of locative property phrases would require a detailed analysis and formalisation of the semantics of spatial prepositions.

3.3 Place-Names

Many proper names apply to things that are normally considered to be 'places'. Such proper names will be called *place-names*. The referent of a place-name, must be an object of a kind that can be classified by a place-like count noun. Since, names in themselves are arbitrary and do not carry any conceptual content, we suggest that the place-like character of a count noun must derive entirely from the count noun with which it is associated.

For example, Leeds is a city, so the place designated by 'Leeds' is of the type associated with cities. Thus we would represent the semantics of (e.g.) "John is in London" as having a form such as

$$\exists x[\mathsf{Name}(x, \texttt{"London"}) \wedge \mathsf{city}(x) \wedge \mathsf{In}(\mathsf{john}, x)] ,$$

where the In relation gets a specific spatial interpretation dependent on the type of place object:

$$\forall x[\mathsf{city}(x) \rightarrow \forall y[\mathsf{In}(y, x) \leftrightarrow \mathsf{P}(\mathsf{footprint}(\mathsf{ext}(y)), \mathsf{footprint}(\mathsf{ext}(x)))]]$$

3.4 Definite Descriptions

Phrases such as 'the library', 'the shed at the end of the garden' function as complex nominal expressions referring to places. These are *definite descriptions* [17], which (within a given context) identify a unique place entity. Most definite descriptions can be paraphrased by an expression of the form 'the κ such that ϕ', where κ is a count noun. Here, it is the count noun κ that determines the place-like character of the definite description, whereas ϕ serves to identify a unique individual. We can represent the logical form of a definite description using the *iota* notation [23].[6] If we separate the count noun from other constraints in the description, the representation will take the form:

$$\iota x[\kappa(x) \wedge \phi(x)] .$$

---

[6] The meaning of the iota operator is defined as follows:
$$\Psi(\iota x[\phi(x)]) \equiv_{def} \exists x[\phi(x)] \wedge \forall y[\phi(y) \rightarrow (y = x)] \wedge \Psi(x)] .$$

There is another form of definite description that is commonly used to refer to places. This is exhibited by phrases such as: 'the top of the hill', 'the side of the mountain', 'the middle of the ocean' and 'the edge of the table'. Such phrases include a particular kind of place-like count noun ('top', 'side', 'middle' *etc*) whose referents are derived by primarily spatial functions from place-like objects. Thus, the form of these phrases could be represented as, for example:

$$\text{the\_top}(\iota x[\kappa(x) \wedge \phi(x)]) \,,$$

where the_top is a function from place objects (of an appropriate kind) to another kind of place object corresponding to their tops.

4   Semantic Attributes of Place Concepts

In this section we present an analysis of the primary semantic attributes that are relevant to the interpretation of place-related concepts: we consider the modes by which a place may be individuated as a subspace of a global space by principles of integration and demarcation, and we also examine the ways in which a place may act as a 'host' for other objects.

4.1   Principles of Integration

As we argued in Section 2.2, continuity and homogeneity are two primary factors in determining the extent of a place and demarcating its boundary. However, when we are dealing with complex geographic or sociological entities, other more subtle principles of integration come into play. We have identified the following five factors as being of particular significance:

- **Homogeneity.** A region individuated on the basis of homogeneity can be formalised as a maximal connected region all of whose points satisfy a characteristic predicate.
- **Control.** This covers any integration principle based on ownership of, or jurisdiction over some subspace of the domain. The control principle is usually only relevant to geographic regions. It is typically applicable in subspaces described by the terms: region, district, domain, demesne, territory.
- **Proximity.** In addition to the purely qualitative topological notion of connectedness, metrical notions of distance also play an important role in the integrating principles used to identify places.
- **Aggregation.** Many place-like count nouns refer to aggregates of similar elements (e.g. trees make up a forest, buildings make up a town).
- **Systemic Grouping.** As noted above, we often consider complex arrangements of heterogeneous parts as constituting an integral whole (e.g. an airport or a neighbourhood within a town/city).

4.2   Principles of Integration Determined by Count Noun

The principle of integration relevant to identifying a place object will normally be determined by a count noun by which the place is described. In the case of a general question about place, one may use an abstract or non-substantive count noun (area, region, location etc.) and the connotations of this term will mean that certain integration criteria are most appropriate. For example, 'region' (in its most common sense) implies a connected geographic scale subspace, which is either homogeneous in some way or is the extent of some jurisdiction. In more specific contexts, a particular physical count noun (e.g. 'city' or 'forest') may either be given in a question or supplied by the answerer. In this case the type of count noun will determine its particular mode of integration.

4.3   Connectedness

All place terms have a very strong connotation that the subspace referred to is *connected*. More precisely, this means that any two points in the subspace can be joined by a line (not necessarily straight) that lies completely within the subspace. Typically the subspace will satisfy the stronger condition of being *interior connected* — i.e. any two interior points can be connected by a line that lies wholly within the interior of the subspace.

In rare cases some place terms (especially primarily spatial terms) may be used to refer to a disconnected (i.e. multi-piece) subspace. This could occur with a 'region', 'district', 'territory' or 'domain', which might consist of two or more separate parts. In particular if the integrating principle employed is one of ownership, control or jurisdiction then multi-piece subspaces may occur. It is a matter of taste whether a word such as 'territory' can be used in this way, or whether a multi-piece area of control should be regarded as several different territories. The word 'area' is also occasionally used to refer to a multi-piece subspace.

4.4   Modes of Partitioning

Partitioning is very similar to integration, but whereas integration focuses on determining the extent of one particular subspace, partitioning starts with the space and divides it up into different subspaces, which will be called *cells*. Certain generic place count nouns (e.g. 'zone' or 'sector') have a strong connotation of referring to a subspace within a partition. In many cases, the criteria by which a space is partitioned are closely related to the integrity criteria mentioned above. For instance, a space may be partitioned into cells such that each is more or less homogeneous, relative to some intrinsic property of points in the space, and such that neighbouring cells are distinguishable relative to this property. A partition may also be made on the basis of control (i.e. ownership and jurisdiction).

Partitioning can also be done in a way that is significantly different from integration. This is where we decompose one subspace into smaller subspaces based on structural properties of the larger subspace. The division may be in relation to the shape of the larger subspace, within which we may identify such features as (e.g.) lobes or necks; or we may divide it purely in terms of the relative positioning of its parts, such as the northern or central part.

4.5 Modes of Locating and Hosting

We noted above that there are a variety of different ways in which a place-like object can spatially constrain the location of other objects. A comprehensive theory of place would have to define these in terms of a theory of spatial relationships, such as the *Region Connection Calculus* [15] or *Region-Based Geometry* [5]. In the current paper we simply list a number of kinds of spatial constraint which can operate as locating relationships:

- Topological inclusion (e.g. 'in the liver'),
- Geometrical containment (e.g. 'in a building'),,
- Containment within a concavity (e.g. 'in a cup'),
- Interposition among elements of aggregate (e.g. 'in a forest' implies 'among trees')
- Location *within or among* elements of aggregate (e.g. 'in a town' implies 'within or among buildings'),
- Containment within a surface demarcation (such as a district or country) — i.e. *footprint* containment.
- Support (such as 'on a table').

5 Classification of 'Place-Like' Count Nouns

In this section we attempt to classify the semantic content of some of the most common place-related count nouns. This is not straightforward, as there are a wide variety of different terms with subtly different meanings. Because there is considerable overlap in the range of applicability of different terms, distinct terms may in some (or most) cases seem to be equivalent. Moreover, the terms themselves are in some cases ambiguous, having a number of distinct though closely related senses. Nevertheless, we shall attempt to tease out the principal semantic elements that underlie their meanings.

We start by making a high-level distinction between three categories:

- *Substantive.* By substantive, we mean those place count nouns that refer to types of entity that, in addition to their place-like characteristics, have essential properties that are non-spatial (e.g. 'town', 'cupboard', 'country', 'planet').
- *Spatial.* In this category we include terms which characterise place entities in terms of purely spatial characteristics. For example, 'region' or 'point'.
- *Abstract.* This category includes the most general place terms (such as 'location', 'position' and 'place' itself), which are themselves used to characterise the semantic nature of more specific place entities and place terminology. Thus these terms may be considered as meta-level place concepts.

Although, in terms of an idealised semantics of place terms, the distinction between these three categories is well-defined, when applied to actual natural language terms, it is often not completely clear cut. One form of blurring arises because terms which seem to be primarily abstract or spatial in nature tend to also have connotations which suggest more specific types of place entity. For instance, 'region' often refers to entities of a geographic or political nature and 'neighbourhood' is typically associated with demarcations within human settlements. Hence, in classifying natural language vocabulary, we shall identify terms that are *primarily spatial*, although they may have non-spatial connotations. The distinction between place-like and non-place-like substantive count nouns is also blurred, since almost any substantive count noun can in certain contexts be regarded as a place.

5.1   Substantive Place-Like Count Nouns

Substantive place count nouns are those whose instances have identity and individuation criteria that are not purely spatial. Although we may regard these instances as places, they also have essential properties that are physical and/or sociological.

In order to give a comprehensive theory of substantive place count nouns one would need to define their individuation and identity criteria and also define the semantics of their role in regard to place-related expressions. In particular one would need to specify how they related to hosting regions and other locating predicates formed by spatial prepositions. This would be a complex task, which is beyond the scope of the present work.

However, we believe that the task may not be quite as large as one might fear. Count nouns can be organised into a subsumption hierarchy starting with very general types and ramifying into more specific. Moreover, it seems that many of the integrating principles and other semantic attributes of place count nouns can be specified at the upper levels of the hierarchy and are thus inherited by more specific count nouns. For instance different kinds of building will share the same modes of association with places, so one does not need to specify completely separate semantics for cinemas and churches, for example.

A possible ontology of the uppermost levels of the hierarchy of substantive place count nouns is as follows:

- Geographic Features (mountains, forests)
- Material Artifacts
  - Static artifacts (buildings, roads)
  - Movable artifacts (containers, vehicles)
- Fiat entities (countries, districts) [18]

Substantive place count nouns operate at many different levels of granularity, enabling us to refer to spatial locations more or less specifically. For instance, we may identify the following sequence, ordered in increasing granularity:

room, building, district, town, county, country, continent

When we ask a 'Where?' question (e.g. 'Where was Susan born?'), we may get an answer at one of several levels of granularity. In this case the range of legitimate answers may vary from a particular room in a building all the way to a country or even a continent (perhaps even a planet). Identifying an appropriate level of granularity is a key consideration in devising mechanisms for automated answering of 'Where?' questions, but is beyond the scope of the present work.

5.2 Primarily Spatial Place Count Nouns

We now consider what we are calling 'primarily spatial' place count nouns. Specifically, we examine following terms:

area	neighbourhood	sector	tract
district	patch	site	zone
domain	point	spot	
locality	region	territory	

We assume that all these terms refer to a type of object that is conceived of as a subspace of a more general spatial universe. Thus, we distinguish them in terms of the character of this subspace and the criteria by which it is individuated. Let us now attempt to explain the connotations associated with the most prominent senses of each of the listed place count nouns:

'Area' is more or less neutral as to the type of subspace being referred to, although it has connotations relating to dimensionality: it is not normally applied to point-like or linear subspaces, but rather to 2D (or $2\frac{1}{2}$D) and occasionally 3D subspaces. Areas are normally self-connected, but the term may occasionally be applied multi-piece subspaces.

Areas are typically demarcated by means of the integrating principles of homogeneity (e.g. 'an arid area'), or proximity to some distinguished object (e.g. 'the area around the church'). They are also sometimes demarcated with respect to ownership or jurisdiction, or with respect to a structural decomposition of a larger subspace (e.g. 'the south eastern area of Australia').

A distinctive aspect of the term 'area' is that it is often used in cases where the boundary of the referenced subspace is poorly defined (e.g. 'a damp area on the wall', 'the loading area').

The term 'area' is strongly associated with the metrical magnitude of a subspace. Indeed, in mathematical or scientific contexts, 'area' normally refers purely to a spatial magnitude, without reference to any particular place. This is really a distinct sense of the term, which is semantically independent from the concept of place. However, when we use the term 'area' to describe a subspace, we may imply that we are also interested in the magnitude of its extension.

'District' almost always refers to geographic regions. We can distinguish two somewhat different senses of the term:

1. One sense of 'district' refers to a unit of jurisdiction smaller than (and contained within) countries.
2. In a more general sense 'district' does not necessarily refer to an actual unit of jurisdiction, but to a region of similar size with some (often vague) geographically related integrating principle.

'Domain' is one of a group of upper-level place terms which have a connotation of ownership or control. ('Domain' is of course also used in a more abstract sense to refer to a field of knowledge or expertise.)

'Locality' is very close in meaning to sense 3 of 'neighbourhood', as described below. Arguably, 'locality' typically refers to a larger geographic subspace than 'neighbourhood'; and, while 'neighbourhood' is most commonly applied to subspaces within an urban environment, 'locality' is equally applicable in rural settings.

'Neighbourhood' has a very general mathematical sense, which can be regarded as abstracting an essential ingredient of the ordinary use of the term:

1. Mathematical context: In topology, a neighbourhood of x is a subspace s of the whole space, such that x is an interior point (or possibly a set of interior points) of s. Neighbourhood is also sometimes given a metrical sense, in which it is determined by spatial proximity.

2. Ordinary langauge context: The ordinary sense of 'neighbourhood' conforms to the abstract topological or metrical sense but is typically applied to geographic or geo-political regions. Common usage also suggests some integrating property such as social unity or uniformity, or systemic integrity. (E.g. the term 'neighbourhood', applied to part of a town, typically denotes an area with a common class of inhabitants or similar standards of buildings; but it is also associated with sharing of amenities, such as a shopping outlets and entertainment venues.)

'Patch' refers to a small subspace of a surface. A patch is generally individuated in terms of its being a maximal (though small) subspace, which is more or less homogeneous with respect to some property of surface points. For instance, one may identify a coloured patch on a surface that is predominantly of a different colour.

'Point' seems to be relatively similar in meaning to the mathematical concept of a point — i.e. a zero-dimensional element of space. Thus, when we refer to something as a point, we do not consider it to have an extension in space, only a location. Nevertheless, 'point' is often used in contexts where it is clear that it cannot refer literally to a mathematical point ('I was standing at this point', 'I sharpened the point of my pencil'). Perhaps it would be more accurate to say that the natural language term 'point' refers to something whose extension in space is negligible with respect to the space (or object) under consideration.

'Point' also has a strong non-spatial connotation, in that it often refers to a sharp protrusion from a physical object. (It is also often applied to time, meaning an instant or perhaps an interval of negligible length.)

'Region' is a general term for an extended subspace (either 2 or 3-dimensional). In most usages it is very close in meaning to 'area', although in mathematical contexts the terms 'region' and 'area' are clearly distinct:

1. Mathematical context: here the term 'region' is used purely spatially to refer to an arbitrary subspace without implying any particular integrating principle, although connotations of being uniformly 2 or 3 dimensional and self-connected usually apply.

2. General context: in most ordinary language usages there is an implication of some additional non-spatial unifying property or principle associated with the region. Appropriate principles are much the same as for 'area' (possibly the connotation of having a vague boundary is less strong for 'region' than 'area').

It is very common that 'region' (more so than 'area') is applied to a subspace whose principle of integration is geographic or geo-political in nature. Structural decomposition of a larger subspace into 'regions' is also a common usage (e.g. 'the central region of Africa').

'Sector' strongly connotes that the subspace is demarcated relative to a *partition* of space. A sector is normally 2 dimensional (sometimes 3). It is often a *fiat* demarcation (in the sense of Smith [18]) imposed to divide a larger space into roughly equal parts.

'Site' is normally used to refer to the place where something is situated. Moreover, it is typically applied to buildings and other large static artifacts. Thus, a site is usually a 2-dimensional subspace of the land surface, where a large artifact is or was situated. Site can also be used with a sense more or less equivalent to sense 2 of 'situation', as described below.

'Spot' typically refers to a subspace with the following characteristics: it is of small scale compared to its context space; its spatial extension approximates a small disc or a point; it is usually on a 2D surface but could also be within a 3D object. A spot need not have any particular intrinsic distinguishing characteristics, but may be identified solely by its relation to other fixed objects.

'Spot' is similar in meaning to 'patch', except that 'spot' implies a roughly disc-like subspace of a surface, whereas a patch can be more irregular.

'Territory' always applies to a geographic scale subspace and also has a strong connotation that the subspace is individuated on the basis of ownership or jurisdiction.

'Tract' refers to a geographic scale subspace, which is typically individuated on the basis of more or less uniform terrain type (often inhospitable). The word derives from the Latin *tractare*, to draw out or drag, which implies extension in space (or possibly in time). Consequently, there may also be a (weak) connotation that a tract refers to an elongated piece of land.

'Zone' like 'sector', refers to a cell of a partition of a reference space. Though there may be slight differences in the connotation of these terms, we assume that they have essentially the same meaning.

Having considered a wide variety of primarily spatial place count nouns, it is evident that certain connotations recur in several cases. We have identified the following as being particularly significant types of connotation:

- **Dimension.** Certain place terms imply that the designated place is (typically) of a particular dimensionality (a point, a line, a 2-dimensional regions, or a 3-dimensional volume).
- **Size.** A term may imply that a subspace is relatively large (tract) or small (spot, patch) relative to the embedding space under consideration.
- **Shape.** A term may imply a subspace of a characteristic shape (e.g. a patch is irregular, although roughly disc shaped).
- **Bounded.** Place terms may be associated with the presence of more or less definite boundaries that demarcate the designated subspace.

- **Partition.** This connotation implies that the designated place is a cell within a partition of the entire space under consideration (e.g. sector and zone).
- **Geographic.** Terms such as 'tract', 'region' (in one sense), 'district', 'domain', 'territory' imply a certain scale, and also a certain relationship with the surface of the Earth. This connotation, being not purely spatial, gives a partially substantive character to such terms.
- **Jurisdiction/Control.** This has been identified as one of the main principles of integration for places, and it is also found to be a connotation of several primarily spatial place-count nouns ('territory', 'district', one sense of 'region'). This connotation also lends a substantive aspect to these terms.

5.3   Abstract Place Count Nouns

By *abstract* place count nouns, we mean those that characterise place in terms of purely logical aspects and do not constrain any physical or spatial properties. In this category we have 'place' itself, as well as 'location', 'position' and 'situation'. Such terms can be regarded as operating at a *meta level*, in that they are often used to describe more specific place terms. For example, one might say 'London is a place' or 'Overlooking the river is a nice location'.

'Place.' We have so far used the term 'place' as if it were the most general of the abstract place concepts, and indeed it does have a very general usage. It can be employed in most contexts where any of the other non-substantive terms are used. There are two distinct modes of linguistic expression that we regard as referring to places.

1. Firstly, the term 'place' may be applied to an *object* of an appropriate type. For instance, one may assert: 'The city of Leeds is a place'. In this sense 'place' is a count noun whose extension covers all objects that are instances of some more specific place-like count noun. This includes, for instance: countries, towns, lakes, hills, buildings, rooms. Although the types of place object are diverse, they have the common feature that they provide a basis for at least one (and often all) of the functions of locating, hosting and anchoring (as identified above in Section 2.3).

2. A second mode of application of the word 'place' is to refer to locative spatial properties rather than place-like objects. Thus, expressions such as 'in England', 'on the table' and 'between the church and the town hall' are often thought of as referring to 'places'. Places in this sense correspond to *reifications* of locative property phrases. Moreover, it is places in this sense that correspond to possible answers to 'Where?' questions.

The ambiguity of 'place' with respect to these different senses is understandable, since they are semantically very closely related. In answering a 'Where?' question, it is common to give simply the name of a substantive place object (such as 'London') instead of a locative property phrase (such as 'in London').

To develop a more comprehensive formal theory of place we believe that it will be necessary to introduce a representation for reified place objects that are abstracted from locative properties. This could be done using the *lambda calculus*: if $\phi(x)$ is a predicate corresponding to a locative property, then $\lambda x[\phi(x)]$ denotes an abstract place object corresponding to the reification of this property.

'Location' is another very general term which has several distinct senses:

1. A spatial property, usually manifest at a geographic scale, which describes the spatial relation of an object to one or more other spatial objects — i.e. a geographic scale locative property phrase.

2. We may also regard the location of an object as being a combination of all spatial properties satisfied by that object. (For example, we may consider the good and bad aspects of a particular location.)

3. The term 'location' is also used to refer to a very explicit designation of the subspace occupied by an object, such as its coordinates within a numerical coordinate system.

'Position' has much in common with 'location' but does have some distinctive connotations:

1. In speaking of 'position' we often refer to the intrinsic form or orientation of an object (or person) as well as its location in space.

2. The term 'position' can be used in a sense where only the form is relevant, not the absolute location in space (e.g. 'Hold your arms in this position').

3. In small and medium scale contexts 'position' is sometimes used just to refer to the location of an object (e.g. the position of a piece on a chessboard).

4. 'Position' is commonly used an a non-spatial sense, to refer to the status of a person in relation to an institutional or social structure.

In so far as 'position' is associated with intrinsic form rather than location, it is not place-like according to the criteria (i.e. locating, hosting, anchoring) that we have suggested apply to typical place count nouns.

'Situation' has two rather different senses:

1. In a general sense, the term 'situation' can be taken as referring to a 'state of affairs' (either purely static or with respect to some on-going event). Although a situation in this sense will generally include spatial aspects, it is only indirectly relevant to the concept of place, and so will not be considered further here.

2. There is another common sense of 'situation' where it does refer to place-like entities. This occurs in contexts such as 'The house was in a beautiful situation, overlooking the lake.' Here, 'situation' refers to where an object is 'situated' and this is normally described by means of locative property phrases — i.e. phrases that locate the object in relation to its surroundings. In this sense, the meaning of 'situation' is very close to that of 'location'. Like 'location', a 'situation' in this sense could be associated with the set of spatial predicates satisfied by an object.

6   Further Work and Conclusions

When we began this work, we believed we could proceed directly to formulate a general logical theory of the concept of place. However, we soon found that the huge variety of different ways in which place enters into language made it impossible to achieve a simple theory that covered all these modes. Thus we were driven to a detailed analysis of the many linguistic expressions of place concepts and their semantic content. We are now still a long way from having a comprehensive formal theory. Nevertheless, we do believe we have delineated a semantic framework that encompasses all the major aspects of the semantics of place and explains many of the ways in which they interact. Hence, we now have a solid basis from which we can proceed to develop a fully formal ontology of place. Clearly this will not be a simple first-order theory over a uniform domain of entities, but a rather complex multi-typed system, which can articulate place-related concepts from a number of different perspectives and levels of abstraction.

Several issues are of particular relevance to the further development of the theory. One is the details of how spatial prepositions should be encoded as spatial relations, in order to formalise the semantics of locative property phrases. Another is the reification of spatial predicates to form abstract place-like entities, which seem to be required in order to model certain natural modes of expression, where locative properties are referred to as if they were a special kind of object. A futher issue concerns the interpretation of place terms in relation to the changing configurations of real physical environments (as studied by Donnelly [10]).

With regard to methodology, although the present work was guided by the more cognitively oriented studies of [1,2,3], the current development has been conducted from a largely theoretical perspective. In order to establish whether the different senses and connotations of place that we have identified correspond well with the intuitions of 'naive' language users, further congnitive experimentation will be required.

A long term goal of this research is to develop a system capable of automated answering of 'Where?' questions. It is evident that these queries are highly unconstrained in that there are a huge number of ways in which answers to such questions can be stated. The current work serves to analyse the forms of possible responses and also provides a basis for a formal specification the truth conditions of place attributions. However, it does not provide a mechanism by which *appropriate* answers can be distinguished from those that are true but uninformative (e.g. Q: 'Where is John?'. A: 'Somewhere in the universe.').

The key to determining an appropriate answer seems to lie in the context within which a 'Where?' question is posed. Pragmatic considerations (such as the epistemic state of the questioner and the topic of the dialogue in which a question arises) appear to provide cues that constrain what would count as an informative answer. Hence, an approach to solving this problem would be to somehow extract semantic constraints and connotations from the context, and use these to determine the mode of expression and level of abstraction most appropriate to a sensible answer. The explication of semantic attributes and connotations of place expressions given in the current paper may provide a useful starting point for devising a mechanism that can achieve this.

Acknowledgements. The support of EPSRC grant EP/D002834 is gratefully acknowledged.

References

1. Agarwal, P.: Contested nature of 'place': knowledge mapping for resolving ontological distinctions between geographical concepts. In: Egenhofer, M.J., Freksa, C., Miller, H.J. (eds.) GIScience 2004. LNCS, vol. 3234, pp. 1–21. Springer, Heidelberg (2004)
2. Agarwal, P.: Sense of place' as a place indicator: establishing 'sense of place' as a cognitive operator for semantics in place-based ontologies. In: Cohn, A.G., Mark, D.M. (eds.) COSIT 2005. LNCS, vol. 3693, Springer, Heidelberg (2005)
3. Agarwal, P.: Topological operators for ontological distinctions: disambiguating the geographic concepts of place, region and neighbourhood. Spatial Cognition and Computation 5(1), 69–88 (2005)
4. Bennett, B.: Application of supervaluation semantics to vaguely defined spatial concepts. In: Montello, D.R. (ed.) COSIT 2001. LNCS, vol. 2205, pp. 108–123. Springer, Heidelberg (2001)
5. Bennett, B.: A categorical axiomatisation of region-based geometry. Fundamenta Informaticae 46(1–2), 145–158 (2001)
6. Bennett, B.: What is a forest? On the vagueness of certain geographic concepts. Topoi 20(2), 189–201 (2001)
7. Bennett, B.: Modes of concept definition and varieties of vagueness. Applied Ontology 1(1), 17–26 (2005)
8. Canter, D.: The Psychology of Place. Architectural Press, London (1977)
9. Canter, D.: The facets of place. Advances in Environment, Behavior, and Design 4, 109–148 (1997)
10. Donnelly, M.: Relative places. Applied Ontology 1(1), 55–75 (2005)
11. Entrikin, J.N.: Place and region 3. Progress in Human Geography 21(2), 263–268 (1997)
12. Harrison, S., Dourish, P.: Replaceing space: The roles of place and space in collaborative systems. In: Proceedings of Computer Supported Collaborative Work, pp. 66–76. ACM, Cambridge, MA (1996)
13. Jordan, T., Raubal, M., Gartrell, B., Egenhofer, M.: An affordance-based model of place in GIS. In: Poiker, T., Chrisman, N. (eds.) Proceedings 8th International Symposium on Spatial Data Handling. International Geographical Union, pp. 98–109 (1998)
14. Massey, D.: Space, Place and Gender. Polity, Cambridge (1994)
15. Randell, D.A., Cui, Z., Cohn, A.G.: A spatial logic based on regions and connection. In: Proceedings 3rd International Conference on Knowledge Representation and Reasoning, pp. 165–176. Morgan Kaufmann, San Mateo (1992)
16. Relph, E.: Place and Placelessness. Pion Press, London (1976)
17. Russell, B.: On denoting. Mind 14, 479–493 (1905)
18. Smith, B.: Fiat objects. Topoi 20(2), 131–148 (2001)
19. Thrift, N.: Steps to and ecology of place. In: Massey, D., Allen, J., Sarre, P. (eds.) Human Geography Today, pp. 295–322. Polity Press, Cambridge (1999)
20. Tuan, Y.-F.: Topophilia: A study of environmental perception, attitudes and values. Prentice Hall, New Jersey (1990)
21. Vandeloise, C.: Spatial Prepositions: A case study from French. University of Chicago Press. Translated by Bosch, A.R.K (1991)
22. Varzi, A.: Vagueness in geography. Philosophy and Geography (2001)
23. Whitehead, A.N., Russell, B.: Principia Mathematica, pp. 1925–1927. Cambridge University Press, Cambridge (1910–1913 second edition 1925–1927)

Spatial Semantics in Difference Spaces

Vlad Tanasescu

Knowledge Media Institute, The Open University,
Walton Hall, Milton Keynes, MK7 6AA, UK
v.tanasescu@open.ac.uk, vladtn@gmail.com

Abstract. Higher level semantics are considered useful in the geospatial domain, yet there is no general consensus on the form these semantics should take. Indeed, knowledge representation paradigms such as classification based ontologies do not always pay tribute to the complexity of geospatial semantics. Other approaches, originating from psychology, linguistics, philosophy or cognitive sciences are regularly investigated to enrich the GIScientist's representational toolbox. However, each of these techniques is often used to the exclusion of others, creating new representational difficulties, or merely as a useful addendum to host theories with which they only superficially integrate. The present work is an attempt to introduce a common ground to these techniques by reducing them to the notion of *differences* or *difference spaces*. Differences are discernible properties of the environment, detected or produced by a computational process. I describe the following semantic frameworks: category-based ontologies, conceptual spaces, affordance based models, image schemata, and multi representation, explaining how each of them can be projected to a model based on differences. Illustrative examples from table top and geographic space are produced in order to show the model in use.

1 Introduction

High level semantics for the geospatial domain are needed in order to (a) help to integrate heterogeneous data sources [1], (b) allow reasoning in GIS [2], and (c) adapt the user's cognitive abilities [3]. Currently, the most studied means of expressing high level semantics is the use of ontologies. An ontology is "a logical theory accounting for the *intended meaning* of a formal vocabulary, i.e. its *ontological commitment* to a particular *conceptualisation* of the world" [4]. The apparent universality and versatility of the definition of ontologies, as well as their academic success in the development of the semantic web [5], has lead many researchers to investigate their use in a geospatial context (e.g. [2] [6] [7]). Ideally, "meaning" expressed by ontologies would provide the long sought for "glue" between geospatial communities by capturing their practices and conceptualisations, and facilitating the alignment of heterogeneous elements expressed at a high semantic level.

However, the actual use of ontologies for spatial knowledge representation has lead to issues which seem difficult to overcome. These issues are related to object representation in geographic space when trying to handle *vagueness* [8] as well as *cultural and subjective discrepancies* [11][3]. Moreover, the absence of large scale

S. Winter et al. (Eds.): COSIT 2007, LNCS 4736, pp. 96–115, 2007.
© Springer-Verlag Berlin Heidelberg 2007

systems based on ontologies is problematic, as is the absence of toy examples fulfilling some of the promises of ontologies and allowing us at least foresee the advent of a new "brain for humankind" [9] and of a world in which machine reasoning would be "ubiquitous and devastatingly powerful"[1].

Efforts in the geospatial community to alleviate these difficulties and build practical solutions have been made by directing ontologies toward a cognitively sounder approach. Indeed, paradigms originating from psychology, cognitive science or philosophy, such as affordances, image schemata, conceptual spaces and multi representation models, have been advocated [12]. However, some of these paradigms seem incompatible with actual knowledge representation techniques and are therefore used in isolation (as for example conceptual spaces in [31] or in [32]), which unfortunately reveals their own expressive limits. Other paradigms seem to integrate better with ontology design practices but this integration appears to be only superficial (e.g. *quale* in DOLCE [10] are inspired by conceptual spaces, but constitute a minimal part of the underlying theory) and tends not to pay tribute to the specificities of the paradigm (Gibson's *affordances* [42], for example, are often assimilated to functions, or their meaning is limited to an HCI environment).

In this paper we present a model of knowledge representation that allows modelling at several levels of semantic detail. To do so we have to depart from the traditional dominance of the notions of categories, objects, properties and relations, common to ontology modelling. We replace these notions with generic *differences*, detectable disturbances of the environment, grounded in the processes which perceive them. Differences aim to be the atomic elements of meaning. Difference spaces constitute networks of meaningfully related discernible differences.

To justify the need for such a move, we first enumerate high level semantic paradigms used to describe the geospatial domain, briefly discussing their advantages with caveats and describing how they represent aspects of the world. Then we show how representations can be based on differences by using three illustrative examples: a toy static example in table top space followed by a more complete one in geographic space and a dynamic example in the way-finding domain. We conclude by giving a summary, relating the framework to other existing frameworks, then discuss future work and issues.

2 Spatial Semantics

The term "conceptualisation" refers to representations whose purpose is to give meaning to the surrounding environment. If the world is believed to be uniform with regard to knowledge representation, it means that a particular conceptualisation can *fully* cover a conceptual region of it. Following this view and according to Smith [13], the final alignment of all conceptualisations should eventually converge to form a global knowledge model for humankind, in the same way as mathematics perennially succeeds in achieving universal consensus.

However, conceptualisations about a similar domain are diverse (as in [52]), dependant on the social or geographic context, and somehow, one could argue,

---

[1] Tim Berners-Lee, 1998, http://www.w3.org/DesignIssues/Semantic.html

seemingly arbitrary. Even the most fundamental usage of spatial concepts or relations, such as the proposition *in* varies across languages, is not the same in English and in Korean [14], related only to the notion of *containment* in the first, but involving notions of *friction* in the other. Given these language discrepancies an eventual unification seems unlikely ever to occur, but they allow us to understand why different conceptualisations of the same domain are possible (without the need to state the rightness of one or the other or to attempt an artificial synthesis), and even unavoidable.

The need to use multiple paradigms for knowledge representation was felt by the geospatial community, not only to match the context in which a concept is used or to accommodate conceptualisations of particular cultures or languages, but simply to respond to the requirements of basic geographic notions such as *place* [15] which are inherently diverse and versatile. We will describe in turn different knowledge modelling paradigms used in a geospatial context, trying to illustrate how these models relate to each other. However, the common link can only be provided by an underlying model, which we believe will be found in the theory of difference spaces.

2.1 Ontologies

Ontologies are an important paradigm of knowledge representation in GIScience [7] as well as in other information science communities. Sowa (in [16], p.15) gives an operational definition of an ontology as "the method to extract a catalogue of things or entities (C) that exist in a domain (D) from the perspective of a person who uses a certain language (L) to describe it". Given this definition, most formal representations used in computer science such as relational database schemata qualify as ontologies. However, the term usually refers to a class of constructs expressed in specific formal languages: *ontology languages*.

Elements of ontology languages allow the description of a domain, usually with *concepts* representing categories and the *relations* between them. Concepts have *attributes* which are themselves sometimes specified by *facets*. The concepts may be hierarchically related by *is-a* relationships to form *taxonomies* allowing reuse through *inheritance*. *Instances*, individuals belonging to the category represented by the concept, are defined by concept. This approach to describing a domain has been introduced into mainstream programming languages, as well as into database design, under the name object-orientation. Moreover, some ontology languages provide *procedures*, *functions*, *axioms* and (production) *rules*. Queries are answered according to *inference* mechanisms which constitute the bulk of the *reasoning* capabilities, allowing database-like queries or determining the category to which a given individual belongs.

However, in the geospatial domain representation problems arise from the use of this semantic apparatus. Indeed, the simple determination of an individual or of a type of individuals in geographic space is problematic, since it seems to have a greater individual or cultural variability than in table top space [17]. This fact naturally leads to seemingly irreconcilable categorization problems between cultures, where categories exist for some objects in a language that are orthogonal to the ones of another [11]. It has been argued that the "what", the "where" and the "what for" of spatial entities are often so intertwined that not only must notions of vagueness be introduced – e.g. one does not know how to define a forest, a mountain or a valley because of their inherent vagueness – but the task of categorization is often arduous and fails to reflect the dynamic nature of the geographic environment. Moreover, depending on the context, there is a need for multi-representation of entities, as nodes in a network, as objects or fields [13] [18], as complex tri-dimensional structures, or as regions on a map as opposed to subjective networks of experiences [15]. Finally, the scientific notion of space – used in maps and GIS – and the *naïve* one, i.e. the one used in everyday life, seems to be irreconcilable, instructive for some (e.g. in [3] naïve geography is deemed useful to the amelioration of GIS user interaction) or epistemologically problematic for others [17].

The origin of this failure to encompass the world's complexity may be found in the actual practice of ontology building. Indeed, categories have been often discussed as problematic even for describing table top space [19] where vagueness and whereness are less obvious. In geographic space 'things' are not crafted to be used for a single purpose, can serve different roles, and afford many other, being strongly dependent of the task at hand as well as other components of the context, most concepts fail to adapt this simplifying filter. Ontology languages are not designed to cope with context other than by building other ontologies and producing mappings between them based on syllogistic reasoning [20]. Gärdenfors has even argued – calling them "free floating island[s] of reeds [with] no anchor in reality" [21] – that symbol based systems such as ontologies are not at all grounded. Looking for grounding is looking for a continuous transition between the world, our means to perceive it and the resulting conceptualization(s). Formal vocabularies alone, by presenting fixed categories accompanied with logical rules, structures and objects, may be unable to emulate the flexibility required by natural reasoning or the imperatives of social interaction.

2.2 Multi-representation

One approach is to enrich the notion of object with context-driven representations. For example, even if we reduce context to scale, an airport will appear as a node in a travel network at world scale, then as a region, then at a smaller scale, showing some roads and inner networks, then present affordance-rich objects such as car parks and zones, then a layer of (interactive) 3D structures allowing shopping and eating. Timpf [24] proposes a framework for a GIS data model for different representations based on directed acyclic graphs, where representations are linked according to zoom level and can become separate objects at certain levels (e.g. a farm may reveal itself being composed of a habitation zone and a barn). However, if we generally agree that the point on a map at world scale *is* the same as the 3D representation of the main airport building, a closer examination of this claim quickly raises problems: it is radically different relative to shape or affordances but has a similar position on the map and is given the same name; do these two facts alone, the "where" and the "what name" justify the fitting of contextually very different objects into the same category? On the other hand, why would categories be dependent only on the notion of scale only, and not task, point of view, etc?

Following these conclusions Spaccapietra et al. [25] show the need of going from multi-scale representation of object geometries to a multi representation of the objects themselves. As a result an object is described as a point in the three dimensions of (1) *spatial resolution* (i.e. the multiple representations of the object in space according to its scale), (2) *viewpoint* (i.e. different viewpoints, or use cases, for which representations are elaborated), and (3) *classification* (what category of object the viewpoint is about). Vangenot [26] gives a formal description of multi-representation in the two first dimensions (resolution and viewpoint) by two means: an *integrated approach* and an *inter-relationship approach*. The former uses a method called *stamping*, where attributes of an object are stamped with a context identifier, which allows attribute filtering. This approach has also been adapted to ontology languages [27].

Another multi-representation model is VUEL [28], which stands for *view element* and represents a graphical component that is explicitly linked to a *semantic*, a *geometric* and a *graphic* dimension. The user action of *drilling* allows the navigation of the representations attached to each of these dimensions, the semantic drill being the option to generalize a layer of representation.

The notion of multi-representation is particularly interesting in its interplay with the notion identity, which stands at the core of any ontology language. The distribution of an object's essence amongst different semantic dimensions for a cognitively sounder approach naturally leads to trying to represent this object using its components, or in other words, projecting it on conceptual spaces. In addition to achieving multi representation, the conceptual spaces approach aims to provide the concept with grounding, by rooting semantic dimensions in perception.

2.3 Conceptual Spaces

The notion of *Conceptual Spaces*, as introduced by Gärdenfors [29], is designed to bridge the gap between symbolic and non-symbolic models of knowledge representation by providing a differentiable notion of an entity. An object becomes the result of the combination of elements representing its qualities or properties, represented as points in their own spaces. These *quality dimensions* ground the final construct in its underlying reality by representing it as points in simpler spaces. For example, a taste could be represented by a point in the four-dimensional quality space composed of the qualities of being *sweet, sour, saline,* or *bitter,* or a colour by its three *red, green* and *blue* components. In this model, categories become regions in a conceptual space organized by voronoi tessellations defined by its typical elements. For example, in a partition of *animal-space* a *robin* would be central in *bird-space* whilst an *emu* would be at the periphery, closer to the spaces related to non flying creatures [30].

In a geospatial context, Raubal [31] gives a formalization of conceptual spaces with application to the urban geographical domain by building the concept space of a "façade". The method resorts to perceptive components – relating to the interpretation of an image – in terms of a system of weights allowing the space to change according to the task at hand or the context (i.e. user or system). If this approach shows convincing results, it is unclear, however, whether it can be generalized to other examples.

Ahlqvist [32] combines conceptual spaces with a formal representation of semantic uncertainty based on rough fuzzy sets, as semantic uncertainty is often recognized in geographic concepts. The model is applied to land cover or use, and the representation (projection) of concepts from different ontologies in additional dimensions allows seamless movement between one representation and another. However, the domains compared are very similar and abstract qualities are not expanded as spaces.

Kuhn [33] provides an example of semantic mapping of wordnet concepts using fine-grained conceptual spaces, and affordances as projection axes during mapping, an operation related to the cognitive science notion of "blending", in order to construct abstract conceptual spaces based on image schemata. The example, boathouses and houseboats, although easy to generalize, shows that the notion of identity can be reinterpreted to achieve elegant semantic mapping between conceptual spaces, used here implicitly as support for image schemata, which represent the final grounding. As this approach involves image schemata and affordances, it will be discussed further in the following chapters.

The notion of conceptual spaces is appealing due to its vision that a concept is a compound that can be manipulated and projected into multiple representations in given contexts. Similarity between instances or concepts becomes very natural once represented as a distance in conceptual spaces. However, in order to keep the natural character of distances, conceptual spaces are given shapes that are intuitive in simple examples (such as colours in [21]) but become hard to imagine for abstract concepts (what would be the shape of e.g. *animality*?). Furthermore, the commendable will to keep concepts grounded has the effect of making abstraction difficult to envision. It is unclear how to represent concepts not directly related to perception or even more complex concrete concepts such as animals, without loosing the grounding. Therefore, although with conceptual spaces it becomes clearer how symbolic and perceptual representations can interact, a middle ground may still be lacking, as stated by Kuhn in [33], between purely geometric conceptual spaces and abstract vocabularies, in order to achieve complexity and abstraction. Image schemata are the effort to bridge this gap by allowing abstract structured elements without losing the ability to compose them in order to form an entity.

2.4 Image Schemata

Image schemata are a model of human perception and cognition initiated by Lakoff and Johnson [35], and further elaborated by Lakoff [36] and even more particularly by Johnson [37]. Johnson claims that mental activities such as perception and cognition are heavily influenced by image schemata as basic cognitive structures. He ([37], p. 29) defines a schema as a "a small number of parts and relations, by virtue of which it can structure indefinitely many perceptions, images, and events" and states that they "operate at a level of mental organization that falls between abstract propositional structure, on the one side, and particular concrete images on the other".

Examples of image schemata can be represented by terms illustrating generic concepts or situations involving physical interaction or relationships ([37], p. 126) such as *container, balance, compulsion, blockage, counterforce, restraint removal, enablement, attraction, mass-count, path, link*, etc. For Johnson these constitute the basic bricks of all possible reasoning and are deeply grounded in bodily and cultural experiences, allowing them to be universally understood. The *image* part is due to the fact that image schema are to a certain extent oriented toward perception, whilst the *schema* part manifests a basic structure not present in the perceptive input.

This allows it to be made sense of and prepares it for higher level concepts, which become grounded at the same time.

Mark [38] introduces image schemata into the geospatial domain by presenting the *platform* and *container* schemata, related to the English prepositions *on* and *in*, respectively. Raubal et al. [39] use a practical approach to acquiring image schematic clues by interviewing participants and then structuring the task at hand (in this case way-finding in airports) based on the extracted schemata. However, the formalization of the obtained schemata as well as their relevance in a context not related to way finding remains unclear. Rüetschi and Timpf [40] provide an insightful approach to image schemata by presenting them as structures ready to be instantiated by *parts* and *induced relations* (e.g. a *container* schemata has an inside, an outside, and a boundary as parts, and induces *part-of* relationships between the elements it contains). This structure allows schemata to be combined, but only for the way-finding domain.

An encompassing formalization is given by Frank and Raubal [41] who use extraction methods borrowed from linguistics for four schemata that are particularly relevant to the geographic space (*location*, *path*, *region* and *boundary*) and three that are equally applicable to the table-top space (*container*, *surface* and *link*). Their approach allows for a variety of combinations, but problems arise from the grounding in linguistic devices (which language should one trust?), as we already know that considerable differences exist (e.g. from [11] and [14]), and from the untyped character of the resulting structures (how to prevent a car from following a footpath instead of a motorway if both can be schematized by a link?). However, the authors point out that some of these issues could be alleviated by replacing the linguistic grounding with a more universal one.

Indeed, if image schemata are supposed to be universally shared structures of human cognition, this still represents a leap of faith. What does not, however, is the fact that individuals present biological similarities and that, given the particular abilities this structure provides to them, they are able to act in a similar way given a specific environment. This concept is represented by affordances, which also could provide a solution to the definition of objects according to what they permit instead of to their position in a hierarchy of categories.

2.5 Affordances

The term and concept of *affordance* was created by J. J. Gibson and used notably in [42] as a basis for *ecological psychology*, which focuses on the interconnections between the subject and its environment. Gibson argues that, when entities are confronting an environment, only affordances immediately matter, i.e. we are immediately aware of the elements of our environment that allow us to express our abilities in terms of action, rather than the physical qualities that do not directly influence our actions (e.g. colour, size, when not related to action, etc). The notion of affordances represents this link between an environment and the agent situated within it. Indeed, what is enabling and therefore meaningful for one agent would be meaningless for another. Simple human affordances such as *climbability* have been verified experimentally: over a certain ratio between the riser height of stairs and the leg length of the subject, or their age, or their degree of fitness, the latter ceases to consider it climbable [43]. Transposed to a geospatial context, fields for example become *cross-able*, as are *roads*, provided the right structure. Affordances, thus, allow the explanation of different notions of space depending on persons and their respective situation.

Kuhn [33] uses affordances as functions provided by objects and uses them to define similarity between them, independently of their application domain. This allows mappings from concept to concept which are orthogonal to the category hierarchy they belong to (e.g. *sheltering* as an affordance, either for humans or for boats, links the concepts of *house*, *houseboat* and *boathouse*). Jordan et al. [15] apply affordances to the notion of *place*, which is most often related to what a particular assemblage of geographic elements allows rather than to attributes attached to coordinate locations and specific borders. The authors state that the environment can be defined as an affordance hierarchy and that the definition of place is the interplay between this environment, the user's capabilities and task requirements. Nevertheless, the task of defining an affordance hierarchy is eventually deemed "formidable". It is indeed doubtful whether affordances are the kind of things that can be easily categorized, since, according to Gibson, "... to perceive an affordance is not to classify an object." ([42], p. 134). Moreover, "... if you know what can be done with a graspable object, what it can be used for, you can call it whatever you please. ... The theory of affordances rescues us from the philosophical muddle of assuming fixed classes of objects, each defined by its common features and then given a name. ... But this does not mean you cannot learn how to use things and perceive their uses. You do not have to classify and label things in order to perceive what they afford." ([42], p. 134). If these quotations are about object categories rather than affordances, Gibson's mistrust of categories is obvious and should be extended to the classification of affordances; indeed, how does one link similar affordances such as *grasp-able* or *take-able* in a hierarchy? To support this claim one can also resort to another important claim of ecological psychology, maybe the most controversial one, which states that the perception of affordances is always *direct*, i.e. it does not involve any (symbolic) inference or other logical process. The direct perception thesis would justify the way our body acts, almost in our absence, as we do not have to think in detail about our actions: our mind / body processes information which does not even reach consciousness, let alone symbolic representation.

2.6 Factoring

Multi-representation highlights the difficulty of ontology modelling, which favours object representation and uses classification as the preferred reasoning paradigm. Although multi-representation may be emulated by using multiple ontologies to represent different point of views, if these ontologies are not reconcilable then this undermines the notion of absolute categories. Indeed, if the same concept is found under different categories in two ontologies using the same top-level categorization, the identity of the two concepts will become problematic.

On the other hand, one can use conceptual spaces as an attempt to deconstruct the notion of object by giving it the status of a point in a multi dimensional space of properties. However, by virtue of the fact that the grounding is limited to perceptual processes and relies only on geometry in order to provide similarity measures, the model is unable to present abstract spaces which do not rely on perception and whose exact shape can hardly be determined.

To solve this issue, image schemata decompose entities into interacting atomic schemata of meaning, including structure, perception and action. Nonetheless, without any grounding other than structural, it is unclear how the combination of schemata can ever reproduce the original concept with its original and very specific affordances. For example both the cup of tea and the lift of my building are containers, yet their respective usage as well as the means of interacting with them differ radically, differences which cannot be solved only by typing (i.e. cup contains coffee, lift contains humans), and it is hard to imagine how they can be reached only by appealing to other schemata.

Affordances ground the conceptualisation of an entity by linking it to the subject and to the actions it allows the latter to perform. However, as for conceptual spaces, this theory is deeply rooted in perception, and examples related to affordances are often related to animal life or to table top object manipulation than to the mix of abstract and perceptual concepts which constitute day to day reasoning.

As there is only one world, these paradigms must somehow describe the same thing, and represent aspects of a complexity that is yet out of reach. If conceptual spaces, image schemata and affordances access objects at a finer level of granularity than categories and attributes, there must be an even lower level at which the differences between these approaches can be reconciled.

3 Thinking with Differences

The following is an attempt to explain discrepancies between the paradigms previously described and to provide a framework that is able to accomodate all of them. The model uses the notion of *difference* as a modelling primitive and it is inspired by the work of philosopher Gilles Deleuze – on the concept of *Difference* in its opposition to *Identity* –, as well as by the recent emergence and generalization of use of folksonomies.

3.1 Deleuze's Difference

Philosopher Gilles Deleuze's early philosophical work, *Difference and Repetition* [47], aimed to provide an alternative take to the traditional debate between identity and difference. Indeed, difference is traditionally seen as a derivative from identity: e.g., saying that "*X* is different from *Y*" implicitly assumes an *X* and a *Y* with stable identities. On the contrary, Deleuze claims that all identities are derivative form differences. Identities do not exist logically or metaphysically prior to any differences "given that there exist differences of nature [even] between things of the same genus." That is to say, things in the world only occur individually and the valuing of classes over instances, in classical philosophy and in ontology design, is irremediably flawed. Things are never "the same"; this affirmation would obscure the difference presupposed by there being two things in the first place. The process of thinking (and therefore learning), is the process of navigating differences, endlessly unfolding some to discover new ones. Indeed, for Deleuze, apparent identities such as *X* are in fact composed of series of differences, where X is "the difference between $x1$ and $x2$ and ...", and Y "the difference between...", and so forth. To confront reality honestly, Deleuze claims, we must grasp beings exactly as they are, which we can only fail to achieve by using concepts of identity (forms, categories, resemblances, predicates, etc.) through which it is too easy to get lost in pure abstraction and lack of grounding.

3.2 Folksonomies

Folksonomies result from tagging systems, which allow a user to enter his or her own metadata [44]. By giving absolute freedom in the elicitation of a label, or *tag*, anyone can draw their own conceptualisation as a "desire line" [45] independently of any pre-established categorisation or object boundary. Therefore, tagging is a bottom-up process which allows folksonomies to grow exponentially, following the social model of the world wide web – building web pages that anyone can link to without caring about the global structure – or more recently the model of wikis – where one can contribute snippets of information only rather than having to write full articles. The absence of structure in tags explains the very low cognitive cost of tagging, as opposed to building fully fledged ontologies or selecting metadata from pre-existing top-down categorizations. Moreover, as metadata accumulates, rather than ending up as vast pools of arbitrary information, basic categorizations emerge from multiple sources from which it is possible to extract hierarchical structures [46] while still leaving space for new elements.

3.3 Interlude: The Juggling Ball Thought Experiments

To get a feeling of how the notion of differences translates into real situations, let us imagine a juggling ball. Basic categorisation suggests that any ball can be used for juggling, as no constraints on the shape of the ball are imposed by this activity (as opposed to in rugby, for example). A finer categorisation however, would attach the object with the conditions of possibility of the activity and decide it should be roughly the size of a hand. However, a more *grounded* look shows that juggling balls are not *any ball*, as a simple web image search reveals. What juggling balls are seems better described by Wittgensteinian "family-resemblances" rather than by ticking a box in a given category hierarchy. As we can see from the pictures, navigating the grounding of the expression reveals objects that are usually multi coloured and soft, covered in fabric or artificial leather. A closer look would reveal that they are often filled with sand-like matter, so that they do not bounce and roll too far away when they fall. This first experiment, imagining a juggling ball, shows that categories sometimes fail to reveal essential characteristics of a family of objects.

For the purpose of another experiment, bearing in mind that a juggling ball is soft to the touch, try for a few seconds to imagine varying its degree of roughness, from a soft leathery material to a texture like pumice stone. When attempting to mentally achieve this variation, the transition is never as smooth as it is theoretically entitled to be: one cannot help but jump from one degree of roughness to another, even after several attempts. Indeed, a difference such as roughness is grounded and appeals to some sense which simulates it. This sense being what it is, and given that imagination appeals to experience, even differences which can be thought of abstractly as continuous are, in reality, discrete.

Moreover, during this exercise, the mental representation of the ball itself may have changed, adopting a representation which is more usually associated with increased roughness such as that of a stone. Additionally, during the process, you may have used a visualisation of your finger to reach for and touch the ball. These elements show the entanglement of different aspects of meaning (it is hard to imagine roughness with a soft material), as well as the need to use a particular sense for each kind of difference we want to detect (it is hard to imagine a feeling related to touch without touching, even virtually). Indeed, as suggested by Kuhn in [49] for spatial measurements, a detectable difference cannot be dissociated from what is used for detecting it in the first place.

With these characteristics in mind, let us draw this interlude to a close and move on to a more generic discussion of how differences can simplify the expression of semantics.

3.4 The Semantics of Differences

The only elements in this theory are called *differences* arranged in *difference spaces*. A difference is anything that emerges from a background and can be isolated: particular colours, a shape, a distinct word, sound, action, event etc. Differences do not exist independently of the process of detecting them. I cannot distinguish as many shades of a colour as a painter: they may exist but have no meaning for me. Some intonations in a conversation, or the speed of a sentence will carry meaning only if I perceive them as different. Children have a very different perception of social differences because of their particular social awareness, itself dependent on other factors in their existence. Etc.

The process that allows a difference to emerge can interchangeably be called *difference space* (emphasizing the reference to conceptual spaces), *sense* or *sensor* (emphasizing the reference to perceptual models), or *simulator*, or *process* (emphasizing the dynamic aspect of the notion). Depending on the point of view, a difference space can be said to *detect* or *produce* differences. In fact, as differences are indistinguishable from the process that detects them, it is simpler to say that there are *only* differences. Even if the detection process is embodied in a biological device, no meaning is produced without the occurrence of the difference, and therefore a difference cannot be isolated from its detection process. Differences are not categories of experience, nor are detectors classification functions. Indeed, as for tags in folksonomies, differences are not elements of pre-existing categories; a detector detects whatever it can detect, which is called a difference.

If senses can be thought of as detectors, they are also, in the full meaning of the term, *processes*. Differences are the result of a process of differentiation operating on other differences, i.e. algorithms operating on data in an information science context, or chemical processes operating on modifications of the physical environment in a biological context, or activity processes operating on an environment in an embodied psychology context. This input originates as other differences, in other difference spaces. Therefore, a difference is itself always the *unfolding* of differences, an endless process of differentiation. For example the difference space detecting the difference *red* may unfold into difference spaces detecting *hue*, *saturation* and *value*, which can in turn be folded together on a sense detecting, for example, the *omnipresent-colour-on-Valentine's-day*. Differences can be similar but are never equivalent: they represent aspects of the world from the perspective of various detectors and their similarity is the result of the differences involved and how they are combined.

There is no hierarchy between difference spaces, although some of them seem more important than others, as they describe large parts of our everyday experience of meaning. For example, differentiable time as well as related notions such as simultaneity or precedence, are further differentiation processes acting on differences. The same occurs with spatial differences. The relative importance of these structures is not essential or structural but due to the fact that many other differences are built on them.

Another example of structuring differences is the distinction we make between objects according to their *separability*, e.g. a book on the table is separable from the table and a building is also separable from the city. In other words, books and buildings *afford* separability – while valleys for example do not. Separability is a difference in itself, something we can notice and isolate; it has to be sensed in its own space distinct from all other difference spaces. In a in given context it could be deduced from the *solid* state (as opposed for example to the state of being liquid) and the presence *crisp-borders* (as opposed for example to fuzzy ones), if and only if there are sensors to capture these differences too. It does not mean that everything that has a clearly defined border and is solid is separable, but that we can build a useful detector of separability by using only these two differences. Here the distinction between *bona fide* and *fiat* boundaries, although useful, simply becomes a matter of relevance to given sensors.

3.5 Table Top Example

As an illustration we analyze the difference space of a juggling ball. The spaces at our disposal, i.e. the sensors or processes that we can use, are here informally described by a name and some differences they may produce. If the name of a sense represents a category, it is just an indication for the reader and has no effect in the model. Suspension points indicate that the set of differences can only be infinite:

{*shape*: roundish, squarish,..},{*dimensions*: 3D, 2D,...},{*coloration*: multicolour, monochrome, black, white,...}, {*3d primitive*: sphere, cube, cylinder,...}, {*hardness*: hard, soft,...}, {*consistence*: liquid, physical, smoke,...}, {*size*: medium, small, big,...}, {*border*: vague-border, hard-border,...}, {*affordance*: roll-a, cut-a, separate-a, take-a,...}, {*words*: ball,...}, {*manifestation*: this-manifestation,...}

Most of these spaces are self-explanatory. The *words* space is constituted by elements of the vocabulary of a language, here English, while the *manifestation* space expresses the difference of something that occurs simultaneously and can be intuitively understood as a timestamp which links differences. Affordances, that are also differences, are suffixed by '-a' (e.g. *cut-a*) to avoid neologisms such as *cutability*. Fig. 1 is a graphical representation of such a space.

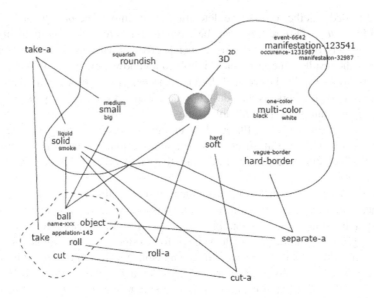

Fig. 1. The Schema of a Difference

In this schema, segments represent difference-folding and -unfolding. Segments indicate which differences are related. For example *roundish*, and *3d* have somehow been determined – the exact process being defined by the simulator itself – from the difference represented by a sphere, and, on this simple schema, only by it, while here, *rollable* (*roll-a*) has been determined by the shape and the physical state. The graphical representation of 3D primitives is a reminder that differences are not "symbols" – although it is convenient to represent them as such – but pure compounds of meaning abstracted from the world by a sensor. The large region links the differences it surrounds to the sensor of its manifestation, a difference allowing to link event occurrences in time. The smaller dashed region indicates that the differences represented there are words in a vocabulary.

By studying this example one can wonder if there is a *top* outside the representation and a *bottom* composed by differences so simple that they can be thought as atomic. There is no *bottom* as a difference can always be unfolded. And there is no *top* because meaning emerges from differences as a new difference, this particular meaning, always foldable into others.

3.6 Restaurants, Bars, Boats, and Houses

In a geospatial context, a typical space is illustrated in Fig. 2. At the bottom left of the figure, *pub* and *restaurant* are linked to the same network of differences, making them indistinguishable – apart if one has access to the given differentiation process – which indeed they are, for most purposes. However we know there are differences between a bar and a restaurant, and these can become explicit by unfolding given differences, i.e. by building sensors for them, for example by refining *eating-out-a*, to reveal the differences in service and kind of offer between the two kind of premises.

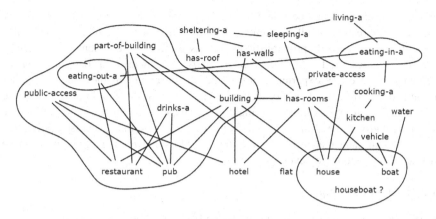

Fig. 2. Difference Space of Geographic Elements

Such an unfolding has been done to obtain *eating-out-a* and *eating-in-a* which both could relate to *eating-a*, which, however, does not allow to differentiate between eating at home and in a pub / restaurant. Similar distinctions can be made between restaurants and type of restaurants without the need of using categories and instances as reasoning as well as grounding is embedded in the difference processes themselves.

Now when new elements are integrated to the difference space, such as *houseboats* one has the choice of linking them to pre-existing elements such as *house* and *boat* to immediately understand the basic purpose of the concept, or to ground them directly to relevant differences, avoiding the relation to house and boat. Therefore, in languages which do not illustrate this similarity at the symbolic (word) level, the meaning can be retrieved anyway, and transmitted, or translated. Moreover, specific differences can be added to the concept to make it sounder, and more grounded, for example *small-kitchen* as opposed to normal sized kitchens generally found in houses.

3.7   Reasoning in Difference Spaces

Reasoning about a situation is achieved by following lines and executing the right algorithm given the observed spaces. By doing so in our given difference spaces we can deduce, for example, that *something physical and small can be taken, something that can be taken and is spherical is sometimes called a ball, a soft ball can be cut in two*, and so forth. There is no *reason* for these facts to stand other that the fact that a process is using them, and consequently, following differentiation lines allows us to always adhere to our grounding. This is *pure* meaning as it allows to make sense of any world in which differences can be sensed. If the grounding changes, i.e. the selection process and/or the related differences, then these propositions are not true anymore, i.e. they cannot be deduced from this particular space. What can be deduced does not only depend on the linked differences but on the execution of the process of differentiation, i.e. on the difference produced, and therefore cannot be deduced in advance, i.e. *a priori*, without a knowledge of the workings of this process.

More generic truths can be found *a posteriori*, based on previous experiences, by comparing the results of all or some simulations that have been made. This also allows reasoning based on probability, as well as qualitative statements such as *most birds have wings*, and *a wingless bird exists*. As mentioned before, *a priori* statements, outside any experience, are based on the nature of the simulator rather than on actual difference data. For example, knowing that the simulator produces the difference *triangle* by counting the angles of a shape, one can say with confidence that *all triangles have 3 angles*, and from the simulator that processes polygonal shapes one can state that a *myriagon* – the 10'000 sided regular polygon – can be constructed without having ever seen one.

3.8 A Dynamic Example

A tourist from Drycountry is visiting Waterland, a part of the world where rivers flow into the sea, while in her home land there are neither rivers nor lakes. She does not know anything about large water bodies although she heard about them. Now imagine the following discussion between her, stopping her car to ask a local for directions:

– How can I go to the capital? – You have to cross the river. – But rivers can't be crossed. – Of course they can be crossed – How? – Don't you know anything, just get on a 渡船[2].

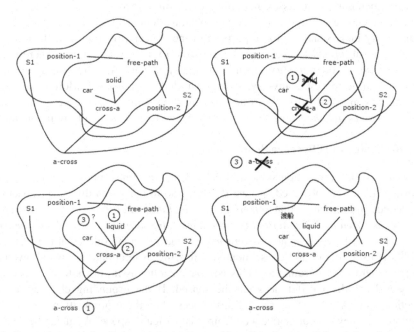

Fig. 3. Dynamic example: a) the original schema; b) a river is not solid and therefore some elements are invalidated; c) it is believed that crossing is possible therefore some new element is expected at (3); d) the meaning of the new element can be deduced from the structure

[2] Translated from English to simplified Chinese using http://translate.google.com

Understanding this sentence for our tourist, although she has never heard of boats nor dreamed about 渡船, is surprisingly easy (cf. Fig. 3, to be read from left to right and from top to bottom). The *action of crossing* (or *a-cross*) can be thought as a process producing a difference between two situations and an enabling action; i.e. the situation before crossing (a field at a given time for example, and a position in this field), the situation after, and the possibility of crossing (i.e. a crossing affordance, or *cross-a*). In other contexts, the a-cross difference, given its simple structure, would be classified as image schema. The fact that the tourist says that crossing the river is impossible, that the river is not *cross-a*, simply indicates that her difference space fails to represent the action of crossing in that situation. However, if she decides to believe her interlocutor she must acknowledge something new to make the crossing possible. That new difference, communicated by a symbol, instantly acquires meaning as it is dropped into the space: it is that which allows crossing when the path is not on solid ground. Even by ignoring "details" about ferries, such as them being boats, etc., the difference can be used and the space allowing river-crossing, completed.

This model of crossing can evolve, and can never be said to be optimal or definitive. As long as it is grounded, i.e. based on computational differences, any operational model can replace it. There are no *bad* difference spaces as there is no *bad* tagging, there are only grounded spaces and non grounded ones, effective ones and less effective ones, depending on a particular intention. Different models can merge in time and produce more and more complex structures, in the same way as folksonomies are progressively acquire organization.

3.9 Summary and Relevance to Other Theories

Below is a summary of a theory of meaning build on differences:

1. *There are only differences* (which can also be called difference spaces, sensors, senses, processes, simulators): categories, classes, or types are not first class elements.
2. *Differences process differences*: a difference can only be deduced from other differences through a given process.
3. *A difference can always be unfolded*: producing other differences.
4. *A posteriori reasoning is always possible*: it operates on traces of the differentiation process.
5. *A priori reasoning is sometime possible*: it operates on the modalities of the differentiation process itself.

The previous examples can be used to illustrate the relationship between difference spaces and ontologies. One could use an ontology to represent *crossing*; it would certainly transform nouns and differences into objects, and affordances, properties and actions into relations, so as to state that one can cross the river if the conjunction of *car*, *solid*, and *free path* is verified. Part *b* of the schema could be explained in this manner, but part *c* would be more difficult to represent, since there is hardly any room in ontologies for the emergence of new elements. Moreover, this implies that the whole process would not be grounded, as there is no guarantee in ontologies that concepts can always unfold themselves, and would not allow other schemas to be merged into it. Having sensors for verbal concepts fully justifies a symbolic comprehension of the world, which can be used by itself as most other differences can be projected (i.e. have a difference computed from them) on the symbolic plane. However, ontologies do not explicitly relate the symbolic plane to other differences, which can be called affordances, perceptive inputs, or image schemata, and which would provide grounding as well as a more interesting meaning to concepts. Instead, ontologies project verbal components onto planes composed of logical differences, with only occasional incursions into non-symbolic or non-logical difference spaces to extract properties and qualities (*quale*). It is this quasi exclusive use of the symbolic plane which forces ontologies to be object-orientated structures, as there is no other means to link symbolic elements whilst keeping the illusion of cognitive soundness.

Difference spaces are related to conceptual spaces, but they are distinct from them by the fact that they are not predefined, and they do not need any particular geometry. Gärdenfor's colour spindle is irrelevant here as we are only concerned about the fact that a given colour can be isolated, which constitutes a difference. The absence of definite shapes in difference spaces, due to the granularity of each space (each difference being absolute and non constitutive of a greater compound), also implies, as already stated, that distance relationships are not automatic in difference spaces and have to be implemented by specific processes. This allows to represent that similarity judgement are not only geometric and also makes the integration of category based similarity judgements possible, when relevant.

4 Conclusion

Advantages of difference spaces and difference-based reasoning are multiple. Firstly, everything that is represented in a difference network is, by definition, grounded. Secondly, difference spaces allow an arbitrary increase in complexity, as senses interact locally, unfolding themselves. Thirdly, the model shows how perceptual symbol systems can be computed from differences provided by perception, and interact with verbal, symbolic, logical and action / affordances planes.

Therefore, when a symbol constructed as a difference and as part of an ontology plane, it still links the origin of the verbal difference to perception and affordance related differences. Image schemata in this model become different choices of abstraction of an object, which allow us to understand schema superposition, i.e. how several schemata can collaborate to the constitution of an object without necessarily interacting with each other as building blocks do. Multiple representations of an object are here automatic as the verbal anchorage has different origins given the process used to sense them, and we have as many representations as we have sensors attached to the name of a manifestation. Each compound can then be given its own name if needed.

Further work includes a more precise analysis of the relationship of ontologies and difference spaces as well as investigating the possibility of devising a formal framework to represent difference spaces.

Acknowledgments. The author would like to thank the anonymous reviewers for their precious feedback. Marie-Kristina Thomson, for her early support, encouragement as well as her patience with reading and commenting the present version of the article. As well as Catherine McKay, for carefully reading and correcting this text, putting considerable effort to give it the appearance of English.

References

1. Bishr, Y.: Overcoming the semantic and other barriers to GIS interoperability. International Journal of Geographical Information Science 12, 299–314 (1998)
2. Fonseca, F.T., Egenhofer, M.J.: Ontology-driven geographic information systems. In: GIS '99: Proceedings of the 7th ACM international symposium on Advances in geographic information systems, pp. 14–19. ACM Press, New York (1999)
3. Egenhofer, M.J., Mark, D.M.: Naive Geography. In: Freksa, C., Mark, D.M. (eds.) COSIT 1999. LNCS, vol. 1661, Springer, Heidelberg (1999)
4. Guarino, N.: Formal Ontology and Information Systems. In: Guarino, N. (ed.) Proceedings of FOIS98 (1998)
5. Fensel, D.: Ontologies: A silver bullet for knowledge management and electronic commerce. Springer, Heidelberg (2001)
6. Winter, S.: Ontology: Buzzword or paradigm shift in GI science? International Journal of Geographical Information Science 15, 587–590 (2001)
7. Agarwal, P.: Ontological considerations in GIScience. International Journal of Geographical Information Science 19, 501–536 (2005)
8. Smith, B., Mark, D.: Do Mountains Exist? Towards an Ontology of Landforms. Environment & Planning B: Planning and Design 30, 411–427 (2003)
9. Fensel, D., Musen, M., Amsterdam, V.: The semantic web: A brain for humankind. Intelligent Systems 16, 24–25 (2001)
10. Gangemi, A., Guarino, N., Masolo, C., Oltramari, A., Schneider, L.: Sweetening Ontologies with DOLCE. In: Proceedings of EKAW, pp. 166–181. Springer, Heidelberg (2002)
11. Mark, D., Turk, A.: Landscape Categories in Yindjibarndi: Ontology, Environment, and Language. In: Spatial information theory: Foundations of geographic information science, pp. 28–45. Springer, Heidelberg (2003)
12. Kuhn, W.: Why Information Science needs Cognitive Semantics - and what it has to offer in return (2003)
13. Peuquet, D., Smith, B., Brogaard, B.: The ontology of fields (1999)
14. Choi, S., McDonough, L., Bowerman, M., Mandler, J.: Early sensitivity to language-specific spatial categories in English and Korean. Cognitive Development 14, 241–268 (1999)
15. Jordan, T., Raubal, M., Gartrell, B., Egenhofer, M.: An Affordance-Based Model of Place. In: GIS 8th Int. Symposium on Spatial Data Handling, SDH, pp. 98–109 (1998)
16. Sowa, J.: Knowledge representation: logical, philosophical, and computational foundations Brooks/Cole (2000)
17. Smith, B., Mark, D.: Ontology with Human Subjects Testing: An Empirical Investigation of Geographic Categories. American Journal of Economics and Sociology 58, 245–272 (1999)
18. Galton, A.: A formal theory of objects and fields. In: Spatial information theory: Foundations of geographic information science, pp. 458–473. Springer, Heidelberg (2001)
19. Wierzbicka, A.: Apples Are Not a Kind of Fruit. The Semantics of Human Categorization. American Ethnologis 11, 313–328 (1984)

20. Shirky, C.: The Semantic Web, Syllogism, and Worldview (2003), http://www.shirky.com/writings/semantic_syllogism.html
21. Gärdenfors, P.: How to Make the Semantic Web More Semantic Formal Ontology in Information Systems. In: Proceedings of the Third International Conference (FOIS 2004), pp. 17–34 (2004)
22. Schoop, M., de Moor, A., Dietz, J.: The pragmatic web: A manifesto. Communications of the ACM 49, 75–76 (2006)
23. Rigaux, P., Scholl, M., Voisard, A.: Spatial Databases: With Application to GIS. Morgan Kaufmann, San Francisco (2001)
24. Timpf, S.: Cartographic objects in a multi-scale data structure. Geographic Information Research: Bridging the Atlantic, 224–234 (1997)
25. Spaccapietra, S., Parent, C., Vangenot, C.: GIS Databases: From Multi-scale to Multi-representation. In: Abstraction, Reformulation, and Approximation, Springer, Heidelberg (2000)
26. Vangenot, C., Parent, C., Spaccapietra, S.: Modeling and manipulating multiple representations of spatial data. In: Proc. of the Symposium on Geospatial Theory, Processing and Applications (2002)
27. Benslimane, D., Vangenot, C., Roussey, C., Arara, A.: Multi-representation in ontologies. In: Kalinichenko, L.A., Manthey, R., Thalheim, B., Wloka, U. (eds.) ADBIS 2003. LNCS, vol. 2798, Springer, Heidelberg (2003)
28. Bédard, Y., Bernier, E.: Supporting multiple representations with spatial databases views management and the concept of VUEL. In: ISPRS/ICA Joint Workshop on Multi-Scale Representations of Spatial Data, Ottawa, Canada (2002)
29. Gärdenfors, P.: Conceptual spaces: The geometry of thought. MIT Press, Cambridge (2000)
30. Gärdenfors, P., Williams, M.: Reasoning about categories in conceptual spaces. In: Proceedings of the Fourteenth International Joint Conference of Artificial Intelligence, pp. 385–392 (2001)
31. Raubal, M.: Formalizing Conceptual Spaces Formal Ontology in Information Systems. In: Varzi, A., Vieu, L. (eds.) Proceedings of the Third International Conference (FOIS 2004), pp. 153–164 (2004)
32. Ahlqvist, O.: A Parameterized Representation of Uncertain Conceptual Spaces Transactions in GIS. Blackwell Synergy 8, 493–514 (2004)
33. Kuhn, W.: Modeling the Semantics of Geographic Categories through Conceptual Integration. In: Egenhofer, M.J., Mark, D.M. (eds.) GIScience 2002. LNCS, vol. 2478, pp. 108–118. Springer, Heidelberg (2002)
34. Barsalou, L.: Perceptual symbol systems. Behavioral and Brain Sciences 22, 577–660 (2000)
35. Lakoff, G., Johnson, M.: Metaphors we live by. University of Chicago Press, Chicago, IL (1980)
36. Lakoff, G.: Women, fire, and dangerous things. University of Chicago Press, Chicago, IL (1987)
37. Johnson, M.: The body in the mind: The bodily basis of meaning, imagination, and reason. University of Chicago Press, Chicago, IL (1987)
38. Mark, D.: Cognitive Image-Schemata for Geographic Information: Relations to User Views and GIS Interfaces. In: Proceedings GIS/LIS 89, pp. 551–560 (1989)
39. Raubal, M., Egenhofer, M., Pfoser, D., Tryfona, N.: Structuring Space with Image Schemata: Wayfinding in Airports as a Case Study. In: Proceedings of the International Conference on Spatial Information Theory, pp. 85–102 (1998)

40. Ruetschi, U., Timpf, S.: Using Image Schemata to Represent Meaningful Spatial Configurations. In: Meersman, R., Tari, Z., Herrero, P. (eds.) On the Move to Meaningful Internet Systems 2005: OTM 2005 Workshops. LNCS, vol. 3762, Springer, Heidelberg (2005)

41. Frank, A., Raubal, M.: Formal specification of image schemata – a step towards interoperability in geographic information systems. Spatial Cognition and Computation 1, 67–101 (1999)

42. Gibson, J.: The Ecological Approach to Visual Perception. Lawrence Erlbaum, Mahwah (1979)

43. Konczak, J., Meeuwsen, H., Cress, M.: Changing affordances in stair climbing: The perception of maximum climbability in young and older adults. Journal of Experimental Psychology Human Percept ion and Performance 18, 691–697 (1992)

44. Mathes, A.: Folksonomies-Cooperative Classification and Communication Through Shared Metadata Computer Mediated Communication. In: LIS590CMC (Doctoral Seminar), Graduate School of Library and Information Science, University of Illinois Urbana-Champaign (December 2004)

45. Merholz, P.: Metadata for the Masses 2004 at http://www.adaptivepath.com/publications/essays/archives/000361.php

46. Mika, P.: Ontologies are us: A unified model of social networks and semantics. In: Proc. ISWC2005 (2005)

47. Deleuze, G.: Difference and Repetition. Columbia University Press (1994)

48. Feynman, R.: The development of the space-time view of quantum mechanics. Nobel Lecture. Physics Today 19, 31 (1966)

49. Kuhn, W.: Geospatial Semantics: Why, of What, and How. Journal on Data Semantics, 2 (2005)

50. Fauconnier, G., Turner, M.: Conceptual Integration Networks. Cognitive Science 22, 133–187 (1998)

51. Coulson, S., Oakley, T.: Blending Basics. Cognitive Linguistics 11, 175–196 (2000)

52. Thomson, M.K., Béra, R.: Relating Land Use to the Landscape Character: Toward an Ontological Inference Tool. In: Winstanley, A.C. (ed.) GISRUK 2007, Proceeding of the Geographical Information Science Research UK Conference, Maynooth, Ireland, pp. 83–87 (2007)

Evaluation of a Semantic Similarity Measure for Natural Language Spatial Relations

Angela Schwering

Institute of Cognitive Science, University of Osnabrueck, Germany
aschweri@uos.de

Abstract. Consistent and flawless communication between humans and machines is the precondition for a computer to process instructions correctly. While machines use well-defined languages and formal rules to process information, humans prefer natural language expressions with vague semantics. Similarity comparisons are central to the human way of thinking: we use similarity for reasoning on new information or new situations by comparing them to knowledge gained from similar experiences in the past. It is necessary to overcome the differences in representing and processing information to avoid communication errors and computation failures. We introduce an approach to formalize the semantics of natural language spatial relations and specify it in a computational model which allows for similarity comparisons. This paper describes an experiment that investigates human similarity perception between spatial relations and compares it to the similarity determined by the our semantic similarity measure.

1 Introduction

Any computer system working with spatial data or interacting with the environment must have the capability to communicate with humans about environments. Spatial human-machine communication via natural language poses great problems: natural language spatial relations are highly ambiguous with only vaguely specified semantics [1,2] and therefore cannot be easily formalized. However, computers require a formal model to understand and process the semantics of spatial relations. For a computer to be able to understand and express spatial configurations in natural language would be the key to eliminate human-machine communication problems and increase usability.

Shariff, Egenhofer and Mark [3,4,5] developed a formal model to describe natural language spatial relations: following the premise *topology matters, metric refines* they investigated several topologic and metric properties of natural language expressions in an experiment conducted on human subjects. Based on their findings, we propose to specify the semantics of spatial relations in natural language by describing all possible formal spatial relations which might apply to the spatial setting described by the natural language term. Hence we use formal spatial relations as qualities to describe natural language spatial relations.

While equality and inequality is very easy to detect for computers, similarity is a vague and relatively undefined measure to compare entities. However, similarity plays an important role in the human cognition: the sense of similarity is the foundation for the human ability to classify similar entities, to reason on similar situations and for learning.

S. Winter et al. (Eds.): COSIT 2007, LNCS 4736, pp. 116–132, 2007.

Gärdenfors proposed conceptual spaces [6,7,8] as a cognitively plausible framework for representing information at a conceptual level. We apply the theory of conceptual spaces to a geometric similarity measure for natural language spatial relations: here the conceptual space is formed by a set of quality dimensions which are grounded in well-defined formal spatial relations. Each natural language spatial relation is specified by values on the quality dimensions and therefore occupies a region within the conceptual space. The geometric space is then used to determine the semantic distance between the natural language expressions.

A semantic similarity measure therefore enables a machine to compare instructions given by humans: when it receives a new instruction it can search for similar instructions in the past and reuse its "experiences". Similarity comparisons are important to simulate human reasoning processes and offer the vagueness and flexibility that formal-logic reasoning is often lacking. The ability of judging similarity is necessary to react adequately to new situations by comparing them to experiences learned in the past. We consider similarity comparisons as an important form of non-classical reasoning.

The remainder of the paper is structured as follows: section 2 gives an overview of the related work concerning our semantic similarity measure between spatial relations, which is described in section 3. Section 4 explains the calculation procedure in our similarity measure with concrete data, the setting of our experiment and the statistical parameters. In section 5 we discuss the results. The final section summarizes the outcome of the paper and outlines directions for future work.

2   Related Work

This section describes how formal spatial relations build the basis for Shariff's model to describe natural language spatial relations. Additionally we describe Gärdenfors' theory of conceptual spaces which is used as framework for our similarity measure.

2.1   Semantics of Natural Language Spatial Relations

Formal spatial relations - topologic [9], metric [10] or direction relations [11,12] - have well-defined semantics, but natural language spatial relations have complex semantics often implying more than one type of formal spatial relation. While people are more familiar with using spatial terms in their natural languages, systems use definitions based on a computational model for spatial relations. To bridge this gap Shariff, Egenhofer and Mark [3,4] developed a model defining the geometry of spatial natural language relations following the premise topology matters, metric refines [13]. This model for spatial relations consists of two layers: first it captures the topology between lines and regions based on the 9-Intersection model. The second layer analyzes the topologic configuration according to a set of metric properties: splitting, closeness and approximate alongness.

Topologic Properties. Based on the Egenhofer's 9-intersection model [14] there can be identified 19 different relations between lines and regions [15]. A conceptual neighborhood graph arranges these relations in a way that it corresponds to the human similarity perception: the nearer two relations are within the graph, the more similar they are (see [15] for discussion on different possible conceptual neighborhood graphs). Shariff et al. used the conceptual neighborhood graph illustrated in figure 1 to describe natural language spatial relations.

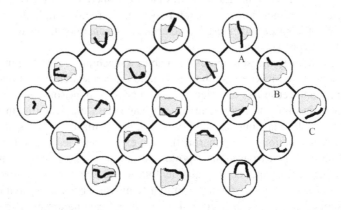

Fig. 1. Conceptual neighborhood graph of topologic relations between line and region [4, p. 301]

A natural language spatial relation can be described by one or several topologic relations: Egenhofer and Mark showed in a human subject test [16] that the relations *cross* and *intersect* were both associated with the topologic relation A in figure 1 and the relation *alongEdge* with relation B and C.

Metric Properties. The metric properties of natural language spatial relations refer to the ratios of certain distances, length or size differences of the region and the line. Splitting determines the way a region is divided by a line and vice versa. Closeness describes the distance of a region's boundary to the disjoint parts of the line. Approximate alongness is a combination of the closeness measures and the splitting ratios: it assesses the length of the section where the line's interior runs parallel to the region's boundary. The following list describes the fifteen different metric configurations proposed by Shariff et al. Several are illustrated in figure 2.

- Interior/Exterior Area Splitting: describes how the line separates the interior/ exterior of the region.
- Interior/Exterior Traversal Splitting: describes the ratio of the line lying inside/ outside the region to the length of the whole line.
- Perimeter/Line Alongness: describes the ratio of the line lying on the region's boundary to the length of the region's boundary or to the line's length.
- Region Boundary Splitting: describes how the boundary of the line splits the boundary of the region.
- Inner/Outer Closeness: the distance between the line's and the region's boundary with the line's boundary being inside/outside the region.

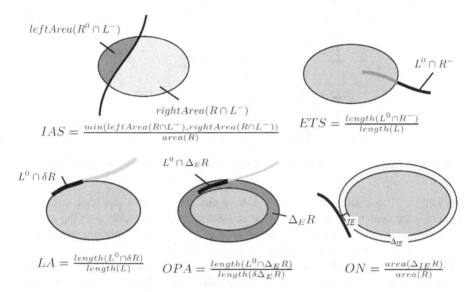

$$IAS = \frac{min(leftArea(R\cap L^-),rightArea(R\cap L^-))}{area(R)}$$

$$ETS = \frac{length(L^0\cap R^-)}{length(L)}$$

$$LA = \frac{length(L^0\cap\delta R)}{length(L)}$$

$$OPA = \frac{length(L^0\cap\Delta_E R)}{length(\delta\Delta_E R)}$$

$$ON = \frac{area(\Delta_{IE}R)}{area(R)}$$

Fig. 2. Illustration of several metric measures: Interior Area Splitting (IAS), Exterior Traversal Splitting (ETS), Line Alongness (LA), Outer Approximate Perimeter Alongness (OPA) and Outer Nearness (ON) [3, pp. 212–214]

- Inner/Outer Nearness: the distance between the line's interior and the region's boundary with the line being completely inside/outside the region.
- Inner/Outer Approximate Perimeter Alongness assesses how the line's interior splits a buffer zone around the region's boundary with the buffer zone being inside/outside the region.
- Inner/Outer Approximate Line Alongness assesses the length of the line's interior falling within a buffer zone around the region's boundary with the buffer zone being inside/outside the region.

We are aware of the fact that this computational model does not include a "complete" semantic description of natural language spatial relations. It does not consider directional properties, neither other aspects such as functional properties of spatial relations [17]. However, our experiment will show that the combination of topologic and metric properties already serves as a good approximation of the semantics inherent in natural language spatial relations and is very useful for the similarity comparison.

2.2 Gärdenfors' Conceptual Spaces

The notion of a conceptual space was introduced by Peter Gärdenfors as a framework for representing information at the conceptual level [6,7,8]. Conceptual spaces can be utilized for knowledge representation and support semantic similarity measurement. A conceptual space is formalized as a multidimensional space consisting of a set of quality

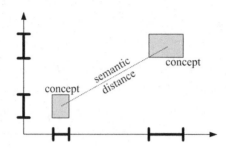

Fig. 3. Two concepts and their semantic distance in a two-dimensional conceptual space

dimensions.[1] The quality dimensions can have any geometric or topologic structure: in figure 3 are shown two linear dimensions. Each concept is described on the quality dimensions with either a single value or an interval. Concepts are therefore represented as n-dimensional convex regions within the conceptual space.

The fact that conceptual spaces have a metric allows for the calculation of spatial distances between concepts in the space. The spatial distance is interpreted as the semantic distance: the nearer two concepts are in the conceptual space, the lower is their semantic distance and the higher is their similarity. The similarity is considered as a decaying function of the semantic distance [18,19]. Gärdenfors proposes to use the Minkowski metrics to calculate the semantic distance. Johannesson and Gärdenfors demonstrated in experimental studies, that the Euclidian metric is more appropriate when stimuli are composed of integral, perceptually fused dimensions and the city-block metric should be preferred when the stimuli are composed of separable dimensions [20,21].

3   A Semantic Similarity Measure for Natural Language Spatial Relations

Our semantic similarity measure for natural language spatial relations combines the above mentioned approaches: Shariff, Egenhofer and Mark showed how natural language spatial relations can be described by their topologic and metric properties. We use their model to determine relevant qualities for the semantic description of spatial relations. For the similarity comparison we apply Gärdenfors' theory of conceptual spaces as a framework to represent the topologic and metric properties and determine the similarity. Via this formalization of natural language spatial relations a machine cannot only "understand" the meaning of the spatial relations, but also compare two relations with respect to their similarity.

---

[1] Gärdenfors' conceptual spaces consist of a more complex structure based on domains represented through a set of integral dimensions, which are distinguishable from all other dimensions. Since this paper focuses on conceptual spaces as a technique for knowledge representation and similarity measurement, we do not focus on the cognitive foundation of conceptual spaces, but concentrate only on the methodology to formalize conceptual spaces. We therefore describe a "simplified" version of conceptual spaces. The complete cognitive foundation of conceptual spaces can be found in [6].

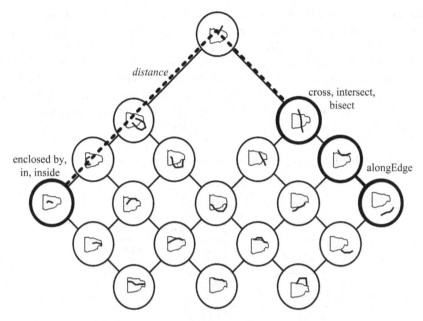

Fig. 4. Examples for the topologic description of spatial relations

3.1   Representation to Formalize Semantics

Shariff et al. proposed 16 different properties to specify the semantics of natural language spatial relations.

Topologic Properties. The first property describes the topologic relation between the region and the line. We use the snapshot model (figure 4), a conceptual neighborhood graph between lines and regions, as one quality dimension of the conceptual space. The topologic properties of natural language spatial relations are described by different values on this dimension. While most qualities are described on a linear dimension, this quality dimension has a network structure. This is necessary to reflect the complex similarity relations between the different topologic configurations. As mentioned above, there exist different ways to construct the conceptual neighborhood graph between lines and regions: the snapshot model organizes the relations according to their topological distance [15,22]. We chose the snapshot model for our calculations, because tests on human subjects confirmed that the conceptual neighborhood identified by the snapshot model corresponds to the way humans conceptualize similarity [15] and corresponds to groupings people made using natural language spatial relations [23]. Shariff et al. use a variant of the snapshot model[2].

---

[2] In [5], Shariff uses a conceptual neighborhood which differs in the connections of the very top relation (figure 1). We discussed our semantic similarity measure using Shariff's variant of the snapshot model in [24]. In our approach it is possible to use different conceptual neighborhoods as long as the distances meet the similarity perception of humans. Nevertheless, the selection of the conceptual neighborhood graph has an influence on the similarity value computed by our model, because the arrangement and therefore also the measured distances differ.

Metric Properties. The 15 different metric properties are represented on separate dimensions: seven for splitting, four for closeness and four for alongness. These dimensions are linear. Based on experiments with human subjects, Shariff et al. determined the value ranges describing each natural language spatial relation according to the formulas given in section 2.1. Table 1 shows the specification for the same six relations as above. Not all metric properties are applicable for each spatial relation: For instance, the dimension Outer Closeness (OC) cannot be determined for the relations *enclosed by*, *in* and *inside*, because the line is completely inside the region.

Table 1. Examples for the metric description of spatial relations

	IAS	IC	OC	IN
crosses	0.03–0.50		0.29–8.62	
enclosed by		0.19–0.79		0.19–0.79
in		0.03–0.78		0.03–0.69
inside		0.02–0.87		0.02–0.77
intersect	0,02–0,50		0,25–9,10	
bisect	0,05–0,48	0,40–8,78		

Grounding. The proposed semantic similarity measure shall serve as a computational model to describe the vague semantics of natural language spatial relations in an unambiguous way. However, this can only be ensured if the dimensions themselves are grounded. Gärdenfors proposes to ground the dimensions of conceptual spaces in our sensory perception such as color, sound or taste. Another possibility is to ground dimensions in mathematically well-specified measurements (compare also [25]). The quality dimensions used in our semantic similarity measure are grounded in formal spatial relations: in mathematics, topologic and metric relations have a clear and unambiguous definition and semantics.

3.2   Determining the Similarity Via the Semantic Distance

The overall semantic distance between two spatial relations is taken as indicator for the semantic similarity. Similarity is a decaying function of the semantic distance [18,19].

Topology Dimension. The semantic distance between two different properties on a network dimension are the shortest distance between the nodes in the network. Figure 4 shows the relations *enclosed by*, *in* and *inside* described by a line lying completely inside the region and the relations *cross*, *bisect* and *intersect* described by a line starting outside of the region, crossing the region and ending outside the region again.

The semantic distance between these relations equals four, because the nodes are four steps away in the network.

$$distance(x_1 \wedge ... \wedge x_k, y_1 \wedge ... \wedge y_m) = \frac{1}{k * m} \sum_{i=1}^{k} \sum_{j=1}^{m} distance(x_i, y_j)$$

If a concept is represented by a conjunction of several nodes, i.e. several topologic configurations are possible for one relation, we compute the average distance as shown in the above formula. The relation $alongEdge$ in figure 4 is specified by two possible topologic relations. The semantic distance between $alongEdge$ and $cross$ equals 1.5. This distance measure was proposed by Rada [26] as semantic distances in a semantic network.

Metric Dimension. The semantic distance on linear dimensions is measured as the spatial distances by the city-block or the Euclidian metric. The Minkowski metrics is a generic formula: $r = 1$ results in the city-block distance and $r = 2$ in the Euclidean distance [27].

$$distance(x, y) = [\sum_{i=1}^{n} |x_i - y_i|^r]^{\frac{1}{r}}$$

If the spatial relations are described via intervals, the semantic distance is computed between the mean values of the intervals. In the multidimensional space, this corresponds to the semantic distance between the center points of the regions describing the spatial relations. The center point of a concept is often interpreted as the prototypical instance of this concept.

Similarity Across Different Dimensions. The similarity measure determines the semantic distance across all dimensions. However, each dimension represents a different quality: to make qualities comparable, e.g. by comparing topologic distance to inner closeness, the values must be standardized and represented in some relative unit of measurement. This is done by calculating the z scores for these values, also called z-transformation [28]. We compute the semantic distance based on the standardized dimension values.

Topology Matters, Metric Refines. According to studies by Egenhofer, Mark et al., the topologic properties of spatial relations are decisive while metric properties serve only as refinement. Our similarity measure accounts for the different importance of dimensions by assigning weighting factors to each quality dimension reflecting the importance of the dimension.

Spaces with Different Quality Dimensions. Moreover, we have to consider that not all spatial relations are described by the same set of quality dimensions: some quality dimensions may not be applicable at all or their values do not matter for a specific configurations. Therefore the similarity comparison must be able to compare descriptions of spatial relations based on different conceptual spaces. In [29,30,31] we proposed a two-step process for computing semantic distances in conceptual spaces based on different dimensions: in the first step we check whether both concepts are described by the same dimension. For this purpose we introduce a boolean dimension with the value 'yes' for dimension existence and 'no' indicating that the dimension is not applicable. If the dimension is applicable, we compare the values for this quality. This way we can compare concepts described in non-identical conceptual spaces. The weighting factors are chosen in a way, that a non-applicable dimension increases the semantic distance more than different values on the same quality dimension.

4   Experimental Evaluation of the Semantic Similarity Measure

We conducted an experiment to evaluate how well the similarities determined by our semantic similarity measure meet the similarity perception of humans. For a set of 15 relations we determined the similarity by our similarity measure and asked human subjects to judge the similarity and compared the answers afterwards.

4.1   Similarity Calculation of the Semantic Similarity Measure

This section outlines how we determined the semantic descriptions for natural language spatial relations as data input for our system and how we transformed semantic distances in different similarity categories.

Data Input. In section 3 we described the dimensions required to specify natural language spatial relations and the algorithm to determine semantic similarity. The values on each quality dimension must be specified for the natural language spatial relations. We used the specification of natural language spatial relations from an experiment by Mark and Egenhofer [16]: they investigated the conceptualization of spatial relations in an experiment. The subjects were presented with a picture of a park and a sentence *The road _____ the park* with the blank substituted by different natural language spatial relations. The subjects had to draw a road in the picture such that the resulting drawing would fit the spatial relation. This way the topologic properties of spatial relations were determined. Figure 5 shows several sentences and examples of these drawings.

Shariff et al. analyzed the drawings provided by the experiment for the metric properties of the investigated spatial relations and constructed a complete topologic and metric description of 59 natural language spatial relations. Figure 4 and table 1 show only a small set of relations as examples; the complete data can be found in [3] and [5].

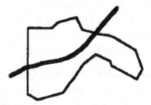

The road crosses
the park.

The road runs along
the edge of the park

The road splits the park.

Fig. 5. Examples from the human subject test [16,5]

Semantic Distances. The semantic measure determines semantic distances ranging between 0.0 and 11.5. These semantic distances need to be interpreted as similarity values. As we will describe later, we used four different categories to describe the similarity of spatial relations: identical, very similar, similar or partially similar and not similar. To compare human judgements with the system's similarity values, we require one common scale: table 2 describes the transformation from semantic distances to those categories.

Table 2. Transformation of semantic distance

Semantic Distance	Similarity Categories
$0.00 = x$	identical
$0.00 < x \leq 1.00$	very similar
$1.00 < x \leq 4.85$	similar or partially similar
$4.85 < x$	not similar

Figure 6 visualizes the results of the similarity measure: multidimensional scaling groups the relations into four major clusters which can be interpreted as follows:

– all inside relations are grouped together, here: *enclosedBy*, *in*, *inside* and *within*,
– all outside or disjoint relations are grouped together, here: *alongEdge*, *avoid*, *bypass* and *near*,
– all cross relations are grouped together, here: *bisect*, *cross*, *intersect*, *split*, *transect*, *traverse*, and
– the *enter* relation is separate from the others and is the fourth cluster.

These results agree with findings from Mark and Egenhofer: they investigated the prototypicality of spatial relations: *inside*, *outside*, *cross*, and *enter* have been identified as the most frequently used group prototypes (compare figure 11 in [23]).

4.2 Similarity Judgements of Human Subjects

Similarity can be measured either directly via similarity judgements or indirectly for example via confusion probability of stimuli (the more similar two stimuli are, the more likely is it to confuse them). The questionnaire used in our experiment is designed to measure similarity in a direct way. In the following we describe the method of the experiment.

Participants. Out of 23 subjects, who took part in the experiment, we could use the results of 22 participants for our evaluation. One participant obviously mixed up the scales and classified identical sentences as 'not similar'. All of them had a university background, but only 6 of them had a background in geographic information science. The participants were between 21 and 57 years old, 8 of them were female. All participants were either native speakers or had an excellent knowledge in English. The participation in the experiment was voluntary. The participants were not payed.

Design of the Questionnaire. The questionnaire was a pdf form which was filled in by the participants in front of the computer. The questionnaire compared 15 natural language spatial relations according to their similarity.

Context has a high influence on similarity judgements [32]. Therefore we chose the same setting as Mark and Egenhofer did in their experiment, because their results are used as data input for our similarity measure: Every comparison was done in the sentence *The road _____ the park* with the blank substituted by different spatial relations. In this manner we were able to eliminate context effects by ensuring that participants in our experiment use the same context as participants in the previous experiments.

In literature it is often argued that similarity is asymmetric [33,34,35,36,37]. Therefore the questionnaire asked for the similarity twice in both directions: we asked subjects to determine the similarity from one spatial relation to all other 15 spatial relations including itself. This was repeated one after the other for all 15 spatial relations. Each comparison of one sentence to all other sentences was done on a separate page and subjects were instructed not to scroll backwards to look up their previous similarity judgements.

The subjects were asked to choose one of the four categories to determine the similarity of sentences: 1 stands for identical, 2 is very similar, 3 is similar or partially similar and category 4 stands for not similar sentences. There have been several reasons for reducing the answer possibilities to four categories: especially for non-supervised internet experiments it is very important that the questionnaire is sufficiently easy for subjects to answer. We found that ranking all 15 sentences according to their similarity is difficult and demands a lot of attention. A questionnaire of this size takes more than 60 minutes to answer. As well, using one continuous scale for similarity judgement expected too much of the participants: subjects complained being unable to assign sensible similarity values to completely dissimilar sentences. The problem of assigning similarity values to such "anomalies" have been discussed in literature: To determine the similarity between two stimuli we usually "presuppose some amount or type of similarity" [32, p. 259]. Medin, Goldstone and Gentner point out that humans would judge the comparison of a relation *avoid* to *avoid* as being identical rather than (very) similar. While the categories "very similar", "similar" and "partially similar" refer to different degrees of similarity, identity and non-similarity do not. It is difficult to determine the similarity degree between two relations *avoid* and *enclosedBy* and it is even more difficult do this consistently across other non-similar relations, i.e. to relate the degree of similarity to the comparison of the relations *near* and *within*. The degree of similarity can only be determined between stimuli which have a common basis, i.e. a minimum amount of similarity. Therefore we did not choose one similarity scale but four different categories in our experiment. In our similarity measure which is based on the similarity of properties, the maximum similarity (i.e. a semantic distance of zero) is interpreted as identical and a a high semantic distance is interpreted as non-similar. For the evaluation of our similarity measure we can therefore assign the four categories the ordinal ranking order "identical", "very similar", "similar or partially similar" and "non-similar".

Procedure. The participants received the questionnaire as pdf document via email. Each of them completed it in front of the computer. The participants had to read the instructions. On the second page the participants had to determine the similarity of one sentence describing a spatial relation between the road and the park to each of the following 15 road-park-sentences. The participants could only use the four categories "identical", "very similar", "similar or partially similar" and "not similar". After completing one page, they could go on with the next page and the next road-park-sentence comparison. They were not allowed to scroll back and look at their previous similarity judgements. It took the participants approximately 20 minutes to complete the questionnaire.

4.3 Statistical Parameters

The system shall be evaluated against the human perception of similarity of natural language spatial relations.[3] This assumes that humans have a common opinion on the similarity of spatial relations. If the subjects themselves disagree about the similarity of two relations we cannot use these similarity comparisons for our evaluation. We demand that more than 50% of the subjects agreed on one similarity categorization. Out of 225 similarity comparison there have been 23 relations where subjects did not sufficiently agree on one categorization. These relations were not included in the evaluation. Therefore the sample size of the evaluation is $n = 202$. Since the similarity categorization has an ordinal scale we computed the average similarity perception of the human subjects by the median of their answers.

The correlation between the system's similarity measurement and the subjects' similarity judgements was measured via the Bravais-Pearson-Correlation Coefficient. According to [38,39] a correlation of $r > 0.5$ has a great magnitude of effect. For a significance level $\alpha = 0.01$ and a power $1 - \beta = 0.8$, the optimal sample size is $n = 36$. Our sample size is greater and therefore is sufficiently big to make a significant statement [4].

5 Results and Discussion

We compared similarity measurements based on our system to similarity judgements of participants in our experiment and computed the correlation. The Bravais-Pearson-Correlation Coefficient equals $r = 0.96$. In general, a correlation greater than 0.8 is already considered as a very strong correlation. Those relations that we had to exclude due to disagreement among the subjects on the similarity would have a negative influence of only 0,01 on the correlation coefficient. To compute the significance we used the t-test: the resulting t-value is $t = 49.55$ which leads to $p < \alpha$. This shows that our semantic similarity measure is very suitable to determine the similarity of natural language spatial relations as humans would perceive it. In almost all cases it was possible

---

[3] The system computes the similarity based on the averaged data of the experiment of Mark an Egenhofer. Therefore it represents the average opinion on the semantics. We do not claim that the system's similarity measurements meet the similarity perception of every human, but only the average opinion. Therefore we did evaluate the system against the average opinion of the subjects and not against every single subject.

[4] The sample size reflects the number of different natural language spatial relations. Since we want to test our similarity measure for many different relations we chose a very big sample size. With a correlation coefficient of r = 0.96, the fact that a sample size much greater than the optimal sample size increases the probability of a type I error should not be a problem.

to find the identical, similar and partially similar relations to one spatial relation. The similarity predictions based on the semantic similarity measure proposed in this paper can be used to predict reliably the similarity perceived by humans.

Table 3. Results of the semantic similarity of relation pairs determined by the system and the average opinion of the subjects (similarity measurement of the system, similarity judgement of human subjects)

	along edge	avoid	bisect	bypass	cross	enclosed by	enter	in	inside	intersect	near	splits	transect	traverse	within
alongEdge	(1,1)	(3,-)	(4,4)	(3,2)	(4,4)	(4,4)	(4,4)	(4,4)	(4,4)	(4,4)	(3,3)	(4,4)	(4,4)	(4,4)	(4,4)
avoid	(3,-)	(1,1)	(4,4)	(2,2)	(4,4)	(4,4)	(4,4)	(4,4)	(4,4)	(4,4)	(2,3)	(4,4)	(4,4)	(4,4)	(4,4)
bisect	(4,4)	(4,4)	(1,1)	(4,4)	(2,2)	(3,-)	(3,3)	(3,-)	(3,-)	(2,2)	(3,4)	(2,2)	(2,2)	(2,2)	(3,-)
bypass	(3,2)	(2,2)	(4,4)	(1,1)	(4,4)	(4,4)	(4,4)	(4,4)	(4,4)	(4,4)	(2,2)	(4,4)	(4,4)	(4,4)	(4,4)
cross	(4,4)	(4,4)	(2,2)	(4,4)	(1,1)	(3,-)	(3,3)	(3,3)	(3,3)	(2,2)	(3,4)	(2,2)	(2,2)	(2,2)	(3,-)
enclosedBy	(4,4)	(4,4)	(3,4)	(4,4)	(3,-)	(1,1)	(4,-)	(2,2)	(2,2)	(3,3)	(4,4)	(3,-)	(3,-)	(3,-)	(2,2)
enter	(4,4)	(4,4)	(3,3)	(4,4)	(3,3)	(4,4)	(1,1)	(4,-)	(4,-)	(3,3)	(4,4)	(3,3)	(3,3)	(3,3)	(4,-)
in	(4,4)	(4,4)	(3,3)	(4,4)	(3,-)	(2,2)	(4,3)	(1,1)	(2,2)	(3,3)	(4,4)	(3,3)	(3,3)	(3,3)	(2,2)
inside	(4,4)	(4,4)	(3,3)	(4,4)	(3,3)	(2,2)	(4,3)	(2,2)	(1,1)	(3,3)	(4,4)	(3,3)	(3,3)	(3,3)	(2,2)
intersect	(4,4)	(4,4)	(2,2)	(4,4)	(2,2)	(3,3)	(3,3)	(3,3)	(3,3)	(1,1)	(3,4)	(2,2)	(2,2)	(2,2)	(3,3)
near	(3,2)	(2,3)	(3,4)	(2,-)	(3,4)	(4,4)	(4,4)	(4,4)	(4,4)	(3,4)	(1,1)	(4,4)	(4,4)	(4,4)	(4,4)
split	(4,4)	(4,4)	(2,2)	(4,4)	(2,2)	(3,3)	(3,3)	(3,-)	(3,3)	(2,2)	(4,4)	(1,1)	(2,2)	(2,2)	(3,3)
transect	(4,4)	(4,4)	(2,2)	(4,4)	(2,2)	(3,-)	(3,3)	(3,3)	(3,3)	(2,2)	(4,4)	(2,2)	(1,1)	(2,2)	(3,3)
traverse	(4,4)	(4,4)	(2,2)	(4,4)	(2,2)	(3,-)	(3,3)	(3,-)	(3,3)	(2,2)	(4,4)	(2,2)	(2,2)	(1,1)	(3,-)
within	(4,4)	(4,4)	(3,3)	(4,4)	(3,3)	(2,2)	(4,3)	(2,2)	(2,2)	(3,3)	(4,4)	(3,3)	(3,3)	(3,3)	(1,1)

Table 3 shows the similarity determined by the system and the similarity judgements by the subjects. For better comparison we show both similarities using the same similarity categorization; "-" marks the cases in which subjects did not sufficiently agree about the similarity value.

While our semantic similarity measure determines only symmetric similarities, the judgements of the subjects can be asymmetric. However, only four relation pairs out of 225 comparisons have been determined as asymmetric: *bisect-enclosedBy*, *enclosed By-enter*, *alongEdge-near* and *bypass-near*. The first two were not taken into the calculation anyway, because of the high disagreement of subjects on their similarity. The participants did not agree very much on the last two as well, but 52% agreed on one categorization and therefore they have been taken into the calculation. Still, four asymmetric similarity judgements out of 225 comparisons is very low and the similarity perception of natural language spatial relations seems to be mainly symmetric. Therefore we think that for the comparison of natural language spatial relations a symmetric similarity measure reflects best the human similarity perception.

Furthermore we did not discover great difference between subjects depending on their expertise in geographic information science (GIS), sex or age. However, for a detailed analysis of possibly influencing facts such as GIS knowledge or language we need to expand our human subject test.

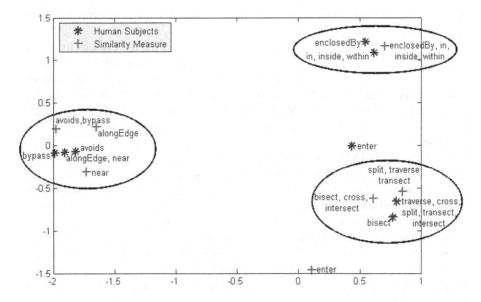

Fig. 6. Plot of the relations

The results of a multidimensional analysis of all similarity values is shown in figure 6. Three clusters can be recognized: each cluster contains those relations grouped by the system and exactly the same relations which were grouped together based on the results of the human subject test. This figure underlines again graphically how well the system and the human similarity perception correlate. Only the relation *enter* differs between the system and the subjects: this is because the system determined higher dissimilarity between *enter* and the cluster of inside-relations than the the human subjects did. [5]

6 Summary and Future Work

In this paper we introduced a semantic similarity measure for natural language spatial relations. The aim of this similarity measure is to ease human-computer communication about spatial configuration: using our system, a computer becomes able to compare natural language spatial relations with respect to their semantic similarity. A computer can "understand" spatial relations in natural language and reason on them.

We combined a framework for semantic similarity measurement, namely Gärdenfors conceptual spaces, with Shariff et al.'s model to describe natural language spatial relations. Shariff et al.'s description is based on topologic properties and various measures for metric properties. Each property is represented on a different quality dimension: the

---

[5] As shown in table 3, the system judges the similarity from *enter* to *in*, *inside*, and *within* as not similar (4). The majority of subjects judged the similarity from *enter* to *in*, *inside* and *within* as similarity or partially similar (3). Since the disagreement was too high they were excluded from the calculation of the correlation coefficient, however had to be included to compute the multidimensional scaling.

topologic properties are modeled on a network dimension while the metric properties are modeled on a linear dimension.

In our experiment we computed the similarity of various spatial relations for which formal descriptions had been produced in advance by a test on human subjects (Mark et al.). Afterwards we compared the results of our semantic similarity measure with human similarity judgements on the same similarity measure. We found that our similarity measure correlates significantly with the answers given by humans. The experimental investigation shows that this similarity measure can predict reliably the similarities perceived by humans.

So far, the similarity measure is restricted to qualities concerning the topologic and metric properties of a spatial relation only. To successfully determine the similarities of natural language spatial relations which also express a certain directional configuration or functional coherences, we need to extend the similarity measure to also model other properties of spatial configurations. Moreover, the current model uses the semantic descriptions resulting from relations between lines and regions in a park-road context. Recently, Nedas et al. [40] and Xu [41] analyzed how line-line natural language spatial relations can be formalized. More work has to be done to test to what extent our approach can be applied to region-region and line-line spatial relations and to other domains as well.

Acknowledgements. I especially thank Isabel Dombrowe and Catharina Riedemann for their help on the design and the statistical evaluation of the questionnaire and the three anonymous reviewers for their valuable comments. Moreover I want to thank the participants who took part in the experiment.

References

1. Montello, D.R., Goodchild, M., Gottsegen, J., Fohl, P.: Where's downtown?: Behavioral methods for determining referents of vague spatial queries. Spatial Cognition and Computation 3, 185–204 (2003)
2. Robinson, V.B.: Individual and multipersonal fuzzy spatial relations acquired using human-machine interaction. Fuzzy Sets and Systems archive 113(1), 133–145 (2000)
3. Shariff, A.R., Egenhofer, M.J., Mark, D.M.: Natural-language spatial relations between linear and areal objects: The topology and metric of English language terms. International Journal of Geographical Information Science 12(3), 215–246 (1998)
4. Egenhofer, M.J., Shariff, A.R.: Metric details for natural-language spatial relations. ACM Transactions on Information Systems 16(4), 295–321 (1998)
5. Shariff, A.R.: Natural-language spatial relations: Metric refinements of topological properties. Phd thesis, University of Maine (1996)
6. Gärdenfors, P.: Conceptual spaces: The geometry of thought. MIT Press, Cambridge (2000)
7. Gärdenfors, P.: Conceptual spaces as a framework for knowledge representation. Mind and Matter 2(2), 9–27 (2004)
8. Gärdenfors, P.: How to make the semantic web more semantic. In: Varzi, A., Vieu, L. (eds.) Formal Ontology in Information Systems, Proceedings of the Third International Conference (FOIS 2004). Frontiers in Artificial Intelligence and Applications, vol. 114, pp. 153–164. IOS Press, Amsterdam, NL (2004)

9. Egenhofer, M., Franzosa, R.: Point-set topological spatial relations. International Journal of Geographical Information Systems 5(2), 161–174 (1991)
10. Hernández, D.: Relative representation of spatial knowledge: The 2-d case. In: Mark, D.M., Frank, A.U. (eds.) Cognitive and Linguistic Aspects of Geographic Space, pp. 373–385. Kluwer Academic Publishers, Dordrecht, The Netherlands (1991)
11. Freksa, C.: Using orientation information for qualitative spatial reasoning. In: Frank, A.U., Formentini, U., Campari, I. (eds.) Theories and Methods of Spatio-Temporal Reasoning in Geographic Space. LNCS, vol. 639, pp. 162–178. Springer, Heidelberg (1992)
12. Moratz, R., Tenbrink, T.: Spatial reference in linguistic human-robot interaction: Iterative, empirically supported development of a model of projective relations. Spatial Cognition and Computation 6(1), 63–106 (2006)
13. Egenhofer, M.J., Mark, D.M.: Naive geography. In: Kuhn, W., Frank, A.U. (eds.) COSIT 1995. LNCS, vol. 988, pp. 1–15. Springer, Heidelberg (1995)
14. Egenhofer, M.J., Sharma, J., Mark, D.: A critical comparison of the 4-intersection and 9-intersection models for spatial relations: Formal analysis. In: McMaster, R., Armstrong, M. (eds.) Autocarto 11, Minneapolis, Minnesota, USA, vol. 1(11) (1993)
15. Egenhofer, M.J., Mark, D.M.: Modeling conceptual neighborhoods of topological line-region relations. International Journal of Geographical Information Systems 9(5), 555–565 (1995)
16. Mark, D., Egenhofer, M.J.: Topology of prototypical spatial relations between lines and regions in English and Spanish. In: Proceedings of the Twelfth International Symposium on Computer- Assisted Cartography, Charlotte, North Carolina, vol. 4, pp. 245–254 (1995)
17. Coventry, K.R., Garrod, S.C.: Saying, seeing, and acting: The psychological semantics of spatial prepositions. Essays in Cognitive Psychology. Psychology Press, Hove, UK (2004)
18. Attneave, F.: Dimensions of similarity. American Journal of Psychology 63, 516–556 (1950)
19. Melara, R.D., Marks, L.E., Lesko, K.E.: Optional processes in similarity judgments. Perception Psychophysics 51(2), 123–133 (1992)
20. Johannesson, M.: Combining integral and separable subspaces. In: Moore, J.D., Stenning, K. (eds.) Proceedings of the Twenty-Third Annual Conference of the Cognitive Science Society, pp. 447–452. Lawrence Erlbaum, Mahwah (2001)
21. Johannesson, M.: Geometric models of similarity. PhD thesis, Lund University (2002)
22. Egenhofer, M.J., Al-Taha, K.K.: Reasoning about gradual changes of topological relationships. In: Frank, A.U., Campari, I., Formentini, U. (eds.) COSIT 1997. LNCS, vol. 1329, pp. 196–219. Springer, Heidelberg (1992)
23. Mark, D.M., Egenhofer, M.J.: Modeling spatial relations between lines and regions: Combining formal mathematical models and human subject testing. Cartography and Geographic Information Systems 21(3), 195–212 (1994)
24. Schwering, A.: Semantic similarity between natural language spatial relations. In: Symposium Spatial Reasoning and Communication on the Conference on Artificial Intelligence and Simulation of Behaviour (AISB07), Newcastle upon Tyne (2007)
25. Kuhn, W.: Geospatial semantics: Why, of what, and how? Journal of Data Semantics 3, 1–24 (2005)
26. Rada, R., Mili, H., Bicknell, E., Blettner, M.: Development and application of a metric on semantic nets. IEEE Transactions on systems, man, and cybernetics 19(1), 17–30 (1989)
27. Suppes, P., Krantz, D.M., Luce, R.D., Tversky, A.: Foundations of measurement - geometrical, threshold, and probabilistic representations, vol. 2. Academic Press, San Diego, California, USA (1989)
28. Devore, J., Peck, R.: Statistics - The exploration and analysis of data. 4th edn. Duxbury, Pacific Grove, CA (2001)
29. Schwering, A., Raubal, M.: Spatial relations for semantic similarity measurement. In: Akoka, J., Liddle, S.W., Song, I.-Y., Bertolotto, M., Comyn-Wattiau, I., van den Heuvel, W.-J., Kolp,

M., Trujillo, J., Kop, C., Mayr, H.C. (eds.) Perspectives in Conceptual Modeling. LNCS, vol. 3770, pp. 259–269. Springer, Heidelberg (2005)

30. Schwering, A.: Hybrid model for semantic similarity measurement. In: Meersman, R., Tari, Z. (eds.) On the Move to Meaningful Internet Systems 2005: CoopIS, DOA, and ODBASE. LNCS, vol. 3761, pp. 1449–1465. Springer, Heidelberg (2005)

31. Schwering, A.: Semantic similarity measurement including spatial relations for semantic information retrieval of geo-spatial data. IfGIprints. Verlag Natur & Wissenschaft, Solingen, Germany (2007)

32. Medin, D.L., Goldstone, R.L., Gentner, D.: Respects for similarity. Psychological Review 100(2), 254–278 (1993)

33. Goldstone, R.L.: Mainstream and avant-garde similarity. Psychologica Belgica 35, 145–165 (1995)

34. Rosch, E.: Cognitive reference points. Cognitive Psychology 7, 532–547 (1975)

35. Rosch, E.: Cognitive representations of semantic categories. Journal of Experimental Psychology: General 104(3), 192–233 (1975)

36. Rosch, E.: Principles of categorization. In: Rosch, E., Lloyd, B. (eds.) Cognition and Categorization, pp. 27–48. Lawrence Erlbaum Associates, Hillsdale, New Jersey (1978)

37. Tversky, A.: Features of similarity. Psychological Review 84(4), 327–352 (1977)

38. Cohen, J.: Statistical power analysis for the behavioral sciences. Erlbaum, Hillsdale, New York (1988)

39. Cohen, J.: A power of primer. Psychological Bulletin 112, 155–159 (1992)

40. Nedas, K., Egenhofer, M., Wilmsen, D.: Metric details of topological line-line relations. International Journal of Geographical Information Science 21(1), 21–48 (2007)

41. Xu, J.: Formalizing natural-language spatial relations between linear objects with topological and metric properties. International Journal of Geographical Information Science 21(4), 377–395 (2007)

Affordance-Based Similarity Measurement
for Entity Types

Krzysztof Janowicz[1] and Martin Raubal[2]

[1] Institute for Geoinformatics, University of Muenster, Germany
janowicz@uni-muenster.de
[2] Department of Geography, University of California at Santa Barbara
raubal@geog.ucsb.edu

Abstract. When interacting with the environment subjects tend to classify entities with respect to the functionalities they offer for solving specific tasks. The theory of affordances accounts for this agent-environment interaction, while similarity allows for measuring resemblances among entities and entity types. Most similarity measures separate the similarity estimations from the context—the agents, their tasks and environment—and focus on structural and static descriptions of the compared entities and types. This paper argues that an affordance-based representation of the context in which similarity is measured, makes the estimations situation-aware and therefore improves their quality. It also leads to a better understanding of how unfamiliar entities are grouped together to ad-hoc categories, which has not been explained in terms of similarity yet. We propose that types of entities are the more similar the more common functionalities their instances afford an agent. This paper presents a framework for representing affordances, which allows determining similarity between them. The approach is demonstrated through a planning task.

1 Introduction

Understanding the interaction between agents and their environment is a fundamental research goal within cognitive science. The theory of affordances [1] describes how agents perceive action possibilities of entities within their environment, arising from both the physical structures of the entities and the agent. A major problem with this theory is that it does not account for cognitive and social processes. As argued by Chaigneau and Barsalou [2], function plays a prominent role in categorization, which also emphasizes the importance of affordances as part of human perception and cognition. The process of categorization itself can be explained in terms of similarity. With the exception of alignment models such as SIAM [3] most similarity theories assume that similarity is a static and decontextualized process. This contradicts the definition of affordances as inseparable constructs of agent and environment where entities are grouped around functionality. Similarity measures, such as MDSM [4] and SIM-DL [5], are context-aware, but at the same time reduce the notion of context to the domain of application, i.e., an unstructured set of entities or entity classes.

Measuring entity (type) similarity with respect to affordances requires their representation. The theory presented in this paper specifies such representation based

S. Winter et al. (Eds.): COSIT 2007, LNCS 4736, pp. 133–151, 2007.
© Springer-Verlag Berlin Heidelberg 2007

on the conceptual design depicted in [6]. It utilizes an extended affordance theory [7], thus incorporating social-institutional constraints and goal definitions. The paper provides a context-aware similarity measure based on the hypothesis that entity types are the more similar the more common affordances their instances offer a specific user for solving a particular task. Hence the presented measurement theory offers a computational approach towards understanding how cognitive processes and social-institutional aspects interact in categorization. This view strongly correlates with the three main components of geographic information science, i.e., cognitive, computational, and social [8]. The presented framework provides additional insights into the grouping of unfamiliar entities to ad-hoc categories [9].

Starting with a review of related work on affordances and similarity measurement, the paper then introduces a formal representation of the extended affordance theory, which supports the separation of perceiving affordances from their execution [6]. Based on this representation similarity measures are developed that determine the similarity between entity types by comparing affordances. For that reason we decompose the language describing the affordances and transform it to conceptual spaces that support similarity measurement by providing a metric [10]. This leads to a representation of functions and actions in conceptual spaces [11]. The approach is demonstrated using a scenario from psychology and AI based planning, where an agent needs to change a light bulb, involving reasoning about what entities offer support for reaching the ceiling. The presented theory focuses on entity types but allows for modification to work on the level of individual entities as well.

2   Related Work

This section introduces the notion of affordance, its extended theory, and a functional representation framework. We then provide an overview of semantic similarity theories—focusing on those related to GIScience—and AI planning.

2.1 Gibson's Theory of Affordances

The term *affordance* was originally introduced by James J. Gibson who investigated how people visually perceive their environment [1]. His theory is based on ecological psychology, which advocates that knowing is a direct process: The perceptual system extracts invariants embodying the ecologically significant properties of the perceiver's world. Animal and environment form an inseparable pair and this complementarity is implied by Gibson's use of ecological physics.

Affordances must be described relative to the agent. For example, a chair's affordance "to sit" results from a bundle of attributes, such as "flat and hard surface" and "height", many of which are relative to the size of an individual. Later work with affordances builds on this *agent-environment mutuality* [12]. According to Zaff [13], affordances are measurable aspects of the environment, but only to be measured in relation to the individual. It is particularly important to understand the action relevant properties of the environment in terms of values intrinsic to the agent. Warren [14] demonstrates that the "climbability" affordance of stairs is more effectively specified as a ratio of riser height to leg length.

Several researchers have believed that Gibson's theory is insufficient to explain perception because it neglects processes of cognition. His account deals only with individual phenomena, but ignores categories of phenomena [15]. According to Eco [16], Gibson's theory of perception needs to be supplemented by the notion of *perceptual judgments*, i.e., by applying a cognitive type and integrating stimuli with previous knowledge.

Norman [17] investigated affordances of everyday things, such as doors, telephones, and radios, and argued that they provide strong clues to their operation. He recast affordances as the results from the mental interpretation of things, based on people's past knowledge and experiences, which are applied to the perception of these things. Gaver [18] stated that a person's culture, social setting, experience, and intentions also determine her perception of affordances. Affordances, therefore, play a key role in an experiential view of space [19, 20], because they offer a user centered perspective. Similarly, Rasmussen and Pejtersen [21] pointed out that modeling the physical aspects of the environment provides only a part of the picture. The overall framework must represent the mental strategies and capabilities of the agents, the tasks involved, and the material properties of the environment.

2.2 Extended Theory of Affordances

In this work we use an extended theory of affordances within a functional model, which supplements Gibson's theory of perception with elements of cognition, situational aspects, and social constraints. This extended theory suggests that affordances belong to three different realms: physical, social-institutional, and mental [7]. In a similar and recent effort, the framework of distributed cognition was used to describe and explain the concept of affordance [22].

Physical affordances require bundles of physical substance properties that match the agent's capabilities and properties—and therefore its interaction possibilities. One can only place objects on stable and horizontal surfaces, one can only drink from objects that have a brim or orifice of an appropriate size, and can be manipulated, etc. Common interaction possibilities are grasping things of a certain size with one's hands or walking on different surfaces. Physical affordances such as the sitability affordance of a chair depend on body-scaled ratios, e.g., doorways afford going through if the agent fits through the opening.

It is often not sufficient to derive affordances from physical properties alone because people act in environments and contexts with social and institutional rules [23]. The utilization of perceived affordances, although physically possible, is frequently socially unacceptable or even illegal. The physical properties of an open entrance to a subway station afford for a person to move through. In the context of public transportation regulations it affords moving through only when the person has a valid ticket. The physical properties of a highway afford for a person to drive her car as fast as possible. In the context of a specific traffic code it affords driving only as fast as allowed by the speed limit. Situations such as these include both physical constraints and social forces. Furthermore, the whole realm of social interaction between people is based on *social-institutional affordances*: Other people afford talking to, asking, and behaving in a certain way.

Physical and social-institutional affordances are the sources of *mental affordances*. During the performance of a task a person finds herself in different situations, where she perceives various physical and social-institutional affordances. For example, a public transportation terminal affords for a person to enter different buses and trains. It also affords to buy tickets or make a phone call. A path affords remembering and selecting, a decision point affords orienting and deciding, etc. In general, such situations offer for the person the mental affordance of deciding which of the perceived affordances to utilize according to her goal.

2.3 Functional Representation of Affordances

Our conceptual framework of affordances uses an adjusted version of the *HIPE theory of function*, which explains how functional knowledge is represented and processed [24]. This theory explains people's knowledge about function by integrating four types of conceptual knowledge: History, Intentional perspective, Physical environment, and Event sequences. Functional knowledge emerges during mental simulations of events based on these domains. The HIPE theory is well suited to the formalization of affordances because of their functional character [6]. Similar to functions, affordances are complex relational constructs, which depend on the agent, its goal and personal history, and the setting. The HIPE theory allows for representing what causes an affordance and therefore supports reasoning about affordances. More specifically, it is possible to specify which components are necessary to produce a specific affordance for a particular agent.

Figure 1 demonstrates the conceptual framework of the relation between the three affordance categories presented in section 2.2 during the process of an agent performing a task. The agent is represented through its physical structure (*PS*), spatial and cognitive capabilities (*Cap*), and a goal (*G*). Physical affordances (*Paff*) for the agent result from invariant compounds (*Comp*)—unique combinations of physical, chemical, and geometrical properties, which together form a physical structure—and the physical structure of the agent[1]. This corresponds to Gibson's original concept of affordance: a specific combination of (physical) properties of an environment taken with reference to an observer.

Social-institutional affordances (*SIaff*) are created through the imposition of social and institutional constraints on physical affordances—when physical affordances are perceived in a social-institutional context *Cont (SI)*. While performing a task the agent perceives various physical and social-institutional affordances within a spatio-temporal environment represented through *Env (S,T)*. This corresponds to HIPE's notion of a physical system and allows for localizing the perception of affordances in space and time.

Mental affordances (*Maff*) arise for the agent when perceiving a set of physical and social-institutional affordances in an environment at a specific location and time. Affordances offer possibilities for action as well as possibilities for the agent to reason about them and decide whether to utilize them or not, i.e., mental affordances. The agent needs to perform an internal operation *Op (Int)* to utilize a mental affordance. Internal operations are carried out on the agent's beliefs (including its

---

[1] The arrows in Figure 1 represent a function that maps *Comp* and *Agent* to *Paff*.

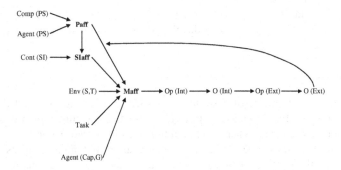

Fig. 1. Functional representation of affordances for an agent

history and experiences) and lead to an internal outcome *O (Int)*. In order to transfer such outcome to the world, the agent has to perform an external operation *Op (Ext)*, which then leads to an external outcome *O (Ext)*, i.e., some change of the external world. This external change, in turn, leads to new physical affordances, situated in social-institutional and spatio-temporal contexts.

2.4 Semantic Similarity Measurement

The notion of similarity originated in psychology and was established to determine why and how entities are grouped to categories, and why some categories are comparable to each other while others are not [25, 26]. The main challenge with respect to semantic similarity measurement is the comparison of meanings. A language has to be specified to express the nature of entities and metrics are needed to determine how (conceptually) close the compared entities are. While entities can be expressed in terms of attributes, the representation of entity types is more complex. Depending on the expressivity of the representation, language types are specified as sets of features, dimensions in a multidimensional space, or formal restrictions specified on sets using various kinds of description logics. Whereas some representation languages have an underlying formal semantics (e.g., model theory), the grounding of several representation languages remains on the level of an informal description. Because similarity is measured between entity types which are representations of concepts in human minds, it depends on what is said (in terms of computational representation) about entity types. This again is connected to the chosen representation language, leading to the fact that most similarity measures cannot be compared. Beside the question of representation, context is another major challenge for similarity assessments. In many cases meaningful notions of similarity cannot be determined without defining in respect to what similarity is measured [26, 27].

Similarity has been widely applied within GIScience. Based on Tversky's feature model [28], Rodríguez and Egenhofer [4] developed an extended model called Matching Distance Similarity Measure (MDSM) that supports a basic context theory, automatically determined weights, and asymmetry. Raubal and Schwering [29, 30] used conceptual spaces [10] to implement models based on distance measures within geometric space. The SIM-DL measure [5] was developed to close the gap between geo-ontologies described through various kinds of description logics, and similarity measures that had not been able to handle the expressivity of such languages. Various similarity theories [31, 32] have been developed to determine the similarity of spatial scenes.

2.5 AI Planning

Planning is the development of a strategy for solving a certain task and therefore a precondition for intelligent behavior. In terms of artificial agents, a plan is a chain of actions, or action sequence, where each action to be performed depends on some pre-conditions, i.e., a certain state of the world. Each action potentially causes effects or post-conditions that affect or trigger subsequent actions in the chain. The plan terminates when the intended goal is reached. A planner in Artificial Intelligence (AI) takes therefore three input variables: a representation of the initial state of the world, a representation of the intended outcome (goal), and a set of possible actions to be performed to reach the goal. Formally, a plan can be regarded as a triple $\langle O, I, A\langle p,q\rangle\rangle$ [33, 34] where O is the intended outcome, I the initial state of the world, and A a set of actions—each defined via pre- and post-conditions p, q. However, after executing actions the state of the world is changed, which impacts the future plan, therefore making planning a non-linear process. One distinguishes between offline and online planning. Offline planning separates the creation of the strategy and its execution into two distinct phases; this requires a stable and known environment. In contrast, online planning is suitable for unknown and dynamic environments where a pre-given set of behavioral rules and models cannot be determined. One of the main challenges within dynamic environments is that one can neither assume complete knowledge of the environment nor the availability of entities (as part of such environment) supporting certain actions.

3 Use Case

Contrary to classical planning, our vision of an affordance-based and similarity-driven planning service executes as follows: The agent determines an intended outcome (goal). Next, the agent selects a possible affordance descriptor (see section 4) from its internal knowledge base that either leads to the intended goal or is part of the chain towards it. The agent then needs to verify whether an entity of the type specified in the affordance descriptor is available within its immediate environment and if not, whether it can be substituted by an entity of a similar type. After that the agent can execute the (similar) action specified by the selected affordance. Thus, reaching a new state, the agent selects the next outcome to be reached towards the final goal and again chooses an appropriate affordance descriptor. The process terminates when the final outcome is reached or no supporting entity can be detected within the current environment.

To illustrate this approach, we introduce a use case, which is derived from the literature on ad-hoc categories [9, 35]. We assume that an agent needs to change a light bulb in an office room. Before doing so, the agent has to fetch an additional entity that raises it up to a certain level in order to reach the ceiling and change the bulb. In terms of affordances, the task is two-fold: the agent must find an entity that affords standing on and has to be movable to be carried or pushed to the required position. If a single entity lacks sufficient height, an additional affordance will come into play, namely that of being stackable. As illustrated in Figure 2 the office room contains several candidate entities, such as a desk, a chair, and books, which could be utilized to fulfill the task. Some entities are movable, stackable, and offer support for standing on them at the same time, while others fulfill these requirements only partially. We assume that the agent has the necessary capabilities to categorize entities accordingly. If an entity is of a certain type, it can be manipulated as specified in the affordance descriptor (section 4) stored in the agent's knowledge base.

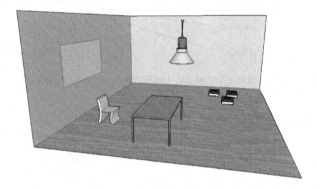

Fig. 2. Candidate entities needed to change a light bulb

Figure 3 shows a simplified representation of the 'changing a light bulb' scenario using the framework discussed in section 2.3. The agent perceives affordances involving the entities *desk, chair,* and *book* in the office environment, where the agent is spatio-temporally located. The task is changing the light bulb, which involves a series of sub goals. The physical structure of the desk affords the agent to move it, stand on it, stack it, and to climb it. The *Paff* of moving the desk is constrained through the following social context (or rule): Moving the desk will lead to scratches on the floor, therefore, one should not slide the desk across the floor, resulting in the *SIaff*(not move).

The chair affords the agent to move, stack, climb, and sit on it. Books in the office afford the agent to move and stack them. Notice that all of this knowledge, which had previously been acquired by the agent, is represented for the entity types desk, chair, and book. Perceived instances in the agent's environment are categorized with respect to these types and therefore the agent can utilize knowledge associated with them. This process is similar to a perceptual cycle [36] where a schema directs exploration and sampled objects modify the schema.

All of these functions result in the top-level *Maff* for the agent, namely to evaluate whether the task of changing the light bulb can be fulfilled with the given constraints represented through the functions. More formally, the (interconnected) sets of physical and social-institutional affordances at a given point in space and time result in a set of mental affordances for the agent: $\{Paff, SIaff\}_{Env(S,T)} => \{Maff\}$. *Maffs* are therefore higher-order functions because *Paffs* and *SIaffs* are functions themselves.

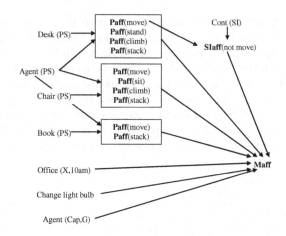

Fig. 3. Functional affordance-based representation of use case

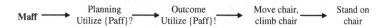

Fig. 4. Functional activity process for the agent

The second part of the process is represented in Figure 4. The agent performs internal operations (within the planning process), deciding whether the task can be performed based on the given functions. This is also where the agent performs affordance-based similarity measurement to find out which entities can be used for the task of changing the light bulb. The outcome of this operation is the decision to utilize a set of affordances. Moving and climbing the chair are external operations, which subsequently lead to the external outcome that the agent stands on the chair and can change the light bulb. This is an abstraction from more complex online planning tasks, as we assume that the same entity can be used for each step (moving, climbing, and standing).

4 Representation of Affordances

This section introduces a representation of affordances built upon the extended theory described in section 2.2, and provides the groundwork for the similarity measurements established in section 5.

Based on previous definitions [37, 38], we specify an affordance A as a triple $(O, E, \{AC\})$. The outcome O is the change of world state after execution of the actions AC with respect to manipulated entities of type E. The same affordance can be realized by several actions—each described by physical PH and social-institutional SI constraints, i.e., pre-conditions. AC is therefore defined as a set of actions $\{ac_1(ph_1,si_1),..., ac_n(ph_n,si_n)\}$. Constraints are tied to a certain action with respect to an entity (of a given type), while the outcome is equal for all actions defined for the affordance A. Therefore the outcome can also be regarded as the post-condition of all actions of *A*. For the representation of outcome, physical, and social-institutional constraints we employ unary and binary predicates, and apply a restricted kind of predicate logic. The predicates (predicate variables) are part of the agent's internal knowledge base, i.e., no semantic problems arise from the question of what a predicate, such as hasPosition, means to the agent. This knowledge base also contains information about the inter-relation between predicates, such as hierarchies or (spatial and temporal) neighborhood models. The predicate logic used to specify actions and outcome can be regarded as a subset of first order logic. Valid operators and quantifiers are:

- Operators (constructors): logical *not* ($\neg$), *and* ($\wedge$), and *implication* ($\rightarrow$)
- Quantifiers: $\forall$, $\exists$, and $\exists!$ *(exactly one)*
- Arithmetic operators: $<, >, \leq, \geq, \neq$ and $=$ applied to P^+

In addition, we use some abbreviations to improve readability. This way, some necessary assumptions are made about the later transformation to conceptual spaces without the need of going into detail about the relation between logics and conceptual spaces (see also [39]), and the problem of mapping.

- $(\forall e \exists! r\ P(e,r) \wedge E(e) \wedge R(r) \wedge R(x) \wedge (r \leq x)) \wedge ... \rightarrow Q(e)$ is abbreviated by Q: $P(e, \leq x) \wedge ...$; where P,Q are predicates, e is an instance of E, and r, x are real numbers. This allows for statements such as the one shown for carry(ability) below.
- The same way $\forall e \exists! f\ P(e,f) \wedge E(e) \wedge F(f)$ is abbreviated by P(e,F(f)); where f is an instance of F such as in hasPosition(e,Position(p)). This states that an entity needs to have a position from where it is moved to another one. The same kind of statement can be made by adding negation, such as on(e, $\neg$Parquet(p)).

The perception and execution of affordances is modeled in terms of mapping statements, i.e., predicates connected via logical *and*, to Boolean values. Physical constraints describe statements about the physical properties of entities that need to be *true* before the specified action can be performed with respect to entities of the type *E*. Social-institutional constraints specify statements about social aspects regarding the interaction with entities (again abstracted to type level) that need to be *true* before the specified action can be performed. Both types of constraints are specific with respect to the agent perceiving the affordance. We claim that entity types *E* are specified only via what their instances afford a given agent and are the more similar the more common or similar affordances they support.

Summarizing, a certain type of entities affords something to a specific agent if the agent can perform actions on such entities; i.e., *A* is *true* with respect to the agent if at least one of the actions of *AC* is performable (its *PH* and *SI* constraints are satisfied, i.e., *true*) and after realizing the affordance *A* the state of the world changes as specified in *O* (i.e., if the predicates specified for the outcome are *true*). Consequently, although an agent can perceive the affordance of something to be moveable, it could fail to move the entity because of external factors not explicitly stated in the action constraints. This reflects both the separation of internal operation and outcome, and external operation and outcome described in section 2.3. Note that because we assume the agent to be fixed it is not part of the affordance definition itself but its physical and social-institutional context is defined via constraints on the actions.

In terms of the light bulb scenario, an affordance descriptor for moveability of desks is specified as follows:

```
Move-ability (
  Outcome (O): hasPosition(e, Position(y)) ∧ y ≠x
  Entity Type (E): Desk
  Actions (AC):
   carry(PH:hasPosition(e, Position(x))∧ WeightKg(e, ≤20)∧
LengthCm(e, ≤100)∧ …)
   push(PH:hasPosition(e, Position(x)))∧ WeightKg(e, ≤100)∧ …
        SI: on(e, ¬Parquet(p))
  …)
```

For our agent desks afford moving if they can be either pushed or carried from a position x to another position (specified as a position *not* being the start location x). Due to the agent's physical capabilities it is able to carry desks with a weight below 21kg and a length of up to 100cm. Pushing the desk is an alternative action and could be even performed with heavier desks (up to 100kg). Pushing though may damage floors. Therefore an entity of type desk is moveable if it either weighs less than 21kg and is not longer than 100cm, or weighs less than 100kg but the supporting floor is resistant to damage caused by sliding heavy entities across it (parquet is not). In this example the restrictions are mostly based on the physical capabilities of the agent and the structure of solid entities. In other ad-hoc categories such as 'things to extinguish a fire' the candidate entities (e.g., water, sand, or a blanket) also differ in their consistency. It is still possible that the agent perceives the affordance and tries to carry the desk but during execution recognizes that for external reasons the desk cannot be moved (e.g., because the desk is mounted to the floor). Depending on these restrictions and the abilities of the agent some desks may not be movable at all. The question whether such entities should still be categorized as desks is not discussed here but will be taken up again in the future work section.

This example also points to the connection between affordances. Depending on the granularity, one may argue that there are explicit pushability and carryability affordances that can be defined as sub-affordances of moveability. In addition, the social-institutional restriction introduced for pushing desks could also be perceived as an affordance (damageability). Damageability of floors is then defined via an outcome O specifying that the state (in this case, the surface) of the floors is changed by several actions AC.

5 Affordance-Based Similarity Measurement

This section describes how similarity between entity types can be measured based on the assumption that types of entities are the more similar, the more common functionalities their instances afford to a given agent. We therefore introduce a framework that specifies what parts of the affordances are compared and how. After defining what makes affordances similar, we use the similarity values determined between affordances to develop a similarity measure for entity types. As we cannot directly handle the expressivity of the affordance representation with respect to similarity measurement, the affordance descriptors need to be transformed to regions within conceptual spaces. Similarity between these regions can be computed using existing measures. Overall similarity is then described as the weighted sum of the individual similarity values measured between affordance descriptors. This step is comparable to a weighted (ω) Tversky Ratio Measure such as used in MDSM. At last the same kind of measure is applied to determine entity type similarity.

5.1 Similarity Between Affordances

Each affordance is specified by the change in world state its execution causes and the actions performed on certain types of entities to achieve this outcome. As the entity types are only described in terms of what they afford, similarity between affordances depends on the action and outcome specification. To evaluate whether an affordance defined for a certain entity type is valid in the context of a specific agent and entity, predicates are resolved to Boolean values. Similarity though rests on the assumption that affordances are the more similar the more similar their descriptors are. As no metric can be defined to reason about the similarity of predicates in general, we define mappings for the predicates to quality dimensions within a conceptual space [10], hence being able to utilize a metric for comparison. Similarity measures are asymmetric, therefore the direction of the comparison must be taken into account. In the following, the index s is used for source while t determines the target, i.e., the compared-to predicate.

First we consider predicates that map entities to non-negative real numbers ($\mathbb{R}^+$). Such predicates can either describe facts about entities of the type specified in the affordance or external entity types. The predicates are transformed to dimensions and the numeric values to upper or lower bounds of the dimensions. If a predicate maps to a single value, lower and upper bounds are equal. If no lower bound is specified it is set to 0 or in the case of upper bound to infinity. Dimensions referring to the entity type E together form a conceptual space while dimensions referring to other entity types form conceptual spaces for those types[2]. This is also the reason why the action and outcome descriptors cannot be directly utilized to determine the similarity between entity types. Physical and social-institutional constraints as well as outcomes may directly refer to the specified entity type or to its environment, e.g., via external types.

In cases where predicates map between entities the transformation to dimensions is more complex. The predicates can still be represented as dimensions but only on a Boolean (i.e., nominal) scale[3]. This means that predicates p_s and q_t are equal (similarity is 1) if $p_s = q_t$, which also includes rewriting rules (such as De Morgan rules for $\wedge$ and $\vee$ as well as for $\exists$ and $\forall$) or if q_t can be inferred from p_s. These

---

[2] In such cases similarity is determined in a recursive way as entity types are again explained in terms of affordances.

[3] While this approach is used in most of the literature on similarity measures within conceptual spaces, one may argue that such mapping violates the notion of dimensions and regions.

inference mechanisms include standard inference rules, such as elimination and introduction but also spatio-temporal reasoning, etc. A typical example can be constructed assuming that p_s: PO(X, Y) and q_t: ¬DC(X, Y) in terms of the Region Connection Calculus [40]. Note that the same example (and other inference rules) does not work in the opposite way, which is consistent with keeping similarity measurement asymmetric. In other cases similarity is 0.

The connection between single predicates using ∧ is preserved within the structure of conceptual spaces by the amalgamation (+) of similarity values (Equation 2). The pre-processing step of turning predicate-based descriptions to conceptual spaces is applied to any physical and social-institutional constraints of all action descriptors ac ∈ AC and the outcome descriptors O of source and target affordance A_s and A_t. This creates at least one conceptual space for each action and outcome of A_s and A_t. The process is computationally expensive, but it is static and therefore easy to cache offline. The process is depicted in Figure 5.

After transformation to conceptual spaces an alignment procedure [5, 41] must be established to determine which conceptual spaces of the descriptors of A_s and A_t are mapped for similarity measurement. Each conceptual space describes either the capabilities (in terms of physical or social-institutional constraints) of the agent to perform an action with respect to an entity type or the desired outcome for all actions specified for the affordance. From this the following alignment rules can be derived.

- Spaces representing the outcome aspects of A_s are mapped to such from A_t.
- Spaces representing physical aspects or social-institutional aspects of an action from A_s are mapped to such from A_t.
- Space specifying dimensions for different types of entities cannot be mapped.
- Physical and social-institutional aspects of an action from A_s are jointly mapped to their counterparts of an action from A_t and not separately to different actions.
- Action names are unique within the agent's knowledge base, therefore action descriptors of A_s and A_t are mapped if both share the same name, or if both are situated within the same hierarchy or neighborhood model. In the latter case, the maximal possible similarity is decreased to the similarity within the hierarchy or neighborhood [5, 42].
- If no counterpart for a conceptual space representing constraints of an action can be found, the similarity value is 0.

Finally, after constructing conceptual spaces for the affordance descriptors, semantic similarity between conceptual regions can be measured. Semantic distances are calculated based on the standardized differences of the values for each quality dimension. The final values are normalized by the number of dimensions used in the calculation. This way, a semantic distance function between two conceptual regions can be established [43]. Here, quality dimensions are represented on either Boolean or interval scale. For Boolean dimensions, the values can take 1 or 0, therefore semantic similarity between two conceptual regions for each Boolean dimension is either 1 (completely similar) or 0 (completely dissimilar). In order to calculate asymmetric similarity for two intervals we consider a simplified version of the *line alongness*

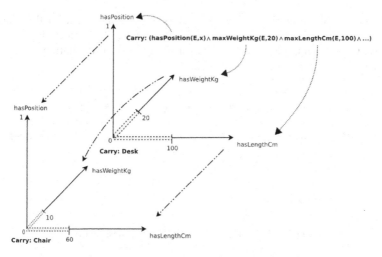

Fig. 5. Creating and mapping dimensions from predicates

ratio for topological line-line relations [44]. Positive similarity values of intervals result only if there exists at least a partial overlap between intervals. If such overlap does not exist, the similarity evaluates to 0, i.e., complete dissimilarity. This makes sense for the described scenario, because minimum and maximum values for dimensions are hard constraints for the agent. The calculation of interval similarity is given in Equation 1.

$$Sim_{Int}(I_i,I_j) = length(I_i \cap I_j)/length(I_i) \text{ with } i,j \in \{1,2\}, i \neq j \qquad (1)$$

The final measure for semantic similarity between two conceptual regions X and Y is depicted in Equation 2, where *Scale* refers to either Boolean or interval, S_i and S_j are the respective values of a quality dimension to be compared, and *n* refers to the number of dimensions.

$$Sim_{CS}(X,Y) = \Sigma(Sim_{Scale}(S_i,S_j)) / n \qquad (2)$$

After being able to determine the similarity within conceptual spaces, the similarity between affordances is defined as depicted in Equation 3.

$$Sim_A(A_s,A_t) = \omega_{ac} * {}^1/_n \Sigma sim_{AC} + \omega_o * sim_O; \text{ where } \Sigma \omega = 1 \qquad (3)$$

Sim_A is specified as the weighted (ω) and normalized sum of similarities for compared actions (sim_{AC}) and outcomes (sim_O) expressed as similarity within conceptual spaces. While sim_O is directly determined from the outcome conceptual space of A_s and A_t, the similarity between actions is determined via the weighted sum of the similarities for the physical and social-institutional aspects (Equation 4). The number of compared actions *n* in Equation 3 represents alignable actions, not the total number of all available actions.

$$Sim_{AC} = \omega_{ph} * sim_{ph} + \omega_{si} * sim_{si}; \text{ where } \Sigma \omega = 1 \qquad (4)$$

Summarizing, as depicted in Figure 6, affordances can be compared by mapping their descriptors to conceptual spaces and expressing predicates as dimensions of such spaces. The overall similarity is then defined as the weighted sum for the individual similarities computed for actions and outcomes, where the former depend on the similarity values computed for their physical and social-institutional constraints (regarding a certain type of entity and agent).

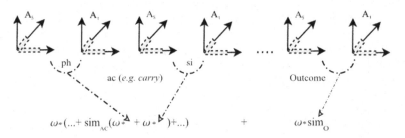

Fig. 6. Comparing affordances via their descriptors

5.2 Similarity Between Entity Types

Based on our core assumption that entity types are the more similar the more common functionalities their instances afford an agent (in solving a given task), similarity between types can be determined as depicted in Equation 5.

$$Sim_E(E_s, E_t) = {}^1/_n * \Sigma Sim_A \tag{5}$$

Again, this raises the question which affordances should be compared. The theory presented in this paper is driven by the idea of solving tasks via the agent's interaction with its environment and hence focus on achieving goals. Therefore, affordances are selected for comparison with respect to their outcome, i.e., the affordances A_s and A_t with the highest outcome similarity (sim_O) are matched.

5.3 Weights

Weighting is a useful method to adjust the similarity measurement process to better fit a given task. However, skeptical readers may argue that weights can be used to manually tweak the numbers until they fit the expected results. This section briefly discusses the role of weights in the presented framework to counteract such argumentation. The weights introduced in section 5.1 are comparable to the commonality and variability weights introduced as part of the context theory of MDSM [4] and can therefore be automatically determined without the need of manual adjustment. While MDSM distinguishes between attributes, functions, and parts to describe and weight entity types, our approach can be weighted with respect to physical and social-institutional constraints as well as outcomes. By using such weights the importance of these aspects can be automatically adjusted to improve the quality of the similarity estimations. Commonality increases the similarity of such kinds of descriptors that are common for most compared entity types and tends to increase overall similarity. In contrast, variability strengthens the weight of such aspects that are unique for certain types of entities and tends to decrease overall similarity. In terms of the light bulb use case, social-institutional aspects may become more relevant for the measured similarity if the physical constraints of the entities are more or less the same (or vice versa).

The proposed weights can be used to define aspects as irrelevant for a certain task (by setting the weight to 0). If a task is of fundamental importance—as can be imagined for the wheelchair example described in [6]—it may be reasonable to ignore social-institutional constraints and focus on the outcome aspects. In other words, the proposed weights can be either automatically determined or used as exclusion factors depending on the task. In both cases the weights do not require manual pre-settings.

6 Application to Use Case

This section applies the presented framework to the light bulb use case. After perceiving the available entities in its environment the agent recognizes that it is not allowed to slide the desk towards the light bulb (modeled as social-institutional constraint). The agent's physical capabilities prevent it from carrying the desk, therefore other entities must be used to perform the task. From previous interaction with its environment the agent knows several facts about certain types of entities. This information is stored as affordance descriptors in the agent's knowledge base. To find out whether other entities (E_t) can be utilized, the agent compares the affordance descriptors (relevant for performing the task) of desks (E_s) with those of books and chairs (figure 3).

As argued in section 4 we assume that the physical constraints on carry and push depend on the abilities of the agent and the structure of the moved entities, leaving other aspects, such as texture, shape, and minimal size (which affect graspability) aside. The following specification for moveability of books does not state that the agent is physically unable to carry heavier books, but represents its current knowledge about the interaction with books[4].

```
Move-ability (
  Outcome (O): hasPosition(e, Position(y)) ∧ y ≠ x
  Entity Type (E): Book
  Actions (AC):
   carry(PH:hasPosition(e, Position(x))∧ WeightKg(e, ≤3)∧
LengthCm(e, ≤30)∧ …)
   push(PH:hasPosition(e, Position(x)))∧ WeightKg(e, ≤ 3)∧ …
…)
```

Contrary to desks, social-institutional constraints are not defined for moving books. The comparison between moveability of desks and books yields a similarity of 0.68. In the case of chairs, where maximal weight and length were set to 10kg respectively 60cm, the resulting similarity is 0.86. This reflects the fact that, with respect to weight and length, the experience of moving desks and chairs is more similar than between desks and books.

---

[4] Instead of assuming the previously acquired knowledge as an upper bound, one may also argue for positive infinity as an upper bound.

The same comparison is also computed for standability, stackability, and climbability, and finally leads to a similarity of 0.36 between desks and books and 0.64 between desks and chairs. Due to the situated nature of similarity and categorization the results cannot be used to argue that chairs or books are similar to desks in general [27]. However, to fulfill the given task, i.e., to change the light bulb the agent can conclude from previously acquired knowledge that chairs are possible candidates (internal operation). While trying to utilize the individual chair (external operation) the agent may fail because—contrary to other chairs—the available chair might not have sufficient stability, and could then utilize the books. If the chair is suitable to solve the task, the agent adds a new affordance (restricted by the physical properties of the used chair) about chairs to its knowledge base. In case of the books, standability and climbability are added. This relates to the fact that humans cannot perceive all information about the physical properties of a certain entity and therefore reason on a category level (based on previous knowledge). This post-processing of entity types can be regarded as a learning process.

7   Conclusions and Future Work

The presented methodology provides a framework for the conceptual affordance representation discussed in [6] and specifies how to measure similarity between affordances and entity types. The formalization captures important aspects of the conceptual design, such as the distinction between physical and social-institutional constraints. Via the outcome specification our approach is able to distinguish between the perception of an affordance and its execution. Similarity measurement is not a static procedure but modeled as a situated process [26, 35] within a context formed by the actor, task, and environment. On the one hand this provides insight into similarity-based categorization of unfamiliar entities (entity types) such as for ad-hoc categories [9]. On the other hand it allows for similarity-based reasoning and planning.

Further work should focus on providing a detailed formal system underlying the presented affordance theory. The important question is not directed to whether a more expressive language can improve the computational representation of affordances but whether such representation still allows for similarity measurement. In this paper simple predicates were transformed to regions in conceptual spaces to determine their similarity. One obvious problem thereby is that all dimensions are regarded as independent from each other, which is normally not the case. For instance, restrictions such as weight and length defined for the actions of the moveability affordance influence each other. In addition, only such predicates were allowed that could be mapped to the conceptual space representation. To overcome these restrictions and allow for more complex logical statements within the affordance definition, one could consider the integration of approaches such as the conceptual spaces logic [39] or similarity measurements a la SIM-DL. Analogy may be an additional tool to compare logical statements that, due to the nature of predicate logic, cannot be compared in terms of similarity. The presented theory would also benefit from the integration of measures focusing on similarity between spatial scenes [31, 32] and case-based reasoning [45]. While we have introduced AI planning to argue for the representation of affordance as a triple $(O, E, \{AC\})$, further work is required to adopt the presented methodology to real planning and learning scenarios. This will involve the interaction with several types of entities to solve a certain task, instead of trying to find one entity (of a given type) that can be used to fulfill all subtasks. Such an extended approach can then be tested with real geographic data sets.

Additional research should investigate the relationship between affordances and how affordances can be combined. While our work focuses on entity types the model can be adapted to entities as well. This raises the question whether an entity is still of some sort independent of whether it offers certain affordances (e.g., a broken cup). Considering temporal aspects, one may argue that an occupied chair (therefore not supporting sitability) should not be categorized as chair anymore. The agent should not reason about entities in terms of categories, such as desks, books, and chairs, but as members of sets determined by affordances. Each 'light bulb changing support'-entity is then defined as a member of the ad-hoc formed set 'moveable *and* standable *and* climbable (*and* stackable)'. The task of similarity is therefore to distinguish between central and radial entities.

Acknowledgments. The comments from three anonymous reviewers provided useful suggestions to improve the content and clarity of the paper. We also thank Johannes Schöning for providing the graphical illustration of the use case. Partial funding for this work came from the SimCat project granted by the German Research Foundation (DFG Ra1062/2-1).

References

1. Gibson, J.: The Theory of Affordances, in Perceiving, Acting, and Knowing - Toward an Ecological Psychology. In: Shaw, R., Bransfor, J. (eds.), pp. 67–82. Lawrence Erlbaum Associates, Hillsdale, NJ (1977)
2. Chaigneau, S., Barsalou, L.: The Role of Function in Categories. Theoria et Historia Scientiarum (forthcoming)
3. Goldstone, R., Medin, D.: Similarity, interactive activation, and mapping: An overview. In: Holyoak, K., Barnden, J. (eds.) Analogical Connections: Advances in Connectionist and Neural Computation Theory, pp. 321–362. Ablex, Norwood, NJ (1994)
4. Rodríguez, A., Egenhofer, M.: Comparing geospatial entity classes: an asymmetric and context-dependent similarity measure. International Journal of Geographical Information Science 18(3), 229–256 (2004)
5. Janowicz, K.: Sim-DL: Towards a Semantic Similarity Measurement Theory for the Description Logic ALCNR in Geographic Information Retrieval. In: Meersman, R., Tari, Z., Herrero, P. (eds.) On the Move to Meaningful Internet Systems 2006: OTM 2006 Workshops. LNCS, vol. 4278, pp. 1681–1692. Springer, Heidelberg (2006)
6. Raubal, M., Moratz, R.: A functional model for affordance-based agents. In: Hertzberg, J., Rome, E., (eds.) Affordance-Based Robot Control. Springer, Berlin (forthcoming 2007)
7. Raubal, M.: Ontology and epistemology for agent-based wayfinding simulation. International Journal of Geographical Information Science 15(7), 653–665 (2001)
8. Goodchild, M., et al.: Introduction to the Varenius project. International Journal of Geographical Information Science 13(8), 731–745 (1999)
9. Barsalou, L.: Ad hoc categories. Memory & Cognition 11, 211–227 (1983)

10. Gärdenfors, P.: Conceptual Spaces - The Geometry of Thought. Bradford Books. MIT Press, Cambridge, MA (2000)
11. Gärdenfors, P.: Representing actions and functional properties in conceptual spaces. In: Zlatev, J., et al. (eds) Body, Language and Mind. Mouton de Gruyter, Berlin (forthcoming)
12. Gibson, J.: The Ecological Approach to Visual Perception. Houghton Mifflin Company, Boston (1979)
13. Zaff, B.: Designing with Affordances in Mind. In: Flack, J., et al. (eds.) Global Perspectives on the Ecology of Human-Machine Systems, pp. 238–272. Lawrence Erlbaum Associates, Hillsdale, NJ (1995)
14. Warren, W.: Constructing an Econiche. In: Flack, J., et al. (eds.) Global Perspectives on the Ecology of Human-Machine Systems, pp. 210–237. Lawrence Erlbaum Associates, Hillsdale, NJ (1995)
15. Lakoff, G.: Women, Fire, and Dangerous Things: What Categories Reveal About the Mind. The University of Chicago Press, Chicago (1987)
16. Eco, U.: Kant and the Platypus - Essays on Language and Cognition. English translation ed. Harvest Book Harcourt (New York) (1999)
17. Norman, D.: The Design of Everyday Things. Doubleday, New York (1988)
18. Gaver, W.: Technology Affordances, in Human Factors in Computing Systems. In: CHI'91 Conference Proceedings, pp. 79–84. ACM Press, New York (1991)
19. Lakoff, G.: Cognitive Semantics. In: Eco, U., Santambrogio, M., Violi, P. (eds.) Meaning and Mental Representations, pp. 119–154. Indiana University Press, Bloomington (1988)
20. Kuhn, W.: Handling Data Spatially: Spatializing User Interfaces. In: Kraak, M., Molenaar, M. (eds.) SDH'96, Advances in GIS Research II, Proceedings. International Geographical Union, Delft 13B.1–13B.23 (1996)
21. Rasmussen, J., Pejtersen, A.: Virtual Ecology of Work. In: Flack, J., et al. (eds.) Global Perspectives on the Ecology of Human-Machine Systems, pp. 121–156. Lawrence Erlbaum, Hillsdale, NJ (1995)
22. Zhang, J., Patel, V.: Distributed cognition, representation, and affordance. Pragmatics & Cognition 14(2), 333–341 (2006)
23. Smith, B.: Les objets sociaux. Philosophiques 26(2), 315–347 (1999)
24. Barsalou, L., Sloman, S., Chaigneau, S.: The HIPE Theory of Function. In: Carlson, L., van der Zee, E. (eds.) Representing functional features for language and space: Insights from perception, categorization and development, pp. 131–147. Oxford University Press, New York (2005)
25. Goldstone, R., Son, J.: Similarity. In: Holyoak, K., Morrison, R. (eds.) Cambridge Handbook of Thinking and Reasoning, Cambridge University Press, Cambridge (2005)
26. Medin, D., Goldstone, R., Gentner, D.: Respects for Similarity. Psychological Review 100(2), 254–278 (1993)
27. Goodman, N.: Seven strictures on similarity. In: Goodman, N. (ed.) Problems and projects, pp. 437–447. Bobbs-Merrill, New York (1972)
28. Tversky, A.: Features of Similarity. Psychological Review 84(4), 327–352 (1977)
29. Raubal, M.: Formalizing Conceptual Spaces. In: Varzi, A., Vieu, L. (eds.) Formal Ontology in Information Systems, Proceedings of the Third International Conference (FOIS 2004), pp. 153–164. IOS Press, Amsterdam (2004)
30. Schwering, A., Raubal, M.: Spatial Relations for Semantic Similarity Measurement. In: Akoka, J., et al. (eds.) Perspectives in Conceptual Modeling: ER workshops CAOIS, BP-UML, CoMoGIS, eCOMO, and QoIS, Klagenfurt, Austria, October 24–28, 2005, pp. 259–269. Springer, Berlin (2005)

31. Li, B., Fonseca, F.: TDD - A Comprehensive Model for Qualitative Spatial Similarity Assessment. Spatial Cognition and Computation 6(1), 31–62 (2006)
32. Nedas, K., Egenhofer, M.: Spatial Similarity Queries with Logical Operators. In: Hadzilacos, T., Manolopoulos, Y., Roddick, J.F., Theodoridis, Y. (eds.) SSTD 2003. LNCS, vol. 2750, pp. 430–448. Springer, Heidelberg (2003)
33. Fikes, R., Nilsson, N., STRIPS,: a new approach to the application of theorem proving to problem solving. Artificial Intelligence 2, 189–208 (1971)
34. Erol, K., Hendler, J., Nau, D.: Complexity results for HTN planning. Annals of Mathematics and Artificial Intelligence 18(1), 69–93 (1996)
35. Barsalou, L.: Situated simulation in the human conceptual system. Language and Cognitive Processes 5(6), 513–562 (2003)
36. Neisser, U.: Cognition and Reality - Principles and Implications of Cognitive Psychology. In: Atkinson, R., et al. (eds.) Books in Psychology, Freeman, New York (1976)
37. Stoffregen, T.: Affordances and events. Ecological Psychology 12, 1–28 (2000)
38. Rome, E., et al.: Towards Affordance-based Robot Control. In: Dagstuhl Seminar 06231 Affordance-based Robot Control, June 5–9, 2006, Dagstuhl Castle, Germany (2006)
39. Fischer Nilsson, J.: A Conceptual Space Logic. In: 9th European-Japanese Conference on Information Modelling and Knowledge Bases. Hachimantai, Iwate, Japan (1999)
40. Cohn, A., Hazarika, S.: Qualitative Spatial Representation and Reasoning: An Overview. Fundamenta Informaticae 46(1–2), 2–32 (2001)
41. Markman, A., Gentner, D.: Structure mapping in the comparison process. American Journal of Psychology 113, 501–538 (2000)
42. Rada, R., et al.: Development and Application of a Metric on Semantic Nets. IEEE Transactions on Systems, Man and Cybernetics 19, 17–30 (1989)
43. Schwering, A., Raubal, M.: Measuring Semantic Similarity between Geospatial Conceptual Regions. In: Rodriguez, A., et al. (eds.) GeoSpatial Semantics - First International Conference, GeoS 2005, Mexico City, Mexico, November 2005, pp. 90–106. Springer, Berlin (2005)
44. Nedas, K., Egenhofer, M., Wilmsen, D.: Metric Details of Topological Line-Line Relations. International Journal of Geographical Information Science 21(1), 21–48 (2007)
45. Stahl, A., Gabel, T.: Optimizing Similarity Assessment in Case-Based Reasoning. In: 21st National Conference on Artificial Intelligence (AAAI-06), AAAI Press, Boston, MA (2006)

An Image-Schematic Account of Spatial Categories

Werner Kuhn

Institute for Geoinformatics, University of Münster
Robert-Koch-Str. 26-28, D-48149 Münster
kuhn@uni-muenster.de

Abstract. How we categorize certain objects depends on the processes they afford: something is a vehicle because it affords transportation, a house because it offers shelter or a watercourse because water can flow in it. The hypothesis explored here is that image schemas (such as LINK, CONTAINER, SUPPORT, and PATH) capture abstractions that are essential to model affordances and, by implication, categories. To test the idea, I develop an algebraic theory formalizing image schemas and accounting for the role of affordances in categorizing spatial entities.

1 Introduction

An ontology, according to Guarino [17], is "a logical theory accounting for the intended meaning of a vocabulary". Ontologies map terms in a vocabulary to symbols and their relations in a theory. For example, an ontology might specify that a house is a building, using symbols for the two universals (house, building) and applying the taxonomic (is-a) relation to them; or it might explain what it means for a roof to be part of a building, using a mereological (part-of) relation [2]. What relations should one use to account for the meaning expressed by the italicized phrases in the following glosses from WordNet[1]?

- a house is a "building *in which something is sheltered* or located"
- a boathouse is a "house at edge of river or lake; *used to store boats*"
- a vehicle is a "conveyance that *transports people or objects*"
- a boat is a "small vessel *for travel on water*"
- a ferry is a "boat that *transports people or vehicles across a body of water* and operates on a regular schedule"
- a houseboat is a "barge that is designed and equipped *for use as a dwelling*"
- a road is an "open way (generally public) *for travel or transportation.*"

These phrases express *potentials* for actions or processes, rather than actual processes. I will use the term *processes* for anything happening in space and time, and the term *affordances* for potentials of processes. This should not be taken as overextending Gibson's notion [13], but as a shorthand covering a collection of phenomena whose best example is an affordance in Gibson's sense (and which have been called telic

---

[1] http://wordnet.princeton.edu/perl/webwn

S. Winter et al. (Eds.): COSIT 2007, LNCS 4736, pp. 152–168, 2007.

relations in [33]). They all pose the same ontological challenge: how can a logical theory capture the categorizing effect of potential processes? For example, how can it express that a vehicle is a vehicle because it affords transportation, even though it may never be used to transport anything, or that a house is a house because it can shelter something? The goal is not to capture peripheral cases of a category (e.g., the vehicle which never transports), but conversely, properties which are central (e.g., the transportation and sheltering affordances). As the above WordNet examples illustrate, affordances are indeed often central for categorization.

What is the relevance of affordances to ontologies of spatial information? Ordnance Survey (the mapping agency of Great Britain) has identified affordance as one of five basic ontological relations to make their geographic information more explicitly meaningful, together with taxonomic, synonym, topological, and mereological relations [19]. Feature-attribute catalogues for geographic information abound with object definitions listing affordances as key characteristics [38]. Recent challenges to semantic interoperability have revolved around use cases that involve affordances. For example, the OGC geospatial semantic web Interoperability Experiment[2] worked on a query retrieving airfields based on their affordance of supporting certain aircrafts. Addressing the ontological challenge posed by affordances is necessary to capture the meaning of vocabularies in spatio-temporal information and to enable semantic interoperability.

Affordances are also increasingly recognized as having a role to play in formal ontology [1]. The enduring core of objects, particularly of artifacts or organisms, has been sought in "bundles of essential functions" [31]. Frank sees affordances as the properties that define object categories in general and uses this idea in his ontological tiers [7]. Smith has proposed an ontology of the environment, and in particular the notion of niche [39], founded on Gibson's affordances and underlying his *Basic Formal Ontology*. The *Descriptive Ontology for Linguistic and Cognitive Engineering* (DOLCE) introduces categories "as cognitive artifacts ultimately depending on human perception, cultural imprints and social conventions" and acknowledges a Gibsonian influence [11].

Recent formal ontology work has attempted to formalize image schemas as abstract descriptions [12]. While this gives an ontological account of the mental abstraction of image schemas, it does not yet allow for using it in category specifications. The latter would require integration with work on roles, particularly thematic roles, to address the categorizing power of objects participating in processes. Thematic roles are essentially a set of types of participation for objects in processes [40]. For example, a vehicle can play the thematic role of an instrument for transportation. The formalization of image schemas given in this paper does capture such thematic roles, but at a more generic level that has not yet been related to the thematic role types discussed in the linguistics literature.

In the context of spatial information theory, affordances and image schemas have been the object of several formalization attempts [5,10,25,34,35,36,37]. At least one of them also made the relationship between the two notions explicit [37]. But none has developed an image-schema based formalization of affordances to the point where it can account for categorizations. The work presented here builds on these attempts

---

[2] http://www.opengeospatial.org/initiatives/?iid=168

and develops them further in the context of spatial ontologies. Its major contribution is a formal algebraic theory of a set of image schemas and their affordances.

What are the basic ontological modeling options for affordances? Unary relations (e.g., "affords-transportation") leave the meaning of their predicates unspecified, produce an unlimited number of them, and suppress the object to which a process is afforded. To model affordances as a binary or ternary relation ("affords"), would require a *kind of* process as argument (e.g., "transportation"). For example, an individual vehicle (say, your car) affords a kind of transportation (i.e., of people), which subcategorizes the vehicle (e.g., as a car). However, first order theories do not permit process types as arguments, as long as these are not treated as individuals. The Process Specification Language PSL [16] takes the latter approach, but cannot state algebraic properties for the process types.

Due to this expressiveness issue, formal ontology has not yet approached affordance relations (see also [2,3,31] for discussions of formal ontological relations). The participation relation comes closest, but relates individual objects to actual processes [31]. It is also much looser than affordance. For example, a boat might participate in a painting process, as well as a transportation event. Both conform to the usual intuition about participation, that there are objects involved in processes, but one is an "accidental participation" and the other an instance of a potential, afforded participation that categorizes the object.

Ontological relations between classes have been proposed for capturing spatial relations [4]. However, they cover only classes of objects (not processes), and the approach would need to be extended to modal logic in order to cope with potential processes, not only with existence and universal quantification. Before taking this route of first order axiomatizations of affordance relations, one has to analyze what relations are necessary. This is the goal pursued here. The main novelty is the combined formalization of affordances and image schemas in a second order algebraic theory, accounting for spatial categories and establishing a framework for ontology mappings. In the remainder of the paper, I will state how image schemas form the basis for the theory (section 2), introduce the formalization method (section 3), present the results of the analysis (section 4), and conclude with implications and directions for future research (section 5).

2   Image Schemas as Theoretical Foundation

The primary source on image-schematic categorization is Lakoff's *Women, Fire, and Dangerous Things* [29]. It presents detailed case studies on image schemas and their role in language and cognition. Its main impact lay in demonstrating the inadequacy of traditional ideas about categorization, which are based on necessary and sufficient conditions. Formalizations, however, were not Lakoff's goal, due to the perceived limitations of formal semantics at the time. Since then, a lot of work in cognitive semantics, including applications to ontology engineering [26] has built on Lakoff's empirical evidence and informal models for image schemas. This section reviews the widely accepted characteristics and attempts to identify "ontological properties" of image schemas.

2.1   Key Characteristics of Image Schemas

Image schemas are patterns abstracting from spatio-temporal experiences. For example, they capture the basic common structures from our repeated experience of containment, support, linkage, motion, or contact. The idea emerged from work in cognitive linguistics in the 1970's and 1980's, mostly by Len Talmy, Ron Langacker, Mark Johnson, and George Lakoff. A recent survey is [32]. The idea gained popularity through Lakoff and Johnson's book on metaphors [28], Lakoff's book on categorization [29], and Johnson's book on embodied cognition [21]. The latter characterizes image schemas as follows (p. 29): "In order for us to have meaningful, connected experiences that we can comprehend and reason about, there must be pattern and order to our actions, perceptions, and conceptions. A schema is a recurrent pattern, shape, and regularity in, or of, these ongoing ordering activities. These patterns emerge as meaningful structures for us chiefly at the level of our bodily movements through space, our manipulation of objects, and our perceptual interactions". Johnson's characterization highlights the spatial aspects of image schemas, their link to activities, and their power to generate meaning and support reasoning. I will briefly discuss each of these aspects and then extract those properties of image schemas that suggest their potential for ontology.

Image schemas are often *spatial*, typically topological (e.g., CONTAINMENT, LINK, PATH, CENTER-PERIPHERY) or physical (e.g., SUPPORT, ATTRACTION, BLOCKAGE, COUNTERFORCE). This makes them obvious candidates for structuring spatial categories in ontologies. Since foundational ontologies still take a rather simplistic view of space, based on abstract geometric locations (for a survey, see [1]), any candidates for more powerful spatial patterns should be tested for their ontological potential. This is what I am doing here with image schemas for the purpose of building ontologies of spatio-temporal phenomena. As Lakoff and Johnson have shown, the spatial nature of image schemas also gives them a role in structuring abstract categories and in constraining conceptual mappings [27]. The scope of this paper, however, is limited to the role of image schemas in categorizing spatial entities.

The second aspect in Johnson's characterization closely relates to the first: image schemas capture regularities in activities and *processes*. Examples are CONTAINMENT (capturing how entering and leaving relate to being inside), SUPPORT (relating getting on and off to being on) or PATH (relating motion to being at places). Since my goal is to account for meaning grounded in affordances (in the sense of process potentials offered by the environment), any regularity observed in processes is interesting. The structuring of processes in foundational ontologies is at least as weak as that of space. Separate branches of foundational ontologies have been devoted to processes, but their internal structure and the participation relations between processes and objects remain underdeveloped [31]. Image schemas provide a special form of such relations.

Thirdly, image schemas support our *understanding* of and *reasoning* about experiences. Thus, they are likely to shape the meaning of vocabularies we use to describe these experiences. After all, information system terminology, like natural language, encodes human experiences in communicable form and as a basis for reasoning. An obvious example are navigation systems [34], guiding human movement through space. Most GIS applications, from cadastral systems and land use databases to tourism services involve categorizations based on human activities and experiences [22]. Exploiting abstract patterns of experiences for the design of spatial (and other) ontologies seems therefore justified.

2.2 Ontological Properties of Image Schemas

A generally accepted precise definition or even formalization of image schemas is still missing, and the evidence for their existence (in whatever form) comes mainly from linguistics. For the purpose of this work, however, I require neither a precise definition nor a broader empirical basis for image schemas. I simply take the notion as a useful theoretical construct with the following ontological properties:

1. Image schemas *generalize over concepts* (e.g., the CONTAINMENT schema abstracts container behavior from concepts like cups, boxes, or rooms);
2. they are internally *structured* (e.g., the CONTAINMENT schema involves behavior associated with an inside, an outside, a contained entity, and possibly a boundary);
3. they can be *nested and combined* (e.g., transportation combines behavior from the SUPPORT and PATH schemas);
4. they are *preserved in conceptual mappings* (e.g., the LINK and PATH schemas are common to all transportation links).

I call these properties ontological, because they relate image schemas to essential questions in ontology, which impact applications to spatial information: what entities should form the upper levels of ontologies? can these have internal structure or should they be atomic? how can ontologies reconcile process and object views of the world? how can ontological primitives be combined? what are useful properties of ontological mappings?

It is far beyond the scope of this paper to address any of these questions in detail. However, the fact that image schemas suggest some answers to them encourages a research program that explores their potential to (1) ground ontologies in human sensory-motor experience, and (2) provide invariants for ontology mappings. In pursuing this longer term goal, I claim here that *an algebraic theory built on image schemas can account for those aspects of meaning that are grounded in affordances*. I test this hypothesis by developing a new style of formalization for image schemas, and applying it to account for empirical data about spatial conceptualizations taken from WordNet. Note that the role of WordNet is only to supply external data on existing categorizations. This avoids the danger of inventing example data to fit the theory, while admitting the inclusion of improvements on WordNet. The somewhat confused upper levels of WordNet have indeed been replaced here by the DOLCE alignment given in [11].

3   Formalization Method

The relation between objects and process *types*, though missing from formal ontology, is a central pillar of algebra and computing. Algebra groups values into sorts based on the kinds of operations they admit. Programming languages model these sorts as data types and classify values or objects based on the kinds of computational processes they offer. These processes are collectively referred to as behavior. For example, a data type Integer offers operations (behavior) of addition, subtraction, and multiplication, but not division or concatenation.

The software engineering technique of algebraic specification exploits this commonality between programs and algebra [6]. It models programs as many-sorted algebras, consisting of sets of values with associated operations. Logical axioms, in the form of equations over terms formed from these operations, define the semantics of the symbols used. For example, an axiom might say that an item is in a container after having been put in. There exists a variety of flavors and environments for algebraic specifications. In the context of ontology engineering, OBJ [15] and CASL [30] are the most popular environments.

For reasons of simplicity, expressiveness, and ease of testing, I use the functional language Haskell as an algebraic specification language. It offers a powerful development and testing environment for ontologies, without the restriction to subsets of first order logic and binary relations typical for ontology languages. It also imposes less overhead for theorem proving than most specification environments and imposes the healthy need to supply a constructive model for each specification. While we have previously published ontological specifications in Haskell (e.g., in [8,23,24,36]), the style has now been refined, clarified, and stabilized. The remainder of this section introduces this style and the associated design decisions, as far as their knowledge is necessary to understand the subsequent formalizations. It is not a syntax-driven introduction to Haskell[3], but one that explains how the language can be used to model individuals, universals, affordances, and image schemas.

3.1 Universals as Data Types

Universals (a.k.a. categories, classes or concepts in ontology) will be modeled here as Haskell data types, and individuals (a.k.a. instances) as values. The standard computing notion of what it means to belong to a type captures the ontological relationship of instantiation between individuals and universals. Note that universals are then not just flat sets of individuals, but are structured by the operations (a.k.a. methods) defined on them.

The Haskell syntax for type declarations simply uses the keyword **data** followed by a name for the type and a right-hand side introducing a constructor function for values, possibly taking arguments. Here is a simple example, declaring a type for the universal Medium with two (constant) constructor functions Water and Air (Haskell keywords will be boldfaced throughout the paper):

```
data Medium = Water | Air
```

Type synonyms can be declared using the keyword type. For example, one defines a synonym House based on a previously declared type Building as follows:

---

[3] Introductions to Haskell, together with interpreters and compilers, can be found at http://www.haskell.org. For beginners, the light-weight Hugs interpreter is highly recommended. It has been used to develop and test all the specifications in this paper. Note that the Hugs option needs to be selected, to allow for multiple class parameters.

```
type House = Building
```

By considering types as theories [14], where operation signatures define the syntax and equational axioms the semantics for some vocabulary, one can now write theories of intended meanings, i.e., ontologies, in Haskell. They introduce a type symbol for each universal (e.g., House), a function type for each kind of process (e.g., enter, not shown here), and equations on them (e.g., stating that entering a House results in being in it).

3.2 Subsumption (and More) Through Type Classes

Organizing categories by a subsumption relation permits the transfer of behavior from super- to sub-categories. For example, houses might inherit their ability to support a roof from buildings. Standard ontology languages define subsumption in terms of instantiation: houses are buildings, if every house is also a building. Haskell does not permit this instantiation of a single individual to multiple types, but offers a more powerful form of subsumption, using type classes. These are abstractions from types and can therefore model something more general than universals or concepts. I use Haskell type classes to represent two forms of abstractions from universals: multiple conceptualizations of the same type (say, multiple ideas of a house) and image schemas.

To declare that houses inherit the behavior of buildings, one first introduces type classes for multiple conceptualizations of each. Class names have only upper case letters here, to distinguish them visually from type names. The sub-category is then derived (=>) from the super-category, and the types are declared instances of their classes:

```
class BUILDING building where
```
   <specify the behavior of buildings here, if any>
```
class BUILDING house => HOUSE house where
```
   <specify additional behavior of houses here, if any>
```
instance BUILDING Building where
```
   <specify how the type Building implements the behavior of type class BUILDING>
```
instance HOUSE House where
```
   <specify how the type House implements the behavior of type class HOUSE>

Note that Haskell's **instance** relation is one between a *type* and a *type class*. The symbols building and house are just variables for the types that can be instances (we could also call them x or a).

While this may look like a lot of overhead for the conceptually simple subsumption relation, it is not only mathematically cleaner and more transparent (every value has exactly one type), but also more flexible than subsumption in standard ontology languages. Type classes allow for behavior inheritance from *multiple* classes (with the same => syntax, denoting a so-called context), without creating dubious cases of subsumption. For example, a class can model the combination of behavior inherited by houseboats from housings and boats, without requiring houseboats to be both, houses and boats. Type classes also allow for *partial* inheritance, so that penguins can be birds which do not implement the flying behavior.

3.3   Image Schemas as Multi-parameter Type Classes

The previous section has shown that Haskell type classes generalize over types (which model universals or concepts), get structured by behavior (in the form of computable functions), and combine behavior from multiple classes. They also define theory morphisms, i.e., mappings from one theory (type) to another. For example, an instantiation of a type `Boathouse` to the `HOUSE` class defined above would map the basic behavior of sheltering (not shown) to the sheltering of a boat. With this mapping capacity, type classes exhibit the ontological properties required to represent image schemas, as listed in section 2.

To make the specification method more powerful, we now extend it to capture generalizations over types belonging to different classes. In the popular Hugs extension of Haskell (and possibly in the next Haskell standard), type classes can have multiple parameters. For example, a class representing the CONTAINMENT image schema might have two parameters (type variables), one for the containing type (called `container` here) and one for the type of the containee (called `for`):

```
class CONTAINMENT container for where

    isIn :: for -> container -> Bool
```

The methods of this class use these parameters in their signatures. Thus, the second line states that an implementation of the `isIn` query requires a containee (`for`) type and a container type as inputs and returns a Boolean value.

Multi-parameter type classes represent relations between types. For example, a container and a containee type are in the relation CONTAINMENT, if an `isIn` method exists that takes a value of each and returns a Boolean. Thus, the device that models subsumption can in fact model any other relations between categories. Image schemas are such a relation, but probably not the only interesting one for ontology.

Relations between types are more abstract than relations between individuals. For example, the CONTAINMENT schema provides a characteristic containment relation for object types (relating, for example, boats and boathouses). As a schema, it relates individual objects more weakly than the topological *isIn* or *contains* relation, but more generally. The *isIn* relation affirms, for example, that a particular boathouse contains a particular boat. The CONTAINMENT schema, by contrast, affirms that a boathouse (i.e., any instance of type boathouse) is a container for boats, no matter whether a particular boathouse contains a particular boat at any point in time.

3.4   Affordances as Behavior Inherited from Image Schemas

The link between objects and process types that characterizes affordances has already been captured in the algebraic notion of a type: The processes afforded by an individual are modeled as the methods offered by the data type representing the universal. For example, the affordance of a house to shelter people is expressed as a set of methods applicable to objects of type house and person (such as the methods to enter and leave a house and to ask whether a person is inside or outside).

Since a lot of this behavior generalizes to other universals (in the example, to all "people containers", including rooms and cars), it makes sense to model it at the most generic level. This level, I claim here, corresponds to image schemas. For example, the behavior required to ask whether somebody is in a house is the `isIn` method of the CONTAINMENT class above. Houses get this affordance from the CONTAINMENT image schema by instantiating the `House` type to the CONTAINMENT type class:

```
instance CONTAINMENT House Person where
  isIn person house =
    (container person = house) || isIn (container person) house
```

The axiom for `isIn` uses variables for individuals of type Person (`person`) and House (`house`) and states that the person is in the house if the house contains her or, recursively, if the house contains a container (e.g., a room) containing her. To state this, I use a representation of state as a labeled field `container`, here assumed to be defined for person types.

3.5 Testable Models

As a logical theory, an ontology per se does not need a formal model. However, if the theory is expressed in a programming language like Haskell, it does, for the code to become executable. The benefit of this is to make the theory testable through its model. Our experience has been that this is enormously beneficial for ontology engineering, as it reveals missing pieces, inconsistencies, and errors immediately, incrementally, and without switching development environments [9]. Ideally, one uses the best of both worlds, executability from a language like Haskell, and the greater freedom of expression that comes with non-constructive languages like CASL or even just Description Logics. With tools like Hets[4], spanning multiple logics, such heterogeneous specifications can now be integrated and tested for coherence.

For the case at hand here, the executability requirement was certainly more beneficial than limiting. Constructing individuals of the categories under study often revealed errors in axioms, and it gave the satisfaction (well known from programming) of seeing a solution "run" (though this is, of course, never a proof of correctness or adequacy). To state the effect in simpler terms: this paper would have been submitted to COSIT two years ago, if Haskell was as patient as paper (or Word) is. It would have had fewer insights, more confused ideas, and certainly more errors than now.

A model of the kind of logical theory presented here consists of a way to record and change state information. It has to satisfy the signatures and axioms of the theory. Haskell offers a simple device to keep track of the state of an object and allow object updates, through its labeled fields (already used above). For example, a parameterized type `Conveyance` can be constructed from two values (a conveying entity of type `Instrumentation` and a conveyed entity of variable type) as follows:

---

[4] http://www.informatik.uni-bremen.de/agbkb/forschung/formal_methods/ CoFI/hets/ index_e.htm

```
data Conveyance for =

  Conveyance {conveyor :: Instrumentation, conveyed :: for}
```

Values of the type can then be constructed and updated by supplying values for these labeled fields to the constructor function `Conveyance`.

4 Formalization Results

The immediate results of applying the above formalization method are, besides testing and improving the method itself, a set of seven image schemas specifications and around twenty category specifications by affordances[5]. Rather than listing the specifications in full length, this section will present results in the form of insights gained in the process, and illustrate these by excerpts from the specifications. It starts with some observations on image schemas and their interactions in general and then shows how an account for affordance-based categorizations is derived from the formalized image schemas.

4.1 Image Schemas and Their Combinations

Image schemas are treated here as purely relational structures, regardless of whether they are also reified as objects. What we consider containers, for example, are objects *playing the role* of containing something else. This role is afforded by the CONTAINMENT schema, but it does not have to turn the object into a container type. For example, a hand can enclose a pebble, instantiating the CONTAINMENT schema, but not becoming a container type; it only plays a containing role. The same object may also take on other roles, such as supporting or covering and it might stop to have them at any time. This distinction is a direct result from the formal difference between a type class (as a relation on types, used to model a role) and a type (used to model rigid properties that an object cannot stop to have [18]).

In formalizing particular image schemas, it became clear that some of them combine simpler patterns which are themselves image-schematic. This generalizes the earlier observation that image schemas have internal structure toward differentiating primitive from complex image schemas. For example, the usual forms of the CONTAINMENT and SUPPORT schemas described in the literature involve a path (at least implicitly), to allow for the contained or supported objects to access and leave containers or supports. It seemed appropriate to factor out the PATH schema to obtain simpler and more powerful CONTAINMENT and SUPPORT schemas. These can then also structure situations where something is contained or supported without ever having (been) moved into or onto something else (e.g., a city in a country, a roof on a building). The resulting specification, as seen previously, is rather simple:

```
class CONTAINMENT container for where

  isIn :: for -> container -> Bool

class SUPPORT support for where

  isOn :: for -> support -> Bool
```

---

[5] The specifications can be downloaded from the Resources tab at http://musil.uni-muenster.de/.

The SUPPORT schema is used here for situations that are sometimes attributed to a surface schema instead. The choice of name reflects the fact that it models support (e.g., against gravity) which may or may not come from a surface. Objects like buildings (supporting a roof), parcels (on which buildings stand), and vehicles (supporting a transported object) obtain this affordance from the schema.

The question remains open how CONTAINMENT and SUPPORT should be formally distinguished (beyond the structurally equivalent class definitions above). The presence of a boundary (as suggested in many characterizations of CONTAINMENT) does not seem to achieve this distinction, as not all containers have explicit boundaries. A more likely candidate to make the distinction is transitivity: containment is transitive (the person in the room is also in the house), but support is not, or at least not in the same direct way (the person on the floor of the room is not supported in the same way by the ground supporting the house).

Though PATH is also a complex schema, the LINK schema cannot be separated, as it is an essential part, i.e., any path necessarily involves a link. PATH is specified here as a combination of the LINK and SUPPORT schemas. It requires a link between the from and to locations and support for the moving entity (for) at these locations. A move changes the support relation to hold between the moving entity and the to location. If the from location also contains the entity (i.e., the two types also instantiate the CONTAINMENT schema), move induces the corresponding change for the isIn relation. All movement furthermore involves a medium in which it occurs [13]. The specification is given by a class context listing the three schemas (the expression in parentheses, constraining the types belonging to this class), and by the signature of the move behavior:

```
class (LINK link from to, SUPPORT from for,
   SUPPORT to for, MEDIUM medium)
   => PATH for link from to medium where
     move :: for -> link -> from -> to -> medium -> for
```

Surfaces, understood as separators of media [13], can be used to further constrain the motion to a surface (water body, road) through a second combination with SUPPORT, where a surface supports the link.

Among the remaining schemas specified until now, the PART-WHOLE schema was modeled using the relations defined in [2]. The COLLECTION schema adds collections of objects of a *single* type. It is useful to capture definite plurals ("the cars"), and it can be combined with the ORDER schema to create ordered collections. The COVER schema is still underdeveloped and structurally not yet distinguished from CONTAINMENT and SUPPORT.

Exploiting this algebra of image schemas further, one arrives at specifications for some fundamental spatial processes. For example, the category of a *conveyance* (WordNet: "Something that serves as a means of transportation") is a combination of the PATH and SUPPORT schemas. The specification becomes more complex, adding constraints on the conveying and conveyed object types and putting the PATH constraint into the transport method, but its basic structure is CONVEYANCE = PATH + SUPPORT:

```
class (INSTRUMENTATION conveyance, SUPPORT conveyance for,
  PHYSICAL_OBJECT for)
  => CONVEYANCE conveyance for where
    transport :: PATH conveyance link from to medium =>
      for -> conveyance -> link -> from -> to -> medium ->
conveyance
```

Conveyances afford support and transportation on paths. Moving a conveyance moves the object it supports, which is captured by the transport method. This affordance structures all derived specifications for transportation-related categories (vehicles, boats, ferries etc.). Note that conveyances may also *contain* transported items (to be expressed by a combination with the CONTAINMENT schema), but in any case need to support them, due to gravity.

4.2 Deriving Affordances from Image Schemas

The hypothesis posited for this paper was that image schemas capture the necessary abstractions to model affordances, or more specifically, that an algebraic theory built on the notion of image schemas can account for aspects of meaning related to affordances. The formalization method discussion in section 3 has shown that the formal instrumentation is up to this task. As a proof of concept and illustration for this claim, the above specification of the CONVEYANCE type class will now be extended to specify vehicle as a "conveyance that transports people or objects" (see its WordNet definition cited in the introduction).

Rather than listing all required instance declarations, the rest of this section focuses on the ones that are essential to specify the transportation affordance. First, a data type representing conveyances is declared. It is the parameterized type already introduced above, where the parameter stands for the type of (physical) object to be transported on the conveyance:

```
data Conveyance for =
  Conveyance {conveyor :: Instrumentation, conveyed :: for}
```

On the right-hand side of this declaration, the constructor function (Conveyance) takes two arguments in the form of labeled fields. The first (conveyor) is of type Instrumentation and contains the state information of the conveyance (e.g., its location). The second (conveyed) has the type of the transported object and is represented by the parameter for this type (for). This parameterized specification allows for having conveyances transporting any kinds of objects. It essentially wraps a conveyor type and adds a parameter for what is conveyed.

Secondly, conveyances have to be declared instances of the PATH schema in order to obtain the affordance to move. This is achieved by instantiating the necessary types of the PATH class. For example, a conveyance for physical objects (generalizing people and objects, as indicated by the WordNet gloss), three particulars (standing for a link, its source, and its goal) and a medium (in which the motion occurs) are together instantiating the PATH relation. This produces an axiom stating that the result of moving a conveyance is to move the conveyor:

```
instance PATH (Conveyance PhysicalObject) Particular
Particular

Particular Medium where

move conveyance link start end med =

conveyance {conveyor = move (conveyor conveyance) link start
end med}
```

The recursion on move is then resolved by instantiating PATH again for the conveyor, i.e., the Instrumentation type. The resulting axiom specifies that the result of moving the conveyor is that it will be supported by the end of the link:

```
instance PATH Instrumentation Particular Particular
Particular Medium

where move conveyor link start end med = conveyor {support =
end}
```

Finally, the transportation affordance results from instantiating conveyances for physical objects to the CONVEYANCE class. The axiom specifies the effect of transport as a move of the conveyance.

```
instance CONVEYANCE (Conveyance PhysicalObject)
PhysicalObject where

transport for con link from to med = move con link from to
med
```

Since each object "knows" its location (in terms of another object supporting it), moving the conveyance automatically moves the object it supports.

The specifications for houses, boats, houseboats and boathouses take the same form. They combine the image schemas required for their affordances into type classes, possibly subclass these further, and then introduce data types inheriting the affordances and specifying the categories. The signatures of these logical theories for categories are those of the inherited behavior. The axioms are the equations stating how the types implement the classes.

5 Conclusions

The paper has presented a method to account for spatial categorizations using a second order algebraic theory built on the notion of image schemas. The method posits image schemas as relations on types which pass behavior on to the types instantiating them. The type classes representing image schemas can be combined to produce more complex behavior and pass this on in the same way. The case study material contains an account for spatial categories from WordNet and has been illustrated here by the category of vehicles.

The contributions of this work lie in several areas. The primary goal and innovation was an integrated formalization of image schemas and affordances with the power to account for *affordance-related aspects of meaning*. The choice of a functional (as opposed to logic) programming language and the use of its higher order constructs (type classes) raise the issue of how to transfer the results into standard ontology languages. At the same time, they suggest an alternative route: The image schematic relations can link domain ontologies. If domain ontologies are specified as Haskell data types (which is in principle straightforward), and these are then declared instances of Haskell type classes, one obtains mappings among domain ontologies. The equations generated by the type class instantiations serve as bridging axioms. This use of the presented theory compromises neither the simplicity of standard ontologies nor the expressiveness of the theory.

With such an ontology mapping goal in mind, this work prepares an important step in realizing semantic reference systems [24], i.e., to establish a *transformation mechanism* between multiple conceptualizations. Consequently, future work will primarily study the ontology mapping capability of the approach. A follow-up hypothesis is that a grounding of ontologies in image schemas can take the form of image-schematic mappings rather than that of some new ontological primitives. Lakoff's invariance hypothesis [27] supports this strategy from a cognitive point of view.

Another direction in which to extend this work is toward a *theory of (relative) location* based on image-schematic relations. With the two central image schemas of SUPPORT and CONTAINMENT, the essential locating relations (isOn, isIn) have already been specified (see also [20]). The approach treats location as a role, allowing for multiple location descriptions for a single configuration, and it is extensible with other image-schematic relations.

The idea of combining image schematic relations to *specify complex behavior* warrants further exploration. Ongoing work extends the specifications to car navigation scenarios (one involving ferry links, the other highway networks). Beyond the case of modeling transportation as a combination of SUPPORT and PATH behavior, other fundamental spatial processes can clearly be modeled in this way, such as diffusion as a combination of CENTER-PERIPHERY and PATH.

The image schema specifications themselves will require some refinement. For example, a function from time to a path position will model *continuous motion*, beyond the current discrete start-to-end motion. Similarly (but not limited to image schemas), it remains to be seen how best to deal in this context with physical and geometrical constraints occurring in category descriptions (such as sizes, weights, and shapes).

Acknowledgments. Discussions with Andrew Frank, Martin Raubal, Florian Probst, Krzysztof Janowicz, David Mark and others over many years have shaped and sharpened my ideas on image schemas and affordances and their role in ontology. Partial funding for this work came from the European SWING project (IST-4-026514). The paper was finished while I was a resident at the e-Science Institute in Edinburgh.

References

1. Bateman, J., Farrar, S.: General Ontology Baseline. Bremen, Germany, SFB/TR8 Spatial Cognition, University of Bremen (2004), Available from http://134.102.58.154/i1/materials/deliverables/del1.pdf
2. Bittner, T., Smith, B.: Individuals, Universals, Collections: On the Foundational Relations of Ontology. In: Varzi, A.C., Vieu, L. (eds.) Proceedings of the Third International Conference on Formal Ontology in Information Systems (FOIS 2004), pp. 37–48. IOC Press (2004)
3. Degen, W., Heller, B., et al.: Contributions to the Ontological Foundation of Knowledge Modelling. Leipzig, Institute for Computer Science (2001), Available from http://lips.informatik.uni-leipzig.de:80/pub/2001-12/en
4. Donnelly, M., Bittner, T.: Spatial relations between classes of individuals. In: Cohn, A.G., Mark, D.M. (eds.) COSIT 2005. LNCS, vol. 3693, pp. 182–199. Springer, Heidelberg (2005)
5. Egenhofer, M.J., Rodríguez, A.M.: Relation algebras over containers and surfaces: An ontological study of a room space. Spatial Cognition and Computation 1(2), 155–180 (1999)
6. Ehrig, H., Mahr, B.: Fundamentals of Algebraic Specification. Springer, Berlin (1985)
7. Frank, A.: Ontology for spatio-temporal Databases. In: Sellis, T., Koubarakis, M., Frank, A., Grumbach, S., Güting, R.H., Jensen, C., Lorentzos, N.A., Manolopoulos, Y., Nardelli, E., Pernici, B., Theodoulidis, B., Tryfona, N., Schek, H.-J., Scholl, M.O. (eds.) Spatio-Temporal Databases. LNCS, vol. 2520, pp. 9–77. Springer, Heidelberg (2003)
8. Frank, A.U., Kuhn, W.: Specifying Open GIS with Functional Languages. In: Egenhofer, M.J., Herring, J.R. (eds.) SSD 1995. LNCS, vol. 951, pp. 184–195. Springer, Heidelberg (1995)
9. Frank, A.U., Kuhn, W.: A specification language for interoperable GIS. In: Goodchild, M.F., Egenhofer, M.J., Fegeas, R., Kottman, C.A. (eds.) Interoperating Geographic Information Systems (Proceedings of Interop'97), pp. 123–132. Kluwer, Norwell, MA (1999)
10. Frank, A.U., Raubal, M.: Formal Specifications of Image Schemata - A Step to Interoperability in Geographic Information Systems. Spatial Cognition and Computation 1(1), 67–101 (1999)
11. Gangemi, A., Guarino, N., et al.: Sweetening WordNet with DOLCE. AI Magazine 24(3), 13–24 (2003)
12. Gangemi, A., Mika, P.: Understanding the semantic web through descriptions and situations. In: Proceedings of Coopis / DOA / ODBASE. LNCS, pp. 689–706. Springer, Heidelberg (2003)
13. Gibson, J.: The Ecological Approach to Visual Perception. Boston, Houghton Mifflin Company (1979)
14. Goguen, J.: Types as Theories. In: Reed, G.M., Roscoe, A.W., Wachter, R.F. (eds.) Topology and Category Theory in Computer Science, pp. 357–390. Oxford University Press, Oxford (1991)
15. Goguen, J.A.: Data, Schema, Ontology and Logic Integration. Logic Journal of IGPL 13(6), 685 (2005)
16. Gruninger, M., Kopena, J.: Planning and the Process Specification Language. In: Olivares, J., Onaindía, E. (eds.) Proceedings of the Workshop on the Role of Ontologies in Planning and Scheduling, ICAPS, pp. 22–29 (2005)

17. Guarino, N.: Formal ontology and information systems. In: Proceedings of Formal Ontology and Information Systems, Trento, Italy, pp. 3–15. IOS Press, Amsterdam (1998)
18. Guarino, N., Welty, C.: Evaluating ontological decisions with OntoClean. Communications of the ACM 45(2), 61–65 (2002)
19. Hart, G., Temple, S., et al.: Tales of the River Bank, First Thoughts in the Development of a Topographic Ontology. In: 7th Conference on Geographic Information Science (AGILE 2004), Heraklion, Greece, pp. 169–178 (2004)
20. Hood, J., Galton, A.: Implementing Anchoring. In: Raubal, M., Miller, H.J., Frank, A.U., Goodchild, M.F. (eds.) GIScience 2006. LNCS, vol. 4197, pp. 168–185. Springer, Heidelberg (2006)
21. Johnson, M.: The Body in the Mind: The Bodily Basis of Meaning, Imagination, and Reason. The University of Chicago Press, Chicago (1987)
22. Kuhn, W.: Ontologies in support of activities in geographical space. International Journal of Geographical Information Science 15(7), 613–631 (2001)
23. Kuhn, W.: Modeling the Semantics of Geographic Categories through Conceptual Integration. In: Egenhofer, M.J., Mark, D.M. (eds.) GIScience 2002. LNCS, vol. 2478, pp. 108–118. Springer, Heidelberg (2002)
24. Kuhn, W.: Semantic Reference Systems. International Journal of Geographic Information Science (Guest Editorial) 17(5), 405–409 (2003)
25. Kuhn, W., Raubal, M.: Implementing Semantic Reference Systems. In: Gould, M., Laurini, R., Coulondre, S. (eds.) AGILE 2003 - 6th AGILE Conference on Geographic Information Science. Presses Polytechniques et Universitaires Romandes, Lyon, France, pp. 63–72 (2003)
26. Kuhn, W., Raubal, M., et al.: Cognitive Semantics and Spatio-Temporal Ontologies. Introduction to a special issue of Spatial Cognition and Computation 7(1) (2007)
27. Lakoff, G.: The Invariance Hypothesis: is abstract reason based on image-schemas? Cognitive Linguistics 1(1), 39–74 (1990)
28. Lakoff, G., Johnson, M.: Metaphors We Live By. The University of Chicago Press, Chicago, IL (1980)
29. Lakoff, G.P.: Women, Fire, and Dangerous Things. What Categories Reveal about the Mind. University of Chicago Press, Chicago, IL (1987)
30. Lüttich, K.: Approximation of Ontologies in CASL. In: International Conference on Formal Ontology in Information Systems (FOIS 2006), Baltimore, Maryland, November 9–11, pp. 335–346. IOS Press, Amsterdam (2006)
31. Masolo, C., Borgo, S., et al.: WonderWeb Deliverable D18 (2004), Available from http://wonderweb.semanticweb.org/deliverables/documents/D18.pdf
32. Oakley, T.: Image Schema. In: Geeraerts, D., Cuyckens, H. (eds.) Handbook of Cognitive Linguistics, Oxford University Press, Oxford (2007)
33. Pustejovsky, J.: The Generative Lexicon. MIT Press, Cambridge, MA (1998)
34. Raubal, M., Egenhofer, M., et al.: Structuring Space with Image Schemata: Wayfinding in Airports as a Case Study. In: Frank, A.U. (ed.) COSIT 1997. LNCS, vol. 1329, pp. 85–102. Springer, Heidelberg (1997)
35. Raubal, M., Gartrell, B., et al.: An Affordance-Based Model of Place. In: Poiker, T., Chrisman, N., eds. 8th International Symposium on Spatial Data Handling, SDH'98, Vancouver, Canada, pp. 98–109 (1998)
36. Raubal, M., Kuhn, W.: Ontology-Based Task Simulation. Spatial Cognition and Computation 4(1), 15–37 (2004)

37. Raubal, M., Worboys, M.: A Formal Model of the Process of Wayfinding in Built Environments. In: Freksa, C., Mark, D.M. (eds.) COSIT 1999. LNCS, vol. 1661, pp. 381–399. Springer, Heidelberg (1999)
38. Rugg, R., Egenhofer, M., et al.: Formalizing Behavior of Geographic Feature Types. Geographical Systems 4(2), 159–180 (1997)
39. Smith, B., Varzi, A.C.: The Niche. Noûs 33(2), 214–223 (1999)
40. Sowa, J.F.: Knowledge Representation. Logical, Philosophical, and Computational Foundations. Brooks Cole, Pacific Grove, CA (2000)

Specifying Essential Features of Street Networks

Simon Scheider and Daniel Schulz

Fraunhofer Institut für Intelligente Analyse und Informationssysteme (IAIS),
Schloss Birlinghoven, 53754 Sankt Augustin, Germany
{Simon.Scheider,Daniel.Schulz}@iais.Fraunhofer.de

Abstract. In order to apply advanced high-level concepts for transportation networks, like hypergraphs, multi-level wayfinding and traffic forecasting, to commercially available street network datasets, it is often necessary to generalise from network primitives. However, the appropriate method of generalisation strongly depends on the complex street network feature they belong to. In this paper, we develop formal expressions for road segments and some essential types of roads, like roundabouts, dual carriageways and freeways. For this purpose, a formal network language is developed, which allows a clear distinction among the geometrical network, its embedding into the Euclidian plane, as well as navigational constraints for a traffic mode.

1 Introduction

Nowadays digital street network data plays an important role in all kinds of businesses related to transport and navigation. One prominent example are navigation devices, which use network graphs together with map matching and shortest path algorithms to navigate a vehicle on a physical network. Another example are topographical maps showing the street infrastructure at different scales, which have to be adapted to different generalisation levels. A third example are traffic forecasting models, which can be used for planning purposes or to assess traffic frequency or traffic quality on a given location in the street network.

There exist only a small number of data providers for digital street data. The data model usually consists of an embedded (usually non-planar) graph, which is a graph whose edges and nodes have some representation in the Euclidian plane and which may be overlapping, and a set of additional street attributes, like e.g. road category and feature identifiers like street name. Of course, each vendor has its own format, and each format has its own set of street attributes, whose quality is often unknown.

However, for probably all kinds of transport network applications, network attributes and feature identifiers are crucial bits of information. This is because transportation modelling obviously is only feasible with complicated relational structures built on the usual network primitives [4]. This is documented by the big amount of research on high-level logical network structures [11]. For example in order to solve the task of wayfinding on a highway network, Timpf et al. [8] [9] distinguished among several task levels (route planning, wayfinding instructions, driving instructions), whose logical primitives are complex network features like e.g. freeway exits. In the research

S. Winter et al. (Eds.): COSIT 2007, LNCS 4736, pp. 169–185, 2007.

area related to transport GIS, Mainguenaud [5] introduced multi-level hypergraph networks that allow master nodes and master edges to represent sub-networks. Additionally, the fact that a given physical network infrastructure can be used by more than one tansport mode, leads to the concept of feature-based virtual networks [10], in which a second graph is built on the primitives of a street network graph. This graph abstracts from unnecessary geometrical and logical details for a certain transport mode.

However, in order to build these high-level data structures from a commercial dataset, given attributes are often not sufficient. Therefore it would be useful to have formal network generalisation techniques available, which could be used to automatically derive certain high-level features from a dataset of primitives. There have been important conceptual contributions to this problem, e.g. Stell and Worboys [7] and Stell [6] introduced a formalisation of simplification methods for spatial networks and graphs in general. As we will illustrate by two examples in section 2 of this paper, the appropriate method of generalisation for a given subset of the network is nevertheless strongly dependent on the type of complex network feature this subset belongs to. These can be different types of roads and different types of junctions.

In this paper, we develop formal expressions for road segments and some essential types of roads, like roundabouts, dual carriageways and freeways. In this scenario it is not sufficient to account only for logical characteristics of a street network, it is also necessary to take into account geometrical properties of its Euclidian embedding, as well as navigational constraints for vehicle traffic.

In the following sections, we will first consider two common examples of generalisation tasks for a street network, and discuss the main challenge they pose to a formal treatment using the existing concept of simplification by Stell and Worboys. Then we will introduce a formal specification of a street network, integrating physical network characteristics with navigational constraints for vehicle traffic on this network. On this basis we introduce definitions for road segment types and types of roads. We conclude with an outlook to future work.

2   Two Examples of Generalisation Tasks

The data model of a spatial street network dataset commonly consists of street segments and their intersections. Their logical equivalents are edges and nodes of a graph. The central question is, if this graph representation is enough for a formal treatment of common generalisation tasks.

Stell and Worboys [7] introduced a formal concept for the simplification of general graphs, which is considered as a combination of "amalgamation" and "selection" operations. A "selection" of a graph $G = (N, E)$ is an operation that takes a subset of nodes $N' \subset N$ and produces a derived graph $G' = (N', E')$, with each edge of E' being a possible path between two nodes of N' in the original graph G. In this way, nodes of the original graph can be removed, but the connecting paths between all remaining nodes still exist as an edge of the derived graph. Amalgamation means to combine nodes of the original graph by a surjective function, so that these nodes and their connecting edges can collapse into a new node, and edges being connected to a remaining node collapse into one new edge.

The first example we consider is the network data model of a roundabout junction linking 3 roads together (figure 1 on the right). If we try to apply the simplification concept to the generalisation of a roundabout junction, it would make sense to have a single amalgamation operation that combines all nodes m, q, r, n, p, o that are part of the loop into exactly one new node. All edges a, b, c, d, e, f belonging to this loop would disappear, and the remaining node would be connected to all three remaining edges entering the junction.

The second example is a dual carriageway road having an intersection with a second road (figure 1 on the left and figure 2). Here, we would rather apply first a se-lection operation, which removes the nodes o and q, leaving the nodes m, n, p and r together their two connecting paths [m a o b p] , [p c n] and [m f q e r] , [r d n]. In a second step, we would like to amalgamate the two nodes p and r to the new node u, so that the graph consists of the three nodes m, u and n and a new linear feature evolves consisting of the new edges j and k. In this case, the generalisation operation trans-lates a dual carriageway to a bidirectional road, consisting of two segments, and one intersection at the new node u.

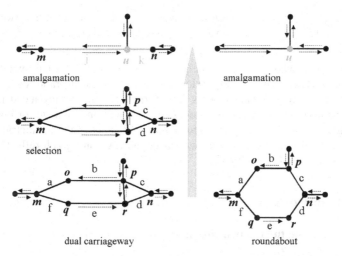

Fig. 1. Simplification of two embedded graphs, one being a common data model of a dual car-riageway, one being a common data model of a roundabout. The simplification method and its result are dependent on the complex feature type: The simplified feature is 1-dimensional in the case of a dual carriageway, and 0-dimensional in the case of a roundabout.

For a formal treatment, the problem now is how to distinguish between these cases? We could apply the appropriate amalgamation and selection operations if we could identify the sub graphs. Unfortunately, there are no appropriate attributes in the origi-nal dataset of the vendor which describe such complex types of roads and intersections. Additionally, note that the logical graph in both cases is nearly equal, so a distinction of nodes and edges on this logical level seems to be impossible. For example how can we distinguish nodes m and n from nodes p and r in the case of a dual carriageway, which is necessary in order to apply the proposed amalgamation operation?

Fig. 2. A dual carriageway road intersecting with two normal roads. Geometries (white lines) of a commercial network dataset are shown against the background of an aerial view.

This example underlines our principal hypothesis that a formal specification of complex network features is not possible only on the logical graph level commonly used to describe network data models. The formal apparatus introduced in the next section is an attempt to provide an integrated specification language for this purpose. In this language we can clearly distinguish among 3 aspects: the geometrical network graph, its embedding into the Euclidian space, and a second graph reflecting navigational constraints for a type of traffic.

3   Street Networks

3.1   Some Notes on the Formal Notation

In the following, data types are expressed by upper case symbols, denoting types of a typed higher order logic. Data structures and operations are algebraic expressions of this language. Syntax and semantics are equivalent to the "Basic Extended Simple Type Theory" (BESTT) approach (compare Farmer [2] and Farmer and Mohren-schmidt [3]), so that each type stands for a set. There are three atomic types with an obvious semantic: "Bool" for boolean values, "Rat" for rational numbers and "INT" for integers, and other atomic types are introduced in section 3.2. Derived types can be constructed from atomic ones by combining them to product, function, set and list types, meaning their appropriate set theoretic combinations (compare table 1). Let A and B be any two types. In this formal approach, every expression "a" has a type "A", which means they can always be written together like this (a:A).

Expressions can be formed by typed variables, constants, function abstraction (written as "λ[variable].[expression]") and function application (written   as "[function] ([expression])"), so that function arguments are of the appropriate type. Some language constants with obvious meaning are used, like:

$\downarrow$: A→Bool (:=definedness), toset: A×...×A→set[A] (:=tuple to set projection), ‖:set[A]→Rat (:=set cardinality), ‖: list[A]→Rat (:=list size), #i: A_0×..., A_i,...×A_j→A_i (:= extracts the ith element of a tuple), [i]: list[A] →A (:=extracts the ith element of a list), if: Bool×A×A→A (:=conditional expression), I: (A→Bool)→A (:= finds a unique object of type A by a predicate (definite description)), as well as well known logical and arithmetic symbols. The function "I" can be used to select unique objects with the help of its predicate argument.

Table 1. The four possible type constructors of a BESTT language. Let A and B be any types.

Type constructors	Comments
A×B	Product type
A→B	Function type
set[A]	Set type
list[A]	List type

The semantics of a BESTT language is strict: Every domain has an error element for undefined functions, and errors are automatically propagated. Therefore expressions are always defined in a classical sense. If an expression cannot be evaluated, it returns the error element, and the application of operator "$\downarrow$" will return false.

3.2 Specifying the Geometrical Street Network

In this section, we introduce an abstract specification of a typical data model of the physical street network infrastructure, as implemented in commercially available navigational maps like "Navteq" and "Teleatlas". In table 2 we introduce street segments (lines) of type L and nodes of type N together with their Euclidian embedding of type set[P] as atomic elements of road geometries, which can be seen as equivalents to "point" and "line" feature types of ISO-GDF.

Table 2. The three basic types of a spatial street network

Basic Types	Comments
N	Network nodes
L	Street segments (lines)
P	2-dim Euclidian space := Rat×Rat

Axiom 1. A street line (type L) has an embedded linear curve (type set[P]) that is not geometrically self-intersecting.

Let P be the type of a tuple of spatial coordinates in the 2-dimensional Euclidian space. According to axiom 1, a geometric representation of a line is a set of points (type set[P]), meaning a geometrical curve that does not cross itself in the 2-D space, equivalent to a "simple line" in [1]. In table 3, we give some basic operations for these types. Their axiomatic specification is omitted in this paper for simplicity reasons.

Table 3. Some basic operations for a spatial street network

Basic operation	Of type	Comments
Line	L→set[P]	A function embedding street segments (lines)
point	N→P	A function embedding street nodes
border	L→N×N	A function returning the two end nodes of a street segment
path	list[L] → Bool	Predicate to identify paths of node-connected street segments
cycle	list[L] → Bool	Predicate to identify closed paths of node-connected street segments
nodes	list[L] → list[N]	A function returning the list of nodes of a path. If the path is a cycle, all nodes will be returned, if the path is unclosed, all nodes without start and end node are returned
next	list[L]×INT →INT	Iterator for looping in a cycle
region$_{inner/outer}$	list[L] → set[P]	A function returning the inner/outer region for a list of lines being a simple cycle

Let "line": L→set[P] be the total function embedding every line as a curve into the 2-dimensional Euclidian plane. Let "border": L→N×N be the total function that produces the tuple of two end nodes for a line, so that their embeddings point(#1(border(h:L))) and point(#2(border(h:L))) are the two bordering points on the line's curve.

Lines can be seen as the atomic spatial entities of street networks, being logically and geometrically connected exclusively at their bordering nodes. They can be considered as edges of an embedded street network graph. The "path" predicate ensures that a given list of lines is a path of this graph. A street network graph must comply with the following axioms:

Axiom 2. A street network is a connected graph of street segments of type L (edges) and nodes of type N (vertices). This means that every pair of vertices of the street network is connected through a path.

Axiom 3. Two lines of the street network geometrically intersect either in the point of their common end node (planar "border" intersection), or by crossing each other non-planarly.

For our purpose, in the remainder we will introduce special operations as typed expressions (numbered formulas) and specify them by definitions. In order to increase legibility, we will sometimes add some comments to the defining expressions in italic.

Axiom 3 has the consequence that the embedding of the street network graph is not necessarily planar. This means there can be "road intersections" as well as "bridge-tunnel" relationships between lines. A "planar" predicate is given by:

$$planar : set[L] \rightarrow Bool .\tag{1}$$

Definition 1

$\forall$ k:set[L]. planar(k) =
    $\forall$(e$_1$:L), (e$_2$:L)∈ k. e$_1$ ≠ e$_2$ $\Rightarrow$
        line(e$_1$) ∩ line(e$_2$) =

if(toset(border(e_1)) ∩ toset(border(e_2))≠∅,
              point(I(n:N).toset(border(e_1))∩toset(border(e_2)) = {n}),
              ∅
)

Additionally, there are two special kinds of paths: simple paths and simple cycles. Simple paths are not closed, which means the end nodes of the path are not equal, and they are "simple", which means not self-intersecting. This is ensured, if they have a planar embedding and their list of nodes is unique, which means there is not any node occurring more than once in the list.

$$\text{simplepath: list[L]} \rightarrow \text{Bool .} \qquad (2)$$

Definition 2

∀ k:list[L]. simplepath(k) =
        path(k) ∧ planar(toset(k)) ∧
        toset(border(k[|k|-1])) ∩ toset(border(k[0])) = ∅ ∧ //unclosed path//
        |toset(nodes(k))| = |nodes(k)| //each node occurs just once in the path//

A simple cycle is a closed simple path.

$$\text{simplecycle : list[L]} \rightarrow \text{Bool .} \qquad (3)$$

Definition 3

∀ k:list[L]. simplecycle(k) =
        cycle(k) ∧ planar(toset(k)) ∧
        |toset(nodes(k))| = |nodes(k)| //each node occurs just once in the path//

For a simple cycle, the function $\text{region}_{inner/outer}$: list[L] → set[P] (compare table 3) is always defined. It returns the expected 2-dimensional region without holes and without bordering lines, equivalent to the interior/exterior of a cell complex [1].

3.3 Traffic on a Street Network

The embedded graph introduced in the last section is undirected. This is because different kinds of traffic on the same geometrical network infrastructure can have multiple directional constraints. In other words, the street network graph provides an infrastructure for different kinds of traffic networks, vehicular, pedestrian or bus transport, with its specific kinds of navigational constraints.

Navigation is a sequence of directional decisions. A directional decision in the proposed street network can only be taken at a common end-node (n:N) of two or more lines $(k_1:L, k_2:L)$, $n \in \text{toset(border}(k_1))$ ∧ $n \in \text{toset(border}(k_2))$. This is because junctions by definition exist only at border nodes (axiom 3).

A transport network on the proposed street network infrastructure therefore can be expressed as a second, directed graph, whose edges ("navigation edges") express navigational rules. They connect one line to another iff turn off at a street network node is allowed. Normally, in navigation applications, it is convenient to specify a single "navigation node" for either side of a street, which means for each direction of driving. For reasons of simplicity, we specify navigation nodes of such a graph to be just street lines of type L, with the consequence that for navigation paths, we ignore the risk of possibly unwanted u-turns at bidirectional lines. Further on, the first line of a navigation edge will be called the "from" line, the second one the "to" line.

In order to express a navigation graph, first a "maximal directed navigation graph" can be extracted from the street network graph. This is the special navigation graph for the case that all geometrically possible turn offs at navigation nodes are allowed.

$$navigraph_{max}:set[L{\times}L{\times}N]$$
$$:=\{t : L{\times}L{\times}N \mid toset(border(\#1(t))) \cap toset(border(\#2(t)))= \{\#3(t)\}\} . \tag{4}$$

In the scope of this paper, we will focus on vehicle traffic. A navigation graph for vehicle traffic is simply the subset of "navigraph_max" for which vehicle turn-off is not prohibited.

$$navigraph :set[L{\times}L{\times}N]$$
$$:= \{t : L{\times}L{\times}N \mid t \in navigraph_{max} \wedge \neg prohibited(t)\} . \tag{5}$$

In the remainder, if we refer to a navigation graph of type set[L×L×N], we always mean the navigation graph for vehicle traffic in formula 5. Let us call two lines (vehicle-) "navigable", if their tuple occurs in the (vehicle-) navigation graph. For navigation graphs, a single axiom is necessary. It ensures that for every line of it, there must be at least two navigation edges, one for incoming and one for outgoing traffic. This means the line is always a "from"-line as well as a "to"-line:

Axiom 4
$$\forall(k:L) \in \{k \mid \exists t \in navigraph. \, k=\#1(t) \vee k=\#2(t)\}.$$
$$\exists t_1, t_2 \in navigraph. \, t_1 \neq t_2 \wedge$$
$$k \in \{\#1(t_1), \#2(t_1)\} \wedge k \in \{\#1(t_2), \#2(t_2)\} \wedge$$
$$(\#1(t_1) = \#2(t_2) \vee \#2(t_1) = \#1(t_2))$$

The axiom prevents so called "graveyards" and "factories", which are lines allowing traffic to enter without giving it a chance to leave them anymore, and vice versa. These anomalies can cause several undesired effects in navigation devices. Graveyards and factories can exist e.g. in the context of spatially separate entering and exiting roads of a parking garage. In these graveyards, traffic exits implicitly exist inside of the garage, but they are not explicitly modelled in the network. These exceptions usually need to be dealt with explicitly. For the sake of simplicity we will not consider them in our specification.

The essential operation for navigation graphs is a shortest path operation. It can easily be formalised without an algorithmic definition. Let "cost": list[L]→Rat be a total function assigning a sum of costs to every path of lines. Let "min": list[L]× set[L]×set[L×L×N]× (list[L]→Rat) → Bool be a predicate that ensures a given path on a given line subset and for a given navigation graph to be minimal according to a given cost function. Then a shortest path operation for two lines k_1 and k_2 and an arbitrary subset "sn" of street network lines can be specified like this:

$$shortest\_path: set[L{\times}L{\times}N]{\times}set[L]{\times}L{\times}L \rightarrow list[L] . \tag{6}$$

Definition 4

$\forall$ ng:set[L×L×N], sn:set[L], k_1:L, k_2:L. shortest_path(ng, sn, k_1, k_2) =
   I(o:list[L]). ($\forall u\in$ toset(o).u$\in$ sn) $\wedge$ k_1 =o[0] $\wedge$ k_2 =o[|o|-1] $\wedge$
       ($\forall i\in$ {0,...,|o|-2}.$\exists t\in$ ng.#1(t) = o[i] $\wedge$ #2(t)=o[i+1]) $\wedge$
       min(o, sn, ng, cost)

An operator complying to definition 4 will return the "undefined" expression, if there does not exist any navigable path between the two lines k_1 and k_2 in the subset "sn", and otherwise, it will return a list of lines being the shortest possible one.

4 Features of a Street Network

4.1 Types of Road Segments

Now, important subtypes of lines can be specified according to their navigational characteristics by introducing predicates accordingly.

Dead Line. Lines of the street network that are not part of the tuple of at least one navigation edge do not belong to the transport network of the respective traffic type. These lines can be specified with the predicate "dead":

$$\text{dead} : \text{set}[L×L×N] \times L \to \text{Bool} . \tag{7}$$

Definition 5

$\forall$ng:set[L×L×N], k:L. dead(ng, k) = $\neg\exists t\in$ ng. k= #1(t) $\vee$ k= #2(t)

Dead End. Lines are called "deadend", if at least one of the two end nodes are not part of any navigation edge. Consequently, "dead" lines are special kinds of dead ends, which means dead(ng, k)→deadend(ng, k):

$$\text{deadend} : \text{set}[L×L×N] \times L \to \text{Bool} . \tag{8}$$

Definition 6

$\forall$ng:set[L×L×N], k:L. deadend(ng, k) = $\exists n\in$ toset(border(k)).$\neg\exists t\in$ ng. n = #3(t)

One-way. If a line is exclusively a "from"- or a "to"- line for all navigation edges at one end node, then it is a "one-way" street segment:

$$\text{oneway} : \text{set}[L×L×N] \times L \to \text{Bool} . \tag{9}$$

Definition 7

$\forall$ng:set[L×L×N], k:L. oneway(ng, k) =
     $\neg$deadend(ng, k) $\wedge$
     ($\forall n\in$ toset(border(k)).
        ($\exists t\in$ ng. k =#1(t)$\wedge$n=#3(t)→$\neg\exists$ t$\in$ ng. k =#2(t)$\wedge$n=#3(t)) $\vee$
        ($\exists t\in$ ng. k =#2(t)$\wedge$n=#3(t)→$\neg\exists$ t$\in$ ng. k =#1(t)$\wedge$n=#3(t))
)
     //both end nodes of a line can be passed in only one direction//

In a common sense, a line is a one-way road segment if it can be passed through in only one direction. This means that one node is only an exit, and the other node is only an entrance. This is the case, if the line is not a dead end, so both nodes can be passed, and if both nodes can be passed in only one direction. Then one node must be exit and the other one must be entrance because of axiom 4, therefore definition 7 is correct.

Bidirectional Line. Bidirectional road segments are lines that are not "dead" or "one-ways".

4.2 Types of Roads and Their Intersections

According to a common understanding, a road is an identifiable route between two or more places. In our formalisation, a road is a complex linear geometrical feature that is a path of lines. Furthermore a road is considered to be a simple path, because it normally does not cross itself. Another essential characteristic of a road is that it must be "identifiable". This means each street segment must uniquely belong to only one road, and therefore roads logically cannot overlap. Definition 8 states that a "road system" is a set of simple paths of lines (roads), and each line must belong to only one road:

$$\text{roadsystem} : \text{set}[\text{list}[L]] \rightarrow \text{Bool} . \tag{10}$$

Definition 8

$\forall$ rs:set[list[L]] . roadsystem(rs) =
　　　$(\forall(r:\text{list}[L]) \in \text{rs.simplepath}(r))$
　　　$\wedge$
　　　$(\forall(k:L). \exists r \in \text{rs. } k \in \text{toset}(r) \Rightarrow (\neg \exists r_2 \in \text{rs. } r \neq r_2 \wedge k \in \text{toset}(r_2)))$

Normal Roads. A normal road can be any simple path of lines in the street network graph that is not a ring. In contrary to a street, it is necessary that one can drive on a road from one start to one end. A normal road allows for every kind of intersection with other roads:

$$\text{normal\_road} : \text{set}[L \times L \times N], \text{list}[L] \rightarrow \text{Bool} . \tag{11}$$

Definition 9

$\forall$ ng:set[L×L×N], k:list[L]. normal_road(ng, k) =
　　　simplepath(k) $\wedge$
　　　$\forall i \in \{0,\dots,|k|-2\}. \exists t \in \text{ng.\#1}(t)=k[i] \wedge \#2(t)= k[i+1]$
　　　//the normal road can be passed through in at least one direction//

Linear Rings. Linear rings are just simple geometrical loops. A simple loop is a simple path of street lines whose end nodes are connected. On a linear ring, it must be possible to drive one way around. Linear rings also allow for every kind of intersection:

$$\text{linear\_ring} : \text{set}[L \times L \times N], \text{list}[L] \rightarrow \text{Bool} . \tag{12}$$

Definition 10

$\forall$ ng:set[L×L×N], k:list[L]. linear_ring(ng, k) =
　　　simplecycle(k) $\wedge$
　　　$\forall i \in \{0,\dots,|k|-1\}. \exists t \in \text{ng.\#1}(t)=k[i] \wedge \#2(t)= k[\text{next}(k,i)]$
　　　//the linear ring can be passed through in at least one direction//

Fig. 3. A roundabout (left) and a linear ring (right) enclosing a housing area

Roundabouts. For a roundabout, further navigational and geometrical constraints have to be considered. A linear ring is a roundabout, if it is a simple loop of one-way lines that can only be passed through in one (e.g. anticlockwise) direction. Whether a roundabout can be passed through clockwise or anticlockwise is dependent on the directionality rule.

So far, a roundabout could not be distinguished from a city block having an appropriate configuration of surrounding one-way streets, or from an appropriate linear ring surrounding a city center. In order to distinguish them, geometrical properties have to be taken into account. The principal idea of a roundabout is that of a special kind of road junction. Therefore the inner region formed by the loop is not usable for street infrastructure or housing infrastructure, because this would complicate the traffic situation inside of the junction. This means that street lines lying "inside" of the loop and being logically connected with the loop are not allowed, as well as buildings lying inside of the loop.

We specify a roundabout to be a linear ring that does not enclose any connected lines inside of the region formed by the loop. Buildings are not considered for simplicity reasons:

$$\text{roundabout} : \text{set}[L \times L \times N] \times \text{list}[L] \rightarrow \text{Bool} . \tag{13}$$

Definition 11

$\forall$ ng:set[L×L×N], k:list[L]. roundabout(ng, k) =

    linear_ring (k) $\wedge$ ($\forall i \in$ toset(k).oneway(ng,i)) $\wedge$

    //the linear ring can be passed through in exactly one

    (counter clockwise or clockwise) direction//

    $\neg\exists$(o:list[L]). path(o) $\wedge$

        ($\exists$(m:L)$\in$ toset(o), (n:L)$\in$ toset(k).

            toset(border(m)) $\cap$ toset(border(n)) $\neq\varnothing$) $\wedge$

        ($\forall$(m:L)$\in$ toset(o). line(m) $\cap$ region$_{inner}$(k)$\neq\varnothing$)

    //there is no logically connected path of lines lying inside of the loop//

A roundabout must have at least one external access point because of axiom 2. A roundabout with exactly one access point can be called a "turnaround circle" (German: "Wendehammer").

Dual Carriageways (closed). If a path is a "closed dual carriageway", then it consists of two parallel, non-crossing one-way passages, each of them carrying traffic in one direction, and ultimately being connected to one node at both ends of the carriageway. The nodes provide exit from as well as entrance to it. This means that all lines of the road must be one-way lines and must form a simple loop. But in contrary to a round-about, there are exactly two special nodes with pairs of connected lines in the loop where drive through is not permitted. We call this intersection feature "diametrical bifurcation" and it has some characteristics which can be used to distinguish a dual carriageway from a roundabout. A diametrical bifurcation (compare figure 5) consists of two one-way lines and their common end node. Navigation is not allowed from one line to the other crossing their common node, although the node is an entrance for traffic to one line and an exit for the other one:

$$\text{diabifurcation} : \text{set}[L \times L \times N] \times N \times L \times L \rightarrow \text{Bool} . \tag{14}$$

Definition 12

$\forall$ ng:set[L×L×N], (n:N), (k_1:L), (k_2:L). diabifurcation(ng, n, k_1, k_2) =
   $k_1 \neq k_2 \land$ oneway(ng, k_1) $\land$ oneway(ng, k_2) $\land$
   ($\neg \exists t \in$ ng.#3(t)=n $\land$ {#1(t),#2(t)}= {k_1, k_2}) $\land$
   ($\exists t_1, t_2 \in$ ng. #3(t_1)=n $\land$ #3(t_2)=n $\land$
        ((#1(t_1)= k_1 $\land$ #2(t_2) = k_2) $\lor$ (#1(t_1)= k_2 $\land$ #2(t_2) = k_1))
)
   //at their common node, driving through both one-way lines is not allowed,
   but exit from one line and entrance to the other line is possible//

Because of axiom 4, a bifurcation node must be connected to at least one third line, allowing for entrance as well as exit to the dual carriageway. In consequence, passage in a dual carriageway is possible from one bifurcation node to the other on all line tuples in between them. These two separately navigable passages are called "lanes" and provide exactly opposite driving directions.

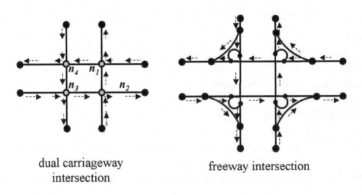

dual carriageway
intersection

freeway intersection

Fig. 4. Two examples of junction types for dual carriageways and freeways

If there were not any junctions between dual carriageways, our specification could be finished now. But of course, dual carriageways can have logical junctions with other ones. In principle, this could have the consequence that a line subset of one dual carriageway could appear also as a part of other dual carriageways (compare figure 4), if they are connected by junctions. This is prevented by the road system definition 8. Nevertheless, one can think of many configurations. In order to ensure the "right" configuration of dual carriageways, it is necessary to formalise the principal types of node intersections at junctions that are allowed for such a road type, taking also into account their geometrical characteristics.

For this purpose it is first necessary to consider all possible intersection nodes that allow to exit from the loop of one dual carriageway and to enter another one. A junction between two closed dual carriageways always means an intersection between two closed loops of one-way roads, so all intersecting lines must be one-way lines. The points of geometrical intersection can be logical, in which case exactly four 4-valued nodes appear, called "crossing intersections" (see nodes 1-4 in figure 4 and node 1 in figure 5). Or there is no direct logical connection, in which case roads are logically connected by "ramps" (all other nodes in figure 4 and node 3 in figure 5). It turns out that in order to "stay on the same road at intersections", three rules for three possible intersection types (compare figure 5) have to be followed, two rules for "diverging bifurcations" (rule 2 and 3) and one for "crossing intersections" (rule 1):

$$\text{crossing\_inters} : \text{set}[L \times L \times N] \times N \rightarrow \text{Bool} . \qquad (15)$$

Definition 13

$\forall \, ng{:}set[L \times L \times N], (n{:}N). \text{ crossing\_inters}(ng, n) =$

$\exists \, (k_{enter1}{:}L), (k_{enter2}{:}L), (k_{exit1}{:}L), (k_{exit2}{:}L) . |\{k_{enter1}, k_{enter2}, k_{exit1}, k_{exit2}\}|{=}4 \, \wedge$

$\quad (\neg \exists \, (k{:}L). \, k \notin \{k_{enter1}, k_{enter2}, k_{exit1}, k_{exit2}\} \wedge n \in border(k)) \, \wedge$

$\quad oneway(k_{enter1}) \wedge oneway(k_{enter2}) \wedge oneway(k_{exit1}) \wedge oneway(k_{exit2}) \, \wedge$

$\quad (\exists t_1, t_2 \in ng.$

$\qquad (\#1(t_1)= k_{enter1} \, \wedge \#2(t_1) = k_{exit1} \wedge \#3(t_1) = n) \, \wedge$

$\qquad (\#1(t_2)= k_{enter1} \, \wedge \#2(t_2) = k_{exit2} \wedge \#3(t_2) = n)$

$\quad)$

//a crossing intersection node is a 4-valued node, in which two one-way roads cross each other, so that there are two entrance lines and two exit lines//

Rule 1: At "crossing intersection" nodes, there exists a second entrance to and an exit from the node, which means external traffic can cross the intersection on an external one-way road (compare node 1 in figure 5). Therefore there must be in total 2 distinct entrance lines and two distinct exiting lines. The clear distinction between the crossing one-way road and the loop road is possible, because if external traffic is allowed to "cross" the road, then it is inevitable that either one external entering line must lie outside and one exiting line must lie inside of the loop, or vice versa. If both lines either lie inside or outside, the external traffic cannot cross the dual carriageway, and therefore this is not a valid configuration.

$$\text{divbifurcation} : \text{set}[L \times L \times N] \times N \rightarrow \text{Bool} . \qquad (16)$$

182   S. Scheider and D. Schulz

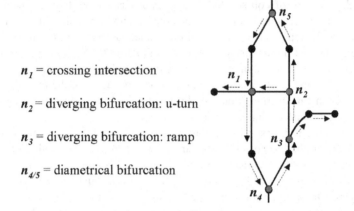

n_1 = crossing intersection

n_2 = diverging bifurcation: u-turn

n_3 = diverging bifurcation: ramp

$n_{4/5}$ = diametrical bifurcation

Fig. 5. Types of one-way intersection nodes for dual carriageways

Definition 14
$\forall$ ng:set[L×L×N], (n:N). divbifurcation(ng, n) =
$\exists(k_{root}:L),(k_{b1}:L), (k_{b2}:L).$ border(k_{root})∩(border(k_{b1})∩border(k_{b2}))={n}∧
    (¬∃ (k_3:L). $k_3 \notin$ {k_{root}, k_{b1}, k_{b2}} ∧ n∈ border(k_3)) ∧
    oneway(ng, k_{root}) ∧ oneway(ng, k_{b2}) ∧ oneway(ng, k_{b2}) ∧
    ($\exists t_1$, $t_2 \in$ ng. #3(t_1)=n ∧ #3(t_2)=n ∧
    #1(t_1)= k_{root} ∧ #2(t_1) = k_{b1} ∧ #1(t_2)= k_{root} ∧ #2(t_2) = k_{b2})
*//at a diverging bifurcation node, it is only possible to drive from one root
line into exactly two branches//*

Rule 2: In the case of a "diverging bifurcation", there is not more than one entrance to
the node, and exactly two exits, and therefore there is no crossing traffic possible.
These nodes exclusively allow exit to ramps or continuation. Because of international
traffic rules aiming to avoid lane crossings, exit to a ramp is only possible to lines lying outside of the loop's inner region. If the principal directionality is right hand
sided, this always means to turn off to one's right. Turning off to one's left at such a
node, towards the inner region, inevitably means to stay on the same road (see node 3
in figure 5). Therefore if there is a possibility to exit towards the inner region of a
dual carriageway loop, this is not a valid configuration.

So far, we have considered all possible intersection nodes at junctions of two dual
carriageways. Of course, a bigger variety of intersection nodes are possible in other
types of junctions. There is one case that seems to be tricky: "U-turn" roads are paths
of planarly embedded one-way lines lying entirely inside of the loop and connecting
the two opposite lanes. The tricky thing is that exit nodes to "u-turn" roads are also
diverging bifurcations, but they allow turning off towards the inner region of the loop
(compare node 2 in figure 5). If we would follow rule 2 at that node, we would not
continue on the road, and therefore we have to deal with this special exception:

Rule 3: We should break rule 2 at a diverging bifurcation node, if it is an intersection with a one-way "u-turn" road. One-way u-turn roads connect two lanes of the same dual carriageway by a diverging bifurcation. So a dual carriageway should allow to exit towards its inner region at a diverging bifurcation, if this exit can be identified as a u-turn road connecting the opposite lane of the dual carriageway.

It turns out that all other intersections with other types of roads do not pose a problem, because their lines cannot belong to a second dual carriageway. Intersections with bidirectional lines therefore do not have to be taken into account.

$$\text{dual\_carriageway} : \text{set}[L{\times}L{\times}N]{\times}\text{list}[L] \rightarrow \text{Bool} . \qquad (17)$$

Definition 15

$\forall$ ng:set[L×L×N], k:list[L]. dual_carriageway(ng, k) =

    simplecycle(k) $\wedge$ ($\forall i{\in}$ toset(k).oneway(ng,i)) $\wedge$

    ($\exists$i,j,g,h${\in}$ {1,...,|k|}. i$\neq$g $\wedge$ j=next(k,i) $\wedge$ h=next(k,g) $\wedge$

        diabifurcation(ng, I(n:N).toset(border(k[i]))∩toset(border(k[j]))={n},

               k[i], k[j]) $\wedge$

        diabifurcation(ng, I(n:N).toset(border(k[g]))∩toset(border(k[h]))={n},

               k[g], k[h]) $\wedge$

        //there are exactly two diametrical bifurcations in the loop//

        ($\forall$f${\in}$ {0,...,|k|-1}. f${\notin}$ {i,g}$\Rightarrow$

               $\exists$t${\in}$ ng. #1(t)=k[f] $\wedge$ #2(t)= k[next(k,f)]

)

        //each pair of connected lines in the loop which is not part of a diabifurcation can be passed through in exactly one (either counter clockwise or clockwise) direction//

) $\wedge$

    ($\forall$(n:N)${\in}$ toset(nodes(k))). $\forall$(to:L)${\notin}$ toset(k). $\forall$t${\in}$ ng.

      n = #3(t) $\wedge$ to = #2(t) $\wedge$ oneway(to) $\wedge$ ($\neg\exists$(f:L). diabifurcation(ng, n, f, to))

      $\Rightarrow$

    //rule 1//   (crossing_inters(ng,n) $\wedge$

                    $\neg\exists$ (to$_2$:L). n${\in}$ border(to$_2$) $\wedge$ to$\neq$to$_2$ $\wedge$

                    line(to) $\subset$ region$_{inner}$(k) $\wedge$ line(to$_2$) $\subset$ region$_{inner}$(k))

              $\vee$

    //rule 2//   (divbifurcation(ng, n) $\wedge$ line(to) $\cap$ region$_{inner}$(k)=$\varnothing$)

              $\vee$

    //rule 3//   (divbifurcation(ng, n) $\wedge$

                ($\exists$(d:list[L]). to=d[0] $\wedge$ planar(toset(k)$\cup$toset(d)) $\wedge$

                    ($\forall$z${\in}$ toset(d). z${\notin}$ toset(k) $\wedge$

                      (line(z) $\cap$ region$_{outer}$(k)=$\varnothing$)

) $\wedge$

                ($\exists$(g:L)${\in}$ k. ($\exists$t$_i{\in}$ ng. #1(t$_i$)=d[|d|-1] $\wedge$ #2(t$_i$)=g) $\wedge$

                    (shortest_path(ng, d, d[0], d[|d|-1])!) $\wedge$

                    $\neg$(shortest_path(ng, k, #1(t), g)!)

)

)

)

)

//intersection nodes allowing exit to one way lines (and not being diametrical bifurcations), are either diverging bifurcations at ramps (allowing exit towards the outside of the loop), or diverging bifurcations at u-turn roads (allowing exit towards the inner region of the loop), or crossing intersections with another one way road//

Dual Carriageways (unclosed). Dual carriageways sometimes can have only one or no diametrical bifurcation node at all. This is the case if the dual carriageway ends at a T-junction. In order to specify unclosed dual carriageways, we have to specify their T-junctions. This is left as a specification task for future efforts.

Freeways. Freeways are dual carriageways with less variety of intersections. Let us first consider the case of one-way intersections. In fact, on a freeway, a crossing intersection is normally not allowed, as well as a u-turn road. This is because at these intersection types, a vehicle has to cross all lanes, which is not a desirable manoeuvre for freeways. It is immediately clear that the only type of one-way exit from a dual carriageway that does not imply lane crossings is the diverging bifurcation at ramps complying with dual carriageway rule 2.

Furthermore, intersections with bidirectional lines are not acceptable for freeways, because each exit or entrance must have an acceleration lane, and the construction of two tangential acceleration lanes for exit and entrance is not possible in one line. In consequence, the specification of freeways is comparably easy: Every dual carriageway that exclusively has got ramp exits (rule 2 exits) or ramp entrances as node intersections is a freeway:

$$\text{freeway} : \text{set}[L \times L \times N] \times \text{list}[L] \to \text{Bool} .\tag{18}$$

Definition 16

$\forall$ ng:set[L×L×N], k:list[L]. freeway(ng, k) =

$\quad$ dual_carriageway(ng, k) $\wedge$

$\quad$ ($\forall$(n:N)$\in$ toset(nodes(k)). $\forall$(to:L)$\notin$ toset(k). $\forall$t$\in$ ng.

$\qquad$ n =#3(t) $\wedge$ to =#2(t) $\wedge$ ($\neg\exists$(f:L).diabifurcation(ng, n, f, to))

$\qquad\Rightarrow$

//rule 2//$\qquad$ (divbifurcation(ng, n) $\wedge$ line(to) $\cap$ region$_{inner}$(k)=$\varnothing$)

)

5 Conclusion

In order to apply advanced high-level concepts for transportation networks, like hyper graphs, multi-level way finding and traffic forecasting, to commercially available street datasets, it is often necessary to generalise from network primitives. The appropriate method of generalisation depends on the complex street network feature these primitives belong to. As was shown in section 2, essential complex features can be e.g. roundabouts and dual carriageways, but also freeways and several types of junctions. In order to specify them, a formal network language was introduced in section 3, which allows a clear distinction among 3 aspects: the geometrical network graph, its embedding into the Euclidian space, and a second graph called "navigation graph", built on the first one, and reflecting navigational constraints for a traffic mode. On this basis, the road segment types "dead", "dead end" and "oneway" as well as the road types "normal_road", "linear_ring", "roundabout", "dual_carriageway" and "freeway" could be specified by formal definitions.

Our future work will focus on the implementation of the proposed predicates, considering efficient algorithms and search strategies for the identification of complex network features. Furthermore, the next step should be to consider some essential features that were not discussed in this paper, e.g. complex junction types.

Acknowledgments. The authors would like to thank Dr. Angi Voss for her valuable support concerning the work on formal expressions.

References

1. Egenhofer, M.J., Herring, J.R.: Categorizing binary topological relationships between regions, lines and points in geographic databases. In: Technical Report, Department of Surveying Engineering, University of Maine, Orono (1991)
2. Farmer, W.M.: A Basic Extended Simple Type Theory. SQRL Report 14 (2003)
3. Farmer, W.M., Mohrenschmidt, M.: Simple type theory: Simple steps towards a formal specification. In: 34th ASEE/IEEE Frontiers in Education Conference, Savannah, GA, F1C-6, IEEE, Los Alamitos (2004)
4. Goodchild, M.F.: Geographic Information Systems and Disaggregate Transportation Modelling. Geographical Systems 5(1–2), 19–44 (1998)
5. Mainguenaud, M.: Modelling the Network Component of Geographical Information Systems. Int. J. Geographical Information Systems 9(6), 575–593 (1995)
6. Stell, J.G.: Granulation for Graphs. In: Freksa, C., Mark, D.M. (eds.) COSIT 1999. LNCS, vol. 1661, pp. 417–432. Springer, Heidelberg (1999)
7. Stell, J.G., Worboys, M.: Generalizing Graphs using Amalgamation and Selection. In: Güting, R.H., Papadias, D., Lochovsky, F.H. (eds.) SSD 1999. LNCS, vol. 1651, pp. 19–32. Springer, Heidelberg (1999)
8. Timpf, S., Volta, G.S., Pollock, D.W., Egenhofer, M.J.: A Conceptual Model of Wayfinding Using Multiple Levels of Abstractions. In: Frank, A.U., Formentini, U., Campari, I. (eds.) Theories and Methods of Spatio-Temporal Reasoning in Geographic Space. LNCS, vol. 639, pp. 348–367. Springer, Heidelberg (1992)
9. Timpf, S., Kuhn, W.: Granularity Transformations in Wayfinding. In: Freksa, C., Brauer, W., Habel, C., Wender, K.F. (eds.) Spatial Cognition III. LNCS (LNAI), vol. 2685, pp. 77–88. Springer, Heidelberg (2003)
10. Zhou, C., Lu, F., Wan, Q.: A Conceptual Model for a Feature-Based Virtual Network. GeoInformatica 4(3), 271–286 (2000)
11. Tang, A.Y, Adams, T.M., Usery, E.L.: A Spatial Data Model Design for Feature-Based Geographical Information Systems. International Journal of Geographical Information Systems 10(5), 643–659 (1996)

Spatial Information Extraction for Cognitive Mapping with a Mobile Robot

Jochen Schmidt[1], Chee K. Wong[1], and Wai K. Yeap[2]

[1] Centre for Artificial Intelligence Research
Auckland University of Technology, Auckland, New Zealand
[2] Department of Artificial Intelligence, University of Malaya,
Kuala Lumpur, Malaysia

Abstract. When animals (including humans) first explore a new environment, what they remember is fragmentary knowledge about the places visited. Yet, they have to use such fragmentary knowledge to find their way home. Humans naturally use more powerful heuristics while lower animals have shown to develop a variety of methods that tend to utilize two key pieces of information, namely distance and orientation information. Their methods differ depending on how they sense their environment. Could a mobile robot be used to investigate the nature of such a process, commonly referred to in the psychological literature as cognitive mapping? What might be computed in the initial explorations and how is the resulting "cognitive map" be used for localization? In this paper, we present an approach using a mobile robot to generate a "cognitive map", the main focus being on experiments conducted in large spaces that the robot cannot apprehend at once due to the very limited range of its sensors. The robot computes a "cognitive map" and uses distance and orientation information for localization.

1   Introduction

Since Tolman [1] suggested that animals (including humans) create a representation of the environment in their minds and referred to it as a "cognitive map", many psychological experiments have been conducted to study the nature of cognitive maps (see [2] for a review). Some of the important characteristics of cognitive maps highlighted by these studies include distorted information about distances and directions, landmarks, places, and paths, a hierarchical organization as well as multiple frames of reference. Many models of cognitive maps have also been proposed and one idea appears to be most prominent, namely that the map begins with some form of a network of "place representations". Several computational theories of cognitive mapping have been published since then, including the works of Poucet [3], Chown, Kaplan and Kortenkamp [4], Kuipers [5], and Yeap and Jefferies [6].

More recently, researchers began to use mobile robots to test ideas about cognitive mapping as opposed to robot mapping. In robot mapping, one is concerned with the development of efficient algorithms for the robot, with its particular sensors, to simultaneously localize and map its environment (SLAM). For some examples of recent work in this area see [7] and the references therein. The maps acquired via this paradigm are precise. By "precise", we refer to the fact that every surface the robot encounters is remembered and its position is known with a certain degree of accuracy. The robot also

S. Winter et al. (Eds.): COSIT 2007, LNCS 4736, pp. 186–202, 2007.

knows its position in the environment. In contrast, the mapping process used by humans (and animals), referred to as cognitive mapping, produces an imprecise map initially, which later turns into a representation laden with one's own interpretations and experiences of the world. The map produced in such a process is known as a cognitive map.

For example, Kuipers and his team have been experimenting with robots to find ways to compute his Spatial Semantic Hierarchy from the ground up [8]. Both the gateway construct in the PLAN model of cognitive mapping [4] and the use of exits in the Absolute Space Representation (ASR) model of cognitive mapping [6] were tested on a mobile robot: refer to [9] for the former and to [10] for the latter. Also, ideas about cognitive mapping based upon neurological findings were being tested using mobile robots. Examples of such work include [11,12]. However, many of these attempts produced algorithms that were more an inspiration from observations about cognitive mapping than a test-bed for theories of cognitive mapping. These researchers were concerned on their robots successfully mapping their environments. Hence, instead of investigating cognitive mapping, they ended up trying to solve the robot mapping problem.

Our goal in this paper differs. Different animals compute cognitive maps using different sensors, and therefore our robot should be treated as a kind of animal with its own peculiar sensing capabilities. For unknown reasons, humans do not remember a precise map after one or two visits to a new environment. We assume animals do not too, and so neither should our robot. To investigate our robot's cognitive mapping process, it is thus best to have our robot compute an imprecise map first and then investigate animal-like strategies for finding its way home using such a map. It is argued that the behavior of such a robot might shed light on cognitive mapping. Refer to [13] for a review on biological navigation approaches implemented on robots.

To do so, we use a robot equipped with sonar sensors to compute a description of each local space visited. The robot's "cognitive map"[1] is thus a network of such local spaces. Following the theory of cognitive mapping as presented in [6], we refer to each local spaces computed as an Absolute Space Representation (ASR). With sonar sensors, the description of each ASR computed (or more precisely, the shape computed) is not accurate enough to allow its identification on its return journey, even more so in large environments. As lower animals (especially rats) have been observed to use distance and direction information encoded in their cognitive map to find their way [14], we implemented similar strategies for the robot.

The algorithms presented here are based on our previous work on robot and cognitive mapping [15,16,17]. They have been considerably extended to be able to cope with "large" spaces (for a robot with sonar sensors), and new experimental results are shown here. The spaces considered are too large for the robot to be apprehended at once due to the limited range of its sonar sensors.

Section 2 describes the way our robot computes its "cognitive map". Section 3 presents the two strategies that our robot uses to localize itself in the environment. Section 4 shows the results of our experiments, the main focus being on large spaces.

---

[1] by definition, a robot cannot have a cognitive map. However, rather than being verbose and say "using a robot to simulate computing a cognitive map", we will simply say the robot computes a cognitive map.

2   Generating an Absolute Space Representation

When exploring the environment for the first time, the robot creates a "cognitive map" of its environment, i. e., a network of local spaces visited, namely ASRs. The process of generating this topological space representation from sonar information gathered while mapping the environment is described in the following.

For mapping, we use a mobile robot equipped with eight sonar sensors and an odometer. The robot acquires sonar readings while moving on a "straight" line (we are not concerned about drift compensation or correction) until it runs into an obstacle. At this point an obstacle avoidance algorithm is used, after which the robot can wander straight on again. A single one of these straight movements will be called *robot path* throughout this paper. The starting point of the algorithm is a geometric map that contains the robot movement path as well as linear[2] approximations of surfaces generated from the original range data. The goal is to split the map into distinct regions, e. g., corridors and rooms. One of the main problems at this stage is that the range of the sonar sensors (for our robot about 4m) is not sufficient for apprehending large rooms at once with range data acquired from a single position. Therefore, the robot has to travel through the environment, which results in distorted maps due to odometry errors. The algorithm presented here is capable of handling these problems, as will be shown in the experiments' section.

Splitting is done along the robot movement path, using an objective function that computes the quality of a region, based on local metric features derived from the geometric map, such as the average room width (corridors are long and narrow compared to rooms) and overall direction (e. g., a corridor is separated from another one by a sharp bend in the wall). No fixed thresholds are used, so the algorithm can balance the influence of the separate criteria used as needed.

2.1   Split and Merge

The basis of the ASR generating algorithm is the well-known split and merge method [18,19,20], which originated in pattern recognition. A classic application of this algorithm is finding piecewise linear approximations of contour points that have been detected in an image. A variety of other applications has been proposed, including segmentation of image regions given a homogeneity criterion, e. g., with respect to color or texture [20]. The split and merge algorithm is the core part of the process resulting in a topological ASR representation of the environment. Therefore, it is described here, based on the classic version for contour approximation.

The input data of split and merge is a sorted set of (contour) points, which is to be approximated. A parametric family of functions $\mathcal{F}$ (e. g., lines) to be used has to be chosen, as well as a metric for computing the residual error ϵ of the resulting approximation (usually root mean square error), or, when used for regions, a homogeneity or quality criterion. The algorithm results in a piecewise approximation of the original points, where every single residual error is below a given threshold θ_s. The single steps of the algorithm are [20]:

---

[2] this is not mandatory; any approximation will be fine as long as the total area of a region can be computed

1. Start with an initial set of points $\mathcal{P}^0$ consisting of n_0 parts $\mathcal{P}^0_0, \ldots, \mathcal{P}^0_{n_0-1}$. Each part $\mathcal{P}^0_i$ is approximated by a function from $\mathcal{F}$. Compute the initial residual error ϵ^0_i for each part of $\mathcal{P}^0$.
2. Split each $\mathcal{P}^k_i$ where $\epsilon^k_i > \theta_s$ into two parts $\mathcal{P}^{k+1}_j$ and $\mathcal{P}^{k+1}_{j+1}$, compute the approximation and residuals ϵ^{k+1}_j, ϵ^{k+1}_{j+1}. Repeat until $\epsilon^k_j \leq \theta_s \; \forall i$
3. Merge two adjacent parts $\mathcal{P}^k_i$, $\mathcal{P}^k_{i+1}$ into one new part $\mathcal{P}^{k+1}_j$ if $\epsilon^{k+1}_j \leq \theta_s$. Repeat until merging is not possible any more.
4. Shift the split point shared by two adjacent parts $\mathcal{P}^k_i$, $\mathcal{P}^k_{i+1}$ to left and right while leaving the overall number of parts fixed. Keep the split that reduces the overall error, repeat until no further changes occur.

2.2   Region Splitting of the Map

Before a region split and merge algorithm on the geometric map can be applied, it is necessary to create an initial split of the map. The easiest way to do so is to treat the whole map as a single large region defined by the start and end points of the journey. More sophisticated initializations can be used as well, e. g., based solely on the robot movement without taking into account range data [15].

After the initialization step, the actual division of the map into distinct regions is performed based on a split and merge that uses a residual error function $h(\mathcal{P}_i, \mathcal{P}_j)$ which compares two regions $\mathcal{P}_i$ and $\mathcal{P}_j$ and computes the homogeneity of the two regions (low values of $h(\mathcal{P}_i, \mathcal{P}_j)$ means homogeneous, high values very inhomogeneous). This function is used during the split phase for deciding whether a region $\mathcal{P}^k_i$ will be split again at a given position into two new regions $\mathcal{P}^{k+1}_j$ and $\mathcal{P}^{k+1}_{j+1}$, and in the merge (or shift) phase to determine whether two adjacent regions can be merged (or the splitting point be shifted). When the homogeneity is above a given threshold θ_r, the region will be split again.

The basic idea is to use the average width of a region in the map as a criterion for splitting, as a width change resembles a changing environment, e. g., a transition from a corridor to a big room. The homogeneity (residual) function used is:

$$h(\mathcal{P}_i, \mathcal{P}_j) = \frac{\max\{f_w(\mathcal{P}_i), f_w(\mathcal{P}_j)\}}{\min\{f_w(\mathcal{P}_i), f_w(\mathcal{P}_j)\}} + s_r r(\mathcal{P}_i, \mathcal{P}_j), \tag{1}$$

where $f_w(\mathcal{P}_i)$ is the average width of region $\mathcal{P}_i$, and $r(\mathcal{P}_i, \mathcal{P}_j)$ is a regularization term that takes care of additional constraints during splitting. The factor s_r controls the influence of $r(\mathcal{P}_i, \mathcal{P}_j)$.

Obviously, the average width is given by $f_w(\mathcal{P}_i) = \frac{A_{\mathcal{P}_i}}{l_{\mathcal{P}_i}}$, where $A_{\mathcal{P}_i}$ is the area of region $\mathcal{P}_i$, and $l_{\mathcal{P}_i}$ is its length. The definition of the length of a region in particular is not always obvious, but can be handled using the robot movement paths, which are part of each region. The length $l_{\mathcal{P}_i}$ is then defined by the length of the line connecting the start point of the first robot path of a region and the end point of the last path of the region, i. e., the line connecting the exits of an ASR, which the robot used while travelling through the environment. This is an approximation of a region's length having the advantage that disturbance caused by zig-zag movement of the robot during mapping does not affect the end result.

For area computation, the gaps contained in the map have to be taken into account, either by closing all gaps, or by using a fixed maximum distance for gaps. Closing a gap is a good approach if it originated from missing sensor data, but may distort the splitting result when the gap is an actual part of the environment. Closing it would make the region appear smaller than it actually is. Our implementation uses a combination of methods: small gaps are closed in a pre-processing step, large ones are treated as distant surfaces.

Depending on how gaps are handled, the algorithm possibly creates a large number of very small regions. This is where the regularization term $r(\mathcal{P}_i, \mathcal{P}_j)$ comes in: it ensures that regions do not get too small. It penalizes small regions but still allows to create them if the overall quality is very good. We use a sigmoid function centered at n, which is the desired minimum size of a region:

$$r(\mathcal{P}_i, \mathcal{P}_j) = \frac{1}{1 + \exp\left(-\frac{\min\{A_{\mathcal{P}_i}, A_{\mathcal{P}_j}\}}{A_{\max}} + n\right)} - 1 . \tag{2}$$

This function can assume values values between -1 and 0. The exponent is basically the ratio of the area of the smaller one of two adjacent regions to the maximum area $A_{\max}$ of the smallest possible region that the algorithm is still allowed to create. Thus, the smallest ratio is 1. It increases when the region gets larger.

The regularization term only has an influence when a region is already small, making it less likely to be split again. As the sigmoid reaches 0 asymptotically, it has virtually no influence when a region is large. The overall influence of the regularization can be controlled by the factor s_r in (1). It is given by $s_r = s\theta_r$, where $0 \leq s \leq 1$ is set manually and defines the percentage of the threshold θ_r that is to be used as a weight. θ_r is the threshold introduced earlier, which determines that a region is to be split into two when the first region is θ_r times larger than the second one.

3   Localization Strategies

In this section we describe the strategies used for localization based on ASR information, which is extracted from two sources, namely a map that has been generated on the outward journey, i. e., while the robot was exploring the environment, and a second map which is being generated during the homeward journey. A data fusion algorithm is applied for merging localization information computed by separate simple strategies with varying reliability. Each strategy by itself may be not sufficient for the robot to localize itself in the environment, but the combined result is. The fusion is based on the *Democratic Integration* technique proposed by Triesch and von der Malsburg [21,22]. Originally it was developed for the sensor data fusion in computer vision, and uses images as input data. The method has been extended and embedded into a probabilistic framework in [23,24], still within the area of machine vision. We have amended the original approach such that instead of images, information extracted from the ASRs is used. A main advantage of the fusion approach is that the extension is straightforward, i. e., more localization strategies can be added easily.

Two different strategies for localization of the robot with respect to the original map generated on its way to the current position are presented in the following. Each method computes a local confidence map that contains a confidence value between zero and one for each ASR of the original map. Note that these confidence values are not probabilities, and they do not sum up to one; the interval has been chosen for convenience, and different intervals can be used as desired. Refer to [21,22] for further details regarding this matter.

As the "cognitive maps" generated are a topological representation enhanced by metric information, there is no need to correct them for odometry drift. It is important to keep this in mind when developing new localization strategies: it necessitates the use of local information only, i. e., information extracted from adjacent ASRs, or information that can be computed without having to worry about any negative influence originating from odometry errors.

3.1 Distance

Just as humans have a rough notion of how far they walked starting at a certain location, so should a robot. Note that we are not talking about exact measurements, but rather about whether the robot has travelled, say, 5m or 10m. Also, we do not use the actual distance travelled as provided by odometry, because the robot often moves in a zig-zag fashion rather than straight, which would result in quite different distances for each journey through the same space. Particularly for large spaces, the odometry readings also depend highly on which path the robot actually took when crossing empty space, and whether it has explored the space wandering around multiple times before exiting. Our proposed solution for this problem is to use distance information obtained from the "cognitive maps" computed during outward and homeward journeys. These maps have been split into distinct ASRs, and the length of each ASR can be computed, which is defined by the distance between the entrance and the exit the robot used when passing through (cf. Sect. 2.2). In the maps shown, e. g., in Fig. 1, start and end points of an ASR are depicted by dark dots (split points) located on a set of connected lines representing the path the robot took. The zig-zag movement of the robot in between two splits is clearly visible, and can be quite different from the line connecting start and end points, in particular if an ASR has been generated from a large empty space. The strategy for computing a local confidence map that can later on be used for data fusion, is to compare the distance d travelled when returning home, measured in ASR lengths taken from the intermediate map computed on the return journey, to the ASR lengths taken from the original map computed during the mapping process.

The local confidence map $c_{\text{Dist}} \in \mathbb{R}^N$ (N being the total number of ASRs in the original map) is computed as follows: Each ASR's confidence is dependent on the overall distance d travelled on the return journey; the closer an ASR is to this distance from the origin, the more likely it is the one the robot is in currently. To model the confidences for each ASR we use a Gaussian, the horizontal axis being the distance travelled in mm, centered at the current overall distance travelled d. Its standard deviation σ is dependent on the distance travelled, and was chosen as $\sigma = 0.05d$. Note that although a Gaussian is used here, we do not try to model a probability density function, but rather make use of the bell-shape it provides. There are a number of reasons making it most suitable for our purpose: It allows for a smooth transition between ASRs, and the width can be easily adjusted by altering the standard deviation. This is necessary as the overall distance travelled gets more and more unreliable (due to slippage and drift) the further the robot travels.

The confidence value for a particular ASR is determined by sampling the Gaussian at the position given by the accumulated (ASR-)distances from the origin (i. e., where the robot started the homeward journey) to the end of this ASR. After a value for each ASR is computed, the local confidence map c_{Dist} is normalized to the interval $[0; 1]$.

3.2 Relative Orientation

The second method for computing local confidence maps containing estimates of the robot's position with respect to the original map is based on using relative orientation information between adjacent ASRs. During its journey, the robot enters an ASR at one location and exits at a different one, usually including zig-zag movements in between. We define the direction of an ASR as the direction of the line connecting the entrance and exit points. As before, direction information varies every time the robot travels through the environment, but the overall shape between adjacent ASRs is relatively stable. Therefore, we propose to use angles between ASR directions as a local measure of the current position of the robot. Note that this information is pretty much useless on its own, because the same angles (i. e., direction changes) can be found in different locations of the environment. However, combining this strategy with others can help to decide between position estimates that would otherwise be indistinguishable.

Firstly, all angles $\alpha_1, \ldots, \alpha_{N-1}$ between adjacent ASRs in the original map are computed. This can be done offline, as this map is fixed during the homeward journey. In the re-mapping process while returning home, new ASRs are computed in the new map based on data gathered while the robot travels. Using the direction information contained in this map, the angle β between the current ASR and the previous one can be computed. Comparing this angle to all angles of the original map gives a clue (or multiple clues) for the current location of the robot.

The comparison of angles is done by computing the difference angle between the current angle β obtained from the newly generated map and all angles α_i of the original map. This difference angle is then mapped linearly to the interval $[0; 1]$:

$$c_{\text{Dir}\,i} = -\frac{1}{\pi}|\alpha_i - \beta| + 1, \quad i = 1, \ldots, N - 1 \quad . \tag{3}$$

Another obvious choice would be to use the cosine of the difference angle instead of the linear mapping. However, this would "compress" confidence values for similar angles, which is not a desired effect.

The mapping given by (3) results in high values for similar angles and low values for dissimilar ones. The confidence map computed this way can already be used for further processing. Since the overall reliability of the relative orientation strategy as described above is rather low compared to the confidence values from other methods (in this case using distance information), we currently reduce the confidence values by a constant factor. As the data fusion method presented in the next section is capable of adjusting the relative weights of the different localization strategies, this is non-critical, because it will change automatically over time anyway, depending on the reliability of the other methods used.

3.3   Fusion of Strategies

The separate local confidence maps are merged into a single global one based on the Democratic Integration method proposed in [21,22]. Fusion itself is done by computing a weighted sum of all local confidence maps, which is straightforward and not a new concept. However, Democratic Integration allows for the weights to be adjusted dynamically and automatically over time, dependent on the reliabilities of the local map.

Given M local confidence maps $c_{1i}(t)$ at time t generated using different strategies, the global map $c_g(t)$ is computed as:

$$c_g(t) = \sum_{i=0}^{M-1} w_i(t) c_{1i}(t) \quad , \tag{4}$$

where $w_i(t)$ are weighting factors that add up to one. In this paper, $M = 2$, namely the local confidence maps based on distance (c_{Dist}) and relative orientation (c_{Dir}).

An estimate of the current position of the robot with respect to the original map can now be computed by determining the largest confidence value in $c_g(t)$. Its position b in $c_g(t)$ is the index of the ASR that the robot believes it is in. The confidence value c_{g_b} at that index gives an impression about how reliable the position estimate is in absolute terms, while comparing it to other ASRs shows the relative reliability.

In order to update the weighting factors, the local confidence maps have to be normalized first. The normalized map $c'_{1i}(t)$ is given by:

$$c'_{1i}(t) = \frac{1}{N} c_{1i}(t) \quad . \tag{5}$$

Recall that N is the total number of ASRs in the original map. The idea when updating the weights is that local confidence maps that provide very reliable data get higher weights than those which are unreliable. Different ways for determining the quality of each local confidence map are presented in [22]. We use the normalized local confidence values at index b, which has been determined from the global confidence map as described above, i. e., the quality $q_i(t)$ of each local map $c_{1i}(t)$ is given by $c'_{1b}(t)$. Normalized qualities $q'_i(t)$ are computed by:

$$q'_i(t) = \frac{q_i(t)}{\sum_{j=0}^{M-1} q_j(t)} \quad . \tag{6}$$

The new weighting factors $w_i(t + 1)$ can now be computed from the old ones:

$$w_i(t + 1) = w_i(t) + \frac{1}{t + 1} \left(q'_i(t) - w_i(t) \right) \quad . \tag{7}$$

This is a recursive formulation of the average over all qualities from time zero to t. Using this update equation and the normalization of the qualities in (6) ensures that the sum of the weights equals one at all times.

4   Experimental Results

For experimental evaluation we have used a Pioneer 3 robot from MobileRobots Inc (formerly Activmedia Robotics), equipped with eight sonar sensors and an odometer. Using sonar sensors only instead of, say, a laser range finder was a deliberate choice, as the restrictions imposed by the limited range of sonar sensors allow us to better test the cognitive mapping algorithms and the behaviour of the robot in large spaces. The robot must first explore and generate a representation of the environment, i. e., a "cognitive map". This is called the "outward journey" further on, the map is called "original map". At some point the robot is stopped and turned around. On its way back home (the "homeward journey") it remaps the environment and computes a new "cognitive map", which is used in conjunction with the original map for localization. Computation of the "cognitive map" and localization is done each time the robot stops, which is normally due to an obstacle in its way. Note that we turn off the robot and change its position slightly before it is allowed to start the remapping process. Therefore, the maps generated during the outward and homeward journeys are recorded in different global coordinate systems, and the robot has no way of aligning them. In the following we show experimental results for generating the spatial representation of the environment based on the split and merge method presented in Sect. 2, as well as experiments on how the robot uses this inexact "cognitive map" for localization on its way back home using the localization method from Sect. 3. The parameters used for computing the maps were the same for all runs, namely $\theta_r = 1.7$ and $s = 0.2$.

The range of the sonar sensors is about 4m; consequently, spaces larger than the sonar range cannot be apprehended with a single scan, but the robot has to move in order to build a spatial representation of the environment. Remember that the "cognitive map" is a metric-topological representation rather than a purely metric one, and the data acquired during mapping is not corrected for odometry errors. This is one of the main advantages of choosing a cognitive mapping approach over more traditional robot mapping techniques like SLAM. Obviously odometry errors will be visible in the generated maps, but as long the algorithm is able to generate a single ASR as opposed to multiple ones from the data acquired, say, in a large room, this does not pose a problem.

We conducted various experiments in an office environment that contains a large room having dimensions of approximately 9m × 15m. In particular when the robot is close to the center of this room, it does not obtain any sensor readings. Depending on the angle of movement, this may even happen at positions farther from the room center. Two experiments have been selected for a detailed discussion in this paper. Each experiment consists of the original map generated on the outward journey, and maps computed on the way back home as well as confidence maps used for localization. The experiments are labelled *Experiment 1* and *Experiment 2*, respectively.

Figures 1(a) and 2(a) show the layout of the environment used, the large room being in the center of the building. The paths that the robot took during mapping (solid line) and going home (dashed line) are both shown as well. For the mapping stage, the robot started at the location marked by 'X' and was stopped at a random position, turned around and started the homeward journey nearby (marked by 'O'). The return journey stopped when the robot believed that it reached home, or, more precisely, the ASR that contains the start location 'X'. The ASRs generated during the outward journey are

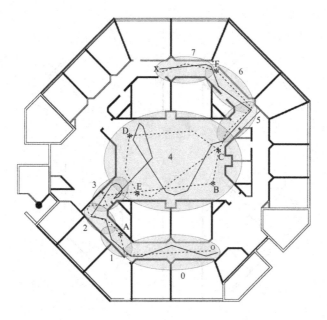

(a) Actual floor plan Experiment 1

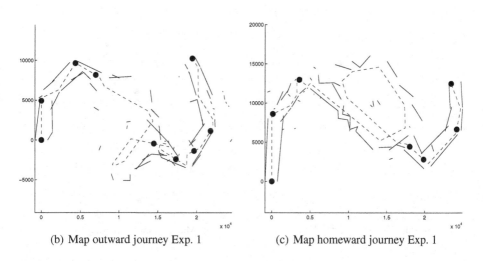

(b) Map outward journey Exp. 1   (c) Map homeward journey Exp. 1

Fig. 1. Experiment 1: (a) Map showing the actual layout of the building. The path the robot took during the outward journey is shown as a solid line, with 'X' marking the starting location. The dashed line visualizes the path for the homeward journey and 'O' marks its starting point. ASRs generated during the outward journey are indicated by ellipses. The labels A - F indicate positions during the homeward journey referred to in the text and in Fig. 3. (b) Map generated during outward journey ("original map"). (c) Map generated during homeward journey.

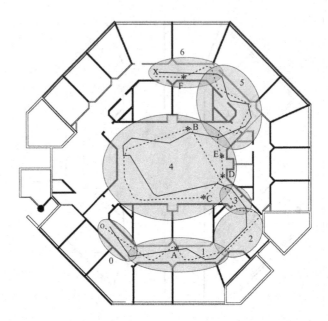

(a) Actual floor plan Experiment 2

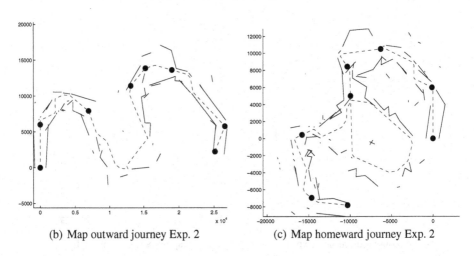

(b) Map outward journey Exp. 2   (c) Map homeward journey Exp. 2

Fig. 2. Experiment 2: (a) Map showing the actual layout of the building. The path the robot took during the outward journey is shown as a solid line, with 'X' marking the starting location. The dashed line visualizes the path for the homeward journey and 'O' marks its starting point. ASRs generated during the outward journey are indicated by ellipses. The labels A - F indicate positions during the homeward journey referred to in the text and in Fig. 4. (b) Map generated during outward journey ("original map"). (c) Map generated during homeward journey.

indicated by ellipses and numbered from zero starting at the *last* ASR (so that during the homeward journey ASRs are visited in ascending order).

The "cognitive map" generated during the outward journey in *Experiment 1* is shown in Fig. 1(b), the map computed on the homeward journey in Fig. 1(c); likewise for *Experiment 2* in Figs. 2(b) and 2(c). Note that the paths the robot took during outward and homeward journey are quite different, particularly when it maps the big room. All units are given in millimeters, black dots indicate the split points between ASRs. Due to the re-initialization of the robot before returning home, the starting point for both, outward and homeward journeys, is the origin of the coordinate system, which also results in the map depicted in Fig. 1(c) to be upside-down with respect to the map in Fig. 1(b). When comparing the maps generated during both journeys, it becomes obvious that different representations may be computed, in particular a different number of ASRs, and splits generated at different locations. This is due to sensory inaccuracies, but it does not pose a problem for the localization method we use. The position of the robot is always given with respect to the ASR representation of the original outward journey map. What can be clearly seen in all maps generated in both experiments is that the algorithm is capable of representing the large room as a single ASR, independent of the path that the robot took while mapping. The split points are nicely located near the exits of the room, as desired. Apart from that, the algorithm separates corridors from bigger rooms (e. g., ASR 1 and 2 in Fig. 2), and corridors from other corridors, most of the times at locations where a human would do so as well. Although robots and humans generating cognitive maps may not necessarily come up with alike representations due to the "hardware" being quite different, this is a desired result as we try to simulate a human cognitive mapping process.

We will now take a closer look at the (obviously incomplete) maps and localization information generated at intermediate stops during the homeward journey. For this purpose we have selected graphs computed at six positions on the homeward journey, which allow for interesting insights into how the localization performs. These positions are marked by '*' and labelled A - F in Fig. 1(a) (*Experiment 1*) and 2(a) (*Experiment 2*).

Figures 3(a) to 3(f) show intermediate maps generated at positions A - F in *Experiment 1* (cf. Fig. 1(a)). The corresponding confidence maps are depicted in Figs. 3(g) to 3(l). In the confidence graphs, the light dotted line shows the ASR estimate using the ASR length information (distance method) and the dark dashed line depicts the ASR estimate using the angles between ASRs (relative orientation method). The solid line is the overall estimate after fusion. The horizontal axis corresponds to the index of the ASR, the vertical axis to the confidence, which can have values between zero and one.

The confidence graph at position A in Fig. 3(g) shows a peak for the overall confidence at ASR 1, signifying the robot is very confident of being in this particular ASR. The same is true for the graph in Fig. 3(h) corresponding to position B, where the robot has moved far into the big room in the center being ASR 4. It can be seen that position C is close to the border between ASRs 4 and 5, which is reflected by high confidence values for both ASRs in the confidence map in Fig. 3(i), where the robot gets more and more unsure about whether it is still in the big room or whether it has entered the next region yet (ASR 5). As it moves further away from the exit and back into the room, reaching position D it is very confident again that it is still in ASR 4, not having actually

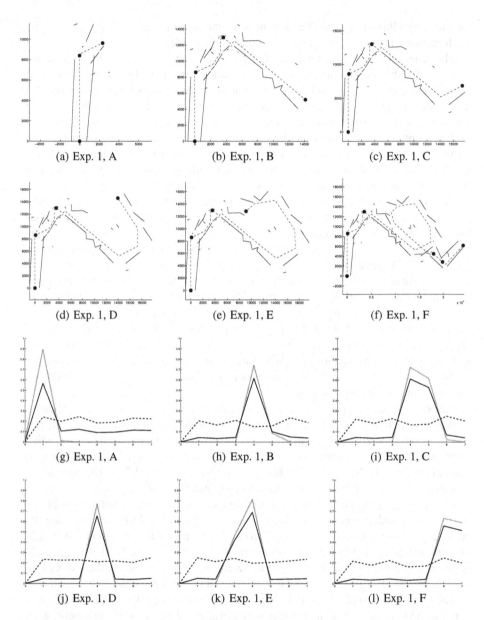

Fig. 3. Experiment 1: (a) - (f) Maps at intermediate stops A - F (cf. Fig. 1(a)); (g) - (l) Confidence maps corresponding to stops A - F: distance (light dotted), relative orientation (dark dashed), and overall confidence (solid). Horizontal axis: ASR number; vertical axis: confidence (0 to 1)

exited the big room (see Fig. 3(j)). Position E is close to where the robot has entered the room, coming through ASR 3. Again, this can be observed in the confidences plotted in Fig. 3(k), which shows that the robot is still quite confident of being in ASR 4, but where the value for ASR 3 has increased considerably, i. e., the robot "knows" that it

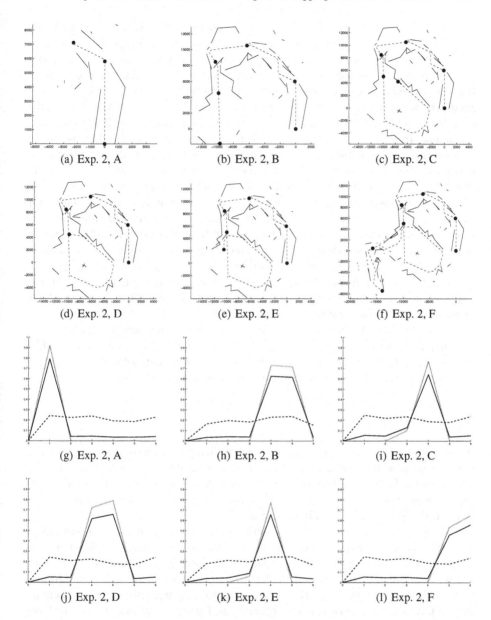

Fig. 4. Experiment 2: (a) - (f) Maps at intermediate stops A - F (cf. Fig. 2(a)); (g) - (l) Confidence maps corresponding to stops A - F: distance (light dotted), relative orientation (dark dashed), and overall confidence (solid). Horizontal axis: ASR number; vertical axis: confidence (0 to 1)

is close to where it was when it entered the room. Finally, the graph in Fig. 3(l) was generated at position F, which is at the border between ASRs 6 and 7, reflected by high confidences for both.

Figures 4(a) to 4(f) show intermediate maps generated at positions A - F in *Experiment 2* (cf. Fig. 2(a)), and the corresponding confidence maps in Figs. 4(g) to 4(l). As in the previous experiment, when reaching position A the robot is very confident about being in ASR 1, indicated by the distinct peak in the graph in Fig. 4(g). At position B it has entered the big room being ASR 4. As can be seen in Fig. 4(h) it is not sure about whether it is still in ASR 4, or whether it has reached the next ASR already, the confidences for both being equally high, with a slight bias towards ASR 4. Taking into account odometry inaccuracies and the information that is available to the robot using the present localization methods, this behaviour makes perfect sense. When it reaches position C after travelling through the big room in a loop, the robot is highly confident (Fig. 4(i)) that it is still in ASR 4 (which is true for the positions in between as well, which are not shown). Being close to where it has entered the room, the confidence for ASR 3 starts to increase at this location. This effect becomes more prominent in Fig. 4(j), when the robot is at position D, which is at the border between ASRs 3 and 4. Again, it is unsure about which ASR it is in, meaning that it can infer that it actually is at the border between two regions, near the entrance to the room. Moving away from the border to position E (Fig. 4(k)), it is again confident about being in ASR 4. At the last position F, the robot is very close to its home position in ASR 6 already, which is reflected in the confidence values shown in Fig. 4(l).

The results show that the method proposed provides a consistent approach for creating and using an inexact "cognitive map" to allow a mobile robot to localize itself. It does not provide the exact pose of the robot in the environment, but rather an approximation, which we believe is sufficient for navigation and new exploration.

5   Conclusion

We have presented algorithms for creating and utilizing a "cognitive map" on a mobile robot equipped with sonar sensors and an odometer. As the range of the sonars is only a few meters, the robot cannot apprehend large spaces at once. The main focus in this paper was on showing that a "cognitive map" can be computed robustly for these spaces, and how the robot localizes itself in these spaces.

The algorithm for creating the "cognitive map" is based on the split and merge method. It divides a given metric map into distinct regions (namely ASRs), thus creating a topological representation on top of the metric one. Features of the ASRs thus computed are then used as part of the localization strategy, which can fuse information from different sources, resulting in a confidence map that tells the robot where its current location in the environment supposedly is. Currently, we exploit basic features only, namely ASR-distance travelled and relative orientation between adjacent ASRs. The purpose of these two simple is strategies is mainly to show how fusion and localization work with our approach. They can easily be replaced or augmented by more sophisticated methods: As long as each method computes a confidence value for each ASR, the integration into the fusion approach is straightforward. Nevertheless, even the basic strategies presented here are powerful enough to allow for a localization that should be accurate enough for many applications.

Much has been discussed with respect to the use of distance information in cognitive mapping. For example, numerous experiments with chickens and pigeons have shown that they are able to use both absolute and relative distance in their search for food (e.g., [25]). Experiments with bees and ants have shown that they can perform internal calculations of the distance and direction travelled to perform path integration (e.g., [26] for a general discussion). Most of these experiments were concerned with the actual distance travelled and how the individual species deal with the errors in their measurements, as do most work on robot mapping to date. Using our robot, we have shown another way of using distance information, namely ASR-distance travelled as opposed to actual distance travelled.

ASR-distance is obtained from the shape of the ASR computed. In the past, there has been scant evidence that humans/animals do pay attention to the shape of each local environment (or, in our terminology, ASR) very early on in their initial exploration of a new environment. However, the debate has now intensified and this is especially true in the animal literature where the problem is commonly referred to as geometry in animal spatial behavior [27]. In many of these experiments, a relocation task utilizing a box-shaped environment is used, and the principal axes of the environment appear to be most useful. Our work here emphasized yet another possibility, namely using a straight line distance between exits of interests in an ASR.

References

1. Tolman, E.C.: Cognitive Maps in Rats and Men. Psychological Review 55(4), 189–208 (1948)
2. Golledge, R.G.: Human wayfinding and cognitive maps. In: Golledge, R.G. (ed.) Wayfinding Behavior: Cognitive Mapping and Other Spatial Processes, John Hopkins University Press, Baltimore (1999)
3. Poucet, B.: Spatial cognitive maps in animals: New hypotheses on their structure and neural mechanisms. Psychological Review 100, 163–182 (1993)
4. Chown, E., Kaplan, S., Kortenkamp, D.: Prototypes, location, and associative networks (PLAN): Towards a unified theory of cognitive mapping. Cognitive Science 19(1), 1–51 (1995)
5. Kuipers, B.: The spatial semantic hierarchy. Artificial Intelligence 119, 191–233 (2000)
6. Yeap, W.K., Jefferies, M.E.: Computing a Representation of the Local Environment. Artificial Intelligence 107(2), 265–301 (1999)
7. Estrada, C., Neira, J., Tardos, J.D.: Hierarchical SLAM: Real-Time Accurate Mapping of Large Environments. IEEE Trans. on Robotics 21(4), 588–596 (2005)
8. Piers, D.M., Kuipers, B.J.: Map learning with uninterpreted sensors and effectors. Artificial Intelligence 92, 169–227 (1997)
9. Kortenkamp, D.: Cognitive Maps for Mobile Robots: A Representation for Mapping and Navigation. PhD thesis, University of Michigan (1993)
10. Jefferies, M., Weng, W., Baker, J.T., Mayo, M.: Using context to solve the correspondence problem in simultaneous localisation and mapping. In: Zhang, C., W. Guesgen, H., Yeap, W.-K. (eds.) PRICAI 2004. LNCS (LNAI), vol. 3157, pp. 664–672. Springer, Heidelberg (2004)
11. Gaussier, P., Revel, A., Banquet, J.P., Babeau, V.: From view cells and place cells to cognitive map learning: processing stages of the hippocampal system. Biol. Cybern. 86, 15–28 (2002)

12. Hafner, V.V.: Cognitive maps in rats and robots. Adaptive Behavior 13(2), 87–96 (2005)
13. Franz, M., Mallot, H.: Biomimetic Robot Navigation. Robotics and Autonomous Systems 30, 133–153 (2000)
14. Gallistel, C.R.: Animal Cognition: The Representation of Space, Time and Number. Annual Review of Psychology 40, 155–189 (1989)
15. Schmidt, J., Wong, C.K., Yeap, W.K.: A Split & Merge Approach to Metric-Topological Map-Building. In: Int. Conf. on Pattern Recognition (ICPR), Hong Kong, vol. 3, pp. 1069–1072 (2006)
16. Schmidt, J., Wong, C.K., Yeap, W.K.: Mapping and Localisation with Sparse Range Data. In: Mukhopadhyay, S., Gupta, G.S. (eds.) Proceedings of the Third International Conference on Autonomous Robots and Agents (ICARA), Palmerston North, New Zealand, pp. 497–502 (2006)
17. Wong, C.K., Schmidt, J., Yeap, W.K.: Using a Mobile Robot for Cognitive Mapping. In: International Joint Conference on Artificial Intelligence (IJCAI), Hyderabad, India (January 2007), pp. 2243–2248 (2007)
18. Duda, R.O., Hart, P.E.: Pattern Classification and Scene Analysis. John Wiley & Sons, New York, USA (1973)
19. Pavlidis, T., Horowitz, S.L.: Segmentation of Plane Curves. IEEE Trans. on Computers C-23, 860–870 (1974)
20. Niemann, H.: Pattern Analysis and Understanding, 2nd edn. Springer Series in Information Sciences, vol. 4. Springer, Berlin (1990)
21. Triesch, J.: Vision-Based Robotic Gesture Recognition. Shaker Verlag, Aachen (1999)
22. Triesch, J., von der Malsburg, C.: Democratic Integration: Self-Organized Integration of Adaptive Cues. Neural Computation 13(9), 2049–2074 (2001)
23. Denzler, J., Zobel, M., Triesch, J.: Probabilistic Integration of Cues From Multiple Cameras. In: Würtz, R. (ed.) Dynamic Perception, pp. 309–314. Aka, Berlin (2002)
24. Kähler, O., Denzler, J., Triesch, J.: Hierarchical Sensor Data Fusion by Probabilistic Cue Integration for Robust 3-D Object Tracking. In: IEEE Southwest Symp. on Image Analysis and Interpretation, Nevada, pp. 216–220. IEEE Computer Society Press, Los Alamitos (2004)
25. Cheng, K., Spetch, M.L.D.M.K, Bingman, V.P.: Small-scale Spatial Cognition in Pigeons. Behavioural Processes 72, 115–127 (2006)
26. Cornell, E.H., Heth, C.D.: Memories of travel: Dead reckoning within the cognitive map. In: Allen, G. (ed.) Human spatial memory: Remembering where, pp. 191–215. Lawrence Erlbaum Associates, Mahwah, NJ (2004)
27. Cheng, K., Newcombe, N.S.: Is there a geometric module for spatial orientation? Squaring theory and evidence. Psychonomic Bull Rev. 12, 1–23 (2005)

Spatial Mapping and Map Exploitation: A Bio-inspired Engineering Perspective

Michael Milford[1] and Gordon Wyeth[2]

[1]Queensland Brain Institute
[2]School of Information Technology and Electrical Engineering
The University of Queensland, Brisbane, Australia
{milford,wyeth}@itee.uq.edu.au

Abstract. Probabilistic robot mapping techniques can produce high resolution, accurate maps of large indoor and outdoor environments. However, much less progress has been made towards robots using these maps to perform useful functions such as efficient navigation. This paper describes a pragmatic approach to mapping system development that considers not only the map but also the navigation functionality that the map must provide. We pursue this approach within a bio-inspired mapping context, and use results from robot experiments in indoor and outdoor environments to demonstrate its validity. The research attempts to stimulate new research directions in the field of robot mapping with a proposal for a new approach that has the potential to lead to more complete mapping and navigation systems.

1 Introduction

The spatial mapping problem has been the subject of much research in the robotics field, resulting in the existence of several well-established mapping algorithms [1]. Sensor and environment uncertainty has caused most robotic mapping and navigation methods to converge to probabilistic techniques. The key strength of probabilistic techniques is their ability to deal with uncertainty and ambiguity in a robot's sensors and environment. Any technique that has some means of handling the uncertainties faced by a mobile robot has an immense advantage over techniques that do not. Probabilistic techniques allow a robot to appropriately use sensory measurements based on their modeled uncertainties. Ambiguous features or landmarks in the environment become useful (albeit less useful than unique features) rather than becoming failure points for the mapping algorithm.

Under appropriate conditions some of these systems can solve the core SLAM (Simultaneous Localization And Mapping) problem in large, real world, static environments. The world representations produced by these systems generally take the form of occupancy grids, or landmark maps. However, while a high resolution (often as small as 10 mm) occupancy grid can faithfully represent the physical structure of an environment, it is not necessarily the most 'usable' representation for general navigation tasks. For environments of any significant size, the amount of data stored in such a map becomes quite large, and requires significant processing before it

S. Winter et al. (Eds.): COSIT 2007, LNCS 4736, pp. 203–221, 2007.

can be used by a robot to perform tasks such as goal navigation. The limitations of such representations are perhaps revealed by the imbalance between a relative abundance of competent mapping algorithms [2–8] and a scarcity of robotic systems that use these maps to perform navigation and other useful tasks [9–11].

This problem has been partially addressed by other mapping techniques that produce topological or hybrid metric-topological representations. Techniques such as the Hybrid Spatial Semantic Hierarchy [12,13] produce maps that are perhaps more suited to route planning and navigation than purely grid-based maps. By embedding navigational concepts into the representation, such as the locations of exits from a room, the maps produced are already somewhat 'navigation-ready' so to speak. A robot need only pick which exit to go through in order to progress towards a goal, rather than process an occupancy grid map to extract possible exit locations.

The concept of map usage shaping the nature of the map applies not only in spatial mapping, but also conceptual and biological mapping domains. In the conceptual domain the structure and form of conceptual maps created by humans is known to change significantly depending on motivation and context [14]. In rodents, place cell firing varies depending on the behavior of the rat at the time [15,16]. Reward location, movement speed, movement direction, and turning angle can all affect place cell firing and even cause remapping of place fields [15]. Furthermore, the place cell maps can gradually change to provide the rat with efficient trajectories between reward sites [17].

Animals are, in fact, excellent examples of the co-evolution of mapping and map usage processes. Foraging honeybees are known as central-place foragers because they have a common point to start and finish their foraging trips. Their 'maps' must provide them with the ability to return to their hive after foraging, a journey typically of two or three kilometers, but stretching as far as 13.5 kilometers [18]. When returning to the nest from a foraging location, the desert ant *Cataglyphis fortis* uses its *home vector*, which is calculated during the outbound trip using weighted angular averaging [19]. Primates possess spatial view cells, that allow them to remember where an object was last seen, even if it is currently obscured, a very useful ability in their natural environment. Each animal or insect represents the environment in a manner which suits the task they must perform in it.

This paper presents a pragmatic bio-inspired approach to developing a robot mapping system that considers both the map and the navigation functionality that the map must provide. The research occurred as part of the RatSLAM project, the aim of which was to use models of rodent hippocampus to produce a robot mapping and navigation system [20–23]. By considering the usability of the maps in navigation tasks, the project developed a robot mapping system with significantly different characteristics to those developed with only the SLAM problem in mind. Furthermore, the bio-inspired approach incorporated concepts from three separate mapping fields; robotic mapping, mapping in nature, and computational models of biological mapping systems.

The hypothesis in this paper is that mapping methods for an autonomous robot must necessarily develop in parallel with mechanisms for map exploitation. Only with awareness of the entire mapping and navigation problem will researchers be able to develop autonomous mobile robot systems that can be deployed across all environments and situations. Since biological systems fulfill both mapping and navigational needs of animals and insects, this hypothesis is pursued from a bio-inspired perspective. We present results demonstrating the capability of a bio-inspired robotic system developed using the proposed approach.

The paper is organized as follows: We start by performing a comparative review of what is known about the mapping and navigation processes in nature and the state of the art robotic mapping techniques. The characteristics of mapping systems in both areas are discussed with respect to solving the entire mapping and navigation problem, rather than just creating a spatial map, resulting in the proposed approach to developing robotic mapping systems. We then describe an example of the pragmatic use and extension of models of biological systems, and the increase in mapping performance that can be achieved by doing so. The parallel development of mapping and navigation processes is illustrated within the context of the RatSLAM system, and results are presented demonstrating the resultant system's effectiveness in indoor and outdoor environments. The paper concludes with a discussion of the significant issues raised by the study.

2 Mapping and Navigation in Nature and Robots

There are many forces driving the diversity that is apparent when examining biological and robot mapping and navigation systems. In nature, creatures and their navigation systems have evolved to suit their environments and their own physical and sensory characteristics. In the natural world, the range of body types, sensory equipment, environments, and lifestyles has produced a myriad of solutions to the problems facing a creature that needs to move around effectively in its environment. Likewise in the research labs of robotics researchers and the domestic home or industrial workplace, mapping and navigation systems have developed in ways to suit the environments and the sensors available, but less so the purpose of the robot.

Given this diversity it seems a challenging task to identify one research direction that is most likely to yield a complete robot mapping and navigation system. The specific mechanisms, representations, and context of each system differ so greatly that direct comparison is not possible. Fortunately, there are a number of fundamental mapping and navigation issues that are common to all these systems. Through consideration of these issues for all these systems it is possible to define goals for future research in this field, and in the process identify one means by which these goals may be met.

The following sections compare and contrast biological systems and models of biological systems with robotic mapping algorithms in a range of areas, with a focus in the biological area on rodents. The comparison highlights the shortcomings of both types of system as well as their complementary characteristics. The shortcomings are particularly relevant when considering where future research into the mapping and navigation problem should concentrate. After a discussion of possible future research approaches, the final section presents a proposed approach for developing complete robot navigation systems.

2.1 Robustness Versus Accuracy

One of the most significant differences between biological mapping systems and probabilistic methods is the world representations that they use. Many probabilistic systems incorporate high resolution occupancy grid maps, such as those shown in Figure 1. In work by Grisetti, Stachniss *et al.* [7] in the Freiburg campus environment, the occupancy grid contained 625 million squares. By contrast the place cells found in the rodent hippocampus appear quite coarse, and likely do not encode information to such a precise degree. The place cells certainly do not represent occupancy, but instead represent the rodent's location in space. The place cells give localization, but the map itself is widely distributed in other regions.

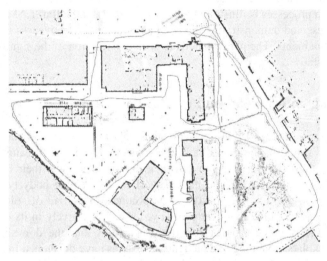

Fig. 1. High resolution grid map of the University of Freiburg campus, measuring 250 × 250 m, with a grid resolution of 0.01 m. The map was generated using the extended Rao-Blackwellized particle filter method [7]. Adapted from Figure 4 of [7] and reprinted with permission of IEEE. © 2005 IEEE.

Furthermore many probabilistic based mapping methods attempt to produce a Cartesian map with direct spatial correspondence to the physical environment. A 2.5 meter wide, 16.7 meter long corridor should appear as an identically sized region of free space in an occupancy grid map, with occupied cells along its edges. Biological systems relax the requirement that there be direct Cartesian correspondence between a map and the environment. In rodents, place field size and shape can vary significantly depending on the type of environment. For instance, the addition of straight barriers to an environment resulted in the destruction of the place fields located where the barrier was added [24]. Barriers can also cause an apparent remapping from the affected place cells into new place cells. Many animals appear to rely more on robust coping strategies than precise representations of their world. One such example is the desert ant; it navigates back to the general vicinity of its nest and then uses a systematic search procedure to find it [19].

2.2 Map Friendliness Versus Map Usability

Maps that are highly accurate, precise, and Cartesian have a number of advantages and disadvantages. One immediate observation is that they are 'human-friendly'. Looking at the map in Figure 1, a human can immediately make a number of high level observations about the environment – there are multiple loops in the environment, several buildings, large open spaces, and so on. It is easy for a human to relate parts of the map back to physical places in the environment, and also to compare the map and environment on a global scale.

In contrast, it is hard to present an equivalent map for any animal or insect. Place fields observed from brain readings in the rodent hippocampus convey only the observed locations in the environment to which a single place cell responds. There is no definitive method of combining the environmental features that a rodent perceives with the place fields of each place cell to produce a single 'conventional' map. An occupancy grid map could be created by using place cell activity and externally determined ranges to obstacles. It is unlikely however such a map would be representative of the rodent's actual mapping process. The extent to which rats rely on "range-to-obstacle" concepts is unknown, as is whether rats use a linear geometric scale for ordering the map.

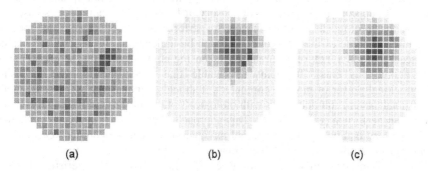

(a)   (b)   (c)

Fig. 2. Sample combined place cell firing rates and area occupancy maps. (a) Time spent by rat at each location in the circular environment, darker squares indicate longer occupancy times. (b) Spike map showing firing rate of one complex-spike cell over the environment, darker squares indicate a higher firing rate. (c) The firing rate array is obtained by dividing the spike map by the time spent in each location, to reveal a single peak of activity.

The maps shown in Figure 1 are human-friendly in that a human can easily understand them. A human could use these maps to navigate efficiently to various locations. However, in doing so a number of high level classification and planning processes would be used. The narrow straight sections of free space in Figure 1 might immediately be classified as pathways, and the large obstacle shapes as buildings. A human might consciously trace along the paths shown in the map, before choosing and following the shortest one. The overall process would be very different to the subconscious navigation that occurs as a person drives to and from work, walks around an office building, or moves around in their house and garden.

This distinction is an important one – the form of most maps created by probabilistic methods is governed by the underlying mathematical algorithms, or human requirements that the map be in a form suitable for performance analysis. In contrast, biological mapping systems have evolved to produce representations that are suited to the navigation tasks animals must perform every day.

2.3 Sensory Differences

Animals and robot platforms have different sensory equipment and movement capabilities. Probabilistic methods exploit the particular characteristics of the sensors they use. Biological sensors have evolved for important activities such as finding food and mating, with biological navigation systems evolving to suit the sensors and activities. Models of biological systems emulate the biological mapping and navigation mechanisms while using robotic sensors. Some researchers have applied compensatory algorithms to the output of the robotic sensors in order to more closely emulate the biological systems. For instance, the use of certain color spaces minimizes the effect of illumination differences [25], a task that many biological vision sensors achieve effortlessly. Other research approaches have involved plausibly modifying the models so that they take advantage of the differences in sensing equipment. For instance, instead of using a camera, a set of sonar sensors can be used to detect bearings and ranges to environment cues [26].

In the search for a biologically inspired mapping and navigation system, there are two ways to approach this problem of sensory differences. One approach is based on the fact that biological systems manage to perform quite well despite the specific limitations of some of their sensors. Although biological vision sensors can be quite advanced, there is a lack of accurate range sensing equipment (with the exception of such animals as bats). It seems reasonable that given the rapidly increasing computational power of modern computers, it should eventually be possible to create an artificial system equipped only with biologically faithful sensors that can match the navigation performance of animals. This approach dictates the meticulous emulation of the sensory equipment and theorized navigation strategies of animals even if this means purposefully handicapping certain aspects of the systems.

The other approach involves trying to exploit the superior characteristics of some robotic sensors by extending the mapping and navigation models. Given the current inability to match the capabilities of many biological sensors, it also seems reasonable that the superior characteristics of some robotic sensors be exploited to make up for other shortcomings in the models. In this approach the focus is on creating functional robotic systems, rather than faithful replication of proposed biological mechanisms. Biological mapping and navigation mechanisms are modified to accommodate the different characteristics of robot sensors. This approach has received relatively little attention, unlike the fields focusing on biologically inspired mechanical robot design. There has been only a limited amount of research into developing practical biologically inspired robot mapping and navigation systems [27].

2.4 Capability in Real World Environments

Biological navigation systems perform well enough to allow the millions of species of animals using them to function and survive every day. These systems combine advanced sensing, clever mapping and robust navigation mechanisms. There is a reasonable amount of knowledge about the capabilities of their sensors, and experiments have gathered a significant amount of knowledge about the navigation capabilities of these animals. Many theories have been devised to account for their capabilities, but in some areas research is only starting to scratch the surface. However, there is no question that animals can navigate well in a wide range of complex, dynamic environments.

In most areas, the state of the art in robotic mapping and navigation systems has not yet come close to matching the abilities of animals. In specific subsets of the problem and under certain conditions, these systems may outperform their biological counterparts, but it is with the sacrifice of robustness and flexibility, and is usually accompanied by a host of assumptions and the use of advanced sensing equipment. Nevertheless the best methods can produce impressive maps of large indoor and outdoor environments [6, 7, 10]. Most of the major mapping and navigation problems, such as closing the loop or coping with ambiguity, can be solved by one or another of these methods with appropriate assumptions.

In contrast to the robotic systems, computational models of the rodent hippocampus have only been used in simulation or on robots in small structured environments, and are yet to solve many of the major mapping and navigation problems. The small size and limited complexity of these environments reduces the major mapping and navigation challenges such as closing the loop, dealing with extended sensory ambiguity, and navigating to goals to trivial or non-existent problems. None of these models have been tested in or successfully applied to the large unmodified environments in which one might reasonably expect an autonomous mobile robot to function, such as an office floor or an outdoor set of pathways.

Biologically inspired models are partly limited because the goal for much of the research is to test navigational theories for a particular animal, rather than to produce a fully functional robot system [28]. The uncertainty about biological systems and subsequent speculation has produced models that may only be partially faithful to the biology, with resulting navigation performance that is inferior to that of the animal. In pursuing the development of a biologically plausible model, it is unlikely that a researcher will stumble upon a better performing model by chance – current biological systems are the product of a long sequence of evolution.

2.5 Proposed Approach

Given the current state of robotic and biological mapping and navigation systems, several conclusions can be drawn. It is unlikely that research in the near future will create a perfect model of the higher level mapping and navigation systems, such as those of a rodent, primate, or human. Animal brains are only partially understood; researchers create theories from the readings of a few dozen electrodes, but the theories are far from being comprehensively proven. Even though models may recreate some aspects of animal navigation behavior, there can be no real confidence that the underlying mechanisms driving the artificial system are the same as those in the real animal. Furthermore, even biological systems do not necessarily possess all the capabilities autonomous robots require to function in the challenging environments earmarked for their deployment.

Conventional robot mapping and navigation research is also facing many challenges. The most impressive recent demonstrations of mobile robotics have been largely made possible by good engineering and incremental improvements in algorithms, sensors, and computational power. The Defence Advanced Research Projects Agency (DARPA) Grand Challenge of 2005 is one prime example of this; while the onboard navigation and mapping systems were state of the art, it was only with an impressive array of expensive, high precision sensing equipment that this specific task could be solved [29, 30]. Some may argue that the continual improvement in sensors and computer power may eventually lead to navigation systems that surpass all the abilities of one of the most adept of navigators – humans. However, it is perhaps more likely that this milestone will be achieved through fundamental methodology changes, rather than steady computational and sensory improvements, although such changes will definitely facilitate the process.

So, where to look for a solution to the mapping and navigation problem? This paper proposes that an eventual solution may be found using a biologically inspired yet completely pragmatic approach, which considers both mapping and navigational requirements. Previous research has investigated bio-inspired robotic mapping methods, although without concurrent consideration of the how the map would be used [31]. However, research on pragmatic models of biological systems to date has still had a heavy emphasis on biological plausibility and has had limited practical success. No research has developed or tested a biologically inspired model under the same conditions and criteria as conventional robot mapping and navigation systems.

For example, the biologically inspired model developed by Arleo [32] can perform localization, mapping and goal recall, but only in a very small artificial arena, with continuous visual input from artificial cues or a distal cue source. Other less biologically faithful models have displayed limited re-localization ability in robot experiments in an artificial arena with visual cues, and have been able to navigate to goals in simple T and two arm mazes in simulation [31]. These approaches have also been fundamentally limited by the rodents on which they are based. The capabilities of rodent mapping and navigation systems do not necessarily fulfill all the desired capabilities of an autonomous mobile robot. If a biologically inspired system is ever to compete with conventional probabilistic methods, it must also contain solutions to the navigational shortcomings of animals. The approach presented in this paper therefore also sought to investigate whether models of the rodent hippocampus can serve as a basis for a complete robot mapping and navigation system.

3 Pragmatic Biological Modeling

To evaluate the potential of current hippocampal models as the basis for a robot navigation system, a model was developed based on the structure shown in Figure 3. Experiments were run on a Pioneer 2DXe mobile robot from Mobile Robots Inc (formerly ActivMedia Robotics). Ideothetic information for path integration is derived from the robot's wheel encoders. Allothetic information for generating the local view is derived from camera images. The testing environment was a flat two by two meter area of a laboratory floor. Colored cylinders were used as visual cues for the robot and were placed just outside the arena in various configurations.

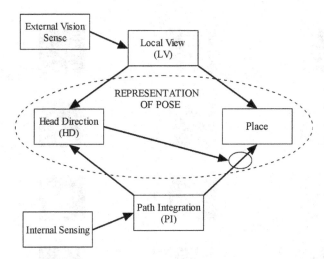

Fig. 3. Overall structure of the hippocampal model

The initial study was based on well known models of the hippocampus [32–34] as shown in Figure 3. The head-direction cells were implemented using a competitive attractor network (CAN) made up of neural units or 'cells' roughly corresponding to biological head-direction cells. Each cell is tuned to be maximally active when the robot's orientation matches the cell's preferred direction. The cell's activity level reduces as the robot orientation rotates away from this preferred direction. The cell arrangement reflects their associated robot orientations – nearby cells encode similar robot orientations. When the ensemble activity of the head-direction cells is viewed as a graph, one can see a 'packet of activity' that resembles a Gaussian curve (see Figure 4 for examples). The center of this 'activity packet' represents the current perceived orientation of the robot.

The place cells were modeled as a two-dimensional CAN, with each cell tuned to be maximally active when the robot is at a specific location. A coarse representation is used, with the path integration system tuned so that each place cell represents a physical area of approximately 250 mm × 250 mm. The cells are arranged in a two-dimensional matrix with full excitatory interconnectivity between all cells. Further details on the implementation can be found in [33], and is similar to the extended implementation described in later sections.

3.1 Experimental Performance

In rotation-only experiments, the robot was able to maintain an accurate estimate of its orientation. However, when the robot moved around the environment, the network's tracking ability proved to be unstable. Over the period of an hour the robot became lost and its perceived location moved well outside its two by two meter arena.

Closer examination revealed that the place cells and head-direction cells developed multiple firing fields in situations of ambiguous sensory input. While activity in the head-direction cells shifted to represent appropriate changes in the robot's orientation, activity in the place cells did not shift appropriately to represent robot translation (Figure 4).

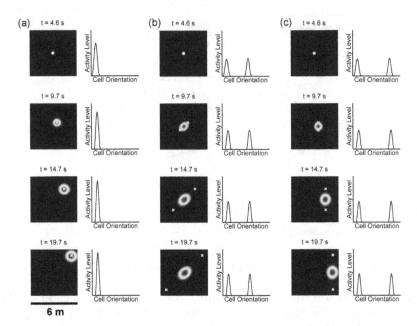

6 m

Fig. 4. Place and head-direction cell firing fields under unique and ambiguous conditions. Crosses show the correct location hypotheses (a) Correct update of place cell activity under a unique orientation hypothesis. (b) With opposing orientation hypotheses, place cell firing smears, and does not shift to represent either possible location estimate. (c) Place cell activity shifts in the net (and consequently incorrect) direction of two orientation hypotheses varying by 90 degrees.

The underlying cause of the failures was the separation of pose (the conjunction of location and orientation) into separate location and orientation representations in the place cells and head-direction cells. Without a strong coupling between the two cell types, allothetic input can incompletely correct errors accumulated during exploration leading to corruption in the associative memory. Most significantly, the separation of the state representation removes the ability to correctly update multiple estimates of pose in perceptually ambiguous situations. Because place cells have no directional attributes, any path integration process, regardless of the specific mechanism, must obtain its orientation information from activity in the head-direction cells. However, each location estimate can only be updated by its one-to-many association with all orientation estimates, rather than updated only by the orientation estimate pertinent to the particular location hypothesis. Consequently, location estimates rapidly become incorrect. The correct robot location estimate can only be reinstated through strong unique visual input.

Separate representations of robot orientation and spatial location are inherently unsuitable for mapping and navigation in large, ambiguous environments, as demonstrated by the work presented in this section. The following section describes the implementation of an extended hippocampal model known as RatSLAM, which combines the concept of head-direction and place cells to form a new type of cell known as a *pose cell*.

3.2   A Model of Spatial Pose

To form a population of pose cells, the competitive attractor network structures that were used to model head-direction and place cells were combined to form a three-dimensional structure (Figure 5). The pose cells are arranged in an (x', y', θ') pattern, allowing the robot to simultaneously represent multiple pose estimates in x', y', and θ'. Primed co-ordinate variables are used because although the correspondence between cells and physical co-ordinates is initially Cartesian, this relationship can become discontinuous and non-linear as the system learns the environment. For example, in indoor experiments, each pose cell initially fires maximally when the robot is in a 0.25×0.25 meter area and within a 10 degree band of orientation. However as an experiment progresses, the pose volume each pose cell corresponds to can grow, shrink, warp, or even disappear under the influence of visual information.

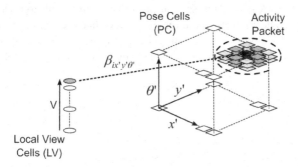

Fig. 5. The local view and pose cell structures. The local view encodes visual information about the environment. The pose cells represent the robot's pose. Co-activated local view and pose cells form associative links. A familiar visual scene activates a local view cell which in turn injects activity into the pose cells associated with it. Re-localization occurs when the activity packet caused by visual input becomes dominant.

The advantage of a unified pose representation is that each cell encodes its own preferred orientation. This allows multiple hypotheses of the robot's pose to propagate in different directions, unlike the more conventional head-direction – place cell model, and enables the mapping of large, ambiguous indoor and outdoor environments (Figure 7c, d) [21, 35].

4   Contextual Mapping

Conventional robot mapping methods typically produce either occupancy grid maps [7] or landmark-based maps [36]. Such maps are appealing from a mathematical perspective because they facilitate the use of probabilistic mapping algorithms. However, such maps are context-less, and store no information regarding the environment that can be used directly by the robot. For example, while an occupancy grid map stores the location of free space and obstacles in the environment, the map does not directly tell the robot about the speeds at which it can traverse different areas of the environment. A cluttered area might require slow movement with much turning, while a long straight corridor offers the potential for a high speed movement. The occupancy map could perhaps be processed using a 'clutter' metric in order to yield this information, but this is a complex and perhaps unnecessary step.

Contextual mapping involves enriching the spatial representation of an environment with additional information relevant to the potential uses of the representation. For example, temporal or speed data provides information about the rate at which different areas of the environment can be traversed, which can be used by a robot to plan the quickest route to a goal. Behavioral data provides information about the appropriate movement behaviors required to move between different places in the environment, which a robot can use to achieve improved movement through the environment. Transition success rate data provides information about how easy it is to cross different areas in the environment, which a robot can use to decide between competing routes to a goal. This section presents a contextual approach to mapping the environment. The map use requirements were that the robot be able to rapidly explore and map its environment, plan and execute the quickest routes to goals, and adapt its map and goal navigation to simple changes in the environment.

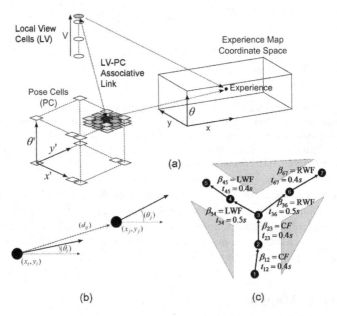

Fig. 6. (a) Experience map co-ordinate space. An experience is associated with certain pose and local view cells, but exists within the experience map's own (x, y, θ) co-ordinate space. (b) A transition between two experiences. Shaded circles and bold arrows show the actual pose of the experiences. d_{ij} is the odometric distance between the two experiences. (c) An example of the behavioral and temporal information. The numbered circles represent experiences, and CF, LWF, and RWF are movement behaviors: CF – Centerline Following; LWF – Left Wall Following; RWF – Right Wall Following.

4.1 Experience Mapping

The experience mapping algorithm was developed to create a contextual representation of the environment containing not only a spatial map, but also movement, behavioral, and temporal information relevant to the robot's motion and higher level behaviors such as goal navigation.

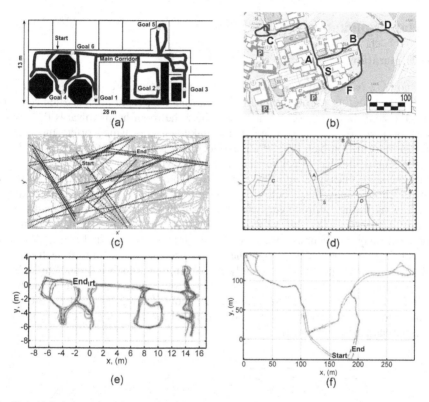

Fig. 7. (a-b) Indoor and outdoor environments. The robot's paths are shown by the thick line. In the outdoor environment the robot started at S and finished at F. An initial loop of the inner loop SABFS was followed by a clockwise traverse of the outer loop SACABDBFS and then a counter-clockwise traverse of the same outer loop. (c-d) The trajectory of the most strongly activated pose cell in (x', y') space, with wrapping in both directions. The thin lines indicate relocalization jumps where the RatSLAM system closed the loop. Each grid square represents 4×4 pose cells in the (x', y') plane. (e-f) Experience maps.

Activity in the pose cells and local view cells drives the creation of experiences. Experiences are represented by nodes in (x, y, θ) space connected by links representing transitions between experiences. Each experience represents a snapshot of the activity within the pose cells and local view cells at a certain time, and in effect represents a specific spatial and visual robot experience (see Figure 6a).

As the robot moves around the environment, the experience mapping algorithm also learns experience *transitions*, which are links between experience nodes.

Transitions store information about the physical movement of the robot between one experience and another, the movement behavior used during the transition, the time duration of the transition, and the traversability of the transition (see Figure 6b, c). This information enables a robot to perform rapid exploration, map correction, route planning and execution, and adaptation to environment change – all with minimal computation [22, 23, 37].

5 Map Exploitation

5.1 Exploration

In a novel environment, an animal or robot must explore in order to acquire information about its surroundings. In classical robotics, exploration is often achieved through greedy frontier-based techniques, where the robot heads towards the nearest unknown area of the environment, or the unknown area which offers the greatest potential information gain [38, 39]. These techniques process occupancy or coverage maps to identify borders of unknown areas in the environment.

By encoding behaviors into the spatial experience map, the RatSLAM system is able to choose movement behaviors that have not yet been used at *behavior intersections* in the environment, where more than one movement is possible, with minimal map processing (Figure 8). Past behavior usage can be detected using the experience map. A recursive tree search function is used to find routes from the currently active experience through the experience map using only transitions of a specific behavior type. If a sufficiently long continuous path is found through the experience map for a certain behavior, that behavior is tagged as having already been followed from the current environment location. During exploration the robot attempts to use previously unused movement behaviors to try to experience new routes through the environment.

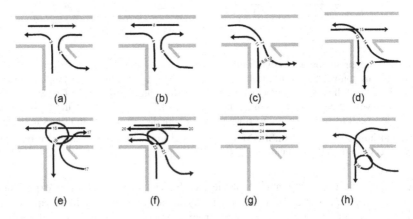

Fig. 8. The robot's local movement at an intersection during an 80 minute experiment. The labels indicate the order in which the paths were traversed.

5.2 Route Planning and Execution

The experience maps provide the spatial, temporal, and behavioral information used to perform goal directed navigation. Temporal information allows the creation of a temporal map, with gradient climbing techniques used to find the fastest (in the robot's experience) route to the goal. Once a route has been found, the robot uses spatial and behavioral information stored in the experience links to navigate to the goal. If the context was different – for instance, if the robot was required to find the straightest routes to the goal, that information could be added to the map.

By creating maps which are both spatial and navigational, a robot is easily able to use them to navigate to goals. The navigation capabilities of the system were tested over a sequence of 12 goal navigation trials in the environment shown in Figure 7a. Figure 9 shows the temporal maps, planned and executed routes to six goals. The robot successfully navigated to all twelve goals, demonstrating the usability of the spatial and navigational experience maps.

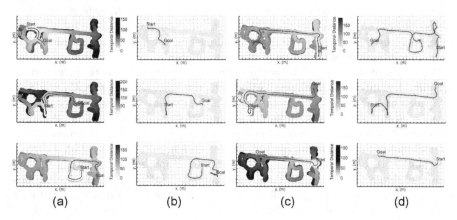

(a)   (b)   (c)   (d)

Fig. 9. Temporal maps, planned and actual routes to the goals for six of the twelve trials. Temporal maps and planned routes are shown in the first and third columns (a, c). The actual routes executed by the robot are shown in the second and fourth column (b, d). Five of the executed routes were near optimal, with major errors in one route.

5.3 Adaptation

The experience maps represent changes in the environment through modification of the inter-experience transition information. As well as learning experience transitions, the system also monitors transition 'failures'. Transition failures occur when the robot's current experience switches to an experience other than the one expected given the robot's current movement behavior. If enough failures occur for a particular transition, indicated by the *confidence* level dropping below a certain threshold, then that link is deleted from the experience map. In this way the robot is able to adapt to simple changes in the environment without any explicit recognition of objects. The system's ability to adapt to changes in the environment was demonstrated in an indoor environment, with the results shown in [22].

6 Discussion

By testing hippocampal models on actual robots in real world environments, it was possible to evaluate their functional navigation performance from a robotics context, rather than just their biological plausibility. The results indicate that functional navigation performance comparable to other more developed navigation systems can be obtained, and demonstrate the potential of a pragmatic bio-inspired approach to the mapping and navigation problem.

However, it might be argued that we have strayed significantly from the original biological models in our research, and that we have, in fact, demonstrated that biological models don't translate well to robot systems. The key finding of the pilot study was that a separate head-direction – place representation system, common to most biological models, could not maintain multiple hypotheses of the robot's pose over time. The separation of the two networks meant that opposing hypotheses of robot orientation were effectively averaged causing the hypotheses of the robot's location to be updated erroneously. To overcome this problem we introduced pose cells, which represent a conjunction of orientation with place. However, even with pose cells, the representation could not be used for meaningful tasks. Discontinuities, ambiguities and collisions in the pose cell representations made path planning impossible.

With consideration of both mapping and navigational requirements, the experience mapping system was developed, which created representations ready-made for navigation and also map adaptation. Surely all of this invention would indicate that the biological models were of little use to translate into a robotic system? We would argue that it is not an error in translation from biology to robotics, but an error in the original biological models themselves. Recent evidence of *grid cells* [40, 41] in the entorhinal cortex (a region of the brain near the hippocampus) indicates that this region of brain integrates ideothetic and allothetic information into a full pose representation – as do the pose cells in the RatSLAM model. Grid cells exhibit many of the same properties of pose cells, such as discontinuity, ambiguity and redundancy. It would seem that the pose cells invented during this research may be closer to biological fact than the previously proposed biological models.

If it is the role of the grid cells to perform the integration of ideothetic and allothetic cues, then the hippocampus may store context-specific episodic sequences combining a number of cues (including the pose information in the grid cells). Perhaps the well defined place representations found in the experience map form a closer analogy to the hippocampal place cells. In future work, we intend to pursue the links from our model to biology. If biological plausibility can be shown, then the RatSLAM model allows predictions to be made regarding neuro-cognitive models of rodent navigation in larger environments than the usual small mazes and arenas.

From a robotics perspective, this research has demonstrated that it is possible to create a bio-inspired robot mapping and navigation system that can function in real-time in large indoor and outdoor environments. Furthermore, the work is one of the few attempts to develop a complete integrated navigation system – one that addresses not only the primary SLAM problem, but also the problems of exploration, navigating to goals, and adaptation. The success of the system can be attributed to the consideration of both mapping and navigation requirements during development, and to the pragmatic adaptation of the biological navigation models on which it is based.

References

1. Thrun, S.: Robotic Mapping: A Survey. In: Exploring Artificial Intelligence in the New Millennium, Morgan Kaufmann, San Francisco (2002)
2. Dissanayake, G., Durrant-Whyte, H., Bailey, T.: A computationally efficient solution to the simultaneous localisation and map building (SLAM) problem. In: 2000 IEEE International Conference on Robotics and Automation, San Franciso, USA, IEEE, Los Alamitos (2000)
3. Newman, P.M., Leonard, J.J.: Consistent, Convergent, and Constant-time SLAM. In: International Joint Conference on Artificial Intelligence, Acapulco, Mexico (2003)
4. Thrun, S.: Probabilistic Algorithms in Robotics. AI Magazine 21, 93–109 (2000)
5. Montemerlo, M., Thrun, S., Koller, D., Wegbreit, B.: FastSLAM: A Factored Solution to the Simultaneous Localization and Mapping Problem. In: AAAI National Conference on Artificial Intelligence, Edmonton, Canada (2002)
6. Montemerlo, M., Thrun, S., Koller, D., Wegbreit, B.: FastSLAM 2.0: An improved particle filtering algorithm for simultaneous localization and mapping that provably converges. In: International Joint Conference on Artificial Intelligence, Acapulco, Mexico (2003)
7. Grisetti, G., Stachniss, C., Burgard, W.: Improving Grid Based SLAM with Rao Blackwellized Particle Filters by Adaptive Proposals and Selective Resampling. In: International Conference on Robotics and Automation, Barcelona, Spain (2005)
8. Thrun, S., Koller, D., Ghahramani, Z., Durrant-Whyte, H., Ng., A.Y.: Simultaneous Mapping and Localization With Sparse Extended Information Filters: Theory and Initial Results. In: Fifth International Workshop on Algorithmic Foundations of Robotics, Nice, France (2002)
9. Yamauchi, B., Beer, R.: Spatial Learning for Navigation in Dynamic Environments. Man and Cybernetics 26, 496–505 (1996)
10. Thrun, S.: Probabilistic Algorithms and the Interactive Museum Tour-Guide Robot Minerva. International Journal of Robotics Research 19, 972–999 (2000)
11. Stentz, A., Hebert, M.: A Complete Navigation System for Goal Acquisition in Unknown Environments. Autonomous Robots 2, 127–145 (1995)
12. Kuipers, B., Byun, Y.T.: A Robot Exploration and Mapping Strategy Based on a Semantic Hierarchy of Spatial Representations. Robotics and Autonomous Systems 8, 47–63 (1991)
13. Kuipers, B., Modayil, J., Beeson, P., MacMahon, M., Savelli, F.: Local Metrical and Global Topological Maps in the Hybrid Spatial Semantic Hierarchy. In: International Conference on Robotics and Automation, New Orleans, USA (2004)
14. Hmelo-Silver, C.E., Pfeffer, M.G.: Comparing expert and novice understanding of a complex system from the perspective of structures, behaviors, and functions. Cognitive Science 28, 127–138 (2003)
15. Kobayashi, T., Nishijo, H., Fukuda, M., Bures, J., Ono, T.: Task-Dependent Representations in Rat Hippocampal Place Neurons. The Journal of Neurophysiology. 78, 597–613 (1997)
16. Bostock, E., Muller, R., Kubie, J.: Experience-dependent modifications of hippocampal place cell firing. Hippocampus 1, 193–205 (1991)
17. Kobayashi, T., Tran, A.H., Nishijo, H., Ono, T., Matsumoto, G.: Contribution of hippocampal place cell activity to learning and formation of goal-directed navigation in rats. Neuroscience 117, 1025–1035 (2003)
18. Frisch, K.v.: The Dance Language and Orientation of Bees. Harvard University Press, Cambridge (1967)

19. Wehner, R., Gallizzi, K., Frei, C., Vesely, M.: Calibration processes in desert ant navigation: vector courses and systematic search. Journal of Comparative Physiology A: Sensory, Neural, and Behavioral Physiology 188, 683–693 (2002)
20. Milford, M.J., Wyeth, G., Prasser, D.: RatSLAM: A Hippocampal Model for Simultaneous Localization and Mapping. In: International Conference on Robotics and Automation, New Orleans, USA (2004)
21. Prasser, D., Milford, M., Wyeth, G.: Outdoor Simultaneous Localisation and Mapping using RatSLAM. In: International Conference on Field and Service Robotics, Port Douglas, Australia (2005)
22. Milford, M.J., Wyeth, G.F., Prasser, D.P.: RatSLAM on the Edge: Revealing a Coherent Representation from an Overloaded Rat Brain. In: International Conference on Robots and Intelligent Systems, Beijing, China (2006)
23. Milford, M., Schulz, R., Prasser, D., Wyeth, G., Wiles, J.: Learning Spatial Concepts from RatSLAM Representations. Robots and Autonomous Systems (2006)
24. Muller, R.U., Kubie, J.L.: The effects of changes in the environment on the spatial firing of hippocampal complex-spike cells. The Journal of Neuroscience 7, 1951–1968 (1987)
25. Tews, A., Robert, J., Roberts, J., Usher, K.: Is the Sun Too Bright in Queensland? An Approach to Robust Outdoor Colour Beacon Detection. In: Australasian Conference on Robotics and Automation, Sydney, Australia (2005)
26. Recce, M., Harris, K.D.: Memory for places: A navigational model in support of Marr's theory of hippocampal function. Hippocampus 6, 735–748 (1998)
27. Gaussier, P., Zrehen, S.: Navigating With an Animal Brain: A Neural Network for Landmark Identification and Navigation. In: Intelligent Vehicles, 399–404 (1994)
28. Franz, M.O., Mallot, H.A.: Biomimetic robot navigation. Robotics and Autonomous Systems, 133–153 (2000)
29. Crane III, C., Armstrong Jr., D., Torrie, M., Gray, S.: Autonomous Ground Vehicle Technologies Applied to the DARPA Grand Challenge. In: International Conference on Control, Automation, and Systems, Bangkok, Thailand (2004)
30. Ozguner, U., Redmill, K.A., Broggi, A.: Team TerraMax and the DARPA grand challenge: a general overview. In: Intelligent Vehicles Symposium, Parma, Italy (2004)
31. Browning, B.: Biologically Plausible Spatial Navigation for a Mobile Robot. PhD, Computer Science and Electrical Engineering, University of Queensland, Brisbane (2000)
32. Arleo, A.: Spatial Learning and Navigation in Neuro-mimetic Systems: Modeling the Rat Hippocampus. PhD, Department of Computer Science, Swiss Federal Institute of Technology, Lausanne (2000)
33. Milford, M.J., Wyeth, G.: Hippocampal Models for Simultaneous Localisation and Mapping on an Autonomous Robot. In: Australasian Conference on Robotics and Automation, Brisbane, Australia (2003)
34. Redish, D.: Beyond the Cognitive Map. Massachusetts Institute of Technology, Massachusetts (1999)
35. Milford, M.J., Wyeth, G., Prasser, D.: Simultaneous Localization and Mapping from Natural Landmarks using RatSLAM. In: Australasian Conference on Robotics and Automation, Canberra, Australia (2004)
36. Nieto, J., Guivant, J., Nebot, E.: Real Time Data Association for FastSLAM. In: International Conference on Robotics & Automation, Taipei, Taiwan (2003)
37. Milford, M.J., Prasser, D., Wyeth, G.: Experience Mapping: Producing Spatially Continuous Environment Representations using RatSLAM. In: Australasian Conference on Robotics and Automation, Sydney, Australia (2005)

38. Stachniss, C., Burgard, W.: Mapping and Exploration with Mobile Robots using Coverage Maps. In: International Conference on Intelligent Robots and Systems, Las Vegas, USA (2003)
39. Yamauchi, B.: A Frontier Based Approach for Autonomous Exploration. In: International Symposium on Computational Intelligence in Robotics and Automation, Monterey, USA (1997)
40. Fyhn, M., Molden, S., Witter, M.P., Moser, E.I., Moser, M.-B.: Spatial Representation in the Entorhinal Cortex. Science 27, 1258–1264 (2004)
41. Hafting, T., Fyhn, M., Molden, S., Moser, M.-B., Moser, E.I.: Microstructure of a spatial map in the entorhinal cortex. Nature Neuroscience 11, 801–806 (2005)

Scale-Dependent Simplification of 3D Building Models Based on Cell Decomposition and Primitive Instancing

Martin Kada

Institute for Photogrammetry, Universität Stuttgart,
Geschwister-Scholl-Str. 24D, 70174 Stuttgart, Germany
martin.kada@ifp.uni-stuttgart.de

Abstract. The paper proposes a novel approach for a scale-dependent geometric simplification of 3D building models that are an integral part of virtual cities. In contrast to real-time photorealistic visualisations, map-like presentations emphasize the specific cartographic properties of objects. For buildings objects, such properties are e.g. the parallel and right-angled arrangements of facade walls and the symmetries of the roof structure. To a map, a clear visual perception of the spatial situation is more important than a detailed reflection of reality. Therefore, the simplification of a 3D building model must be the transformation of the object into its global shape. We present a two-stage algorithm for such an object-specific simplification, which combines primitive instancing and cell decomposition to recreate a basic building model that best fits the objects original shape.

1 Introduction

The acquisition and presentation of 3D city models has been a topic of intensive research for more than 15 years. In general, such data sets include digital representations of the landscape, the buildings and more frequently also of the vegetation and the street furniture. A number of commercial software products and service companies exist nowadays for the reconstruction of buildings. For an efficient data collection of large areas, the objects are measured from aerial images or laser data. Therefore there is no façade information available in the source data which results in building models where the ground plan is simply extruded and intersected with the interpreted roof structure.

Besides the traditional analysis applications of 3D city models, which are e.g. the planning of mobile antennas, alignment of solar installations and noise propagation, the presentation of urban areas gains in importance. Real-time and web-based visualisation systems offer nowadays graphics of near photorealistic quality. To limit the amount of data that needs to be transferred over the network and to increase rendering performance, objects are represented in different levels of detail depending on their distance to the viewer. For 3D city models, the following classification of building objects in three discrete levels of detail is very common: block models with flat roofs and no facade structure, models with roof structures and architectural models with detailed roofs and facades. So far, cities have mostly collected data in the second level

S. Winter et al. (Eds.): COSIT 2007, LNCS 4736, pp. 222–237, 2007.

of detail with only a few selected landmarks being of higher detail. Because of the high costs involved in the acquisition, there are efforts to facilitate the exchange and interoperability between data and application providers. The Special Interest Group 3D (SIG 3D) of the initiative Geodata Infrastructure North-Rhine Westphalia (GDI NRW), Germany, e.g., proposed the application schema CityGML to the Open Geospatial Consortium (OGC) for standardisation [1]. Therein, the properties of five levels of detail are defined to support a broad variety of applications like car navigation systems, flight, driving and nautical simulators, tourism information systems, etc. A preliminary survey lists applications of 3D city models and their specific levels of detail requirements [2].

A photorealistic visualisation is not always the most adequate tool to communicate spatial information. Architects and designers often produce sketch like hardcopy outputs to make their objects appear more alive or to express the preliminary status of their designs. Recent works on interactive visualisations of 3D city models (e.g. [3]) explore non-photorealistic rendering techniques that imitate this style so that spatial situations are easier to perceive and comprehend. Such techniques, however, rely on information about the characteristic edges that best reflect the global shape of a building. This is basically what results from a cartographic simplification.

Another field of application for 3D city models are location based services or context-aware applications. Their users rely heavily on a location- or situation-dependent presentation of the information that is most relevant to their current task. To be useful anywhere at all times, such systems run on mobile devices like digital personal assistants (PDA) or mobile phones. As their screen size and resolution will always be a limiting factor, a geometric simplification of 3D objects is necessary to guarantee the graphical minimum feature size required by maps or map-like presentations. Otherwise the high line density makes it impossible to recognize important aspects of the building object.

Because it is not reasonable to collect and store data for all requested levels of detail, an automatic process is necessary that transforms 3D building models towards more simplified shapes. Object features that are under a minimum size, which can be determined from the scale parameters of the map projection, should be removed without disturbing the global shape. Properties that are specific for the object itself as well as the object type, however, must be preserved. In the case of 3D building models, these are the parallel and right-angled arrangements of facade walls and the symmetries of the roof structures. Object specific features are especially important for landmarks. The simplified model of a church or cathedral, e.g., must not miss its towers after generalisation as otherwise the object is hardly recognisable anymore.

A simplification of solitary objects under these spatial constraints is one of the elemental operators of cartographic generalisation. In cartography, both the spatial objects themselves as well as their arrangement are transformed with the goal to create maps or map-like presentations that help to communicate a spatial situation. Other generalisation operators omit or emphasise objects depending on their importance, aggregate semantically similar objects, replace a number of objects by fewer entities or displace them to relax the spatial density in areas with many objects. The generation of a situation- and context-dependent abstraction level of the spatial data is therefore possible to help viewers apprehend the presented spatial information.

2 Previous Work

The automatic generalisation of building models has been a research topic ever since Staufenbiel [4] proposed a set of generalisation actions for the iterative simplification of 2D ground plans. Several algorithms have been developed that remove line segments under a pre-defined length by extending and crossing their neighbour segments and by introducing constraints about their angles and minimum distances (e.g. [5, 6, 7, 8, 9]). Other approaches use vector templates [10, 11], morphological operators like opening and closing [12, 13], least-squares adjustment [14] or techniques from scale space theory [15].

A few algorithms also exist by now for the generalisation of 3D data. Forberg [16] adapts the morphology and curvature space operators of the scale space approach to work on 3D building models. Thiemann and Sester [17] do a segmentation of the building's boundary surface with the purpose of generating a hierarchical generalisation tree. After a semantic interpretation of the tree's elements, they can selectively be removed or reorganized to implement the elemental generalisation operators for simplification, emphasis, aggregation and typification. Another aggregation approach is proposed by Anders [18]. It works for linearly arranged building groups. Their 2D silhouettes, which are the results of three projections from orthogonal directions, are simplified, extruded and then intersected to form the generalised 3D model. With a strong focus on the emphasis of landmarks do Thiemann and Sester [19] present adaptive 3D templates. They categorise building models into a limited number of classes with characteristic shapes. A building model is then replaced by the most similar 3D template that is best fit to the real object. Because the semantics of the template is known, the object itself or specific features of the model can be emphasised at will.

The simplification of 3D models has been a major topic in the field of computer graphics. See e.g. the survey of Luebke et al. [20] for an up-to-date summary of the most important work. However, these algorithms are designed for general models that approximate smooth surfaces and therefore typically do not perform well on 3D building models. The main reason is that building models consist of considerably fewer planar faces, but many sharp edges. Coors [21], Rau et al. [22] and Kada [23] show that the simplification operators and metrics can be modified so that the characteristic properties of the building models can be preserved during their simplification.

Despite the number of available 3D generalisation approaches, a continuous difficulty seems to be the simplification of the roof structure. Most algorithms avoid this problem by simply generating flat or pent roofs or assume that the roof type is already available as the result of a preceding interpretation. In this paper, we describe a generalisation approach for 3D building models and concentrate on a new procedural method to generate reasonable roof geometries.

3 Generalisation of 3D Building Models

We propose a two-stage generalisation algorithm for the geometric simplification of solitary 3D building models. As can be seen from the intermediate results of the example in Fig. 1, the two stages consist in a total of five steps. The first stage generates

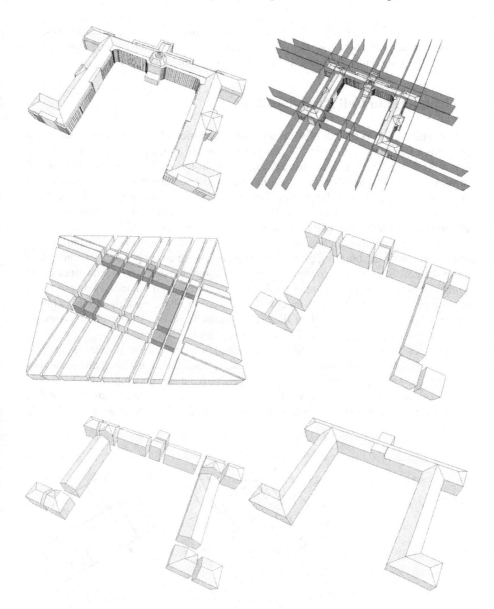

Fig. 1. Original 3D building model (top left) and the five generalisation steps

a 2D decomposition of space that approximates the ground plan polygon by a disjoint set of quadrilateral primitives. We accomplish this by deriving plane equations from the major façade walls (1), subdividing the infinite space along these planes (2) and identifying the resulting cells that feature a high percentage of overlap with the original ground plan polygon (3). The second stage reconstructs the simplified geometry of the roof. Here, a primitive instancing approach is shown where the roof parameters are determined individually for each cell so that they best fit the original model under distinct adjacency constraints (4). By altering those parameters, the simplification of the roof can be properly adjusted. A union operation of the resulting primitives composes the final 3D building model and concludes the generalisation (5).

4 Ground Plan Cell Decomposition

Cell decomposition is a form of solid modelling in which objects are represented as a collection of arbitrarily shaped 3D primitives that are topologically equivalent to a sphere. The individual cells are usually created as instances from a pre-defined set of parameterized cell types that may even have curved boundary surfaces. Complex solids are then modelled in a bottom-up fashion by "gluing" the simple cells together. However, this operator restricts the cells to be nonintersecting, which means adjoining cells may touch each other but must not share any interior points [24].

In our algorithm, the cell decomposition serves two purposes: First, it is build as an approximation of the building ground plan and is consequently per se also a generalization thereof. Second, it provides the basic building blocks for the reconstruction of the roof geometry. Since the input models are provided as 3D data, all computations are also performed in 3D, even though the dimension of the resulting cells is really 2D; or 2.5D if a height is applied like in the example of Fig. 2. For clarity reasons, however, the accompanying Fig. 2, 3 and 4 are given as 2D sketches.

The faces in a polyhedral building representation are always planar. If the real building facade features round or curved elements, then they must be approximated in the model by small polygons. We therefore generate the cell decomposition by subdividing a finite 3D subspace by a set of vertical planes. Fig. 2 e.g. shows a building and the cell decomposition which results from subdividing space along the facade segments.

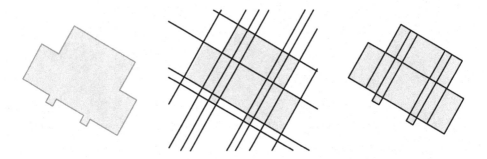

Fig. 2. Building ground plan (left), overlaid decomposition of space along its façade segments (middle) and resulting cell decomposition (right)

As it can be seen, the union of the cells is not yet a simplification of the original shape and the small cells complicate the reconstruction of the roof geometry. So instead of using each individual façade polygon, we cluster them together with a special buffer operation for the purpose of generating fewer planes that in turn produce a decomposition of fewer cells. However, these planes should correspond with the most important facade segments so that the decomposition reflects the characteristic shape of the object. The importance of a plane is measured as the surface area of all polygons that are included in the generating buffer and that are almost parallel to the created plane. Polygons with a different orientation are not counted.

4.1 Generation of Decomposition Planes

We implemented a greedy algorithm that generates the plane of highest importance from a set of input facade polygons. At this point, we ignore all roof polygons and only use polygons with a strict horizontal normal vector. By repeatedly calling the algorithm, new planes are added to the result set and all polygons inside the buffer are discarded from further processing. The generation of planes ends when no input polygons are left or when the importance of the created planes falls under a certain threshold value.

At the beginning of the algorithm, buffers are created from the input polygons (see Fig. 3. Each buffer is defined by two delimiting parallel planes that coincide with the position and normal direction of a generating polygon. These planes may move in opposite directions to increase the buffer area until a generalisation threshold is reached. The buffers are first sorted by their importance and then merged pair wise to create larger buffers. Starting with the buffer of highest importance, the buffers of lower importance are tested for their inclusion in this buffer. If all polygons of a buffer can be included into the one of higher importance without increasing the distance between their delimiting planes above the generalisation value, then the merge is valid and is executed. The algorithm stops when no more buffers can be merged and the averaged plane equation of the polygons of the buffer of highest importance is returned.

Fig. 3. Initial buffer from facade segments (left), delimiting planes of the maximised buffer (middle) and resulting averaged plane (right)

In order to enforce parallelism and to support right angles of the facade segments, the resulting planes are analysed in a last step. If the angle of the normal vectors from two or more planes is found to be below a certain threshold, these planes are made parallel or rectangular. If the deviation is only a small angle, this can be done by changing the normal vector of the plane equation and adjusting the distance value. For larger values, a rotation of the planes around their weighed centroids of the polygons is chosen.

For our computations, we use four threshold values. The most important one is the generalisation distance that the buffer planes may move apart. As this value also determines the distance of the planes used for the decomposition, it is also approximately the smallest ground plan feature length of the resulting set of cells. Another threshold value determines the lowest importance of a plane that is still a valid result. Here, the square of the generalisation distance is used. Buffers below that value probably do not contain polygons with a side length of the generalisation distance and are therefore not important. The last two threshold values are angles. As it is important for the roof construction that the cells are parallelograms, the angle for enforcing parallelism is rather large. We chose 30° for parallelism and 10° for right angles.

See Fig. 4 for the set of buffers that result in a simplified cell decomposition.

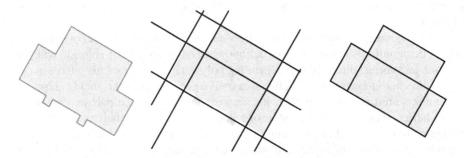

Fig. 4. Building ground plan (left), overlaid simplified decomposition of space along its façade segments (middle) and resulting cell decomposition (right)

4.2 Cell Decomposition

Once the planes have been determined, they are then used to generate the cell decomposition of the building model. Theoretically, an infinite 3D space should be subdivided brute force by the planes. However, as an infinite space is unpractical, a solid two times the size of the building's bounding box is used. Because the plane equations were averaged from facade segments and therefore have no horizontal component, the space is only divided in two dimensions. The resulting cells are therefore 2D polygons extruded into the third dimension.

The decomposition consists of building and non-building cells. Only the building cells are of interest for further processing. The other cells should be discarded. However, these cells can not directly be identified from the decomposition process. Therefore, a further step is necessary.

For that reason, a percentage value is calculated that denotes the overlap of the cell with the original building ground plan. Cells that result in a high overlap value are considered building cells whereas the other cells are considered as non-building cells. A precise value can be computed by intersecting the cell with the ground plan polygon and dividing the resulting area by the area of the cell. As the cells are rather big, an overlap threshold of 50% is able to correctly distinguish between building and non-building cells.

5   Roof Simplification by Cell Decomposition

The roof structure for general 3D building models can be very complex. We therefore present two methods for their simplification. Both recreate a simplified version of the original roof structure for the previously generated ground plan cell decomposition. The first method extends the cell decomposition approach to the third dimension. It is general enough to recreate all roof shapes. As it will be shown in section 6, however, limiting the possible 3D shapes of the cells to a subset of common roof types can lead to a more suitable roof structure for a subset of common buildings.

So far, the roof polygons have been neglected. Now they are used to determine the decomposition planes of arbitrary orientation in order to generate 3D cell decompositions from the ground plan cells. Although the decomposition is done per cell, the planes are determined globally from all roof polygons to ensure that neighbouring cells fit well against each other. We use the buffer approach as previously described. The subdivision process is then done with the subset of planes that has polygons in their buffer that intersected the respective cells. This avoids a heavy fragmentation of the cells.

The resulting cells are now real 3D solids, so the classification in building and non-building cells has to be done in 3D space. Consequently, a percentage value that denotes the volume of the original building model inside each respective cell is computed. Fig. 5 shows the decomposition of the example building of Fig. 1 by the roof planes and the resulting building cells after their identification.

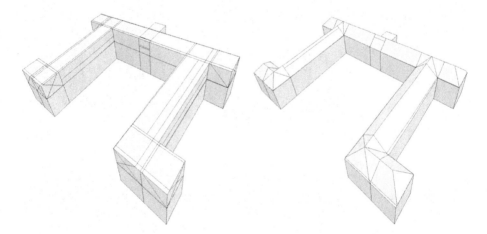

Fig. 5. Decomposition of the roof before (left) and after (right) identification of building cells

As can be seen in Fig. 5, there are some inaccuracies in the resulting model. These are caused by planes that do not cut the 2.5D cells at exactly the same location in space. We remove these inaccuracies by a vertex contraction process that pulls the roof vertices to the closest ground cell corner point, edge or cell centre if they are within close distance. Fig. 6 and 7 show results of the generalisation algorithm for simple example models as well as rather complex landmarks.

Fig. 6. Original (left) and generalised (right) 3D building models and their overlays (middle)

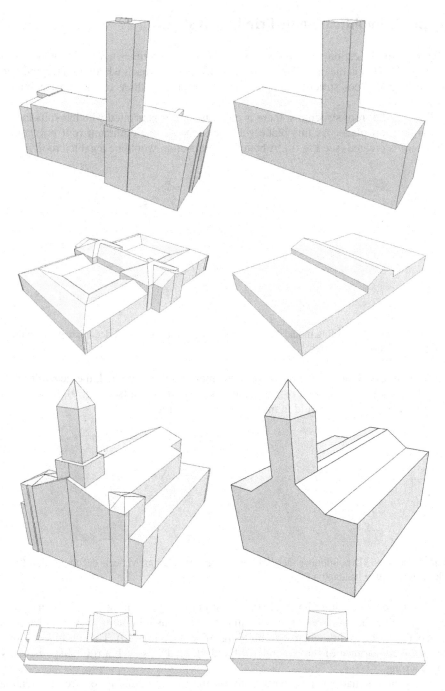

Fig. 7. Original (left) and generalised (right) 3D landmarks

6 Roof Simplification by Primitive Instancing

The roof simplification via cell decomposition does sometimes not lead to good look-ing models. This is the consequence of the universal approach where no interpretation of the original roof structure is performed. We present three of the most common shortcomings.

For very flat roof structures, there is only one buffer generated which results in one decomposition plane. As this plane gets the slope of one dominant roof polygon, a shed roof is created (see Fig. 8). A better generalisation would be a gabled roof.

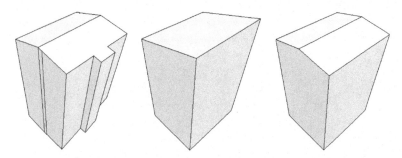

Fig. 8. Original 3D building model (left) and its generalisations via cell decomposition (middle) and primitive instancing (right)

Because the slopes of the decomposition planes are not aligned, the generalisation of hipped roofs often results in an asymmetric roof structure (see Fig. 9). However, a symmetric hipped roof would in most cases be preferred.

Fig. 9. Original 3D building model (left) and its generalizations via cell decomposition (mid-dle) and primitive instancing (right)

Due to different ridge heights, some roof cells may not have a high enough per-centage value that is necessary to classify it reliably as a building cell. This happens especially at the valley where two buildings meet (see Fig. 10). The missing cell dis-turbs the appearance of the generalised building model as such a roof shape is likely to be wrong.

In all three situations, an interpretation of the roof structure is required to create a simplified roof that best resembles the original model, is symmetric and has a realistic shape. Because the height discontinuities of the roof structure have already been in-corporated into the cell decomposition, the interpretation can be done per cell.

Fig. 10. Original 3D building model (left) and its generalizations via cell decomposition (middle) and primitive instancing (right)

The interpretation of the roof type is performed via a cell based primitive instancing approach. Here, every cell is tested against all possible primitive types that are parameterised in terms of the roof properties. So far, we support the eight roof types that are shown in Fig. 11. These are flat, shed, gabled, hipped roof and some connecting elements. The gabled and hipped roof elements need a ground plan in the shape of a parallelogram. Other shapes are not possible and must result in a flat or shed roof. However, the cell decomposition for most buildings with a gabled or hipped roof will usually provide an adequate set of cells. This prerequisite of the cells can be ensured during the generation of the ground plan decomposition by using only approximating planes parallel and rectangular to the general orientation of the building. Otherwise if the prerequisites can not be met, the primitives must be shaped as flat or shed roofs. Alternatively, the roof can be generalised via the cell decomposition approach.

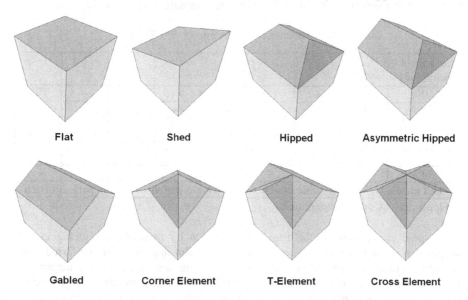

Fig. 11. The eight primitive types supported by the roof simplification

The selection of the roof type and the parameter estimation works by discretising the bounding box of the cell's ground plan. For example, if the 2D space is divided in ten times ten subspaces, then 100 samples are the result. For each subspace inside the cell area, a length and an angle are computed. The length denotes the distance of the original model to the instanced primitive and the angle is computed as the horizontal difference of their normal directions. The vertical components of the normal vectors are ignored, which makes the angle independent from the eaves and ridge heights.

In order to determine the roof type, all eight primitives are instanced and the angle values for all samples computed. An angle below 30° is considered a match. For the hipped roof type, changes in the ridge length results in a different number of matches. Therefore, several parameter values are used. The primitive type with the highest number of matches is taken as an intermediate result for the cell's roof type. Afterwards the ridge and eaves heights are initially set to the highest and lowest value of the original roof polygons inside that cell. The geometric error is determined as the sum of all squared length values. Both parameters are then individually altered until the error is at a minimum.

Once the roof types and their parameters have been determined, the type of each cell is validated against a preference table (see Table 1). Therefore, the cells with approximately the same eaves and ridge heights are first grouped together. For the validity check of a cell, only the cells in the same group are of interest. For each cell, the number of neighbour cells and their arrangement are considered and compared with the preference table. The roof types of the cells of a group are altered until the best overall match is found and the roof parameters that are shared among a group of cells are estimated for the whole group again. This concerns mainly the eaves and right heights, so that smooth roof polygons are created for neighbouring cells.

Table 1. Preference table for primitive roof types

neighbour primitives								
0	+	+	++	−	++	−	−	−
1	+	+	o	++	++	−	−	−
$2^{(1)}$	+	+	o	−	++	−	o	−
$2^{(2)}$	+	+	−	−	−	++	−	−
3	+	+	−	−	−	−	++	−
4	+	+	−	−	−	−	−	++
− bad match		o possible match			+ good match		++ perfect match	

[1] Opposite. [2] Corner arrangement.

In some circumstances, the roof simplification by primitive instancing creates shapes that do not conform to reality. Rather a valid appearance is preferred. This would be a problem in building reconstruction where similar techniques are applied. Here the aim is to generate a true-to-life representation of the building. In generalisation, as long as multiple cells have a high probability that results from the type and parameter estimation and are good matches in the preference table, then the roof shape is very likely also a good overall simplification. If the overall deviation in the length, angle and preference table is too high, then a fallback to the cell decomposition approach is always possible.

The interpretation of the roof structure, however, has some major advantages compared to the cell decomposition approach. First, symmetries in the roof structure can easily be maintained by adjusting the slopes of gabled and hipped roof elements. Second, a special simplification for uniform roof elements is also possible. For example, we experimented with parallel gabled and hipped roofs that are quite common for factory or shopping halls. Once the parameters of a building are known, which includes the number of the uniform elements, a typification of the roof structure is possible. Typification is an elementary generalisation operator that replaces n features by a lower number of features. The seven uniform hipped roof elements in the building of Fig. 12 are e.g. replaced by five elements. In this example, the parameterised rim can also either be retained or removed by the generalisation operator.

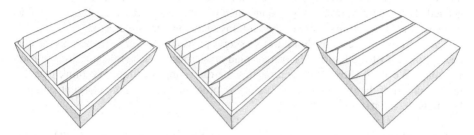

Fig. 12. 3D building model with uniform parallel hipped roof elements in its original shape (left) and before (middle) and after (right) typification

Another example is the simplification of round and curved building elements. For the palace in Fig. 13, the three tower elements were first identified from the ground plan polygon by their circular arranged facade segments. After their parameters were determined, all tower polygons were removed and the simplification of the remaining building model was performed by the primitive instancing approach as described. Afterwards, the towers were added again to the final model as simplified versions. Without an interpretation of the towers, these elements would be eliminated by the simplification or might even interfere with the generalization process.

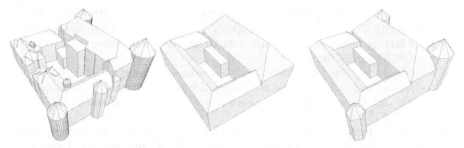

Fig. 13. 3D building model with circular tower elements in its original shape (left), after generalisation of the main building (middle) and with simplified towers (right)

7 Conclusion

Map and map-like presentations are essential to communicate spatial information. As 3D city models are becoming standard products of surveying offices, map-like 3D presentations are only a matter of time until they become available for a wide audience.

Because maps need to be mobile, such applications will run on mobile devices with all their limitations. As 2D generalisation operators are already a common tool to prepare data to the scale of maps, such a scale-depending transformation of 3D data will require new operators.

This paper proposes a new algorithm for the simplification of solitary 3D building models. It is based on cell decomposition and primitive instancing. Geometric properties that are specific to buildings like the coplanarity, parallelism and rectangularity of facade segments are preserved during simplification or can even be enforced if needed. The generalisation is solely controlled by an intuitive distance threshold value that specifies the minimum size of the building elements that are created.

The partition of the algorithm into two stages proved to be very effective as the cell decomposition of the building's ground plan simplifies the generalisation of the roof structure. We have shown two approaches for roof simplification. We think that the interpretation of the roof shape is necessary in order to execute more elaborate simplification operations.

Acknowledgements. The research described in this paper is founded by "Deutsche Forschungsgemeinschaft" (DFG – German Research Foundation). It takes place within the Collaborative Research Centre No. 627 "NEXUS – SPATIAL WORLD MODELS FOR MOBILE CONTEXT AWARE APPLICATIONS" at the Universität Stuttgart. The 3D building models are provided by Stadtmessungsamt Stuttgart.

References

1. Gröger, G., Kolbe, T.H., Plümer, L.: City Geographic Markup Language. Approved Discussion Paper of the Open Geospatial Consortium (2006)
2. Albert, J., Bachmann, M., Hellmeier, A.: Zielgruppen und Anwendungen für Digitale Stadtmodelle und Digitale Gelandemodelle – Erhebung im Rahmen der Arbeitsgruppe Anwendungen und Zielgruppen der SIG3D im Rahmen der Initiative GDI-NRW (2003)
3. Buchholz, H., Döllner, J., Nienhaus, M., Kirsch, F.: Real-Time Non-Photorealistic Rendering of 3D City Models. In: Proceedings of the 1st International Workshop on Next Generation 3D City Models, Bonn (2005)
4. Staufenbiel, W.: Zur Automation der Generalisierung topographischer Karten mit besonderer Bercksichtigung großmaßstäbiger Gebäudedarstellungen, Wissenschaftliche Arbeiten der Fachrichtung Vermessungswesen der Universität Hannover (51), Ph.D. Thesis (1973)
5. Powitz, B.-M.: Zur Automation der Kartographischen Generalisierung topographischer Daten in Geo-Informationssystemen, Wissenschaftliche Arbeiten der Fachrichtung Vermessungswesen der Universität Hannover (185), Ph.D. Thesis (1973)
6. Regnauld, N., Edwardes, A., Barrault, M.: Strategies in Building Generalization: Modelling the Sequence, Constraining the Choice. In: Progress in Automated Map Generalization – ACI (1999)
7. Van Kreveld, M.: Smooth Generalization for Continuous Zooming. In: Proceedings of the ICA, 4th Workshop on Progress in Automated Map Generalization, Peking, China (2001)
8. Harrie, L.E.: The Constraint Method for Solving Spatial Conflicts in Cartographic Generalisation. In: Cartography and Geographic Information Systems (1999)

9. Weibel, R.: A Typology of Constraints to Line Simplification. In: Proceedings of the 7th International Conference on Spatial Data Handling, pp. 533–546 (1996)
10. Meyer, U.: Generalisierung der Siedlungsdarstellung in digitalen Situationsmodellen. Wissenschaftliche Arbeiten der Fachrichtung Vermessungswesen der Universität Hannover (159), Ph.D. Thesis (1989)
11. Rainsford, D., Mackaness, W.A.: Template Matching in Support of Generalisation of Rural Buildings. In: Richardson, D., van Oosterom, P. (eds.) Advances in Spatial Data Handling, 10th International Symposium on Spatial Data Handling, pp. 137–152. Springer, Berlin (2002)
12. Camara, U., Antonio, M., Lopez, A., Javier, F.: Generalization Process for Urban City-Block Maps. In: Proceedings of the XXII International Cartographic Conference, La Coruna, Spain (2005)
13. Li, Z.: Transformation of Spatial Representations in Scale Dimension. International Archives of Photogrammetry and Remote Sensing 31(B3/III), 453–458 (1996)
14. Sester, M.: Generalization based on Least Squares Adjustment. International Archives of Photogrammetry and Remote Sensing 33(B4/3), 931–938 (2000)
15. Mayer, H.: Scale-Space Events for the Generalization of 3D-Building Data. International Archives of Photogrammetry and Remote Sensing 32(3/1), 520–536 (1998)
16. Forberg, A.: Generalization of 3D Building Data based on a Scale-Space Approach. In: Proceedings of the XXth Congress of the IRPRS, vol. 35, Part B, Istanbul, Turkey (2004)
17. Thiemann, F, Sester, M.: Segmentation of Buildings for 3D-Generalisation. In: Working Paper of the ICA Workshop on Generalisation and Multiple Representation, Leicester, UK (2004)
18. Anders, K.-H.: Level of Detail Generation of 3D Building Groups by Aggregation and Typification. In: Proceedings of the XXII International Cartographic Conference, La Coruna, Spain (2005)
19. Thiemann, F., Sester, M.: 3D-Symbolization using Adaptive Templates. In: Proceedings of the GICON, Wien, Austria (2006)
20. Luebke, D., Reddy, M., Cohen, J.D: Level of Detail for 3D Graphics. Morgan Kaufmann, USA (2002)
21. Coors, V.: Feature-Preserving Simplification in Web-Based 3D-GIS. In: Proceedings of the 1st International Symposium on Smart Graphics. Hawthorne, NY, USA, pp. 22–28 (2001)
22. Rau, J.Y., Chen, L.C., Tsai, F., Hsiao, K.H., Hsu, W.C.: Automatic Generation of Pseudo Continuous LoDs for 3D Polyhedral Building Model. In: Innovations in 3D Geo Information Systems, Springer, Berlin (2006)
23. Kada, M.: Automatic Generalisation of 3D Building Models. In: Proceedings of the Joint International Symposium on Geospatial Theory, Processing and Applications, Ottawa, Canada (2002)
24. Foley, J., van Dam, A., Feiner, S., Hughes, J.: Computer Graphics: Principles and Practice (2nd Edition), 2nd edn. Addison-Wesley, Reading (1990)

Degradation in Spatial Knowledge Acquisition When Using Automatic Navigation Systems

Avi Parush[1,2], Shir Ahuvia[2], and Ido Erev[2]

[1] Department of Psychology, Carleton University,
B552 Loeb Building, 1125 Colonel By Drive, Ottawa, ON, Canada, K1S 5B6.
Avi_Parush@Carleton.Ca
[2] Faculty of Industrial Management and Engineering
Israel Institute of Technology, Haifa, Israel, 32000

Abstract. Over-reliance on automated navigation systems may cause users to be "mindless" of the environment and not develop the spatial knowledge that maybe required when automation fails. This research focused on the potential degradation in spatial knowledge acquisition due to the reliance on automatic wayfinding systems. In addition, the impact of "keeping the user in the loop" strategies on spatial knowledge was examined. Participants performed wayfindings tasks in a virtual building with continuous or by-request position indication, in addition to responding to occasional orientation quizzes. Findings indicate that having position indication by request and orientation quizzes resulted in better acquired spatial knowledge. The findings are discussed in terms of keeping the user actively investing mental effort in the wayfinding task as a strategy to reduce the possible negative impact of automated navigation systems.

1 Introduction

Wayfinding and navigation become increasingly easier in the age of automated systems that continuously compute and display users' position in the environment. Although automating parts of the wayfinding task results in reducing the workload associated with it, other problems may arise. Over-reliance on the automated system may cause users to be "mindless" of the environment and not develop wayfinding and orientation skills nor acquire the spatial knowledge that maybe required when automation fails. This research was focused on the potential degradation in spatial knowledge acquisition due to the reliance on automatic wayfinding assistance. It was specifically aimed at exploring ways of engaging the user in the wayfinding task and thus facilitating spatial knowledge acquisition concurrent with the use of automated navigation assistance.

1.1 Wayfinding and Spatial Knowledge Acquisition

Spatial knowledge can be acquired by exploration or with the assistance of wayfinding artifacts and devices. Many studies compared active navigation and exploration in the environment to learning it from a map or route descriptions. The active navigation in the

S. Winter et al. (Eds.): COSIT 2007, LNCS 4736, pp. 238–254, 2007.

environment was classified according to the level by which people were familiar both with the environment and the destination [1, 2]. In general, findings indicated that map learning produced better performance with orientation tasks such as direction pointing, map drawing, and relative location estimation [3, 4, 5]. In comparison, direct navigation produced better performance with navigation tasks such as orienting to unseen targets, route distance estimation and route descriptions [3, 6]. Studies also indicated that spatial cognition changes as a function of the direct experience in the environment: more experience led to more survey knowledge, and less exposure still produced route knowledge [7, 3]. Taken together, many studies have demonstrated the importance of direct and active experience in effective spatial knowledge acquisition.

Another approach was to examine how people actively explore or learn a new environment or even navigate in a familiar environment with a navigation aid and not only by exploration. Two major navigation aids have been recognized and studied [8, 9, 10, 11]: route description or list which is primarily verbal in the egocentric frame of reference (e.g., turn right or left) and a survey description which is more often graphic, pictorial (e.g., a map) in the exo-centric frame of reference (e.g., turn north). In general, maps and survey descriptions produced better survey knowledge as compared to route descriptions [12]. The common aspect of these studies is that the navigation assistance artifact still required the user to be actively engaged in using it effectively. In other words, the user is required to continuously be aware of where she is in the environment and align it with the navigation aid in order to orient and make the subsequent wayfinding decisions [e.g., 13, for a review].

Along with such traditional and actively engaging navigation assistance, the last couple of decades were characterized by the introduction of automated navigation systems. The most common are those based on the Global Positioning System (GPS) which can continuously compute current position and display it within a variety of applications (e.g., route planning). These can replace some of the cognitive tasks the user did while wayfinding and orientating in familiar and unfamiliar environments. Most published research addresses the technological aspects of such automated navigation systems. Studies have examined and demonstrated the positive impact of such automated assistance on wayfinding and orientation performance [e.g., 14]. Another study [15] reported that such assistance reduces workload of operators executing wayfinding along performing other tasks. In addition, many studies have demonstrated the efficacy of GPS-based navigation systems for people with visual impairments who cannot rely on visual cues [16].

In contrast with the many studies that examined the impact of traditional navigation assistance (e.g., map or route directions) on spatial knowledge acquisition, it is hard to find studies addressing the impact of using automated navigation assistance on the human acquisition of spatial knowledge. This is surprising because many studies in various domains of automation have been concerned with and demonstrated the negative impact of automation on the performance of the human operator and the overall performance of the system.

1.2 The Downside of Automation

Wickens [17] functionally categorized automation by human processes: Perception – replacing human reception of information; Cognition – replacing human information processing, memory, and decision making; and Control – replacing human actions. Another categorization of automation [18] is also according to information acquisition, processing, decision making, action selection and execution. Based on such categorizations of automation in human-machine systems, an automatic orientation and navigation device such as a GPS-based system can: 1. replace human perception by eliminating the need to gather information from the environment; 2. replace human cognition by eliminating the need to integrate, comprehend the information, and process it (e.g., compare it to previous information or to information in the memory); and finally, 3. Eliminate the need for wayfinding decision making and problem solving.

While automation can alleviate human workload, there is an ironical aspect to it since it also creates problems. The introduction of automation to cars has the potential of degrading driving skills [19]. With automation, operators have degraded understanding of how the system works, and consequently have difficulties dealing with problems and new situations [20, 21]. Automated systems may reduce the cognitive effort required for decision making, but in turn may cause the users to make decisions based on error-prone heuristics [22, 23]. Users' reliance and trust in automated systems may induce the tendency not to monitor the system's performance [24]. Such tendency can reduce situation awareness, degradation in skill acquisition and skill maintenance, and poor vigilance. Users have reduced situation awareness if they are not actively involved with the system and if another agent (human or machine) is responsible to monitor and control changes [24, 18]. In addition, using automation may have longer term impacts. If information acquisition, information processing, decision making, or actions, are done by a system, the human user looses the skill to do all that when required [18, 25, 26]. The implication is that if automation fails to some extent, it maybe impossible for the human operator to take over. In a study [27] examining the transition from passive monitoring to active participation when working with a system, it was shown that operators who just had to monitor passively the system had many more errors and poorer performance when required afterwards to perform tasks requiring selective attention and rapid information processing.

In summary, the main problem of humans with automated system is "being out of the control loop". A passive operator will have reduced vigilance, reduced situation awareness, and be less skilled. Such an operator will have significant difficulties in making the transition to perform some of the tasks when automation fails. Such difficulties may have detrimental implications depending on the system. Based on this, the following question arises: Can the use of automated navigation system to alleviate the workload in wayfinding also have a negative impact on the acquisition of spatial knowledge? The very few studies on the impact of automated navigation systems on spatial knowledge have shown that the development of survey knowledge was impaired when using a mobile guiding navigation system [28] or was even affected negatively by the use of vehicle navigation systems [29]. The basic hypothesis of this study was that spatial knowledge acquisition was degraded while performing wayfinding with automatic navigation assistance.

1.3 This Study

The objective of this study was to assess the possible negative impact of navigation systems on spatial knowledge acquisition by contrasting it with situations where the user was "kept in the loop" while wayfinding. First, we simulated the reliance on automated navigation assistance with participants having continuous present position indication to support search and wayfinding tasks in a virtual building environment. Then, to address the possible impact of automated navigation assistance, we employed two strategies of keeping the user "In the Loop". The first was to eliminate the continuous availability of spatial information but let users access this information only they chose to. The assumption was that this will force the user to be more active and "mindful" to the cues in the environment and not rely continuously on the navigation system. The second "Keep the user in the loop" strategy was to get users more actively involved in the wayfinding task by introducing periodic orientation quizzes. The assumption was that if people have to indicate their position once in awhile, they will be more "mindful" of the environment, monitor their position, compare and update their cognitive map.

The hypothesis was that participants with the continuous position indication will exhibit superior wayfinding performance. Based on the many findings with automation it was expected that once this assistance was removed, the performance of those participants will be degraded relative to participants who were actively engaged ("Kept in the Loop") with the wayfinding task by one of the two strategies described above. To test these hypotheses, the study design consisted of four between-participant experimental conditions based on the combination of two factors: Position indication, consisting of a continuous position indication or a position indication by request; Orientation quizzes, consisting of wayfinding with or without orientation quizzes. The impact of these strategies on wayfinding performance was tested with a third factor of wayfinding trials.

2 Methods

2.1 Participants

The experiment sample included 103 participants, 45 men and 58 women. All participants were students from the Industrial Management and Engineering Faculty, Israel Institute of Technology. Their ages ranged from 21 to 40. Participants were assigned randomly to one of the four conditions of this experiment. All participants were proficient with MS Windows and use of a game joystick. All participants, according to their self-report, had normal or corrected-to-normal vision and no known manual dexterity or eye-hand coordination problems. All participants were native Hebrew speakers.

2.2 Experimental Design and Tasks

Design. The experimental design was a fully factorial mixed design based on three factors: 1. Position Indication (between-participants) consisting of two levels: position displayed continuously vs. position displayed only by request; 2. Orientation quizzes (between-participants) consisting of two levels: with random orientation quizzes and without quizzes; and 3. Search and wayfinding trials (within-participants) consisting of 16 trials for each participant.

Experimental Tasks. The experimental tasks included the following:
Target Search and Wayfinding. This task was assigned to all participants in all four experimental conditions. Each participant viewed a picture of the target (e.g., a cabinet, house plant, sculpture, etc.) and was then asked to navigate in the experimental environment to find that target. Tasks were counter-balanced across conditions and participants in terms of the following parameters: distance from the starting point of each search trial, the scene at the starting point (within a corridor or in an open space), the distinctiveness of the target (non-distinct such as a table vs. a sculpture), and the need to change between floors in the simulated building in order to find the target. As was mentioned above, each participant performed 16 search and wayfinding tasks in one of the four experimental conditions. There was one additional search and wayfinding trial (trial 17) not crossed with any of the between-participant factors; i.e., there was no present position indication, neither continuously nor by request, and there were no orientation quizzes. This trial was the cost of transition test trial.

Position Display Request. Participants assigned to the Position by Request experimental condition could, at any time, request a map display showing their present position. This display was on for 5 seconds during which participants could not proceed with their wayfinding.

Response to Orientation Quiz. Participants assigned to the experimental condition with orientation quizzes were required, at random times and distances during the wayfinding trial, to indicate their present position by pointing and clicking on a displayed map. Participants could not view the virtual environment until they indicated their position.

Measurements. The following measurements were taken and computed in order to reflect wayfinding performance and level of acquired spatial knowledge:
Wayfinding Performance. The main index for wayfinding performance was the excess distance, in meters, to the target. This measure was computed as the difference between actual distance traveled by the participant and the minimal distance required to reach the target from the starting point.

Present position requests. This measure was simply the number of times the display of the present position was requested by participants in the relevant experimental condition.

Orientation quiz performance. This measure was the difference, in meters, between the position indicated by the participant and the correct present position.

Level of acquired spatial knowledge. This measure was based on a Judgment of Relative Direction. Participant was shown a given scene and was asked to point in the direction of one of the targets (also shown) by clicking on one of the possible directions displayed as buttons. Performance was computed as precentage of correct responses in the judgment of relative direction. In addition, performance was computed as the difference between the direction pointed by the participant and the correct direction.

2.3 Experimental Setup

The experimental environment consisted of a simulated four-story building, each story spanning approximately 9000 square meters. The building consisted of closed offices, corridors and hallways, open spaces, doors and elevators. In addition, various objects were placed at various locations in the corridors and open spaces. The simulated visual field was 45 degrees laterally and 60 degrees vertically. A cross displayed in the center of the visual field aided with pointing at the target. The general experimental display is presented in figure 1.

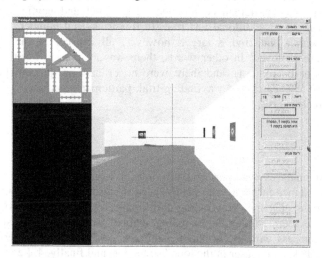

Fig. 1. The experimental environment. The virtual first-person perspective of the building is in the center. Left-hand side includes the map with position indication, and the right-hand side includes a control panel for performing various operations.

The left hand side included a map display of the current floor the participant was in. A yellow dot indicated the current position with an arrow pointing to the direction the participant was looking at. A blue dot indicated the starting position and a red dot marked the target location. This map was displayed either continuously or upon request depending on the experimental condition. The right hand side consisted of a control panel for various operations. These included: start trial, request the present position indication, indicating finding the target, and starting a new trial.

The simulation was developed using the program Vega in an MFC environment. It ran on a desktop PC with a 1024X768 17" monitor. A Microsoft Sidewinder Joystick was used for movement, point and click.

2.4 Procedure

Each participant was assigned randomly to one of the four experimental conditions. Participants filled in a questionnaire with information regarding age, experience with virtual games and orienteering, and then received the instructions relevant to their experimental condition and one practice trial. The purpose was to ensure that participants were capable of navigating with the joystick, request a map, or respond to the quiz (depending on their experimental condition), find the target, move between floors, terminate a trial, and start a new one. Once the experimenter verified that participants understood the tasks and interacted properly with the experimental environment, the first of the 16 search and wayfinding trials began.

Participants could view and read the trial instructions on the right-hand side of the screen throughout the trial. Participants could also consult written instructions with respect on how to use the joystick. Once participants decided they have located the target, they clicked on it and terminated the trial. The transition to the next trial was self-paced to ensure participants had some rest between trials and not develop any cyber-sickness. Once participants completed the 16 search and wayfinding trials, they were introduced to an additional test trial. In this trial (trial number 17) they were required to search and find a target, however, all aspects of the experimental conditions were removed. In other words, there was no present position indication (continuously or by request), and there were no orientation quizzes. Finally, after completing the test search and wayfinding trial, participants performed the judgment of relative direction task.

3   Results

The following analyses were performed in order to examine the impact of automatic navigation assistance in contrast to using "keep the user in the loop" strategies with the automatic assistance: 1. Wayfinding performance; 2. Performance with the "keep the user in the loop" strategies, including number of requests to display present position and success with the orientation quizzes; 3. Performance after removal of automation and "keep the user in the loop" strategies; and finally, 4. Level of acquired spatial knowledge.

3.1   Wayfinding Performance

Excess Distance. The mean excess distance traveled in the search and wayfinding trials was computed as a function of blocks, each consisting of 5 trials. The means are presented in figure 2. It can be seen that the mean excess distance traveled by participants with continuous position indication was consistently lower than the excess distance traveled by participants who had present position displayed by request only. In addition, it can be seen that there was no visible change in the performance of participants with continuous position display as a function of the blocks. In contrast, performance of participants with position display by request improved (excess distance decreased) as a function of the blocks.

A three-way ANOVA with two between-participant variables (2 quiz conditions X 2 position indication conditions) and repeated measures on the block factor was performed on the data to explore whether the above observations were statistically significant. There was a significant main effect for the block factor ($F_{2,170}=3.49$, p<.05). This effect was associated with a marginally significant two-way interaction between the block and the position indication factor ($F_{2,170}=2.71$, p=.07). The interaction between the blocks and the position indication implies that the differences between the blocks are due to the interaction with the position indication. The main change as a function of the block took place with the participants who had position indication by request. In other words, with each block, the mean excess distance was reduced from 69m in the first block, to 53m in the second, and 47 in the last block.

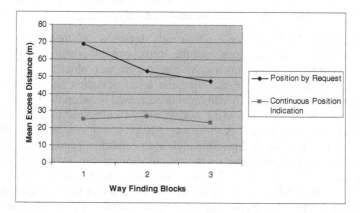

Fig. 2. Mean excess distance traveled by participants as a function of the wayfinding blocks. The top curve is for the group with position display by request, and the bottom curve for the group with continuous position indication.

Position Display Requests. The mean number of requests to display the present position in the relevant experimental groups was computed as a function of the experiment blocks. The means are presented in figure 3.

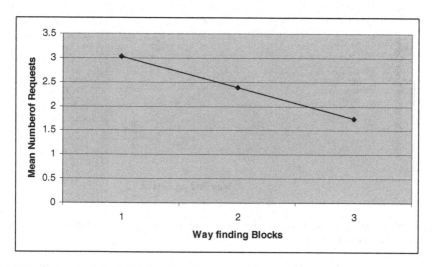

Fig. 3. Mean number of requests to display the present position as a function of the wayfinding blocks

It can be seen that the number of requests to display present position decreased consistently as a function of experiment blocks (from a mean = 3 in the first block to a mean of 1.75 in the third block). A three-way ANOVA with two between-participant variables (2 quiz conditions X 2 position indication conditions) and repeated measures on the block factor was performed on the data to explore whether the observed trend was statistically significant. A main effect was found for the repeated block factor ($F_{2,170}$=4.0, p<.05) indicating that the decrease in number of requests was significant.

Orientation Quiz Performance.   The mean deviation, in meters, between the position indicated by the participant and the correct present position for the relevant groups, was computed as a function of the experiment blocks. The means are presented in figure 4. It can be seen that the mean deviation between the correct position and position indicated by participants decreased between the first and second block (from a mean of 37 meters to 31 meters) and then remained unchanged between the second and the third block. A three-way ANOVA with two between-participant variables (2 quiz conditions X 2 position indication conditions) and repeated measures on the block factor was performed on the data to explore whether the observed trend was statistically significant. A main effect was found for the repeated block factor ($F_{2,170}$=48.11, p<.01) indicating that the drop in magnitude of the deviation in present position indication by participants was significant. A series of pair-wise Bonferroni t-tests were performed as post-hoc tests. These tests confirmed that the mean deviation in the first block was significantly higher than the respective means in the second and third blocks. There was no significant changes in the deviation between the second and third blocks.

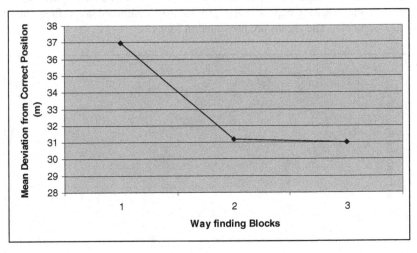

Fig. 4. Mean deviation from correct position in responses to the orientation quizzes as a function of the wayfinding blocks

Summary of Search and Wayfinding Performance.   Performance, as measured by the excess distance taken by participants, improved significantly with more search tasks only for participants who did not have a continuous present position indication. This should be viewed in conjunction with the additional finding that the mean number of requests to display present position decreased with more search tasks.

Participants who had orientation quizzes also improved the accuracy of their responses with more search tasks. Taken together, the findings imply that participants who either requested the display of present position or responded to orientation quizzes may have acquired some spatial knowledge about the environment as a function of practice.

3.2 Cost of Transition

The assumption that participants having the "keep the user in the loop" strategies (request the present position or respond to orientation quizzes) may have acquired more spatial knowledge was tested by examining the cost of transition to performing wayfinding without any of the navigation assistance mechanisms and strategies. The mean excess distance traveled in the search and wayfinding was computed for the additional trial (trial 17, the cost-of-transition test trial). The means for trial 16 (last trial with all mechanisms and strategies still in place) and trial 17 (with all mechanisms and strategies removed) are presented in figures 5 (for the present position display factor) and 6 (for the orientation quiz factor).

It can be seen in figure 5 that mean excess distance for both groups, with continuous position display and with display by request, increased in the cost-of-transition test trial. It can also be seen that, while the group that had continuous present position display during the preceding 16 trials had a better overall performance, the increase in excess distance was more pronounced relative to the group with position display by request.

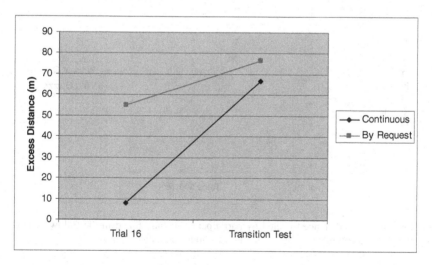

Fig. 5. Mean excess distance traveled by participants in trial 16 and in the following cost-of-transition test trial for the two groups with continuous position display and display by request

A three-way ANOVA with two between-participant variables (2 quiz conditions X 2 position indication conditions) and repeated measures on a trial factor (trial 16 and trial 17) was performed to explore whether this trend is significant. A significant two-way interaction was found between the position display and the trial factor

($F_{1,85}$=4.34, p<.05). A series of pair-wise Bonferroni t-tests were performed as post-hoc tests to explore the source of the interaction. These tests showed that the mean excess distance of the group which had a continuous present position display was significantly lower than the group that had present position display by request, in trial 16 (pre-transition trial). There were no differences between the groups in the cost-of-transition trial. In other words, there was a greater performance degradation reflecting a higher cost of transition for the group that had continuous present position display.

Figure 6 shows that the mean excess distance for both groups, with and without orientation quizzes, was identical in trial 16 and then increased in the cost-of-transition test trial. However, the increase in excess distance was more pronounced with the group that had no orientation quizzes during the preceding 16 trials. Based on the same ANOVA described in the previous paragraph, a significant two-way interaction was found between the orientation quiz and the trial factor ($F_{1,85}$=5.29, p<.05). A series of pair-wise Bonferroni t-tests were performed as post-hoc tests to explore the source of the interaction. These tests showed that the mean excess distance in the cost-of-transition trial for the group which did not have the orientation quizzes was significantly higher that the group that responded to orientation quizzes,. There were no differences between the group in trial 16 (pre-transition trial). In other words, there was a higher cost of transition for the group that did not respond to orientation quizzes.

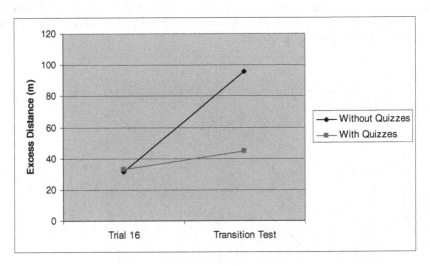

Fig. 6. Mean excess distance traveled by participants in trial 16 and in the following cost-of-transition test trial for the two groups with and without orientation quizzes

It is important to note that the transition to trial 17 did not actually introduce any changes in terms of the lack of quizzes for the two groups without orientation quizzes. However, the position display, either continuous or by request, was removed. Taken together, the above results indicate that the removal of position indication for the groups with the orientation quizzes was not associated with any performance regression, whereas there was a significant degradation of performance for the groups without orientation quizzes. This implies a stronger impact of the orientation quizzes as a mechanism for keeping the user in the loop.

3.3 Spatial Knowledge

Mean percentages of correct responses in the judgment of relative direction test were computed for all participants as a function of the four experimental groups. The mean percentages for each of the groups are presented in figure 7.

It can be seen that the group of participants who performed the search and wayfinding task with orientation quizzes and position display by request had the highest percentage of correct responses. A χ^2 test confirmed that there was a significant association between the orientation quiz and position display factors with respect to proportion of correct responses in the judgment of relative direction (χ^2 = 5.3, p<.05).

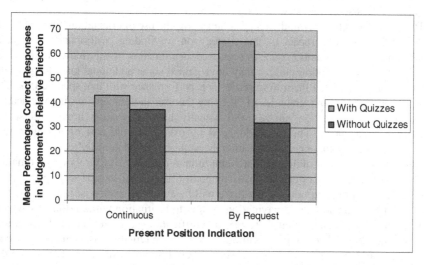

Fig. 7. Mean percentage of correct responses in the judgment of relative direction for the four experimental groups

A two way ANOVA ((2 quiz conditions X 2 position indication conditions) was performed on the mean difference between the direction pointed by the participant and the correct direction. There was a significant main effect for the orientation quiz factor ($F_{1,85}$=5.01, p<.05). The mean difference between the direction indicated by the participant and the correct direction was significantly higher for the group without orientation quizzes (Mean= 77.14) than the group with orientation quizzes (Mean = 49.09).

Taken together, the two measures reflect a strong combined impact of the two "keep the user in the loop" mechanisms on the acquisition of spatial knowledge. These findings are also in line with the cost of transition findings that emphasized the strong positive impact of the orientation quizzes.

4 Discussion

4.1 Summary of the Findings

The basic hypothesis of this study was that participants with continuous position indication will exhibit superior wayfinding performance. The findings show that wayfinding performance (i.e., the excess distance traveled by participants to their targets) was consistently better for participants with continuous position indication. In comparison, participants with position indication by request (one "keep the user in the loop" strategy) traveled more excess distance to the target. However, this excess distance decreased consistently as a function of more wayfinding trials. In addition, the mean number of times those participants requested the display of their position also decreased consistently as a function of more wayfinding trials. Taken together, both findings imply that those participants may have acquired knowledge of the environment that enabled them to improve their wayfinding performance and need less and less assistance.

The other "keep the user in the loop" strategy was the use of orientation quizzes. While there was no apparent impact of this strategy on wayfinding performance, the actual performance in the orientation quizzes improved as a function of wayfinding trials. Thus, participants with orientation quizzes may have gained some additional knowledge of the environment that enabled them to improve their performance with that specific task. The question remained if both strategies of "keeping the user in the loop", position indication by request and orientation quizzes, did actually have a positive impact on spatial knowledge acquisition. This was addressed by examining the cost of transition to a situation without automation, and by testing the acquired spatial knowledge.

Automation research indicates that when human operators fully rely on automation and are then required to perform tasks manually, they tend to perform poorly. This was the basis for our hypothesis that there will be a performance cost in the transition from relying on the automatic navigation system to a situation without that assistance. The findings here indicate that there was a significant cost, performance degradation, with the transition to a situation without automated navigation assistance. This cost was particularly strong for participant who had continuous position indication and did not have orientation quizzes. In other words, the two "keep the user in the loop" strategies, position indication by request and orientation quizzes, actually reduced or moderated the cost of transition to having no navigation assistance.

Finally, the combination of the two "keep the user in the loop" strategies resulted in the highest level of acquired spatial knowledge. The group of participants with the combination of position indication by request and orientation quizzes during the wayfinding trials had the best judgment of relative direction performance (i.e., highest percentage of correct responses and lowest magnitude of deviation from the correct direction).

4.2 Theoretical Implications

Many studies in the field of spatial cognition have demonstrated the importance of direct and active experience in effective spatial knowledge acquisition. The study reported here has also demonstrated that when using strategies which actively engage the user in the wayfinding task, it can have a positive impact on spatial knowledge acquisition. Specifically, the strategies used in this study "forced" participants to actively monitor and track their location in the environment in order to make wayfinding and route planning decisions. This active engagement in the wayfinding task resulted in better spatial knowledge. The following theoretical account of the findings builds upon some theoretical principles of learning and transfer of learning.

One of the aspects of the continuous position indication in terms of learning and transfer to new situations is the concept of Knowledge of Results (KR). KR is important at the learning stage as a guiding mechanism. If there is immediate KR at every step, the learner achieves very good performance during the acquisition phase because they have continuous feedback on their performance. However, since learners focus on the KR to guide their actions, they may not be attentive to various cues that maybe critical for the learning, and thus learning maybe degraded. Consequently, when participants reach the stage of transfer, i.e., performing without KR, performance is degraded [30, 31, 32]. Continuous display of present position acted as a continuous knowledge of results. This was associated with superior performance during the acquisition phase (the search and wayfinding trials). However, the continuous knowledge of results had a negative impact on transfer, the transition to wayfinding without position indication.

Continuous KR may have some implications on the extent to which a learner is passive or active during the learning phase. It is conceivable that a learner with continuous KR maybe more passive during the learning phase. Research on automation showed that a passive operator will have reduced vigilance, reduced situation awareness, and be less skilled. Such an operator will have significant difficulties in making the transition to perform some of the tasks when automation fails. Another study [33] showed that when participants did search and wayfinding based on route directions, their acquired level of spatial knowledge was inferior compared to participants wayfinding with a map. It was hypothesized that participants using a map were more actively engaged with monitoring, tracking, updating their position, and making wayfinding decisions, as opposed to participants who followed the instructions and were relatively passive and "mindless" of the environment. In other words, using artifacts or strategies that "force" the user to invest more mental effort in their task tends to result in better learning.

The construct of invested mental effort during learning [34] assumes intentional, meta-cognitively guided, non-automatic, effortful processes. The impact of such an approach is particularly strong in the abilities to transfer to new situations beyond the learning context [also 35]. A further explanation of this can be found in some of the earlier learning theorists that distinguished between the "habit strength" - long-term retention and performance - and the "momentary response potential" [36] or "response strength" [37] which represent more the performance during learning. In a more recent "new theory of disuse" [38], following Thorndike's "Law of Disuse", a distinction was made between "storage strength" - representing the degree of learning - and "retrieval strength" - representing the ease of accessing the learned material. Retrieval strength weakens in the course of time if the storage strength was not built properly during the learning process. In our study, the use of "keeping the user in the loop" strategies may have acted to build storage strength which prevented or at least slowed significantly the decay of the response strength. This was reflected in the reduced cost of transition and the better spatial knowledge.

4.3 Practical Implications

In many circumstances, it may seem somewhat unreasonable to force users of automatic navigation systems to be more active in the wayfinding task in order to gain better spatial knowledge. For example, the occasional travelers using GPS-based navigation systems cannot be expected to actually increase their workload and not benefit fully from the guidance the navigation system provides. The practical implications of the findings are more relevant to users that wayfinding is an inherent, critical, and frequent part of their duties. This may include professional drivers (e.g., cab drivers), military personnel, pilots, search and rescue operators, and others. For such populations it is conceivable to develop instruction and training in such a way as to encourage the investment of mental effort in the learning process. This in turn may induce users to develop strategies of not completely and continuously rely only on the automatic navigation assistance but keep their own positional awareness. This will enable them to cope better with unexpected situations of automation failure.

The research reported here was carried out in a desktop virtual building. It should be emphasized that navigation and orientation may differ between virtual and the real world [e.g., 39]. The research paradigm utilized here should also be carried out in the real world, in comparison with a virtual world, to further verify the findings and theoretical implications of this study. Furthermore, direction estimation was used here to assess spatial knowledge. It should be noted that there are several other metrics for assessing spatial knowledge such as distance estimation, map drawings, and more. Such various metrics should be measured in future studies to further confirm and generalize the finding reported here.

The research reported here opens additional research questions. There are various levels of automatic navigation assistance such as route planning, route guidance, or "You Are Here" assistance. The impact of these different levels of automatic assistance on spatial knowledge acquisition needs to be studied in a controlled manner. Various non-interrupting ways of engaging the user in the wayfinding task along with automatic assistance need to be developed and tested. Finally, the spatial cognition research community should consider developing and testing various strategies for people with visual impairments to use various non-visual cues along with the automatic wayfinding assistance.

Acknowledgments. The study reported here was conducted while the first author was a visiting scientist at the Israel Institute of Technology.

References

1. Allen, G.L.: Spatial abilities, cognitive maps and wayfinding: Bases for individual differences in spatial cognition and behavior. In: Golledge, R.G. (ed.) Wayfinding behavior: Cognitive mapping and other spatial processes, pp. 46–80. Johns-Hopkins Press, Baltimore, MD (1999)

2. Darken, R.P., Sibert, J.L.: Navigating in large virtual worlds. International Journal of Human Computer Interaction 8(1), 49–72 (1996)
3. Thorndyke, P.W., Hays-Roth, B.: Differences in spatial knowledge acquired from maps and navigation. Cognitive Psychology 14, 560–589 (1982)
4. Ruddle, R., Payne, S.J., Jones, D.M.: Navigating buildings in desk-top virtual environments: Experimental investigations using extended navigational experience. Journal of Experimental Psychology: Applied 3, 143–159 (1997)
5. Moeser, S.D.: Cognitive mapping in a complex building. Environment and Behavior 20(1), 21–49 (1986)
6. Hintzman, D.L., O'Dell, C.S., Arndt, D.R.: Orientation in cognitive maps. Cognitive Psychology 13, 149–206 (1981)
7. Golledge, R.G., Spector, N.A.: Comprehending the urban environment: Theory and practice. Geographical Analysis 14, 305–325 (1978)
8. Taylor, A.H., Tversky, B.: Descriptions and depictions of environments. Memory & Cognition 20, 483–496 (1992)
9. Taylor, A.H., Tversky, B.: Spatial mental models derived from survey and route descriptions. Journal of Memory and Language 31, 261–292 (1992)
10. Taylor, A.H., Tversky, B.: Perspectives in spatial descriptions. Journal of Memory and Language 35, 371–391 (1996)
11. Wickens, C.D., Hollands, J.G.: Engineering Psychology and Human Performance, 3rd edn. Prentice-Hall, Englewood Cliffs (1999)
12. Schneider, L.F., Taylor, H.A.: How do you get there from here? Mental representations of route descriptions. Applied Cognitive Psychology 13, 415–441 (1999)
13. Lobben, A.K.: Tasks, Strategies, and Cognitive processes Associated with Navigational Map Reading: A Review Perspective. The Professional Geographer 56(2), 270–281 (2004)
14. Willis, K.: Mind the Gap: Mobile Applications and Wayfinding. In: Workshop for User Experience Design for Pervasive Computing, Pervasive 05 (2005)
15. Leggat, A.P., Noyes, J.M.: Navigation Aids: Effects on Crew Workload and Performance. Military Psychology 12(2), 89–104 (2000)
16. Marston, J.R., Loomis, J.M., Klatzky, R.L., Golledge, R.G., Smith, E.L.: Evaluation of Spatial Displays for Navigation without Sight. ACM Transactions on Applied Perception 2(3), 110–124 (2006)
17. Wickens, C.D.: An introduction to human factors engineering. Addison-Wesley, Reading (1997)
18. Parasuraman, R.: Designing automation for human use: Empirical studies and quantitative models. Ergonomics 43(7), 931–951 (2000)
19. Stanton, N.A., Young, M.S.: Vehicle automation and driving performance. Ergonomics 41(7), 1014–1028 (1998)
20. Sarter, N.B., Woods, D.D., Billings, C.E.: Automation surprises. In: Salvendy, G. (ed.) Handbook of human factors and ergonomics, 2nd edn. pp. 1926–1943. John Wiley Sons, Chichester (1997)
21. Sarter, N.B., Woods, D.D.: Pilot interaction with cockpit automation II: An experimental study of pilot's model and awareness of the flight management and guidance system. International Journal of Aviation Psychology 4(1), 1–28 (1994)
22. Tversky, A., Kahnman, D.: Judgment under uncertainty: Heuristics and biases. Science 185, 1124–1131 (1984)
23. Parasuraman, R.: Humans and automation: use, misuse, disuse, abuse. Human Factors 39(2), 230–253 (1997)

24. Endsley, M.R., Kiris, E.O.: The out-of-the-loop performance problem and level of control in automation. Human factors 37(2), 381–394 (1995)
25. Johanson, G.: Stress, autonomy, and the maintenance of skills in supervisory control of automated systems. Applied Psychology: an international review 38(1), 45–56 (1989)
26. Wiener, E.L.: Cockpit Automation. In: Wiener, E.L., Nagel, D.C. (eds.) Human factors in aviation, pp. 433–461. Academic, San Diego (1988)
27. Johansson, G., Cavalini, P., Pettersson, P.: Psychobiological Reactions to Unpredictable Performance Stress in a Monotonous Situation. Human Performance 9(4), 363–384 (1996)
28. Aslan, I., Schwalm, M., Baus, J., Kruger, A., Schwartz, T.: Acquisition of Spatial Knowledge in Location Aware Mobile Pedestrian Navigation Systems. In: MobileHCI'06, pp. 105–108. ACM Press, New York (2006)
29. Burnett, G.E., Lee, K.: The effect of vehicle navigation systems on the formation of cognitive maps. In: Underwood, G. (ed.) Traffic and Transport Psychology: Theory and Application, pp. 407–418. Elsevier, Amsterdam (2005)
30. Jarus, T.: Is more always better? Optimal amounts of feedback in learning to calibrate sensory awareness. The Occupational Therapy Journal of Research 15(3), 181–197 (1995)
31. Salmoni, A.W., Schmidt, R.A., Walter, C.B.: Knowledge of results and motor learning: A review and critical reappraisal. Psychological Bulletin 95(3), 355–386 (1984)
32. Sparrow, W.A., Summers, J.J.: Performance on trials without knowledge of results (KR) in reduced relative frequency presentations of KR. Journal of Motor Behavior 24(2), 197–209 (1992)
33. Parush, A., Berman, D.: Orientation and navigation in 3D user interfaces: The impact of navigation aids and landmarks. International Journal of Human Computer Studies 61(3), 375–395 (2004)
34. Salomon, G.: Television Is Easy and Print Is Tough The Differences Investment of Mental Effort in Learning as a Function of Perceptions and Attributions. Journal of Educational Psychology 76(4), 647–658 (1984)
35. Cuevas, H.M., Fiore, S.M., Bowers, C.A., Salas, E.: Fostering constructive cognitive and metacognitive activity in computer-based complex task training environments. Computers in Human Behavior 20, 225–241 (2004)
36. Hull, C.S.: The principles of behavior. Appleton-Century-Crofts: New York (1943)
37. Estes, W.K.: Statistical theory of distributional phenomena in learning. Psychological Review 62, 369–377 (1955)
38. Bjork, R.A., Bjork, E.L.: A new theory of disuse and an old theory of fluctuation. In: Healy, A., Kosslyn, S., Shiffrin, R. (eds.) From learning process to cognitive processes: Essays in honor of William K. Estes, vol. 2, pp. 35–67. Earlbaum, Hillsdale, NJ (1992)
39. Richardson, A.E., Montello, D.R., Hegarty, M.: Spatial knowledge acquisition from maps and from navigation in real and virtual environments. Memory Cognition 27(4), 741–750 (1999)

Stories as Route Descriptions

Volker Paelke and Birgit Elias

ikg – Institute of Cartography and Geoinformatics, Leibniz University of Hanover,
Appelstr. 9a, 30167 Hannover, Germany
{Volker.Paelke,Birgit.Elias}@ikg.uni-hanover.de

Abstract. While navigation instructions in terms of turn instructions and distances are suitable for guiding drivers on roads, a different context of use - like pedestrian navigation - requires extended routing data and algorithms as well as adapted presentation forms to be effective. In our work we study alternative forms of navigation instructions for pedestrians in city environments. In this paper we explore the usefulness of directions given in the form of a short story. To aid retention of navigation instructions and recognition of decision points along the route we have expanded a landmark-based navigation system to present navigation instructions as a sequence of story elements. In this paper we introduce the concept of stories as route descriptions, describe the current prototype implementation, and present preliminary evaluation results from user tests that will guide further development.

1 Introduction

Current pedestrian navigation systems mainly use maps combined with positioning and routing information to represent geographical knowledge. Research that examined the specific context of use in pedestrian navigation has suggested that landmarks are well suited to identify important places along a route [1] and thus can form a valuable extension of route descriptions. Experiments with the use of landmarks in route descriptions have shown to increase the perceived quality of maps [2]. It is generally acknowledged that landmarks represent a central concept in successful route descriptions for pedestrians [2,3,4,5,6]. However, experience shows that some users still have difficulties in memorizing route descriptions, even when they are augmented with landmarks. Experiments with existing systems [7] have also shown that different usage scenarios (e.g. guidance during a sightseeing tour through a city vs. guidance with the aim of reaching a destination on the most efficient route) result in completely different expectations and requirements of users. In inner city navigation a particularly common task is to reach a specific location from a given starting point by the most effective route. In contrast to sightseeing systems and many proposed location-based services the users in this scenario are not interested in additional information on their surroundings or targeted advertisements. Instead, they want a route description that is easy to memorize and can be executed without repeated interaction with a navigation device. Since this use case is very common it seems useful to examine how memorable route description can be synthesized and presented.

S. Winter et al. (Eds.): COSIT 2007, LNCS 4736, pp. 255–267, 2007.

The approach to memorable route description presented in this paper builds on digital storytelling techniques and is inspired by the mnemonic techniques associated with the "Method of Loci" attributed to Cicero [8] and the system of "songlines" across the Australian continent, as popularized by the author Bruce Chatwin [9]. The method of loci is a technique used to remember large amounts of information by associating each "unit" of information with a (virtual) place. Forming a story connecting these places and mentally revisiting these places can help to recall the information "units" associated with them. "Songlines" are part of the aboriginal Australian creation myth of the dreamtime, in which each place has its own creation story captured in the form of a song. The practical use of songlines as a navigation aid has been explored by ethnographers (e.g. Lewis [10]). Helbling [11] provides an overview of songlines as a system of interconnected paths across the Australian continent and how cultural practices served to maintain accurate distribution of the songs. In this sense songs can be viewed as route descriptions, with the strophes of a song following the linear structure of one of these paths. Strophes of the song are based on places, e.g. landscape features like rocks, mountains, springs, groups of trees, crossings etc. From a navigation perspective the system of songlines can thus be thought of as a map with interconnected routes that is passed on from generation to generation in the oral form of songs. Myers [12] provides an interesting overview of the specific cultural traditions and practices of the Pintupi tribe. While we have taken inspiration from the concept of songlines we do not try to mimic specific characteristics of Aboriginal songs.

2   Related Work

2.1   Route Descriptions for Wayfinding

It is everyday experience that verbal information is a common means for conveying wayfinding information. Linguistic and psychological studies are conducted to identify the types of transactions that take place and determine their basic components [2,3,13,14]. Due to the relevance for the design of computer-aided driving assistance systems the communication types and structures of driving directions has been analysed intensely [15,16,17,18]. Also the nature of spatial mental representations induced by verbal descriptions has been explored [19,20].

From these studies it follows that directions are a specific kind of spatial discourse used to communicate spatial relations to other people. The purpose of directions in navigation tasks is to convey all information that is necessary to guide someone from a starting point to a destination. Therefore the route is subdivided into sequential instructions. An analysis of conventional verbal route descriptions reveals that the most important elements in these instructions are the reference to the direction of movement and landmarks [2,3,14].

To generate appropriate route directions automatically it is necessary to understand the underlying structure of human generated instructions and how landmarks are incorporated in them. Following [21] directions are given in wayfinding narratives. The understanding of directions is guided via narrative structures in which landmarks are embedded.

But route descriptions are not only communicated in verbal form. According to [22] the communication of spatial relations is appropriate both in verbal and graphical form. The general structure and content of directions and route maps is the same and it is assumed that an automatic translation between both representation forms is possible [23]. Since techniques for automatic map generation are comparatively well developed route maps are commonly used in navigation instructions.

2.2 Digital Storytelling

In recent years techniques and approaches that originate in the entertainment domain have found widespread use in serious applications. The most prominent example is the exponential growth of 3D graphics performance in personal computers which was driven by the 3D game market and has enabled a multitude of 3D visualization applications in other domains. In the geo context examples include the use of 3D technologies originating in games to communicate 3D geo-information (e.g. [24]) and to train users in geographic analysis tasks (e.g. [25]). Digital storytelling is an emerging discipline in which digital presentation media (which might be generated automatically) are used to relate a complex story. Digital storytelling and game techniques have been used successfully to communicate relevant information to users, e.g. by requiring users to solve quests for which information has to be discovered in a virtual environment [26]. Cartwright [27] considers entertainment technologies as helpful to deliver geographical information efficiently by exploiting the familiarity of users with spatial concepts in games. Thus new users, who are experienced with web and game technologies, can be made 'geographically aware' of map based information. First experiments with this approach suggest that game like interfaces can improve the way new users interact with geographical information spaces.

3 Generating Stories from Route Directions

In the "KuGeRou" ("short story routing", in German "Kurz-Geschichten-Routing") system we aim to use short memorable stories as route descriptions. The hypothesis is that stories related to the sequence of places that a user encounters when following a route can be used as memorable descriptions of it. The use of a story instead of a map could reduce complexity for the user. While the data space of a map contains the complete network of possible routes and thus easily overwhelms the memory capacity of a user, the linear path of a route and the corresponding linear experience of following it contain far less information. By mapping this experience to a similarly linear story taking place at the decision points along a path we aim to create a mental representation of the route that could be easier to memorize than both the complete map and abstract route descriptions. In our approach route descriptions are mapped to general story elements: a route is related as a story, a decision point / landmark location is represented as a scene in the story, a navigation (e.g. turn) instruction corresponds to a similar action within the story, and objects (e.g. landmarks) and agents in a route description are mapped to actors and items in the story.

In the system we use a conventional route planning system to generate position information, route actions, landmarks and distances to decision points on a route and encode this in a data structure, similar to existing systems. A navigation tuple is placed at each decision point and consist of a landmark, the street/place name and a direction as well as the distance to this decision point. We augment this with information for the storytelling system in a second data structure. Each decision point is also associated with a story tuple, consisting of three pre-authored presentation sequences, a main actor, a set of additional actors and objects determined dynamically from the navigation tuple, a set of actions, and possible user interactions. Figure 1 shows an example of both data sets for a specific decision point where the user should turn right into "Georgstrasse" at a "H&M" Shop. The navigation tuple contains the relevant information from the route planner. The story tuple encodes the "story" in a sequence of three templates that can be interpreted as follows: The mouse "George" (name determined by the street name) meets the main actor, gives him a hat (object determined by H&M landmark) and disappears to the right (animation direction determined by turn instruction). Figure 3 (left) shows how this sequence is rendered by the Flash player. The first sequence template also contains the instructions to display standard information (e.g. turn arrow, street-name).

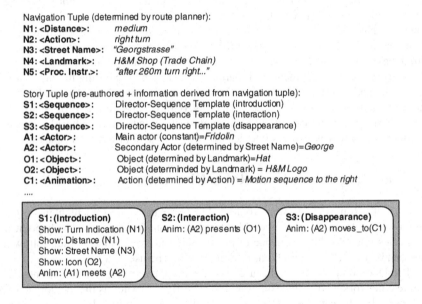

Fig. 1. Composition of navigation tuple and story tuple

3.1 Story Design

While the design of navigational stories can be informed to some degree by existing research on spatial cognition, cartography and digital storytelling, the complex interdependencies between design components make predictions with regards to the performance of a proposed design difficult. As the basis for future systematic developments insights are required that generalize and apply to large user groups. Valid statements with regards to complex information presentation designs (such as story based navigation instructions) can only be established by systematic testing with users under controlled conditions. Our initial goal is to establish the viability of story based navigation instructions and to provide a foundation for such evaluations in a systematic way. To achieve this, we have implemented a test system using a scenario based design approach with an iterative development process following the ISO 13407 cycle [28]. For the initial development we have focused on a single use-case in inner-city pedestrian navigation, where the user arrives by train at the central station of Hanover (giving him time to familiarize himself with the instructions prior to arrival) and than navigating to different destinations in the inner city from memorized instructions. For the technical implementation we have reused existing components for the navigation and storytelling functionality and used standardized techniques (e.g. Macromedia Flash) for the presentation where possible.

A central design decision with regards to navigational stories is the type of story and the relation of its content to the underlying navigation instructions. It is therefore necessary to study different story types and constructions. For our approach we have selected the following story design variables for further study:

Story Type: Different types of stories could result in different recall properties. Due to limitations of the currently used storytelling system certain restrictions are placed on the story type and content. In particular we are currently restricted to use a set of pre-authored story elements for each decision point that are then adapted to the current route description by a template mechanism. We have therefore limited the initial study to two types of stories: The first approach is the most simple and uses non-coherent scenes for each decision point. Thus, there is no logical connection in the sequence of story elements and the main actor is confronted with a unique event at each decision point where new actors, objects and actions are introduced as appropriate. This approach corresponds to the story types invented for the method of loci and lends itself to automatic story generation. However, it may be harder to memorize the sequence of events.

The second approach presents a coherent story line throughout subsequent scenes. The story concept that we have used for this approach is a detective story where a suspect is followed through different places with corresponding clues. This story type follows the ideas of the prominent computer game "Where in the world is Carmen Sandiego" [29] and also allows to generate scenes based on spatially located story templates. More complex storylines would demand a more advanced storytelling system as their basis and are therefore beyond the scope of the current prototype.

Connection Between Geographic Location and Story Elements: The story segment at each decision point should establish a reference to the real world location so that the user can recognize it and apply the corresponding navigation action. In our prototype each story segment follows the simple structure of introduction, interaction, and disappearance. During various iterations of our prototype development we have examined the following connections between geographic location and story elements:

Connection Based on Landmarks: If a landmark is available at a decision point it can be used as a story element, either directly or by symbolic reference. Figure 3 (left) shows an example where a landmark (a shop from the H&M chain) is used both directly (as a shop) and indirectly by verbal allusion (Head&Mouse = H&M).

Connection Based on Street Names: Street names can be used as story elements. Again the reference can be direct or indirect as shown in Figure 3, where the street name ("Georgstrasse") is used directly in the scene and mirrored by the name of the main actor in the scene (a mouse called "Georg").

Connection Based on Action: Locations that are characterized by specific actions (e.g. railway gates) can also be used to establish a reference between story and real world environment. Another connection based on action was inspired by the concept of minimally nonintuitive narratives as proposed by Norenzayam [30]. Here an unusual aspect (object, actor or behaviour) is combined with an otherwise very normal situation to create a memorable event. E.g. in the case of a place called "Brühlstraße" phonetic similarities like yelling ("brüllen" in German) are exploited to combine statues (landmarks at this location) with a memorable action (yelling statues).

Connection Based on Actor Names and Objects: Actor names and objects can be chosen to reflect arbitrary geographical information that could be relevant to the user and are especially suitable for minimally nonintuitive narratives. E.g. in a story segment where the location "Appelstraße" is relevant, apples are picked up.

Connection Based on Direction: The direction of movement of the main actor in a scene can be used to convey the direction in which the route continues. In our experiments we have found that a useful convention is for the main actor to appear at a decision point from the same direction as the user and to leave in the intended direction of travel. E.g. in the situation of Figure 3, where the user should turn right at the H&M Shop onto the Georgstraße, the actors in the scene disappear to the right.

Static vs. Dynamic vs. Interactive Stories: Animation and the possibility for user interaction could improve retention of story content. While static scenes are easiest to define, animated actions can help to present more information in a scene and were found to be essential in early pilot tests to convey the direction of travel. Interaction could help to further improve instruction retention, e.g. by forcing users to select correct objects and directions during the learning phase. The current prototype only supports simple user interaction like moving between scenes.

Realistic vs. Virtual/Abstracted Content Presentation: Both realistic and abstracted visual styles have benefits and shortcomings when it comes to the presentation of story segments. Guided by the results of a study of visual presentation styles for landmarks [31], we have focused our initial study on the use of an abstracted cartoon-like presentation style. Benefits of a cartoon style include it's suitability for a wide range of mobile devices ranging from mobile phones over PDAs to laptops as the performance and memory requirements are much low than with photorealistic styles; the simplification of the authoring process and the abstraction from possibly changing temporal detail. In the future alternative presentation styles should also be examined.

3.2 KuGeRou Prototype

The initial prototype was build using existing components for the navigation and storytelling functionality and used standardized techniques (e.g. Macromedia Flash [32]) for the presentation where possible. Although current capabilities are limited the

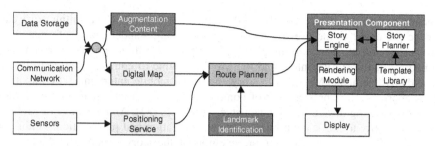

Fig. 2. Prototype system architecture

modular construction was chosen to be extensible for future experiments with a wider scope of design options.

The principle structure of system (see Fig. 2) is similar to conventional navigation systems, with modifications and extensions shown in dark grey. In the first step digital map data is used by the route planner to generate a route, consisting of segments and decision points that connect them, e.g. in the form of turning instructions. In the second step landmarks along the generated route are identified using the techniques described in [33] The process identifies objects with unique features that could serve as landmarks. A digital surface map (DSM) is then used to ensure the visibility of the identified potential landmarks from the user's direction of approach. Objects with limited visibility are discarded at this step. For details of the landmark identification process, see [6]. Both the landmarks and the route itself are then passed on to the presentation component. The presentation component then uses a story planner to generate a story based route description from this data. For our initial prototype we use an existing route planning systems for offline generation of the route and landmarks that is then transferred to the mobile navigation device, e.g. a PDA. Only the presentation component is currently implemented on the PDA for interactive mobile use. While interactive story generation in general is a very challenging research problem, the task in our system is significantly simpler as the stories are typically short, non-branching and are restricted to motion related storylines. In the initial prototype we use a simple mechanism based on pre-defined templates for the overall story structure. For each decision point a suitable story template is selected from a library, according to the instructions provided by the route planner. The template contains slots for all elements of the information tuple that is associated with the decision point. By filling the slots with pre-authored multimedia content according to the identified landmark and the turn instructions the story segment is constructed. Individual scene description templates for decision points with the corresponding actors are stored in the augmentation content database. According to the turn instructions provided by the route planner the story planner thus identifies a suitable template and composites the corresponding scene elements accordingly (E.g. in the scene described by the structures in Fig. 1 and shown in Fig. 3 the H&M landmark is represented in a template in which the head and the mouse are pre-authored elements selected by the navigation instructions). Dynamic element (slots) that are filled according to the chosen route include the name of the mouse and the direction of travel. The name of the mouse depends on the chosen street and the direction of the animation is adapted to the direction of travel). The resulting story can then be presented to the user by a story engine, using standard media formats and players for output. The current prototype employs Flash to present cartoon like animations of the story content and the MobEE engine to compose stories from the pre-authored elements. MobEE is a story engine for mobile entertainment computing that support run-time presentation of content on a wide variety of mobile hardware platforms by using device-independent story structures [34]. An advantage of MobEE is the possibility to use standard tools for media creation and presentation within the system. The device-independent representation of the story structure in MobEE is implemented by hierarchical finite-state automatons that communicate by a common variable pool. This hierarchical structure is well suited to support the composition of stories from predefined elements as in our prototype. The adaptation mechanisms of MobEE was extended to implement the template filling mechanism described above. Figure 3 shows different scenes in a story generate by the current prototype for a trip from Hanover's main station to the university.

3.3   Chosen Test Routes

In addition to digital map data (suitable for pedestrian routing) pre-authored story templates for each possible decision point in the test area and corresponding multimedia elements are required. We have therefore restricted our test system to the inner-city area of Hanover, extended with two longer routes, both starting at the central station and leading to the stadium and the university, respectively. Most story templates were realized for the non-coherent story type. A smaller sample was also realized for the coherent detective story type. All presentations are realized in an abstracted cartoon style. In the current test routes each story segment starts with a short audio commentary that provides an overview of the coming events. Then the story segment itself is related, consisting of the introduction, interaction, and disappearance of a main actor. Figure 3 shows exemplary story segments for the test route shown in Fig. 4. After the story is completed a summary is presented in which all scenes are first presented in a table with the main geographical information and story elements that is followed by visualization in which the geographical context of story elements and landmarks is repeated.

Fig. 3. KuGeRou Prototype

 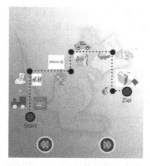

Fig. 4. Alternative representations of the test route

3.4 Evaluation

During the iterative development of the prototype several evaluations of the approach were conducted. The main goal was to guide the development of the system. While the results also provide some insight into the usefulness of the approach it should be kept in mind that these tests were not designed to test the general hypothesis that a story can be used as a route description. To test the general hypotheses requires not only a well designed system but also a carefully designed tests approach under controlled conditions as will discussed in the following sections. The test approach was two-fold:

The primary objective in the initial development phase was to validate that story-based descriptions aid the recall of route descriptions of significant length. The secondary objective was to ensure that our story-based approach produces viable descriptions for a concrete route.

To simplify the test approach a pure recall test was used for the majority of test-users that they could complete using a PC at an arbitrary location. For this purpose a between-subject design was chosen in which test-users were provided either with the story-based route description or a description using conventional guidance instructions plus landmarks. In the test user were asked to envision that they were travelling to Hanover by train and planning to walk to a specific university building approximately three kilometres from the station. The instructions then asked the users to familiarize themselves with the route description (either story-based or conventional) during a five to ten minute period. Following this, users had to perform a task unrelated to navigation, namely to search for a train connection, to simulate some distraction from the recall task after arrival. Users were then asked to write down what they recalled from the route description. Finally, users were asked to evaluate this against the true description and asked for further feedback on different aspects of the system. The key benefit of this approach is that it allows conducting frequent and fast evaluations with user who are unfamiliar with the real environment to guide the development process.

During the development of the initial prototype 22 participants (12 male, 10 female; age 20-50) were asked for feedback on different aspects of the system, with 8 completed overall evaluations. However, this approach does not provide a real check for the suitability of the description for navigation purposes. Therefore, a second test approach was used in which the suitability of the description was evaluated. Here both an expert critiquing approach was used in which test-users who are familiar with the area were asked to critique the route description against there real-world experiences and unfamiliar users were asked to check the use of the system in practice.

Due to the prototypical state of the "KuGeRou" system we did not yet conduct a formal experiment to verify the hypothesis that a story-based presentation improves retention of instructions and used only a limited number of participants. Thus, the results of this pretest are only informal and used to refine the system and give advices for the design of the final evaluation. Due to these limitations no statistical significant performance improvement of a story-based navigation approach can be asserted at this time, but users tend to better memorize the route if it was presented in a story-line format. Many test users reported that they enjoyed the multimedia presentation but some complained that it was at first difficult to accept an "unserious" story description as an aid to memorize a route. Most participants agreed that once they accepted this way of information representation, it was easy to memorize the route. The expert reviews confirmed that the idea of using stories to memorize route descriptions is suitable for pedestrian navigation. Test users who checked the system in practice reported that they found their way easily. Even if it was not possible to memorize the complete information tuple for a navigation point, landmarks or street names detected on the route helped to recall the complete information tuple.

4 Discussion and Future Work

From our initial experience we think that the use of stories is an interesting and potentially relevant format for guidance information. It is obvious that our initial experiments related here are only a first step in this direction, posing many new questions. The limited scope and simple structure of the required stories make this domain an interesting field for experimentation in virtual storytelling. As an example, a simplistic generator often produces absurd stories. It is however unclear if this is really problematic as it is non obvious whether users prefer simple, coherent, sense-making stories. From a mnemonic perspective absurd stories may even be preferable to some users. However, the testing of navigation instructions is subject to a number of difficulties that preclude the direct application of established test methods as they are regularly conducted in usability laboratories. The key problem is to conduct tests under controlled conditions: While outdoor tests allow to cover the complete use of a navigation systems the complex outdoor environment can not be controlled and the results depend on prior location knowledge as well as navigation and spatial orientation skills of the participants. In addition to the problems of repeatability and control it is also difficult to capture all relevant test data in a mobile outdoor use situation. Indoor laboratory tests, on the other hand, allow repeatable tests under controlled conditions and more complete recording of tests data, but fail to cover the important aspects of unequivocal recognition of decision points and the enactment of the instructions.

The initial results seem to warrant further investigation of the "KuGeRou" concept of navigational storytelling. Obviously, the development of stories for navigation purposes is still in very early stages. Based on our experiences with the initial prototype several research directions seem worth pursuing , including:

- The systematic examination of the design space of story designs (and the corresponding implications on story generation) for navigation systems. This includes an extension of the possible environment for navigation from cities to a wide variety of possible environments.
- The creation and implementation of different story types. The extension of the system to larger areas and more diverse content will probably require a more sophisticated story generation and management system. Existing storytelling systems should be evaluated to judge their suitability for this application context.
- The integration of route planning into the mobile system as well as dedicated authoring tools to create scene descriptions and possibly new story templates must be considered to make the inclusion of larger areas possible as a prerequisite for realistic use.

The extension of system scope will then enable relevant user tests under more realistic conditions and the comparison against conventional baseline systems.

Since the evaluation of different types of navigation instruction is hindered by a number of aspects that are difficult to control, including spatial knowledge of test users and the impact of outdoor situations we are also interested in the use of virtual environments for such test. In particular we aim to examine the use of virtual environments based on digital 3D city models for the tests. The use of virtual environments allows to repeatedly generate environments with prescribed features (e.g. cityscapes of a certain type and complexity) that are unknown to test users, to test navigation in specific situations and for the complete use cycle from instruction recognition to enactment, and to automatically record large portions of the desired test data. A central aspect of the work will be to establish the relevant characteristics of the virtual test environments and to validate them against real outdoor test data.

Acknowledgments. We would like to thank Ivonne Gansen (University of Applied Science Harz) and her advisor Christian Geiger (University of Applied Science Duesseldorf) for implementing the initial prototype of KuGeRou. Christian Reimann (University of Paderborn) provided support for the MobEE system. Holger Reckter and Martin Kreyßig (University of Applied Science Harz) helped to design the look of the representation and the interface. Dirk Zimmermann (Siemens Usability) provided valuable advise for the evaluation experiments. We would also like to thank all test participants and the anonymous reviewers for their valuable comments.

References

1. May, A., Ross, T., Bayer, S., Tarkianinen, M.: Pedestrian Navigation Aids: Information Requirements and Design Implications. Personal and Ubiquitous Computing 7, 331–338 (2003)
2. Denis, M., Pazzaglia, F., Cornoldi, C., Bertolo, L.: Spatial Discourse and Navigation: An Analysis of Route Directions in the City of Venice. Applied Cognitive Psychology 13, 145–174 (1999)
3. Daniel, M.-P., Denis, M.: Spatial Descriptions as Navigational Aids: A Cognitive Analyses of Route Directions. Kognitionswissenschaft 7(1), 45–52 (1998)

4. Lovelace, K., Hegarty, M., Montello, D.: Elements of Good Route Directions in Familiar and Unfamiliar Environments. In: Spatial Information Theory: Cognitive and Computational Foundations for Geographic Information Science, pp. 65–82. Springer, Heidelberg (1999)
5. Golledge, R.: Human Wayfinding and Cognitive Maps. In: Wayfinding Behavior, pp. 5–45. John Hopkins Press (1999)
6. Elias, B.: Extraktion von Landmarken für die Navigation, Dissertation, University of Hanover, DGK, series C, number 596 (2006)
7. Paelke, V., Elias, B., Hampe, M.: The CityInfo Pedestrian Information System. In: Proceedings of 22nd International Cartographic Conference, La Coruña/Spain, July 9-16 (2005)
8. Rossi, P.: Logic and the Art of Memory. University of Chicago Press, Chicago (2000)
9. Chatwin, B.: The Songlines, Penguin (1988)
10. Lewis, D.: Route finding by desert aborigines in Australia. The Journal of Navigation 29, 21–38 (1976)
11. Helbling, J.: Die Organisation des sozialen und natürlichen Raumes bei den australischen Aborigines. In: Michel, P. (ed.) Symbolik von Ort und Raum, Bern: P. Lang, pp. 281–303 (1997)
12. Myers, F.: Pintupi Country, Pintupi Self: Sentiment, Place and Politics Among Western Desert Aborigines, University of California Press; Reprint (1991)
13. Mark, D.M., Gould, M.D.: Wayfinding as discourse: A comparison of verbal directions in English and Spanish. Multilingua 11(3), 267–291 (1992)
14. Allen, G.: From Knowledge to Words to Wayfinding: Issues in the Production and Comprehension of Route Directions. In: Frank, A.U. (ed.) COSIT 1997. LNCS, vol. 1329, pp. 363–372. Springer, Heidelberg (1997)
15. Streeter, L., Vitello, D., Wonsiewicz, S.: How to tell people where to go: comparing navigational aids. International Journal Man-Machine Studies 22, 549–562 (1985)
16. Deakin, A.: Landmarks as Navigational Aids on Street Maps. Cartography and Geographic Information Systems 23(1), 21–36 (1996)
17. Mark, D.: A Conceptual Model for Vehicle Navigation Systems. In: Proceedings First Vehicle Navigation & Information Systems Conference (VNIS '89), IEEE Vehicular Technology Section, Toronto, pp. 448–453. IEEE Computer Society Press, Los Alamitos (1989)
18. Freundschuh, S., Mark, D., Gopal, S., Gould, M., Couclelis, H.: Verbal Directions for Wayfinding: Implications for Navigation and Geographic Information and Analysis Systems. In: Proceedings, 4th International Symposium on Spatial Data Handling, pp. 478–487 (1990)
19. Tversky, B., Franklin, N., Taylor, H., Bryant, D.: Spatial Mental Models from Descriptions. Journal of the American Society for Information Science 45(9), 656–668 (1994)
20. Taylor, H., Tversky, B.: Spatial Mental Models Derived from Survey and Route Descriptions. Journal of Memory and Language 31, 261–292 (1992)
21. Weissensteiner, E., Winter, S.: Landmarks in the Communication of Route Directions. In: Egenhofer, M.J., Freksa, C., Miller, H.J. (eds.) GIScience 2004. LNCS, vol. 3234, pp. 313–326. Springer, Heidelberg (2004)
22. Schlender, D., Peters, O., Wienhöfer, M.: The effects of maps and textual information on navigation in a desktop virtual environment. Spatial Cognition and Computation 2(4), 421–433 (2000)

23. Tversky, B., Lee, P.: Pictorial and Verbal Tools for Conveying Routes. In: Spatial Information Theory: Cognitive and Computational Foundations of Geographic Information Science, pp. 51–64. Springer, Heidelberg (1999)

24. Herwig, A., Kretzler, E., Paar, P.: Using games software for interactive landscape visualization. In: Bishop, I., Lange, E. (eds.) Visualization in landscape and environmental planning, pp. 62–67. Spon Press, London (2005)

25. Katterfeld, C., Paelke, V.: Interoperable Learning Environment. In: Geosciences - A Virtual Learning Landscape. In: Proc. of the ISPRS Technical Commission VI Symposium on "E-Learning And The Next Steps For Education" Tokyo, Japan (2006)

26. Tosca, S.: The Quest Problem in Computer Games. In: Proceedings of TIDSE 2003, Darmstadt, Germany, March (2003)

27. Cartwright, W.E.: Using the web for focused geographical storytelling via gameplay. UPIMap 2004 In: The First international Joint Workshop on Ubiquitous, Pervasive and Internet Mapping, Tokyo, Japan. International Cartographic Association Commission on Ubiquitous Cartography pp. 89–109 (2004)

28. ISO 13407: Human-centred design processes for interactive systems (1999)

29. Broederbund: Where in the world is Carmen Sandiego? Game accessed 27.2.2007 (1985), available at: http://free-game-downloads.mosw.com/abandonware/pc/ educational_games/ games_s_z/where_in_the_world_is_carmen_sandiego_deluxe_edition.html

30. Norenzayan, A., Atran, S., Faulkner, J., Schaller, M.: Memory and mystery: The cultural selection of minimally counterintuitive narratives. Cognitive Science 30, 531–553 (2006)

31. Elias, B., Paelke, V., Kuhnt, S.: Concepts for the cartographic visualization of landmarks. In: Gartner, G. (ed.) Location Based Services & Telecartography, Proceedings of the Symposium 2005, Geowissenschaftliche Mitteilungen, Nr., vol. 74, pp. 149–155 (2005)

32. Adobe (2007), http://www.adobe.com/support/documentation/en/flash/

33. Elias, B., Brenner, C.: Automatic Generation and Application of Landmarks in Navigation Data Sets. In: Fisher, P. (ed.) Developments in Spatial Data Handling - 11th International Symposium on Spatial Data Handling, pp. 469–480. Springer, Heidelberg (2004)

34. Reimann, C., Paelke, V.: Adaptive Mixed Reality Games. In: Proc. ACM SIGCHI International Conference on Advances in Computer Entertainment Technology (ACE 2005), 15th - 17th June 2005, Polytechnic University of Valencia (UPV), Spain (2005)

Three Sampling Methods for Visibility Measures of Landscape Perception

Gerd Weitkamp[1], Arnold Bregt[1], Ron van Lammeren[1],
and Agnes van den Berg[2]

[1] Wageningen UR, Centre for Geo-Information, PO Box 47, 6700 AA Wageningen,
The Netherlands
[2] Wageningen UR, PO Box 47, 6700 AA Wageningen, The Netherlands
{Gerd.Weitkamp,Arnold.Bregt,Ron.vanLammeren,
Agnes.vandenBerg}@wur.nl

Abstract. The character of a landscape can be seen as the outcome of people's perception of their physical environment, which is important for spatial planning and decision making. Three modes of landscape perception are proposed: view from a viewpoint, view from a road, and view of an area. Three sampling methods to calculate visibility measures simulate these modes of perception. We compared the results of the three sampling methods for two study areas. The ROPE method provides information about subspaces. The road method enables the analysis of sequences. The grid point method calculates visibility measures at almost every location in space, providing detailed information about transitions and pattern change between original and new situations. The mean visibility values for the study areas reveal major differences between the sampling methods. Combining the results of the three methods is expected to be useful for describing all the facets of landscape perception.

1 Introduction

The concept of landscape more closely matches the way people think about the natural world than the abstract notions of nature and biodiversity [1]. People identify themselves with landscapes [2]. The appearance of the landscape is important to people, which is why it is important for policy makers and planners to take the human perception of landscape into account when developing and protecting the landscape. Visual perception of both the urban and non-urban landscape is used in numerous studies to describe landscape characteristics [3-7]. Unfortunately, landscape perception is poorly implemented in landscape policies and spatial planning. One important reason why landscape perception is not a common criterion in environmental decision making is the lack of well-validated geo-referenced models that connect spatial landscape characteristics to perception values.

Currently, there are two main approaches to describing the spatial characteristics of landscape perception. A common approach is a perception-driven approach where perceived landscape characteristics are derived from people's self-reported judgments

S. Winter et al. (Eds.): COSIT 2007, LNCS 4736, pp. 268–284, 2007.

of photographs or field settings [8]. Although this approach generates direct information about landscape perception, the lack of geo-referenced spatial information makes it difficult to use the findings of these studies in decision-making on interventions in the physical landscape. Another approach is the object-driven approach in which expert judgments are used to derive perception values from data about the physical landscape, such as land cover data, which provide spatial information about the landscape [9]. This is, however, not a valid method since land cover data do not provide information about how people perceive the landscape.

Visibility measures may provide a good method for bridging the gap between these two approaches to describing spatial characteristics of landscape perception [10-12]. This method measures characteristics of the *visible* space and is based on Gibson's notion of direct perception, which regards perception as a direct result of the affordances provided by the physical environment [13]. Drawing on Gibson's theory, Thiel describes spatial experience as an inherited biological function derived from the ecological relationship between humans and the environment [14].

Several methods are available to calculate visibility measures. One method is called viewshed analysis, in which visibility is calculated using a terrain model. The work of De Floriani [15] and Llobera [16] are examples of the use of viewshed analysis. A similar method, which is used in this study, is isovist analysis. In this method, calculations are based on vertical landscape elements, which are of interest of spatial planners because they are a useful instrument for monitoring and changing the landscape openness.

The concept of *isovists* was introduced by Tandy [17] and further developed by Benedikt [18]. The isovist technique describes and analyses visible space. Benedikt defines an isovist as the set of all points visible from a given viewpoint in space with respect to an environment [18]. The development of powerful analytic GIS tools has led to great advances in measuring the visible space. Increasing computing power and the availability of high resolution topographic data open up opportunities to enhance visibility analyses.

One advantage of visibility analyses is the ability to calculate the visibility of a large area from many points of view. Many sampling methods are available for locating suitable viewpoints, the most appropriate method depending on how people are expected to perceive the landscape in any given case. For example, a landscape will be perceived differently when viewed from one or two viewpoints (e.g. a bench) than if viewed from a vehicle travelling along a road, with a sequence of viewpoints. Policy makers and planners require different sampling methods depending on the task at hand, e.g. planning a road for recreational purposes or preserving the open character of an area.

In general, we can distinguish three modes of landscape perception: views from a single viewpoint, views from a route, and views of an area [19]. These modes of perception correspond with the spatial descriptors for constructing an image of a city as defined by Lynch [20, 21]: nodes, paths and districts. Moreover, the three modes are closely related to the relative scale of the landscape compared to the human body [22, 23]: vista spaces are comprehensible from a single viewpoint; environmental spaces can only be viewed by moving through them (locomotion); geographical spaces are too large to comprehend through locomotion and need to be learned by symbolic representation, such as the use of maps [23].

The scale of the landscape relative to the human body is related to the type of perception required to comprehend the landscape, and therefore characterize the landscape as a whole. Both *how* people perceive the landscape and *what* they perceive (the relative scale of the landscape) determine the choice of sampling method for locating the viewpoints. Three sampling methods are proposed which simulate the three modes of perception. We investigate the possibilities and limitations of the three sampling methods by analysing the results of the visibility calculations for two areas. We compare the specific results of each sampling method for the two areas and compare the mean value characteristics for these areas. The results of the three sampling methods for the present situation are then compared with the results obtained for a proposed future situation of one area.

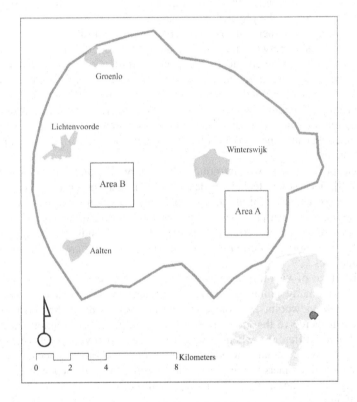

Fig. 1. Locations of study areas within national landscape the Achterhoek

2   Materials and Methods

2.1   Study Area

The two study areas are located in the Achterhoek, one of the twenty national landscapes of the Netherlands (Fig. 1). These national landscapes have a number of scenic core qualities, including landscape openness. The openness of the Achterhoek is classified as 'small-scale openness' [24]. 'Small-scale' refers to open spaces which are closer in scale to the size of the human body than the sizes of large-scale spaces [25]. The geomorphology of the Achterhoek is diverse and visible in the spatial configuration of landscape elements. This geomorphologic diversity was used to select the two study areas for comparing visibility measures of openness. The first area is located to the south-east of the city of Winterswijk (area A) and is characterized by creek dell soils and cover-sand ridges. The second area is located to the west of Winterswijk (area B) and is characterized by a plateau-like terrace influenced by land ice. Area A is visually dominated by patches of forest, while area B is visually dominated by tree rows.

2.2 Data

Landscape openness in the Netherlands is primarily defined by vertical landscape elements because the lack of relief [26]. Vertical landscape elements are space dividers and easy to control [27]. For spatial planners, vertical elements are a useful instrument for monitoring and changing the landscape openness. High resolution topographic maps provide information about these elements and are therefore a suitable starting point to analyse landscape perception. However, we do not want to describe the perception of elements, but the perception of space. Although classes such as 'space' do not occur in topographic maps, by knowing where the vertical elements are located one can analyse the perception of space using topographic maps because vertical elements and spaces are mutually exclusive [28]. This is valid for landscapes where the terrain does not make a substantial contribution to shaping the space, as in the landscapes of the Netherlands [26].

The base dataset for performing visibility analyses is TOP10vector, a Dutch high resolution topographic dataset. This needs preprocessing to calculate the visibility measures. Since only the vertical elements above eye level, which is fixed at 1.6 m, are assumed to be relevant, the other elements were removed. The result is a one-class topographic dataset with vertical landscape elements higher than 1.6 m.

In order to illustrate the effect of modifying vertical landscape elements on the perception of space, one of the original locations (area A) was altered by randomly removing 50% of the forest area and 50% of the length of the tree rows. The modified area is referred to as *area A changed*.

Spatial analyses performed with vector data in GIS, which treat properties of space as scale-independent, do not match perception of the landscape by people, who treat space as scale-dependent [22]. Calculating visibility based on vector-based topographic data without taking human perception into consideration would imply that the extent of an area does not influence the number of viewpoints needed to perceive this area. For instance, a square room of 10 m^2 without any vertical landscape elements would need one viewpoint to cover the whole space, as would a similar room of 10 km^2. However, human perception of space changes with distance; the maximum distance at which a person can distinguish a space-shaping vertical landscape element is about 1200 m, depending on the type of landscape [29, 30]. Montello states that if an area is larger than can be observed from one point of view, locomotion is needed to comprehend space [23]. Hence, more than one viewpoint is needed to perceive a square room of 10 km^2. The dimensions of the two study areas as described were set at 3 km long and 3 km wide. This size is larger than vista space and assumed to be comprehensible by locomotion (environmental space).

The accuracy of the topographic data is a decisive factor in deciding the sampling intensity of the viewpoints. To cover an area with viewpoints, the distance between viewpoints does not need to be smaller than the accuracy value. The accuracy of the Top10vector is approximately 5–10 m if geometric precision and fuzziness of boundaries are combined [31, 32].

In addition to the extent of an area and the accuracy of the topographic data, the number and the shape of landscape elements are also relevant for the sampling intensity of viewpoints. For example, more viewpoints are required to comprehend an area with many irregular patches of forest than an area with equal size and one regular shaped patch of forest. Landscape metrics have been studied thoroughly with software like Fragstats [33, 34]. These kinds of approaches to analysing landscape metrics are based on landscape elements, not on space. Moreover, these approaches focus on the physical landscape, not on human perception of the physical landscape. Since the focus of this paper is on the perception of space, the number and shape of landscape elements is not taken into consideration when determining the number of viewpoints needed to perceive space. However, results derived from an approach based on landscape element metrics could be useful in deciding the intensity of viewpoints needed to perceive space.

A number of software packages exist to calculate visibility measures. We chose to work with Isovist Analyst 1.2, an extension of Arcview 3.x, because it is a vector-based program and able to calculate the selected visibility measures [35]. The visible area is represented by isovist polygons, which are calculated from an obstacles layer – a representation of vertical landscape elements higher than 1.60 meters – and an observer layer – a representation of the locations from which a person views the landscape. The extension uses a combination of Binary Space Partition Trees (BSP) and a ray-tracing approach to compute the isovists [35] from each viewpoint. One parameter is the maximum ray length, the default value for which is infinite. As explained, this distance was set at 1200 m, based on human perception. With this condition, the maximum longest line of sight is 1200 m. Another parameter for calculating isovist polygons is the ray interval (in degrees). The smaller the value of the ray, the more accurate the calculation of the visible area. With this ray-tracing approach, objects in the foreground are detected more accurately than objects in the background. This is in keeping with human perception theory, which states that objects in the foreground have a bigger impact than similar objects in the background [11, 36, 37]. The ray interval was set at 5 degrees, which means that, for instance, a house with a width and length of 12 m will always be detected within a distance of 135 m from the observer. The area within 135 m from an observer is considered to be a human-scale space, according to Lynch [29]. nevertheless, a ray interval of 5 degrees is still rather arbitrary and open to discussion. The main argument for not using a smaller ray interval value is the greater computing time.

As the ray interval and maximum ray length were related to the nature of human perception, the locations of the viewpoints should also be related to the modes of human perception. Therefore, three modes of perception were studied: view from a point, view from a road, and view of an area. These perception stimuli types were translated into three types of viewpoint sampling methods.

2.3 Sampling Methods

The view from a point has been commonly used in landscape paintings since the 17th century and, more recently, photographs record an expression of a landscape from one point of view. The observed space from a single viewpoint is comparable with a vista space [23] or with the space around the body [38], the latter implying a view in all directions. Isovist polygons represent the space from one point in all directions. To locate a viewpoint in an area which shows the character of the area, in this case landscape openness, the point with the largest view was selected. However, it is not possible to perceive the whole area from one point of view (since the study areas have a length and width of 3 km and a maximum longest line of sight of 1200 m). Therefore, the minimum number of viewpoints needed to cover all space was calculated with an extension of Arcview 3.x., the Art Gallery Problem Solver, which is usually applied to select the minimum number of cameras needed to observe all paintings in a gallery. The Rank and OverlaP Elimination (ROPE) technique was used to select the viewpoints. The ROPE technique is a greedy-search method, which iteratively selects the most visible dominant observer with minimum overlapping vistas [39]. The viewpoints were ranked according to the size of the visible area.

Thiel was the first to try to analyse explicitly the visual properties of spatiotemporal paths through the built environment [14]. He acknowledged the importance of sequences in perceiving the environment. Landscape perception is about wholes [40], not about isolated elements. Therefore, locomotion is needed to comprehend these wholes [22] because it allows the observer to know the shape and scale of obstacles [41]. In this study, existing roads located across the area were selected from the topographic map. These line attributes were converted to a sequence of points every 10 m along the road. Isovist polygons were calculated from each point and, together with a sequence number, variations in the shape and size of spaces could be analysed to characterize openness.

When monitoring the change of landscape character it is important to regard the landscape as a whole, taking into account information about perception from single viewpoints, from the road, and of an area. The most detailed information about landscape openness is obtained by calculating visibility measures from infinite viewpoints covering the whole space. However, the reasonable minimum distance between viewpoints with regard to human perception of the landscape is one walking step, or approximately 1 m. A reasonable distance with regard to the accuracy of topographic data is 5–10 m. Considering the accuracy of the topographic data, human perception and computation time, a sampling grid of viewpoints with a distance of 10 m in a horizontal and vertical direction was created.

2.4 Openness Variables

The Isovist Analyst extension derives many geometric variables from the isovist polygons. Four variables were selected for the indicator of openness, the first three of which provide information about size characteristics: (1) the visible area (*Size*) – the size of the area of the isovist polygon – which provides information on how much of the open space is visible; (2) the longest line of sight (*LoS*), calculated by the maximal radial of the isovist (the maximum was set to 1200 m); (3) the distance to the closest object (*ClO*), calculated by the minimal radial of the polygon. Because not only the size, but also the shape can influence the perception of space, a fourth variable was used to calculate the shape of the visible space: (4) the compactness (*Comp*), a measure of the radial compactness of the isovist polygon. If one of two isovists with equal size variable values is compact, this indicates that relatively more space is close to the observer than in the second, less compact isovist. Therefore, the perceived openness is greater.

The values of the four variables were calculated for the three cases (area A, area A changed, and area B), based on the modes of perception (perception from a viewpoint, from a road, and of an area). The sampling methods for the three modes of perception were compared for the kind of openness information they provide and the possibility of detecting differences between landscapes.

3   Results and Discussion

3.1   Views from a Point

The results of the visibility measures based on the ROPE technique are isovist polygons which cover all the space and which have the minimum possible overlap. The viewpoints are ranked by the visible area value. The viewpoint with the highest value in area A has a visible area value of 470,565 m^2 and covers 6% of the total space. The largest visible area in area B is 715,036 m^2, or 8.9% of the total space of area B. Since the overlap between isovist polygons is minimal – viewpoints are not mutually visible and therefore not directly visual accessible – a single isovist polygon can be defined as a subspace. The value of the visible area size only gives information about the subspace, not the total space. Gradual changes of openness between subspaces cannot be detected, and mean values of a space are calculated by single values for subspaces.

The creation of subspaces simplifies the complex structure of the continuous space of landscapes. Space syntax research uses this idea to first split urban space into subspaces, and then analyse the connections between the spaces by using the graph-theory notation to describe the subspaces and their connections [42, 43]. Refinements of the calculation methods are needed to study the possible advantages.of the configuration of these subspaces.

In total, 541 viewpoints were needed to cover all the space in area A, whereas 420 viewpoints were needed to cover all the space in area B. These numbers are higher than expected in reality and result from the use of the ray technique. For example, when calculating the isovist polygon from a viewpoint in a circle with a ray interval of 5 degrees, 72 small spaces were not covered by the isovist polygon, while in reality the whole space is visible from that single viewpoint. Nevertheless, the number of viewpoints is a potential indicator of openness: fewer viewpoints are needed in more open landscapes than in enclosed landscapes. The number of viewpoints that are needed to obtain a representative description of openness is expected to be between one and the number needed for total visibility.

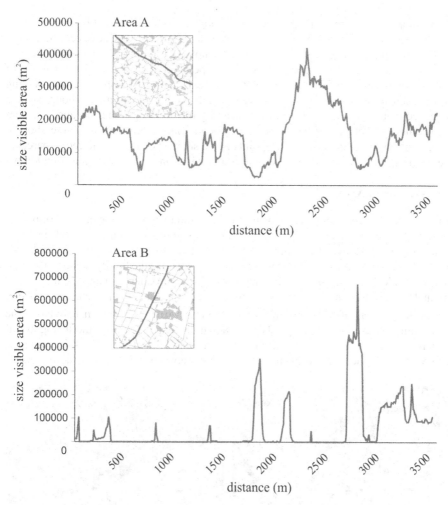

Fig. 2. Visibility sequences from road sampling in area A and B

3.2 View from the Road

The principal difference between the road sampling method and the ROPE sampling method is that the first is based on a landscape element, the road, while the latter is based on space characteristics. In contrast to the ROPE-based viewpoints, adjacent viewpoints are mutually visible and physically directly accessible. This sequence of openness values represents the view from a road.

The visible area size was calculated for points along a road in area A and B (Fig. 2). The size of the visible area was plotted against the location of the viewpoint along the road. The roads are shown in the map of the two areas. The graph of area A shows gradual changes, while the graph of area B shows very abrupt changes. The reason for this difference is the presence of tree rows along the roads in area B, which are characteristic for this area. More abrupt changes in visible area along the road are found in area B and the visible area size is more constant. The visible area from the road in area A changes every 10 m, which indicates a greater variety of openness.

The information derived from the graphs not only indicates openness, but also expresses the variety and complexity of the perception of space. The analysis of values in a sequence of viewpoints along the road is directly applicable to planning issues such as recreational route design. For the other sampling methods, such a sequence would not have such a direct meaning for planning issues because adjacent viewpoints are not necessarily directly physically accessible. However, the shortest route connecting all the ROPE viewpoints is the shortest distance required to perceive the whole space. The length of this route is an indication of the landscape complexity.

3.3   View of an Area

The distance between viewpoints using the grid point sampling method is 10 m. The mapped viewpoint values show the gradual transitions between open and enclosed areas. Figure 3 shows the visible area size values for area A and B. While the descriptive statistics show similar values for both areas, with slightly higher values for area B (Table 1), the classes on the map show different patterns in area A and B. The gradual transitions in area A and the more abrupt changes in area B are clearly visible.

The histograms of the (logarithm of) openness variables values usuallyhave one maximum because the landscape shows gradual changes. However, the histogram of area B shows two maxima, one around 5.1 and one around 5.5 (Fig. 4A), which is also visible on the map in Figure 3B. Values between 5.2 and 5.4 hardly occur.

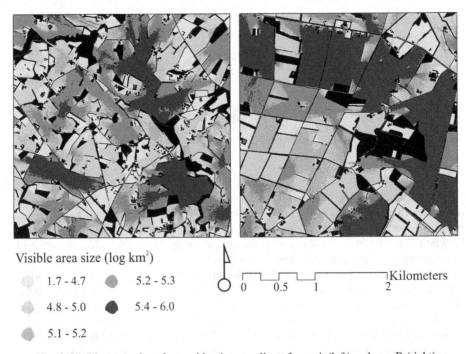

Visible area size (log km^2)

1.7 - 4.7	5.2 - 5.3	
4.8 - 5.0	5.4 - 6.0	
5.1 - 5.2		

0   0.5   1   2   Kilometers

Fig. 3. Visible area values from grid point sampling of area A (left) and area B (right)

These openness values can be used to predict whether the landscape character is changing due to physical changes in the landscape.

The values for compactness (Comp) and closest object (ClO) in Table 1 show the most distinct values for areas A and B, although all values for area A and area B lie within a small range. These similarities are due to the fact that both areas are located within one national landscape with one openness class. More distinct values are expected if areas in different national landscapes with different openness classes were compared.

Table 1. Mean values of openness variables of areas A and B

		Size	ClO	LoS	Comp
Maximum value		6.66	3.08	3.08	1.00
Area A	ROPE	3.86	0.51	2.38	0.16
	Road	5.13	1.33	2.86	0.24
	Grid point	5.01	1.34	2.76	0.26
Area B	ROPE	3.76	0.37	2.40	0.15
	Road	3.65	0.37	2.41	0.11
	Grid point	5.16	1.49	2.77	0.32

3.4 View of an Area Derived from Three Sampling Methods

The three histograms of the visible area size of area B are shown in Figure 4. The histograms for all three sampling methods show two maxima, but the maxima of the ROPE sampling and the road sampling have different values from the point grid sampling. The maxima values for both the ROPE and the road sampling method are 2.8 and 5.0, while the maxima for the grid point sampling are 5.1 and 5.5. The reason for the maximum value of 2.8 in the ROPE sampling results (Fig. 4C) is that smaller subspaces have the same weight as larger subspaces. The ROPE technique produces many very small spaces, which are emphasized in the results. The maximum value of 2.8 for the road sampling method (Fig. 4B) can be explained by the view blocked by tree rows. Histogram B in Figure 2 confirms this. The histograms show similar trends, resulting from three different types of perception.

Table 1 shows the mean values of the four openness variables. The value range of each of the variables is from zero to the maximum value shown in the table, based on a maximum line of sight of 1200 m. The values for the three sample methods have been calculated for area A and B. Since the grid point method provides the most representative mean values of an area, the values calculated by the other sampling methods were compared with the values of the grid sampling method. The mean values of the road sampling are similar to the mean values of the grid point sampling in area A. The results indicate that the perception of the landscape from the road in area B does not reflect the openness character of the area according to the values in Table 1. This can be explained by the presence of tree rows located along the roads, which is characteristic for this area. These tree rows block the view, which results in very low openness variable values (Table 1). The difference between the view from the road and the view of an area is valuable for spatial planning and monitoring.

Based on the values in Table 1 it is difficult to determine whether an area is open or enclosed; there are no guidelines as to what size makes an area open or enclosed. Nonetheless, issues of openness are important to consider when making decisions for spatial planning and monitoring [44]. The values show, for example, whether the perceived openness from the road reflects the openness of the whole area.

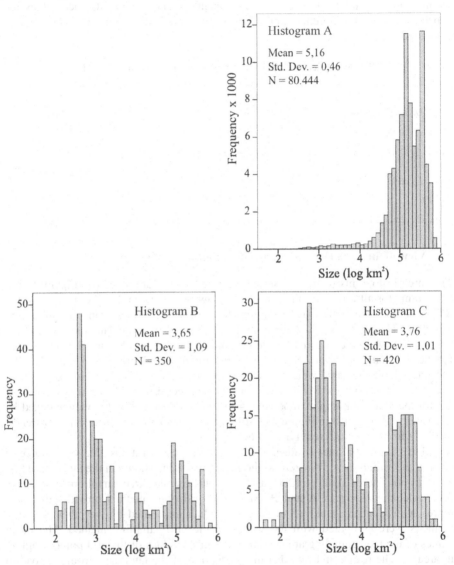

Fig. 4. Visibility area values for the three sampling methods (A: grid point sampling; B: road sampling; C: ROPE sampling) of area B

The difference between the view from the road and the view of an area is smaller in reality than when derived from the values in Table 1 because tree rows in the topographic dataset are presented as solid lines, which is different from how they are perceived in the landscape. The transparency of tree rows is important for visual perception [45], but this is not indicated in the topographic data, which is why the openness variable values from the road in area B underestimate the openness of the area. The representation of tree rows is an example of the difficulties involved in representing natural landscape elements in topographic data in general. One of the ways of gaining insight into this matter is to validate the openness variable values in the field.

The results in Table 1 show that mean values from the ROPE method are much lower than the values from the grid point method. When using the ROPE method, a large visible space has the same weight as a small visible space, whereas in the grid point method, a large space contains more viewpoints than a small space. The weights of the values of the areas differ, which results in different mean values for the two sampling methods.

As suggested, the number of viewpoints needed to perceive the whole space has the potential to be an indicator of openness. However, the results of the three sampling methods for the three areas show that this is not valid for this study area.

3.5 Perceived Change of Openness

The openness variable values of the proposed future situation of area A (with 50% of the forest area and 50% of the tree rows removed) were compared with the values derived from the original situation. The viewpoint with the highest value has a visible area value of 809,602 m^2 and 9.7% of the space is visible from this viewpoint. The first 10 viewpoints cover 44% of all the space. This is an increase of 15% compared with the original situation, whereas the area of total space only increased by 5.4%. 280 viewpoints were needed to cover the whole space, which is 52% of the number needed in the original situation.

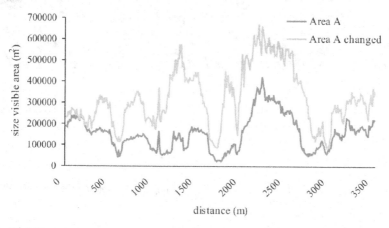

Fig. 5. Visibility sequence based on road sampling, comparing the original (area A) and the proposed future situation of area A (area A changed)

The values of the visible area from the road are two times higher (323,491 m^2) than in the original situation. Figure 5 shows that the general trend of the visible area along the route is similar to the original situation. However, the difference between the highest and lowest value is larger and the changes over a short distances are more abrupt.

The mapped viewpoints of the point grid method show pattern changes. Figure 6 shows the longest lines of sight values of the original situation (Fig. 6A) and the new situation (Fig. 6B). The maximum length of the longest line of sight is 1200 m and the viewpoints with this value are shown in dark grey. In the new situation, 32% of the viewpoints have a maximum line of sight. The increase in visual connectivity between points in the altered situation in area A indicates the decrease in enclosure. This spatial pattern, which is not visible using the other sampling methods, is similar to the representation of subspaces with axial lines [46]. Methods like the visibility graph analysis can calculate local and global connectivity values of axial lines based on graph theory [41, 47]. In this paper all values derived from each viewpoint are local, while landscape perception is about wholes [48, 49]. To calculate these global values, information derived from the ROPE method (creating subspaces), from the road method (sequences, locomotion) and from the grid point method (connect every location) should be considered.

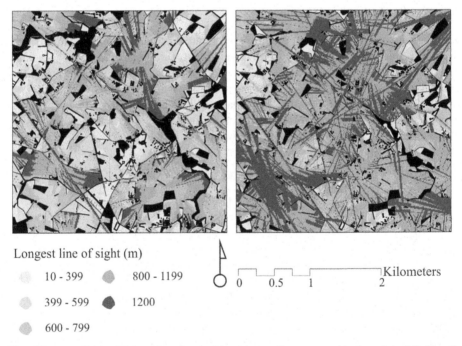

Longest line of sight (m)

  10 - 399   800 - 1199

  399 - 599   1200

  600 - 799

0   0.5   1   2   Kilometers

Fig. 6. Longest lines of sight values from grid point sampling, comparing the original (left) and the new (right) situation of location A

The results show that visibility calculations are a useful method for bridging the gap between the object-driven approach and the perception-driven approach. The results indicate possibilities for distinguishing between different modes of perception by using different sampling methods. However, the results do not prove that the different modes of perception are meaningful for perceived openness in the field. A study by Palmer [26] illustrates, through empirical research, that openness is meaningful for landscape perception, and, accordingly, research has to be done to find out how effective it is to use the three sampling methods to differentiate the different modes of perception.

4 Conclusions and Directions for Further Research

The quantification of landscape characteristics with visibility measures based on geo-data enables spatial planners and policy makers to include visual perception in landscape research. The three proposed methods for sampling viewpoints, representing three modes of perception, produce visibility measures specifying what people perceive in the landscape (visibility) and how they perceive it.

The ROPE sampling method resulted in openness variable values for individual viewpoints. Those values are valid for corresponding subspaces, which are defined by the isovist polygons. Future studies should be undertaken to investigate the advantages of connecting the subspaces in order to derive both local and global characteristics with space syntax measurements.

The road sampling method generated a sequence of openness variable values, which is important for obtaining information about variation in the landscape when moving. The use of an existing road for locating the viewpoints allows the method to be used for spatial planning and landscape design. Movement is important for perceiving whole areas, which is important in landscape perception.

The grid point sampling method provided a high number of openness variable values and covers the whole area with viewpoints. Mapped values show openness patterns and the characteristics of transitions between open and enclosed areas. Grid point sampling is assumed to be the most representative for the view of an area. Therefore, the comparisons between the mean values from the grid point sampling method and the values from the other sampling methods indicate how well the openness of an area is perceived from single viewpoints or from the road. In view of the fact that the mean openness variable values of the two areas A and B are very similar, further research is needed to find out whether indicators such as openness, and their variables, are related to specific scales.

The differences between the mean openness variable values for the three sampling method indicate that the sampling methods are not replaceable, but complement each other. The results for the proposed future landscape (area A changed) and the original situation (area A) are promising for spatial planners and decision makers. They show that the effect of a change in the physical landscape is translated into a change in the character of openness, which can be quantified, mapped and specified for three modes of perception.

The emphasis of this paper is on the comparison of perceived openness by three modes of perception represented by different sampling methods, rather than on the openness variable values as such. Nonetheless, the relevance of the openness variables and landscape perception needs to be clear. Further research should therefore be conducted into validating the calculated openness variable values with people's judgments of openness in the field.

Acknowledgments. The authors wish to thank three anonymous reviewers for their useful comments, and Sanjay Rana for providing the Isovist Analyst 1.2 and the Art Gallery Problem Solver 1.0 software.

References

1. Buijs, A.E., Pedroli, B., Luginbühl, Y.: From hiking through farmland to farming in a leisure landscape: changing social perceptions of the European landscape. Landscape Ecology 21, 375–389 (2006)
2. Nohl, W.: Sustainable landscape use and aesthetic perception - preliminary reflections on future landscape aesthetics. Landscape and Urban Planning 54, 223–237 (2001)
3. Granö, J.G.: Pure Geography. The Johns Hopkins University Press, London (1997)
4. Higuchi, T.: The Visual and Spatial Structure of Landscapes. MIT Press, Cambridge (1983)
5. Kaplan, S., Kaplan, R., Wendt, J.S.: Rated preference and complexity for natural and urban visual material. Perception & Psychophysics 12, 354–356 (1972)
6. Tuan, Y.-F.: Topophilia: A study of Environmental Perception, Attitudes, and Values. Prentice-Hall, Englewood Cliffs, New Jersey (1974)
7. Zube, E.H.: Perceived land use patterns and landscape values. Landscape Ecology 1, 37–45 (1987)
8. Berg, A.E.v.d.: Individual Differences in the Aesthetic Evaluation of Natural Landscapes. Psychologische, Pedagogische en Sociologische Wetenschappen, Vol. PhD. Rijksuniversiteit Groningen, Groningen (1999)
9. Brabyn, L.: Solutions for characterising natural landscapes in New Zealand using geographical information systems. Journal of Environmental Management 76, 23–34 (2005)
10. Miller, D.R.: Spatial Modelling of the Visibility of Land Use Change. In: 4th International Conference on Integrating GIS and Environmental Modeling Banff, Alberta, Canada (2000)
11. Smardon, R.C., Palmer, J.F., Felleman, J.P.: Foundations for Visual Project Analysis. John Wiley & Sons, New York (1986)
12. Bishop, I.D.: Assessment of visual qualities, impacts, and behaviours, in the landscape, by using measures of visibility. Environment and Planning B: Planning and Design 30, 677–688 (2003)
13. Gibson, J.J.: The Ecological Approach to Visual Perception. Houghton Mifflin Compagny, Boston (1979)
14. Thiel, P.: A sequence-experience notation: for architectural and urban spaces. Town Planning Review 32, 33–52 (1961)
15. De Floriani, L., Magillo, P.: Algorithms for visibility computation on terrains: A survey. Environment and Planning B: Planning and Design 30, 709–728 (2003)
16. Llobera, M.: Extending GIS–based visual analysis: the concept of visualscapes. International Journal of Geographical Information Science 17, 25–48 (2003)
17. Tandy, C.R.: The isovist method of landscape survey. In: Murray, H.C., (ed.) Methods of Landscape Analysis. Landscape Research Group, London (1967)

18. Benedikt, M.L.: To take hold of space: isovists and isovists fields. Environment and Planning B: Planning and Design 6, 47–65 (1979)
19. Dijkstra, H.: Het visuele landschap: onderzoek naar de visuele kwaliteit van landschappen. Landschap 8, 157–175 (1991)
20. Lynch, K.: Image of the city. MIT Press, Cambridge (1960)
21. Dalton, R.C., Bafna, S.: The syntactical image of the city: a reciprocal definition of spatial elements and spatial syntaxes. In: 4th International Space Syntax Symposium, London (2003)
22. Freundschuh, S.M., Egenhofer, M.J.: Human conceptions of spaces: implications for Geographic Information Systems. Transactions in GIS 2, 361–375 (1997)
23. Montello, D.R.: Scale and multiple psychologies of space. In: Frank, A.U., Campari, I. (eds.) Spatial Information Theory: a Theoretical Basis for GIS, pp. 312–321. Springer, Heidelberg (1993)
24. Ministerie van VROM: Nota ruimte. Ministerie van VROM, Den Haag (2004)
25. Kuipers, B.: Modeling spatial knowledge. Cognitive Science 2, 129–153 (1978)
26. Palmer, J.F., Lankhorst, J.R.-K.: Evaluating visible spatial diversity in the landscape. Landscape and Urban Planning 43, 65–78 (1998)
27. Simonds, J.O.: Landscape Architecture: a Manual of Site Planning and Design. McGraw-Hill, New York (1998)
28. Laan, H.V.D.: Architectonic Space: Fifteen Lessons on the Disposition of the Human Habitat. Brill Academic Publishers, Leiden (1983)
29. Lynch, K.: Site Planning. MIT Press, Cambridge (1984)
30. Van der Ham, R.J.I.M., Iding, J.A.M.E.: Landscape Typology System based on Visual Elements: Methodology and Application. [s.n.], Wageningen (1971)
31. Mücher, C.A., Kramer, H., Thunnissen, H.A.M., Clement, J.: Monitoren van kleine landschapselementen met IKONOS satellietbeelden. Alterra, Wageningen (2003)
32. Van Buren, J., Westerik, A., Olink, E.J.H.: Kwaliteit Top10vector: De geometrische kwaliteit van de Top10vector van de topografische dienst. Kadaster (2003)
33. Giles Jr., R.H., Trani, M.K.: Key elements of landscape pattern measures. Key elements of landscape pattern measures 23, 477 (1999)
34. Manderller, F., Wrbka, T.: Functional and structural landscape indicators: Upscaling and downscaling problems. Ecological Indicators 5, 267–272 (2005)
35. Rana, S.: Isovist Analyst Extension. Centre for Advanced Spatial Analysis, London (2002)
36. Lynch, J.A., Gimblett, R.H.: Perceptual values in the cultural landscape: a spatial model for assessing and mapping perceived mystery in rural environments. Journal of Computers, Environment and Urban Systems 16, 453–471 (1992)
37. Hanna, K.C., Haniva, K.C.: GIS for Landscape Architects ESRI, Inc. (1999)
38. Tversky, B., Morrison, J.B., Franklin, N., Bryant, D.J.: Three spaces of spatial cognition. Professional Geographer 51, 516–524 (1999)
39. Rana, S.: Use of GIS for planning visual surveillance installations. ESRI Homeland Security GIS Summit ESRI, Denver, Colorado (2005)
40. Vroom, M.J.: The perception of dimensions of space and levels of infrastructure and its application in landscape planning. Landscape Planning 12, 337–352 (1986)
41. O'Sullivan, D., Turner, A.: Visibility graphs and landscape visibility analysis. International Journal of Geographical Information Science 15, 221–237 (2001)
42. Hillier, B.: Space is the Machine: A Configurational Theory of Architecture. Cambridge University Press, Cambridge (1996)
43. Hillier, B., Hanson, J.: The social logic of space. Cambridge University Press, London (1984)

44. Kaplan, R., Kaplan, S., Ryan, R.L.: With People in Mind: Design and Management of Everyday Nature. Island Press, Washington (1998)
45. Hendriks, K., Stobbelaar, D.J.: Landbouw in een leesbaar landschap: hoe gangbare en biologische landbouwbedrijven bijdragen aan landschapskwaliteit. Vol. PhD. Wageningen University, Wageningen, 268 (2003)
46. Jiang, B., Claramunt, C.: Integration of space syntax into GIS: new perspectives for urban morphology. Transactions in GIS 6, 295–309 (2002)
47. Turner, A., Doxa, M., O'Sullivan, D., Penn, A.: From Isovists to visibility graphs: a methodology for the analysis of architectural space. Environment and Planning B: Planning and Design 28, 103–121 (2001)
48. Arthur, L.M., Daniel, T.C., Boster, R.S.: Scenic assessment: an overview. Landscape Planning 4 (1977)
49. Bourassa, S.: Toward a theory of landscape aesthetics. Landscape and Urban Planning 15, 241–252 (1988)

Reasoning on Spatial Semantic Integrity Constraints

Stephan Mäs

AGIS - Arbeitsgemeinschaft GIS, Universität der Bundeswehr München,
Werner Heisenberg Weg 39, 85577 Neubiberg, Germany
{Stephan.Maes}@unibw.de

Abstract. Semantic integrity constraints specify relations between entity classes. These relations must hold to ensure that the data conforms to the semantics intended by the data model. For spatial data many semantic integrity constraints are based on spatial properties like topological or metric relations. Reasoning on such spatial relations and the corresponding derivation of implicit knowledge allow for many interesting applications. The paper investigates reasoning algorithms which can be used to check the internal consistency of a set of spatial semantic integrity constraints. Since integrity constraints are defined at the class level, the logical properties of spatial relations can not directly be applied. Therefore a set of 17 abstract class relations has been defined, which combined with the instance relations enables the specification of integrity constraints. The investigated logical properties of the class relations enable to discover conflicts and redundancies in sets of spatial semantic integrity constraints.

1 Introduction

Integrity, sometimes also called consistency, is a term originally used for the property of database systems of being free of logical contradictions within a model of reality. This model also contains defined integrity constraints that must hold on the database to grasp the semantics intended by the model [6]. Integrity constraints play a major role when the logical consistency of a data set has to be evaluated. For spatial data in particular constraints which comprehend the spatial peculiarities are of interest. While for database modelling a universally valid classification of integrity constraints is established and the constraint types are supported by most database systems, at present there is no sufficient integration of spatial integrity constraints and not even a theoretical basis for the formalisation of their contents and restrictions existing. This paper tries to contribute to a solution of these problems. It focuses on the formalisation of spatial semantic integrity constraints and the identification of conflicts and redundancies in sets of such constraints. Therefore we start with a categorisation of integrity constraints and point out how spatial integrity constraints integrate into this classification. Further the categorisation is used to outline the definition of spatial semantic integrity constraints. Since integrity constraints are defined at the level of entity classes the paper reviews the application of class relations in section 3. Based on that a set of 17 abstract class

S. Winter et al. (Eds.): COSIT 2007, LNCS 4736, pp. 285–302, 2007.

relations is defined which particularly supports the specification of spatial semantic integrity constraints. Another focus is on the investigation of the logical properties of the defined class relations. The reasoning algorithms investigated in section 4 enable to discover conflicts and redundancies in sets of spatial semantic integrity constraints. The practical value and the usability of the researched concepts are demonstrated in section 5, where a possible user interface for the definition of spatial semantic integrity constraints is designed.

2 Integrity Constraints

The restrictions defined by integrity constraints can be manifold. This paper focuses only on spatial semantic integrity constraints. The aim of this section is to categorize integrity constraints and to outline spatial semantic integrity constraints.

2.1 Categories of Integrity Constraints

In Elmasri and Navathe (1994) [7], pp. 638–643, the properties of integrity constraints for data modelling have been analysed. They propose the following classification of integrity constraints according to the type of the specified conditions:

1. **Domain constraints** restrict the allowed types of values of an attribute.
2. **Key and relationship constraints** refer to key values of entity classes, cardinalities of relationships between entity classes and participation requirements of relationships defined in the data model.
3. **General semantic integrity constraints** are explicitly specified and usually more complex. They refer to the semantics of the modelled entity classes. Therefore they can not be specified as domain or key and relationship constraints.

Spatial integrity constraints fit into two categories of this classification. Currently, a variety of database systems is already capable to handle the particular requirements of spatial data and provides predefined spatial data types. The constraints on this data types and the corresponding constraints on geometric and topological primitives are domain constraints. Such spatial integrity constraints are not discussed in this paper.

The second category of spatial constraints is semantic integrity constraints. This paper focuses on the definition of such spatial semantic integrity constraints. Therefore the following subsection will give a closer look at the restrictions which are specified by these constraints.

2.2 What Do Semantic Integrity Constraints Restrict?

Following the definition of Elmasri and Navathe semantic constraints are based on relations between the involved entities or on specific properties of a single entity. The validity of the relations is based on the semantics of the entities. Semantic integrity constraints are defined at the level of the entity classes and have to be explicitly defined. The restrictions defined by semantic integrity constraints can be manifold and complex, what makes a differentiation of the kinds of defined restrictions necessary. An approach to a categorisation of integrity constraints according to the restricted data model elements was made by Ditt et al. [1] and later extended by Friis-Christensen el al. [9]. They differentiated the following integrity constraint categories:

1. constraints referring to an attribute of a single entity
2. constraints referring to at least two attributes of a single entity
3. constraints referring to all entities of a single entity class
4. constraints referring to an entity and its associated entities of various classes
5. constraints referring to operations of entities.

All five categories include semantic integrity constraints. In this paper we only consider the categories three and four, leaving out constraints restricting single entity's attributes and operations of entities. We investigate integrity constraints on relations between the entities of a single or of two entity classes. As a further restriction we only consider binary relations which are not explicitly modelled. Relations which are explicitly defined in the data model are restricted by key and relationship constraints. Implicit relation can be deduced from the corresponding attributes of the involved entities. A typical example of such implied relations are topological relations (e.g. figure 1) between spatial entities [3]. Usually they are not explicitly stored since they can be derived from the entity's geometries.

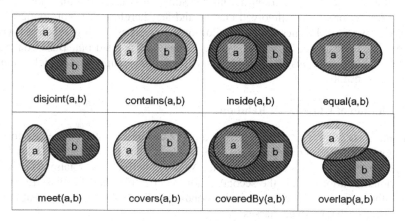

Fig. 1. Set of topological relations between areal entities

All examples of class relations and integrity constraints given throughout the paper are based on the binary topological relations between areal entities shown in figure 1. However, the defined class relations can also be combined with any other instance relation. To enable for reasoning on the integrity constraints (section 4) the instance relation must be part of a limited, jointly exhaustive and pairwise disjoint (JEPD) set of relations. This requirement is necessary when the consistency of the class relations shall be checked making use of reasoning algorithms. Most spatial relations are part of such a JEPD set of relations.

3 Integrity Constraint Definition Based on Class Level Relations

Integrity constraints are defined at the level of entity classes since they are always restricting entire classes or subsets of classes. When a database is checked against a spatial semantic integrity constraint, the checking procedure proves spatial relations between the involved instances. Thus a formalised description of such an integrity constraint must be linked to the instance relations the quality checking procedure applies. But as the following example illustrates instance relations are not suitable for integrity constraint formalisation. A natural language statement about two instances could be: "the watermill is overlapping the river". Since *overlap* is a symmetric relation it also implies "the river is overlapping the watermill". A corresponding semantic integrity constraint for the classes "watermill" and "stream" could be: "A watermill must overlap a stream". Applying the symmetry of the instance relation again it becomes: "A stream must overlap a watermill". These two statements about the classes obviously don't have the same semantic and since not every stream is flowing through a watermill the second is not true. This example shows that instance relations can not clearly represent the semantic of statements about classes and that the formalisation of such statements requires specific class relations. Donnelly and Bittner [2] also identified this problem and provided an approach for the definition of class relations. The following subsections will review their solution with regard to an application for integrity constraint definition. Some class level relations don't define violable restrictions on the involved classes and are therefore also not applicable as integrity constraints. Further on, some additional properties of class relations, which have not been considered by Donnelly and Bittner, are pointed out. Based on that a new set of class relations is defined which particularly supports the definition of integrity constraints.

3.1 Definition of Class Level Relations

Before the class relations can be applied it must be ensured that the classes conform to the following two requirements. First, the involved classes must have at least one instance, i.e. empty classes are not feasible. As stated before class relations are linked to individual relations. Thus the second condition specifies that if a class relation is defined, there must be at least one corresponding individual relation existent among the instances of the involved classes.

x,y,z	Denote variables for individuals / instances. Every instance must be associated to a class.
A, B, C	Denote variables for classes. Every class must have at least one instance.
Inst(x,A)	Means individual x is an instance of class A.
r(x,y)	Means individual x has the relation r to individual y; x and y are said to participate on the relationship instance r. The meta-variable r can stand for any relation of individuals (e.g. topological relations). Every relationship instance r can be associated to a class relation R.

R(A,B)   Denotes that R relates the classes A and B. The meta-variable R can stand for any class relationship. Every R is related to an individual relation r. If a class relation R(A,B) is defined at least one r must be existent between the instances of A and B.

Based on these variables and functions, Donnelly and Bittner [2] define the following class relations:

$$R_{some}^{D\&B}(A,B) := \exists x \exists y (Inst(x,A) \cap Inst(y,B) \cap r(x,y)). \tag{D\&B1}$$

$$R_{all-1}^{D\&B}(A,B) := \forall x (Inst(x,A) \rightarrow \exists y (Inst(y,B) \cap r(x,y))). \tag{D\&B2}$$

$$R_{all-2}^{D\&B}(A,B) := \forall y (Inst(y,B) \rightarrow \exists x (Inst(x,A) \cap r(x,y))). \tag{D\&B3}$$

$$R_{all-12}^{D\&B}(A,B) := R_{all-1}^{D\&B}(A,B) \cap R_{all-2}^{D\&B}(A,B). \tag{D\&B4}$$

$$R_{all-all}^{D\&B}(A,B) := \forall x \forall y (Inst(x,A) \cap Inst(y,B) \rightarrow r(x,y)). \tag{D\&B5}$$

$R_{some}^{D\&B}(A,B)$ holds if at least one instance of A stands in relation r to some instance of B. $R_{some}^{D\&B}(A,B)$ relations are very weak, but nevertheless useful for example when class relations are defined in an ontology. Integrity constraints which are only based on such relations are not expedient, since they only specify that a relation universally exists in reality without any concrete cardinalities. Within a data set the relation is in principle possible, but does not necessarily occur within the modelled part of reality. This means that a data set, which is usually representing parts of the reality, can either contain individuals that have the relation or it doesn't; both cases are conform to the integrity constraint. Since a violation against constraints which are only specifying $R_{some}^{D\&B}(A,B)$ relations is not possible, such constraints are not useful for quality assurance. This changes if $R_{some}^{D\&B}(A,B)$ relations are specified in conjunction with a defined set of entities, like for example a relation r holds for some entities of A and B within a certain area (possibly defined by an individual entity of C). Therewith the constraint is violable by the subsets of A and B and useful for quality assurance. Since the definition of such subsets can be manifold and complex the analysis in this paper is restricted to binary relations between entire entity classes, leaving out subsets of classes.

$R_{all-1}^{D\&B}(A,B)$ holds if every instance of A has the relation r to some instance of B. In set theory such relations are called left-total. This class relation can be used to define the integrity constraint of the above mentioned windmill / stream example: $OVERLAPS_{all-1}^{D\&B}(Windmill, Stream)$ specifies the *overlap* relation for all windmills but it doesn't include all streams.

$R_{all-2}^{D\&B}(A,B)$ holds if for each instance of B there is some instance of A which stands in relation r to it. This means that every instance of B has the inverse relation of r to some instance of A. $R_{all-2}^{D\&B}(A,B)$ is right-total / surjective.

$R_{all-12}^{D\&B}(A,B)$ combines the definitions of (D&B2) and (D&B3). It holds if every instance of A stands in relation r to at least one instance of B and for each instance of B there is at least one instance of A which stands in relation r to it. R is left-total and right-total.

This differentiation of class relations according to the totality of the involved individuals of the entity classes is very useful for the definition of integrity constraints. The class relations define constraints on all individuals of A (D&B2), all individuals of B (D&B3) or on all individuals of both arguments A and B (D&B4). In data modelling such definitions are called participation constraints on the relation. They specify whether the existence of an entity depends on its relation to another entity via the relationship type [7]. $R_{all-1}^{D\&B}(A,B)$, $R_{all-2}^{D\&B}(A,B)$ and $R_{all-12}^{D\&B}(A,B)$ define total participation constraints on their relationship instances, since at least one of the classes is totally effected. $R_{some}^{D\&B}(A,B)$ defines a partial participation constraint since not necessarily all instances of the classes A and B have the relationship instance.

A specific case of $R_{all-12}^{D\&B}(A,B)$ is defined by the $R_{all-all}^{D\&B}(A,B)$ relation, which holds if all instances of A have an relationship instance of R to all instances of B. This relation is very strong, since it defines restrictions on all relations between all individuals of the arguments A and B. Therewith the corresponding integrity constraints are very restrictive but for example useful when all instances of two classes are not allowed to intersect: $DISJOINT_{all-all}^{D\&B}(Streets, Lakes)$.

Beside the total participation constraint the $R_{all-all}^{D\&B}(A,B)$ relationship defines a so called cardinality ratio constraint, which specifies the number of relationship instances an entity can participate in [7]. In this case the number of B entities (i.e. "all" instances of B) defines in how many relationship instances each entity of A is participating and vice versa.

In data modelling total participation and cardinality ratio constraints are well established, for example when using the Entity-Relationship Model as a notation. In such models a total participation is represented by a double line for the relation and cardinality ratio for example by a N:1 next to the relation signature (see figure 2). In this example all buildings are restricted to be contained by only one parcel, while the parcels are allowed to contain an undefined number of buildings.

The number of different cardinality ratio constraints of such a notation is infinite. Thus it is impossible to represent them all by separate class relations. But since some of them are indispensable for the definition of integrity constraints we decided to extend the framework of class relations of [2] by the concept of unambiguousness. Correspondingly the following class relations are defined:

$$R_{Left-D}(A,B) := \forall x,y,z(Inst(x,A) \cap Inst(y,B) \cap Inst(z,A) \cap$$
$$r(x,y) \cap r(z,y) \rightarrow x = z) \cap R_{some}^{D\&B}(A,B). \tag{1}$$

$$R_{Right-D}(A,B) := \forall x,y,z(Inst(x,A) \cap Inst(y,B) \cap Inst(z,B) \cap$$
$$r(x,y) \cap r(x,z) \rightarrow y = z) \cap R_{some}^{D\&B}(A,B). \tag{2}$$

$R_{Left-D}(A,B)$ relations are left-definite / injective and specify that for no instance of B there is more than one instance of A which stands in relation r to it. This relation restricts the number of R relations an instance of B can participate; the instances of A are not restricted. The last term $R_{some}^{D\&B}(A,B)$ ensures that at least one instance relation r does exist between the instances of A and B.

The right-definite relations $R_{Right-D}(A,B)$ specify that no instance of A participates in a relationship instance of R to more than one instance of B. When this relation is defined all instances of A are restricted while the instances of B are not affected.

Both relations are very useful for the definition of integrity constraints, since they restrict the number of possible relations of the involved individuals to a maximum of one.

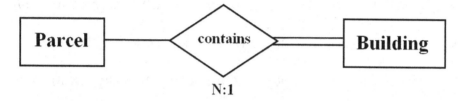

Fig. 2. Total participation and cardinality ratio constraints in an entity-relationship model

3.2 Class Level Relations for Integrity Constraint Definition

The main properties of the class level relations used in the previous section are left-definite, right-definite, left-total and right-total. These properties are independent of each other, what means that no property implies or precludes one of the other properties. If a class relation is only defined as right-total there is no information about its left totality and the cardinality ratio available. For the definition of integrity constraints this situation is insufficient since the constraint relations should allow for combinations of properties as well as their negations. Therefore we define a new set of class relations which implies adjustments to achieve sets of pairwise disjoint relations.

$$R_{some}(A,B) := R_{some}^{D\&B}(A,B) \cap \neg R_{all-1}^{D\&B}(A,B) \cap \neg R_{all-2}^{D\&B}(A,B) \cap$$
$$\neg R_{Left-D}(A,B) \cap \neg R_{Right-D}(A,B). \tag{3}$$

$R_{some}(A,B)$ is similar to the definition (D&B1) of Donnelly and Bittner [2] but while $R_{some}^{D\&B}(A,B)$ contains all other defined class relations these are now excluded. R_{some} is defined as not left-total and not right-total, what implies that some instances of A/B participate in a relation r to an instance of B/A and some don't. Furthermore the exclusions of $R_{Left-D}(A,B)$ and $R_{Right-D}(A,B)$ specify that some A/B participate in a relation r to at least two instances of B/A. Hence R_{some} is only valid for classes with more than two instances.

$$R_{LD}(A,B) := R_{Left-D}(A,B) \cap \neg R_{Right-D}(A,B) \cap \neg R_{all-1}^{D\&B}(A,B) \cap \neg R_{all-2}^{D\&B}(A,B). \tag{4}$$

$$R_{RD}(A,B) := \neg R_{Left-D}(A,B) \cap R_{Right-D}(A,B) \cap \neg R_{all-1}^{D\&B}(A,B) \cap \neg R_{all-2}^{D\&B}(A,B). \tag{5}$$

$$R_{LT}(A,B) := \neg R_{Left-D}(A,B) \cap \neg R_{Right-D}(A,B) \cap R_{all-1}^{D\&B}(A,B) \cap \neg R_{all-2}^{D\&B}(A,B). \tag{6}$$

$$R_{RT}(A,B) := \neg R_{Left-D}(A,B) \cap \neg R_{Right-D}(A,B) \cap \neg R_{all-1}^{D\&B}(A,B) \cap R_{all-2}^{D\&B}(A,B). \tag{7}$$

The definitions (4) to (7) specify class relations which are either left-definite, right-definite, left-total or right-total. The corresponding other properties are excluded.

$$R_{LD.RD}(A,B) := R_{Left-D}(A,B) \cap R_{Right-D}(A,B) \cap \neg R_{all-1}^{D\&B}(A,B) \cap \neg R_{all-2}^{D\&B}(A,B). \tag{8}$$

$$R_{LD.LT}(A,B) := R_{Left-D}(A,B) \cap \neg R_{Right-D}(A,B) \cap R_{all-1}^{D\&B}(A,B) \cap \neg R_{all-2}^{D\&B}(A,B). \tag{9}$$

$$R_{LD.RT}(A,B) := R_{Left-D}(A,B) \cap \neg R_{Right-D}(A,B) \cap \neg R_{all-1}^{D\&B}(A,B) \cap R_{all-2}^{D\&B}(A,B). \tag{10}$$

$$R_{RD.LT}(A,B) := \neg R_{Left-D}(A,B) \cap R_{Right-D}(A,B) \cap R_{all-1}^{D\&B}(A,B) \cap \neg R_{all-2}^{D\&B}(A,B). \tag{11}$$

$$R_{RD.RT}(A,B) := \neg R_{Left-D}(A,B) \cap R_{Right-D}(A,B) \cap \neg R_{all-1}^{D\&B}(A,B) \cap R_{all-2}^{D\&B}(A,B). \tag{12}$$

$$R_{LT.RT}(A,B) := \neg R_{Left-D}(A,B) \cap \neg R_{Right-D}(A,B) \cap R_{all-1}^{D\&B}(A,B) \cap R_{all-2}^{D\&B}(A,B) \cap \neg R_{all-all}^{D\&B}(A,B). \tag{13}$$

The definitions (8) to (13) combine pairs of the four defined class relation properties and exclude the corresponding others. A special case is (13) which additionally excludes $R_{all-all}^{D\&B}(A,B)$.

$$R_{LT.RT-all} := R_{all-all}^{D\&B}(A,B). \tag{14}$$

Definition (14) is equivalent to (D&B5). $R_{LT.RT-all}$ is left-total and right-total and holds if all instances of A have a relationship instance of R to all instances of B.

$$R_{LD.RD.LT}(A,B) := R_{Left-D}(A,B) \cap R_{Right-D}(A,B) \cap R_{all-1}^{D\&B}(A,B) \cap \neg R_{all-2}^{D\&B}(A,B). \tag{15}$$

$$R_{LD.RD.RT}(A,B) := R_{Left-D}(A,B) \cap R_{Right-D}(A,B) \cap \neg R_{all-1}^{D\&B}(A,B) \cap R_{all-2}^{D\&B}(A,B). \tag{16}$$

$$R_{LD.LT.RT}(A,B) := R_{Left-D}(A,B) \cap \neg R_{Right-D}(A,B) \cap R_{all-1}^{D\&B}(A,B) \cap R_{all-2}^{D\&B}(A,B) \cap \neg R_{all-all}^{D\&B}(A,B). \tag{17}$$

$$R_{RD.LT.RT}(A,B) := \neg R_{Left-D}(A,B) \cap R_{Right-D}(A,B) \cap R_{all-1}^{D\&B}(A,B) \cap R_{all-2}^{D\&B}(A,B) \cap \neg R_{all-all}^{D\&B}(A,B). \tag{18}$$

The definitions (15) to (18) combine three of the four defined class relation properties respectively and exclude the corresponding fourth. Particular attention must be given to the class relations which are left-total and right-total ((17) and (18)). In case only one instance of A or B exists left-total and right-total class relations are always left-definite or right-definite, respectively. Furthermore they will also hold (D&B5). Thus it is necessary to separate the relations (17) and (18) from (14), which is done by the exclusion of $R_{all-all}^{D\&B}(A,B)$. Therewith the relations (17) and (18) are not possible if class A or class B has only one instance.

$$R_{LD.RD.LT.RT}(A,B) := R_{Left-D}(A,B) \cap R_{Right-D}(A,B) \cap R_{all-1}^{D\&B}(A,B) \cap$$
$$R_{all-2}^{D\&B}(A,B) \cap \neg R_{all-all}^{D\&B}(A,B). \tag{19}$$

Definition (19) specifies class relations which are left-definite, right-definite, left-total and right-total. Similar to the definitions (17) and (18) $R_{all-all}^{D\&B}(A,B)$ is excluded to distinguish the relation from (14) for the case that A and B have only one instance. In this case the relation can't occur.

All together the definitions (3) to (19) specify 17 class relations which can be combined with any binary instance relations to associate classes. With the exception of $R_{some}(A,B)$ all of these relations specify restrictions which can be used as integrity constraints for quality assurance of the data.

Depending on the relations it is possible to define more than one class relation between two classes, even when the applied instance relations are part of the same JEPD set of relations. Figure 3 illustrates such an example, where two classes can be restricted by three topological class relations. In this scene every instance of A meets one and contains another instance of B. One instance of B meets two instances of A and some are disjoint from all instances of A. Other than those three relations are not occurring. The corresponding integrity constraints are the class relations defined in (11), (15) and (14):

$$\text{MEET}_{RD.LT} \text{ (Entity Class A, Entity Class B)}$$
$$\text{CONTAINS}_{LD.RD.LT} \text{ (Entity Class A, Entity Class B)}$$
$$[\text{MEET} \cup \text{CONTAINS} \cup \text{DISJOINT}]_{LT.RT-all}(A,B)$$

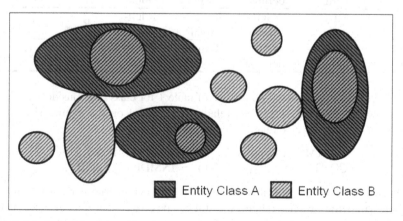

Fig. 3. Scene of two entity classes which could be defined by three class relations

The possibilities of such combinations of class relations are limited if they rest on instance relations which are part of a JEPD set of relations. For example if $R_{LT.RT\text{-all}}$ is defined no second $R_{LT.RT\text{-all}}$ relation and no class relation based other instance relations of that domain is possible between the two classes.

4 Reasoning on Semantic Integrity Constraints

This section investigates the logical properties of the class relations defined in the previous section. For the work with integrity constraints these properties can be very useful; for example they enable to discover inconsistencies and redundancies in a set of integrity constraints.

4.1 Reasoning on the Symmetry of the Class Relations

The transfer of logical properties of instance relations to class relations, like their symmetry and transitivity, has been researched by Donnelly and Bittner [2]. The purpose of this subsection is to deepen the analysis of symmetry properties of the defined class relations. Spatial relations between instances are usually either symmetric or have a well defined inverse relation. Table 1 shows the correlation between symmetry properties of instance level relations and those of the corresponding class level relations.

r^i Inverse instance relation.
R^i Inverse class relation.

Table 1. Symmetry properties of the class relations

Individual Relation r is...	Class Relation R is...					
	left-definite	right-definite	left-total	right-total	R_{some}	$R_{LT.RT\text{-all}}$
symmetric	R right-definite	R left-definite	R right-total	R left-total	R_{some}	$R_{LT.RT\text{-all}}$
Not symmetric	R^i right-definite	R^i left-definite	R^i right-total	R^i left-total	R^i_{some}	$R^i_{LT.RT-all}$

The following examples illustrate the use of table 1 for class relations defined for the scene shown in figure 3. The class relations are based on the symmetric instance relation *meet* and the inverse relations *contains* and *inside*:

$$(\text{MEET }_{RD.LT} (A, B))^i = \text{MEET }_{LD.RT} (B,A).$$
$$(\text{CONTAINS }_{LD.RD.LT} (A, B))^i = \text{INSIDE }_{LD.RD.RT} (B,A).$$

The examples show that not all class relations are symmetric, even when they are based on symmetric instance relations. But it can be proven that if an instance relation is symmetric or has an inverse relation there exists also an inverse relation for each of the corresponding class relations.

4.2 Correlation Between Class Relations and the Number of Individuals

For many entity classes the number of existing individuals is unknown or variable. For these classes the dependency between class relations and the number of individuals of a class is irrelevant. But for classes with a small and well defined number of individuals the designer of a data model is in many cases aware of these numbers. Such classes are for example earth surface or continents. Another example is the class "capital" which can only have one instance if the area of interest is restricted to a single "country". The knowledge about these numbers and their correlation to the class relations should be included when reasoning on class relations.

As already stated in section 3.2 some class relations are not valid if one or both of the involved classes have less than three instances. The only class relation that is possible if both classes have only one instance is $R_{LT.RT\text{-all}}$. If class A has one instance the only possible class relations are $R_{LD.LT}$, $R_{LT.RT\text{-all}}$ and $R_{LD.RD.LT}$; if B has one instance only $R_{RD.RT}$, $R_{LT.RT\text{-all}}$ and $R_{LD.RD.RT}$.

The definition of class relations can also be restricted if the number of instances is more than one. If the number of instances of one of the classes is known, some class relations allow for conclusions about the number of instances of the other class. These reasoning properties are shown in the following list of theorems:

Count(A)   Denotes the number of individuals of the class A.

T1 $$R_{LD.RD.LT}(A, B) \rightarrow Count(A) < Count(B).$$

T2 $$R_{LD.RD.RT}(A, B) \rightarrow Count(A) > Count(B).$$

T3 $$R_{LD.LT.RT}(A, B) \rightarrow Count(A) < Count(B).$$

T4 $$R_{RD.LT.RT}(A, B) \rightarrow Count(A) > Count(B).$$

T5 $$R_{LD.RD.LT.RT}(A, B) \rightarrow Count(A) = Count(B).$$

Furthermore the number of instances can restrict the possible combinations of class relations between two classes. For example if exactly two instances of A exist, only a combination of two $R_{LD.RT}(A,B)$ class relations can be defined for one set of JEPD instance relations. A third $R_{LD.RT}(A,B)$ would require at least one more instance of A.

4.3 Composition of Class Relations

The composition of binary relations enables for the derivation of implicit knowledge about a triple of entities. If two binary relations are known the corresponding third one can potentially be inferred or some relations can be excluded. Examples of composition tables of instance relations can be found in [5] and [10] for topological relations between areal entities and in [11] and [8] for directional/orientation relations. Many other sets of binary spatial relations also allow for such derivations.

A transfer of this reasoning formalism to the class level would be very useful for the work with integrity constraints and other applications of class relations. In general the composition of class relations is not independent of the composition of instance relations. The composition of class relations is possible if the applied instance relations belong to the same set of JEPD relations and this set allows for compositions at the instance level. Using for example the 17 class relations together with the 8 topological relations between regions (see figure 1) would result in 136 topological class relations and almost 18500 compositions. Since such an amount of compositions is hardly manageable we propose a two level reasoning formalism, which separates the compositions of the abstract class relations from those of the instance relations. For lack of space we don't derive all compositions in this paper, but the following three examples shall illustrate the general approach. For the compositions of the applied instance relations we refer to the composition table of binary topological relations between areal entities of Egenhofer [5].

The first example derives the composition from the two abstract class relations $R1_{LT.RT-all}(A,B)$ and $R2_{LT.RT-all}(B,C)$. Therewith all instances of A have a relationship instance of R1 to all instances of B and all instances of B have a relationship instance of R2 to all instances of C. Since all instances of A have the same kind of relation to all instances of B and all instances of B participate in same kind of relation to all instances of C, it is obvious that all instances of A must have the same relation to all instances of C. In other words every possible triple of instances of A, B and C is related by the same relations. Thus the composition of the abstract class relations must be:

$$R1_{LT.RT-all}(A,B) \cap R2_{LT.RT-all}(B,C) \Rightarrow R3_{LT.RT-all}(A,C).$$

For the combination of this result with the instance level compositions two cases have to be distinguished. If the composition of the instance relations is unique (i.e. it results in only one relation) the combined composition is also based on that single relation like in the following example:

$$equal(a,b) \cap disjoint(b,c) \Rightarrow disjoint(a,c).$$

$$EQUAL_{LT.RT-all}(A,B) \cap DISJOINT_{LT.RT-all}(B,C) \Rightarrow DISJOINT_{LT.RT-all}(A,C).$$

If the instance relation composition results in a disjunction of instance relations the combined composition also leads to a disjunction in the class relation, for example:

$$meet(a,b) \cap covers(b,c) \Rightarrow disjoint(a,c) \cup meet(a,c).$$

$$MEET_{LT.RT-all}(A,B) \cap COVERS_{LT.RT-all}(B,C) \Rightarrow [DISJOINT \cup MEET]_{LT.RT-all}(A,C).$$

In this example it is derived that all instances of A have one of the relations *disjoint* or *meet* to all instances of C.

For the second example the $R1_{LT.RT-all}$ relation between the classes A and B is kept and the relation between B and C is $R2_{RD}$. Therewith no instance of B participates in a relationship instance of R2 to more than one instance of C and some instances of B and C are not related by a relationship instance of R2. A possible scene that implements the two abstract class relations for the instance relations *meet* and *contains* is shown in figure 4. As the figure illustrates, for every possible triple of instances the relation between the instances of A and B is *meet* (instance of R1) but only triples which include the instances b1 and c1 have *contains* (instance of R2) as relation. In general R3 can only be derived for the triples of instances which have an R2 relation. In figure 4 this are only the relations a1 to c1 and a2 to c1 with the instance composition:

$$meet(a,b) \cap contains(b,c) \Rightarrow disjoint(a,c).$$

Figure 4 also shows that there are many other relations possible between instances of A and C. Thus it is not possible to derive a unique class relation; all of the abstract class relations (3) to (19) are possible. The only implication is that some (but not necessarily all) of the instances of A and C participate in the relation that results from the instance composition. Thus the composition of the abstract class relations is

$$R1_{LT.RT-all}(A,B) \cap R2_{RD}(B,C) \Rightarrow R3_U(A,C).$$

$R_{\mathcal{U}}$ denotes the universal disjunction of all class relations of (3) to (19) of the corresponding instance relation r.

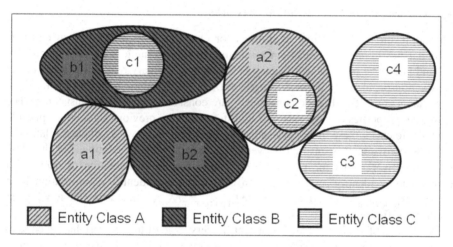

Fig. 4. Scene that implements MEET$_{LT.RT-all}$ (A, B) and CONTAINS$_{RD}$(B,C)

It might be possible that the number of instances of the classes allow for an exclusion of some of the class relations from the disjunction (see section 4.2). For the example in figure 4 the combined composition is:

$$MEET_{LT.RT-all}(A,B) \cap CONTAINS_{RD}(B,C) \Rightarrow DISJOINT_U(A,C).$$

Instance compositions which don't result in a unique relation are treated in analogy to the first example.

In the third example R1$_{some}$ relates the classes A and B and R2$_{RD}$ the classes B and C. This implies that some (not all) instances of A and B are related by a relationship instance of R1 and some (not all) instances of B and C are related by a relationship instance of R2. This doesn't mean that a triple of A, B and C instances exists, which includes both instance relations r1 and r2. Thus the instance composition is not possible and also the composition of the abstract class relations leads to no restriction of possible relations. The combined composition is undetermined:

$$R1_{some}(A,B) \cap R2_{RD}(B,C) \Rightarrow U_U(A,C).$$

$\mathcal{U}$ denotes the universal disjunction of instance relations of the corresponding set of relations, for example for topological relations the disjunction of all 8 relations shown in figure 1.

These examples show that the composition of the defined class relations is possible. It can be extended with similar derivations to all possible compositions. The two levels of compositions can be separately analysed and therewith the reasoning formalism can be used with any spatial or non-spatial set of instance relations. In general the composition of class relations is not independent of the composition of instance relations. The class level composition is only possible if the corresponding instance relation can be derived.

4.4   Consistency of Class Relation Networks

The application of reasoning algorithms for checking consistency and discovering redundancies in networks of instance relations has for example been demonstrated in [4] and [12]. The proof of consistency of a network of binary relations is a constraint satisfaction problem. In a consistent network of JEPD relations the following three constraints are fulfilled: node consistency, arc consistency and path consistency. The reasoning properties of class relations investigated in the previous subsections provide the basis to check these three consistency requirements in networks of class relations.

Node consistency is ensured if every node has an identity relation. For the class relation networks this means that every class must have a relation to itself. If a corresponding identity instance relation is available the identity class relation is in general $R_{LD.RD.LT.RT}$; for example $EQUAL_{LD.RD.LT.RT}(A,A)$ when using the topological relations of figure 1.

A network of relations is arc consistent if every edge of the network has an edge in the reverse direction, i.e. every relation has an inverse relation. It has been shown in section 4.1 that if an instance relation is symmetric or has an inverse relation there is also an inverse relation for each of the corresponding class relations. For instance relations this is the only requirement to proof the arc consistency. As exemplified in section 3.2 it is possible to define more than one class relation between two classes, even when the applied instance relations are part of the same JEPD set of relations. This is a fundamental difference to the instance relations and has to be considered when checking the arc consistency at the class level. If there are combinations of class relations defined their consistency has to be proven, because not all class relations can be combined (see section 4.2). This also includes the available knowledge about the number of instances. If the number of instances of more than one class is known also the theorems (T1) – (T5) have to be checked for those classes. This shows that the arc consistency of networks of class relations is more complex to prove than for networks of instance relations, but it is possible to exclude inconsistencies.

For the proof of path consistency the compositions of all possible node triples must be checked. Therefore the composition of the class relations investigated in the previous subsection can be used.

5 Stepwise Definition of Semantic Integrity Constraints

The logical properties of class relations investigated in the previous section can be used to check the consistency and to find redundancies in a set of integrity constraints. Thus the corresponding reasoning algorithms should be applied when the semantic integrity constraints are defined to discover inconsistencies as early as possible. Figure 5 shows a possible user interface for the definition of semantic integrity constraints between two classes.

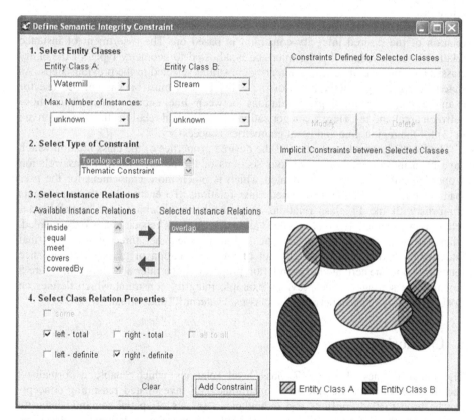

Fig. 5. User interface for the definition of semantic integrity constraints between two classes

Therewith the definition of an integrity constraint based on the defined class relations is compiled in four main steps:

Firstly the user selects the entity classes which shall be restricted by the constraint. After the selection previously defined and implied semantic integrity constraints between these two classes are displayed on the right side of the window. For the definition of spatial integrity constraints the geometry types of the entity classes must be known, because some spatial relations are only valid for certain geometry types. Hence the geometry types should be either known by the system or can be read from available data model or schema information like UML models (Unified Modeling Language) encoded in XMI (XML Metadata Interchange) or GML application schema (Geography Markup Language) documents. If there is no information about the geometry types of the entity classes available or if some entity classes have more than one geometry a corresponding listbox for each entity class should be added to the interface. If the user is aware of the number of existent instances he can enter them after the selection of entity classes.

Secondly, the user selects the type of semantic constraint he wants to define for example topological, metric, directional or other non-spatial constraints like temporal. These types are classified according to the semantic domains of the individual relations.

As a third step the user selects one or more instance relations which the class relation of the desired integrity constraint is based on. The assortment of instance relations made available by the interface is adjusted to geometry types of the entity classes and the type of semantic integrity constraint selected in the previous steps. As stated before the geometry types of the entity classes must be known, because for example the valid topological relations between line entities differ from those between areal entities. Here a categorisation of topological relations like the one given in [3] for the region, line and point geometries is necessary.

The fourth step is the selection of the desired properties of the class relation which have been introduced in the previous sections of the paper. The four class relation properties can be separately activated which is much more convenient for the user than selecting one of the 17 defined class relations. To ensure that the user inputs conform with the 17 class relations only check boxes which lead to valid class relations are enabled. For example the "all to all instances" check box is only enabled, when left-total and right-total are checked while the other three aren't. The final integrity constraint relation results out of the combination of the selected instance relation(s) and the activated class relation properties. The interface shown in figure 5 contains the settings of the previous example integrity constraint which defines an $OVERLAP_{RD,LT}$ relation between the classes "watermill" and "stream".

6 Conclusion

The paper defines abstract 17 class level relations which enable a formalised specification of semantic integrity constraints. The investigated reasoning concepts can be used to find conflicts and redundancies in sets of spatial semantic integrity constraints. The definitions and reasoning rules of the class relations are described independently of a concrete set of instance relations, what makes them applicable for many spatial and non-spatial relations. The only requirements on the instance relations are that they are part of a JEPD set of relations and have defined inverse relations and compositions. Further work will be on the implementation of the introduced approach for checking consistency of sets of integrity constraints to prove the introduced reasoning concepts.

The formalised specification of integrity constraints is improving their management and usability, which will finally result in an improvement of data quality. If an integrity constraints can be composed of other integrity constraints a dataset automatically complies with this constraint if it has been checked against the composing constraints. Hence the exclusion of redundant integrity constraints minimises the number of integrity constraints which have to be verified during a quality check and therewith it is reducing calculation costs.

The investigation of the categories of integrity constraints revealed that there are many different kinds of semantic integrity constraints, but not all of them can be covered by this approach. Nevertheless this framework provides a basis that can be extended by other, possibly more complex types of semantic integrity constraints. As one possible next step semantic constraints on attributes could be included.

The defined class relations are not restricted to applications as integrity constraints. As originally suggested by Donnelly and Bittner [2] they can also be useful for the definition of relations between classes in an ontology. Moreover the reasoning concepts can be used to check the consistency of the relations in such ontology or to discover conflicts in the concepts defined in different ontologies.

The use of the introduced concepts is currently restricted by the unavailability of composition tables for many of the spatial or non-spatial relations. For a broader application at least composition tables for topological relations between entities with simple geometries like points or linestrings must be available. The application of other spatial relations is mostly hampered by the lack of a common understanding of their concepts.

References

1. Ditt, H., Becker, L., Voigtmann, A., Hinrichs, K.H.: Constraints and Triggers in an Object-Oriented Geo Database Kernel. In: 8th International Workshop on Database and Expert Systems Applications (DEXA '97), pp. 508–513 (1997)
2. Donnelly, M., Bittner, T.: Spatial Relations Between Classes of Individuals. In: Cohn, A.G., Mark, D.M. (eds.) COSIT 2005. LNCS, vol. 3693, pp. 182–199. Springer, Heidelberg (2005)
3. Egenhofer, M., Herring, J.: Categorizing Binary Topological Relationships Between Regions, Lines, and Points in Geographic Databases. Technical Report, Department of Surveying Engineering, University of Maine, Orono, ME (1991)
4. Egenhofer, M., Sharma, J.: Assessing the Consistency of Complete and Incomplete Topological Information. Geographical Systems 1(1), 47–68 (1993)
5. Egenhofer, M.: Deriving the Composition of Binary Topological Relations. Journal of Visual Languages and Computing 5(2), 133–149 (1994)
6. Egenhofer, M.J.: Consistency Revisited. GeoInformatica 1, 323–325 (1997)
7. Elmasri, R., Navathe, S.B.: Fundamentals of Database Systems, 2nd edn. (Addison-Wesley), The Benjamin/Cummings Publishing Company Inc. (1994)
8. Freksa, C.: Using Orientation Information for Qualitative Spatial Reasoning. In: Frank, A.U., Formentini, U., Campari, I. (eds.) Theories and Methods of Spatio-Temporal Reasoning in Geographic Space. LNCS, vol. 639, pp. 162–178. Springer, Berlin (1992)

9. Friis-Christensen, A., Tryfona, N., Jensen, C.S.: Requirements and Research Issues in Geographic Data Modeling. In: Proceedings of the 9th ACM international symposium on Advances in geographic information systems, Atlanta, Georgia, USA, pp. 2– 8 (2001)
10. Grigni, M., Papadias, D., Papadimitriou, C.: Topological inference. In: Proceedings of the International Joint Conference of Artificial Intelligence (IJCAI), pp. 901–906 (1995)
11. Hernandez, D.: Qualitative Representation of Spatial Knowledge. In: Hernández, D. (ed.) Qualitative Representation of Spatial Knowledge. LNCS, vol. 804, Springer, Heidelberg (1994)
12. Rodríguez, A., Van de Weghe, N., De Maeyer, P.: Simplifying Sets of Events by Selecting Temporal Relations. In: Egenhofer, M.J., Freksa, C., Miller, H.J. (eds.) GIScience 2004. LNCS, vol. 3234, pp. 269–284. Springer, Heidelberg (2004)

Spatial Reasoning with a Hole

Max J. Egenhofer and Maria Vasardani

National Center for Geographic Information and Analysis
and
Department of Spatial Information Science and Engineering
University of Maine, Boardman Hall, Orono, ME 04469-5711, USA
{max,mvasardani}@spatial.maine.edu

Abstract. Cavities in spatial phenomena require geometric representations of regions with holes. Existing models for reasoning over topological relations either exclude such specialized regions (9-intersection) or treat them indistinguishably from regions without holes (RCC-8). This paper highlights that inferences over a region with a hole need to be made separately from, and in addition to, the inferences over regions without holes. First the set of 23 topological relations between a region and a region with a hole is derived systematically. Then these relations' compositions over the region with the hole are calculated so that the inferences can be compared with the compositions of the topological relations over regions without holes. For 266 out of the 529 compositions the results over the region with the hole were more detailed than the corresponding results over regions without holes, with 95 of these refined cases providing even a unique result. In 27 cases, this refinement up to uniqueness compares with a completely undetermined inference for the relations over regions without holes.

1 Introduction

Some spatial phenomena have cavities (Figure 1a-d), which require geometric representations of regions with holes when they are modeled in geographic information systems. Most prominent geographic examples are the territorial configurations of South Africa (which completely surrounds Lesotho), the former East Germany (completely surrounding West Berlin), and Italy (which completely surrounds San Marino and the Vatican City). Many other spatial configurations with holes have been thought of [1]. This paper focuses on the topological relations involving regions with a single hole.

Although such regions with holes resemble visually regions with indeterminate (i.e., broad) boundaries (Figure 2a), their topologies differ conceptually, since a region with a broad boundary is an open set (Figure 2b), whereas a region with a hole is a closed set (Figure 2c). Therefore, the various versions of topological relations between regions with broad boundaries [2,3,13] do not apply immediately to regions with holes.

While geometric models of spatial features have matured to capture appropriately the semantics of regions with holes [11,17], models of qualitative spatial relations over regions with holes have essentially stayed in their infancy. Some models of

S. Winter et al. (Eds.): COSIT 2007, LNCS 4736, pp. 303–320, 2007.
© Springer-Verlag Berlin Heidelberg 2007

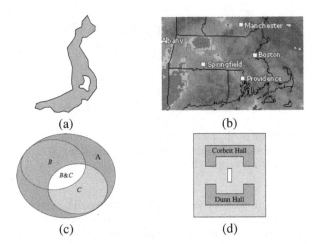

(a)   (b)

(c)   (d)

Fig. 1. Regions with a hole: (a) the Lago Iseo (with Monte Isola, Italy's largest island in a lake), (b) the area of Massachusetts with precipitation on December 23, 2006 at 7:30am, (c) the part of region A that cannot be reached by both persons B and C when they were to travel from their current locations for a set amount of time, and (d) the smoke-free zone around Corbett Hall and Dunn Hall on the UMaine campus implied by a 30ft buffer zone around each building

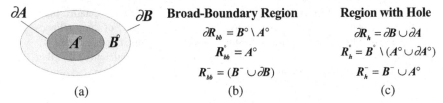

∂A ∂B	**Broad-Boundary Region**	**Region with Hole**
A° B°	$\partial R_{bb} = B^\circ \setminus A^\circ$	$\partial R_h = \partial B \cup \partial A$
	$R_{bb}^\circ = A^\circ$	$R_h^\circ = B^\circ \setminus (A^\circ \cup \partial A^\circ)$
	$R_{bb}^- = (B^- \cup \partial B)$	$R_h^- = B^- \cup A^\circ$
(a)	(b)	(c)

Fig. 2. Two closed discs (A and B): (a) their topological components boundary (∂R), interior (R°), and exterior (R^-) that contribute to the formation of (b) a region with a broad boundary (R_{bb}) and (c) a region with a hole (R_h)

topological relations have addressed regions with holes [8,14,20], distinguishing varying levels of details about the placements of the holes, however, qualitative spatial reasoning with such relations has been either discarded or treated like reasoning without holes. The 4-intersection [9] and 9-intersection [10], for instance, exclude explicitly as the relations' domain and co-domain any regions with holes, so that the comprehensive body of inferences over topological relations based on the 9-intersection does not apply directly to regions with holes. On the other hand, the region-connection calculus (RCC) [19] makes no explicit distinction between regions with or without holes so that this model of topological relations, and their inferences from compositions, applies to regions with holes as well. While the 9-intersection composition table is extentional [15], RCC's applicability to regions with or without holes has given rise to a non-extentional composition table. These differences in the composition of topological relations are the key justification for the need to differentiate relations for regions with holes from relations for regions without holes when making inferences over topological relations.

The following example highlights the need for such an explicit distinction. Given a region B such that it *overlaps* with a region A and also *overlaps* with a region C. From the composition A *overlaps* B and B *overlaps* C [6] one can deduce the possible relations between A and C (Figure 3a-h), yielding in this case the universal relation U_8 (i.e., all eight topological relations are possible).

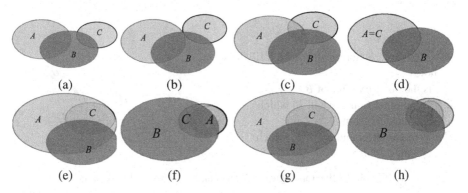

Fig. 3. The eight possible configurations if region A *overlaps* region B and B *overlaps* region C: (a) A *disjoint* C, (b) A *meet* C, (c) A *overlaps* C, (d) A *equal* C, (e) A *covers* C, (f) A *coveredBy* C, (g) A *contains* C, and (h) A *inside* C

If one starts, however, with a region with a hole (E) that overlaps with two other regions, D and F, each without holes, the hole may play a significant role in constraining the possible relations between D and F. For instance, let E *overlap* with D such that E's hole is completely contained in D, and let E overlap with F such that F meets E's hole. Then D could *overlap* with F (Figure 4a), D could *cover* F (Figure 4b), and D could *contain* F (Figure 4c). Therefore, the insertion of a hole into the first region results in a composition scenario that is more constrained than the one without the hole. Treating both cases with the same (less constrained) composition would offer some incorrect choices in case the region has a hole.

Is this example an anomaly? Or maybe even the only case in which relation inferences differ for regions and regions with a hole? Or are there so many more cases that typically the reasoning over a region with a hole differs from the well known topological inferences of regions without holes? This paper provides answers to these questions through a systematic study of the topological relations involving a region with a hole, the derivation of the composition inferences of these relations, and a quantitative comparison of these compositions with the compositions of topological relations between regions without holes [6]. The topological relation between a region and a region with a hole is denoted by t_{RR_h} (and its converse relation by t_{R_hR}), while t_{RR} refers to the topological relation between two regions (each without a hole).

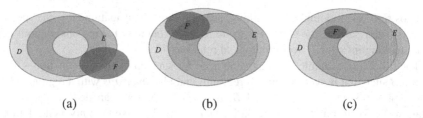

Fig. 4. The three possible configurations if *E overlaps* with *D* such that *E*'s hole is *inside D*, and *E* also overlaps with *F* such that *E*'s hole *meets F*: (a) *D overlaps F*, (b) *D covers F*, and (c) *D contains F*

$B°$ is B's interior
B^{-1} is the inner exterior of B, which fills B's
   hole
B^{-0} is the outer exterior of B
$\partial_i B$ is the inner boundary of B, which
   separates $B°$ from B^{-1}
$\partial_o B$ is the outer boundary of B, which
   separates $B°$ from B^{-0}

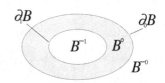

Fig. 5. B's five topologically distinct and mutually exclusive parts

Throughout this paper, the qualitative model of a region with a hole (B) is based on B's five topologically distinct and mutually exclusive parts (Figure 5).

The elements of the qualitative description of a region with a hole are (1) its hole B_H ($B^{-1} \cup \partial_i B$) and (2) the generalized region B^* ($B_H \cup B° \cup \partial_o B$). B^* and B_H are spatial regions, that is, each region is homeomorphic to a 2-disk so that the eight t_{RR} [9] apply to B^* and B_H (but not to B, because B with the hole is not homeomorphic to a 2-disk). The topological relation between B^* and B_H is *contains*, therefore, this is a more restrictive model than the generic region-with-holes model [8], where B_H also could have been *coveredBy* or even *equal* to B^*, thereby leading to somewhat different semantics of a region with a hole.

The remainder of this paper is organized as follows: Section 2 specifies the canonical model used for modeling a region with a hole as well as such a region's topological relation with another region. Section 3 presents a method to derive the t_{RR_h} that are feasible between a region and a region with a hole. Section 4 presents the 23 relations that can be found between a region and a region with a hole, followed by an analysis of these relations' algebraic properties in Section 5. Section 6 derives the qualitative inferences that can be made with t_{RR_h} and $t_{R_h R}$, focusing on compositions over a common region with a hole. Section 7 analyzes these compositions, comparing their reasoning power with the compositions of topological relations between regions without holes. The paper closes with conclusions and a discussion of future work in Section 8.

2 Qualitative Model of a Region with a Hole

The topological relation between a region (A) and a region with a hole (B) is modeled as a *spatial scene* [5], comprising A, B^*, and B_H together with nine binary topological relations among these three regions (Figure 6).

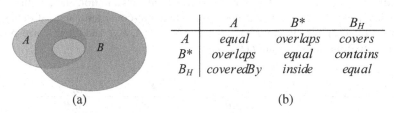

	A	B^*	B_H
A	*equal*	*overlaps*	*covers*
B^*	*overlaps*	*equal*	*contains*
B_H	*coveredBy*	*inside*	*equal*

(a)   (b)

Fig. 6. Topological relation of a region with a hole: (a) graphical depiction of a configuration and (b) the corresponding symbolic description as a spatial scene

In such a spatial scene, five of the nine binary topological relations are implied for any configuration between a region and a region with a hole: each region is *equal* to itself, B^* *contains* B_H, conversely B_H is *inside* B^*, and for the two relations between A and B^* and A and B_H, their converse relations (from B^* to A and from B_H to A) are implied by the arc consistency constraint [16]; therefore, a model of such a spatial scene only requires the explicit specification of the two relations between A and B^* and A and B_H to denote t_{RR_h} (Eqn. 1). These relations between A and B^* and A and B_H are called the *constituent relations* of a topological relation between a region and a region with a hole. Their horizontal 1x2 matrix is a direct projection of the top elements in the two right-most columns of the spatial scene description.

$$t_{RR_h} (A, B) = \left[t(A,B^*) \quad t(A,B_H) \right] \tag{1}$$

The principal relation $\pi(t_{RR_h})$ is then the first element of t_{RR_h} (Eqn. 2).

$$\pi(t_{RR_h} (A,B)) = r(A,B^*) \tag{2}$$

Section 3.3 shows that some configurations actually only require the principal relation in order to specify t_{RR_h} completely.

3 Deriving the Topological Relations Between a Region and a Region with a Hole

The spatial scene can also be used for the derivation of what topological relations actually exist between a region and a region with a hole. Since two of the scene's nine topological relations are subject to variations (the relations between A and B^* and A and B_H), a total of $8^2=64$ t_{RR_h} could be specified. But only a subset of these 64 relations is feasible. For example, t_{RR_h} [*contains   disjoint*] is infeasible, because B^*

cannot be *inside* A at the same time as B_H (which is inside B^*) is *disjoint* from A. Therefore, a topological relation from a region with a hole to another region is feasible if (1) that scene's representation is consistent and (2) there exists a corresponding graphical depiction.

The binary topological relation between a region (A) and a region with a hole (B) is established as a 3-region scene comprising A, B^*, and B_H with the constraint that B^* *contains* B_H (Figure 7). The topological relation between a region and a region with a hole holds if this 3-region scene is node-consistent, arc-consistent, and path-consistent (Macworth 1977) for the four values $t(A, B^*)$, $t(A, B_H)$ and their corresponding converse relations $t(B^*, A)$ and $t(B_H, A)$.

	A	B^*	B_H
A	*equal*	$t(A, B^*)$	$t(A, B_H)$
B^*	$t(B^*, A)$	*equal*	*contains*
B_H	$t(B_H, A)$	*inside*	*equal*

Fig. 7. A 3-region spatial scene that captures the constituent relations of a binary topological relation between a region (A) and a region with a hole (B)

The range of these four relations is the set of the eight t_{RR}. With four variables over this domain, a total of $8^4 = 4,096$ configurations could be described for the topological relations between a region and a region with a hole. Only a subset of them is feasible, however. These feasible configurations are those whose 3-region scenes are consistent. Since in the feasible configurations $t(A, B^*)$ must be equal to the converse of $t(B^*, A)$, the enumeration of the relations in the feasible configuration can be reduced. The same converseness constraint also holds for $t(A, B_H)$ and $t(B_H, A)$; therefore, for a feasible t_{RR_h} two of the four relations are implied. Thus, only two of the four unknown relations are necessary to completely describe a feasible t_{RR_h}, reducing the number of possible configurations to $8^2 = 64$.

4 Twenty-Three Relations Between a Region and a Region with a Hole

In order to determine systematically the feasible t_{RR_h}, a scene consistency checker has been implemented, which iterates for each unknown (i.e., universal) relation over the eight possible relations and determines whether that spatial scene is node-consistent, arc-consistent, and path-consistent [16]. Only those configurations that fulfill all three consistency constraints are candidates for a valid t_{RR_h}. Twenty-three spatial scenes representing a region and a region with a hole have been found to be consistent (Figure 8).

The remaining 64–23=41 candidate configurations for t_{RR_h} have been found to be inconsistent. Therefore, the 23 consistent cases establish the 23 binary topological relations between a region and a region with a hole.

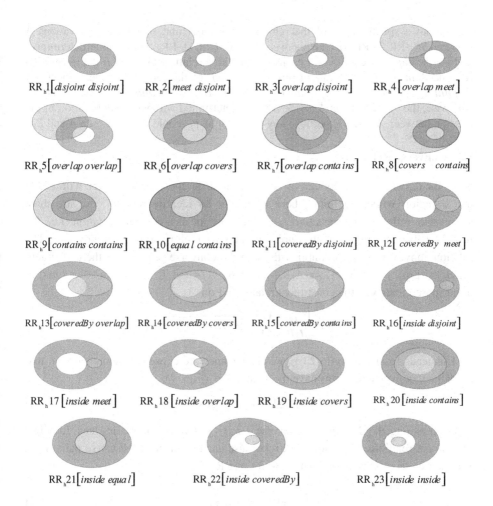

Fig. 8. Graphical depictions of the 23 topological relations between a region and a region with a hole

5   Properties of the Twenty-Three Relations

These 23 t_{RR_h} can be viewed as refinements of the eight t_{RR}. Five of the eight t_{RR}—*disjoint, meet, covers, contains, equal*—do not reveal further details if region B has a hole, because in each of these cases the relation between A and B^* is so strongly constrained that only a single relation is possible between A and B's hole B_H. The remaining three t_{RR}—*overlap, coveredBy, inside*—are less constraining as each offers multiple variations for the topological relations between A and B_H: *overlap* and *coveredBy* each have five variations for $t(A, B_H)$, while *inside* has a total of eight variations.

Without a specification of the relation between A and B_H, the configurations A *overlap* B^* (RR_h5– RR_h8) and A *coveredBy* B^* (RR_h11–RR_h15) are underdetermined, that is, one can only exclude for each case the three relations A *equal* B_H, A *coveredBy* B_H, and A *inside* B_H, but cannot pin down which of the remaining five choices—A *disjoint* B_H, A *meet* B_H, A *overlap* B_H, A *covers* B_H, A *contains* B_H—actually holds. Likewise, the configuration A *inside* B^* is undetermined without a specification of the relation between A and B_H, because any of the eight t_{RR} could hold between A and B_H.

5.1 Converse Relations

Since the domain and co-domain of t_{RR_h} refer to different types—a region with a hole and a region without a hole—there is neither an identity relation, nor are there symmetric, reflexive, or transitive t_{RR_h}. The concept of a converse relation (i.e., the relation between a region with a hole and another region) still exists, however. The relation converse to t_{RR_h} is implied through the converse property of the constituent relations (Eqn. 1)—$t(B^*,A)=\overline{t(A,B^*)}$ and $t(B_H,A)=\overline{t(A,B_H)}$—which is captured in a transposed matrix of the constituent relations (Eqn. 3).

$$t_{R_hR}(B,A)=\begin{bmatrix} t(B^*,A) \\ t(B_H,A) \end{bmatrix}=\begin{bmatrix} \overline{t(A,B^*)} & \overline{t(A,B_H)} \end{bmatrix}^T \tag{3}$$

This leads immediately to 23 t_{R_hR}. Their names are chosen systematically so that all pairs of converse relations have the same index (Eqn. 4).

$$\forall x:1...23: \quad R_hRx = \overline{RR_hx} \tag{4}$$

From among the 23 pairs of converse t_{RR_h} t_{R_hR}, five relation pairs have identical constituent relations (Equations 5a-e), because each element of these five pairs has a symmetric converse relation, that is, $R_hRx = (RR_hx)^T$.

$$\begin{bmatrix} disjoint \\ disjoint \end{bmatrix}^T = [disjoint \; disjoint] \tag{5a}$$

$$\begin{bmatrix} meet \\ disjoint \end{bmatrix}^T = [meet \; disjoint] \tag{5b}$$

$$\begin{bmatrix} overlap \\ disjoint \end{bmatrix}^T = [overlap \; disjoint] \tag{5c}$$

$$\begin{bmatrix} overlap \\ meet \end{bmatrix}^T = [overlap \; meet] \tag{5d}$$

$$\begin{bmatrix} overlap \\ overlap \end{bmatrix}^T = [overlap \; overlap] \tag{5e}$$

5.2 Implied Relations

The dependencies among a region's relations to the generalized region and the hole reveal various levels of constrains (Figure 9). While five $t(A, B^*)$ imply a unique relation for $t(A, B_H)$, two other $t(A, B^*)$ restrict $t(A, B_H)$ to five choices. Only one $t(A, B^*)$—inside—yields the universal relation U_8, imposing no constraints on $t(A, B_H)$.

Known Relation $t(A, B^*)$	Implied Relation $t(A, B_H)$
disjoint	disjoint
meet	disjoint
covers	contains
contains	contains
equal	contains
overlap	not {equal, coveredBy, inside}
coveredBy	not {equal, coveredBy, inside}
inside	U

Fig. 9. Constraints imposed by a specified $t(A, B^*)$ on $t(A, B_H)$

Reversely, knowledge of the relation $t(A, B_H)$ implies in three cases—if $t(A, B_H)$ is equal, coveredBy, or inside—a unique relation between A and B^*; has three choices for three relations between A and B^* (if $t(A, B_H)$ is meet, overlap, or covers); five choices in one case (if $t(A, B_H)$ is disjoint); and six choices if $t(A, B_H)$ is contains (Figure 10).

Known Relation $t(A, B_H)$	Implied Relation $t(A, B^*)$
equal	inside
coveredBy	inside
inside	inside
disjoint	not {equal, covers, contains}
meet	{overlap, coveredBy, inside}
overlap	{overlap, coveredBy, inside}
covers	{overlap, coveredBy, inside}
contains	not {disjoint, meet}

Fig. 10. Constraints imposed by a specified $t(A, B_H)$ on $t(A, B^*)$

The dependencies may be seen as an opportunity for minimizing the number of relations that are recorded. For example, if one of the two implications were such that all known relations implied a unique relation, then it would be sufficient to record only the known relation, thereby cutting into half the amount of relations to be stored for each t_{RR_h}. Such a simple choice does not apply, however. Since five $t(A, B_H)$ are implied uniquely by $t(A, B^*)$, $t(A, B_H)$ needs to be recorded only in three cases to fix a complete t_{RR_h} specification. Reversely only three $t(A, B^*)$ are implied uniquely by their $t(A, B_H)$. Therefore, the common-sense choice of favoring the relation with respect to the generalized region over the relation to the hole gets further support.

6 Compositions over a Region with a Hole

A key inference mechanism for relations is their composition, that is, the derivation of the relation A to C from the knowledge of the two *relations* $t(A, B)$ and $t(B, C)$. A complete account of all relevant compositions considers first all combinatorial compositions of relations with regions (R) and regions with a hole (R_H). Since all compositions involve two binary relations (i.e., $\_\,\_$; $\_\,\_$), each over a pair of R and R_H, there are $2^4 = 16$ possible combinations (Figure 11). Eight of these sixteen combinations specify invalid compositions (C 3–6 and C 11–14), because the domain and co-domain of the composing relations' common argument are of different types (i.e., trying to form a composition over a Region and a Region with a hole). Among the remaining eight combinations, C 1 is the well-known composition of region-region relations. Two pairs of combinations capture converse compositions—C 2 and C 9, as well as C 8 and C 15—while three combinations capture symmetric compositions—C 7, C 10, and C 16.

C 1	$t_{RR}; t_{RR}$	C 5	—	C 9	$t_{R_hR}; t_{RR}$	C 13	—
C 2	$t_{RR}; t_{RR_h}$	C 6	—	C 10	$t_{R_hR}; t_{RR_h}$	C 14	—
C 3	—	C 7	$t_{RR_h}; t_{R_hR}$	C 11	—	C 15	$t_{R_hR_h}; t_{R_hR}$
C 4	—	C 8	$t_{RR_h}; t_{R_hR_h}$	C 12	—	C 16	$t_{R_hR_h}; t_{R_hR_h}$

Fig. 11. The 16 combinations of compositions of binary relations with regions (R) and regions with a hole (R_H)

From among these combinations of compositions involving a region with a hole, we focus here on Comp 7, the inferences from t_{RR_h} ; t_{R_hR}. A spatial scene serves again as the framework for a computational derivation of all compositions. Objects A and C are two regions without a hole, whereas object B is a region with a hole. The corresponding spatial scene has four regions (A, B^*, B_H, and C) with their sixteen region-region relations (Figure 12). The pair of relations $t(A, B^*)$ $t(A, B_H)$ must be a subset of the 23 valid t_{RR_h}, while the pair of relations $t(B^*, C)$ $t(B_H, C)$ must be a subset of the 23 valid t_{R_hR}. Furthermore, $t(B^*, A)$ and $t(B_H, A)$ must be the respective converse relations of $t(A, B^*)$ and $t(A, B_H)$. The same converse property must hold for the pair $t(C, B^*)$ $t(C, B_H)$ with respect to $t(B^*, C)$ $t(B_H, C)$. With 23 pairs for each t_{RR_h} and t_{R_hR}, there are 529 compositions. The range of the inferred relation $t(A, C)$ is the set of the eight t_{RR}. This composition of $t(A, B)$; $t(B, C)$ is specified for any spatial scene that is node-consistent, arc-consistent, and path-consistent. To determine systematically all consistent compositions, we have developed a software prototype of a consistency checker that evaluates a spatial scene

	A	B^*	B_H	C
A	equal	U	U	U
B^*	U	equal	contains	U
B_H	U	inside	equal	U
C	U	U	U	equal

Fig. 12. The spatial scene over four regions used for the derivation of the composition $t\,(A, B)$; $t\,(B, C)$

for the three consistencies. All compositions where found to be valid (i.e., none of the compositions resulted in the empty relation).

Figures 14a and 14b summarize the result graphically, using for the composition the iconic representation of the region-region relations based on their conceptual neighborhood graph [12]. A highlighted relation in that graph indicates that that relation is part of the particular composition. The universal relation U_8 is then an icon with all relations highlighted (Figure 13a), while a unique inference has a single relation highlighted (Figure 13b). The composing relations t_{RR_h} and t_{R_hR} are also captured by the same neighborhood graph, in which the relation to the generalized region is superimposed over the relation to the hole (Figure 13c).

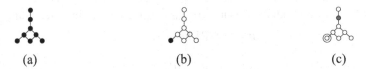

(a)	(b)	(c)

Fig. 13. Iconic representation of relations and compositions: (a) universal relation of region-region relations, (b) unique composition result (*inside*) of region-region relations, and (c) unique t_{RR_h} with the large circle identifying the relation between region A and the generalized region B^* and the black disc highlighting the relation between A and B_H

7 Analysis of Compositions

The 64 compositions of t_{RR} ; t_{RR} [6] form the benchmark for the assessment of the reasoning power of compositions involving regions with holes.

Finding 1: The composition table t_{RR_h} ; t_{R_hR} (Figure 14a and 14b) shows that all 529 compositions are valid (i.e., there is no empty relation as the result of any of the compositions). This means none of the 529 4-object scenes considered to calculate the compositions (Figure 12) is inconsistent. The same level of consistency was also found for the t_{RR} ; t_{RR} composition table.

Finding 2: All compositions are compatible with the composition results of their principal relations (Eqn. 6), that is, the inferences from the principal relations provide an upper bound for the reasoning over regions with a hole.

$$\forall a:1\ldots23, \forall b:1\ldots23: \quad RR_ha ; R_hRb \subseteq \pi(RR_ha) ; \pi(R_hRb) \qquad (6)$$

Finding 3: Among the 529 compositions there are 263 (49.7%) whose results are identical to the compositions of the relations' principal relations (Eqn. 7). Therefore, for slightly less than half of the inferences the hole is immaterial, while it matters for the remaining 266 inferences.

$$\exists a,b \mid a \neq b: \quad RR_h a ; R_h Rb = \pi(RR_h a); \pi(R_h Rb) \tag{7}$$

Finding 4: Among the 266 compositions whose results are more refined than the compositions of their principal relations, 95 compositions are refined to uniqueness (Eqn. 8). If one were to resort in these cases to the compositions of their principal relations, one would incorrectly infer that these compositions are underdetermined.

$$\exists a,b \mid a \neq b: RR_h a ; R_h Rb \subset \pi(RR_h a); \pi(R_h Rb) \wedge \#(RR_h a ; R_h Rb) = 1 \tag{8}$$

To further assess the inference power of the compositions, we use the composition's *cardinality* (Eqn. 9a), which is the count of relations in that composition result, and the *composition table's cardinality* (Eqn. 9b), which is the sum of the cardinalities of all compositions in a table. This yields the *composition table's normalized crispness* (Eqn. 9c), whose lowest value of 0 stands for compositions that result in the universal relation and whose value increases linearly for composition results with fewer choices. The latter measure also applies to subsets of a composition table to assess and compare the inferences of particular groups of relations. The corresponding measures for t_{RR} ; t_{RR} can be defined accordingly.

$$card_{23}^{ij} = \#(RRhi; RhRj) \tag{9a}$$

$$\gamma_{23} = \sum_{i=1..\#(U_{23})}^{j=1..\#(U_{23})} card_{23}^{ij} \tag{9b}$$

$$\overline{\Gamma}_{23} = 1 - \frac{\gamma_{23}}{\#(U_8)*\#(U_{23})*\#(U_{23})} \tag{9c}$$

Finding 5: While the cardinality of the t_{RR_h} ; $t_{R_h R}$ composition table is over seven times higher than that of the t_{RR} ; t_{RR} composition table (γ_{23} =1389 vs. γ_8 =193), the overall inferences from t_{RR_h} ; $t_{R_h R}$ are crisper, because the average composition cardinality is approximately 8% higher for all t_{RR_h} ; $t_{R_h R}$ than for all t_{RR}; t_{RR} ($\overline{\Gamma}_{23}$ =0.67 vs. $\overline{\Gamma}_8$ =0.62).

Finding 6: The increase in crispness is primarily due to a decrease in the relative number of compositions with a cardinality of 5 (and to a lesser degree cardinalities 6 and 8), while simultaneously the relative numbers of compositions with cardinalities 3, 2, and 4 (and to a miniscule amount those of compositions with cardinality 1) increase (Figure 15). Overall 239 ambiguities of pure topological reasoning are reduced, but not fully eliminated, when considering the holes in the regions.

Finding 7: In absolute numbers the count of compositions with unique results goes up from 27 in t_{RR} ; t_{RR} to 224 in t_{RR_h} ; $t_{R_h R}$. Since—for a different set of relations, though—people have been found to make composition inferences more correctly if the result is unique [19], this increase augurs well for people's performance when reasoning over relations with holes.

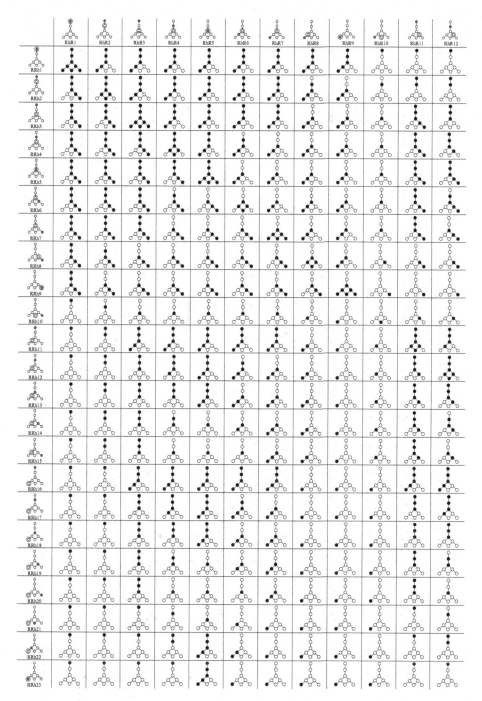

Fig. 14a. Composition table t_{RR_h} ; t_{R_hR} (for $t_{R_hR} = [R_hR1...R_hR12]$)

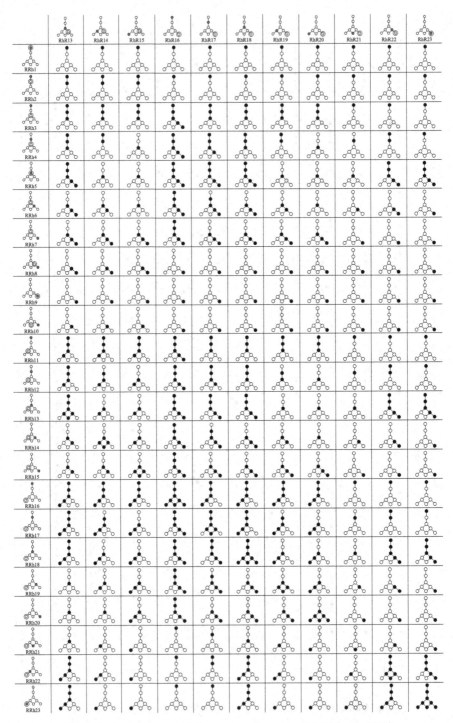

Fig. 14b. Composition table t_{RR_h} ; t_{R_hR} (for $t_{R_hR} = [R_hR13...R_hR23]$)

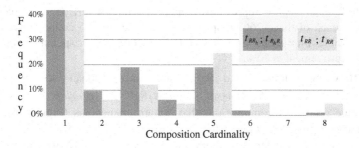

Fig. 15. Comparison of the frequencies of compositions results with cardinality 1 (unique inference) through 8 (universal relation) for composition table t_{RR_h} ; t_{R_hR} with t_{RR} ; t_{RR}

	RhR1	RhR2	RhR3	RhR4	RhR5	RhR6	RhR7	RhR8	RhR9	RhR10	RhR11	RhR12	RhR13	RhR14	RhR15	RhR16	RhR17	RhR18	RhR19
	d/d	m/d	o/d	o/m	o/o	o/cB	o/i	cB/i	i/i	e/i	cv/d	cv/m	cv/o	cv/cB	cv/i	ct/d	ct/m	ct/o	ct/cB
RRh1 [disjoint disjoint]																			
RRh2 [meet disjoint]																			
RRh3 [overlap disjoint]				3	3	3	3					2	2	2	2		2	2	2
RRh4 [overlap meet]			3	2	3	4	5					1	2	3	4		1	2	3
RRh5 [overlap overlap]			3	3		5	5							4	4				4
RRh6 [overlap covers]			3	4	5	4	5					1	2	3	4		1	2	3
RRh7 [overlap contains]			3	5	5	5	2					2	2	2	2		2	2	2
RRh8 [covers contains]																			
RRh9 [contains contains]																			
RRh10 [equal contains]																			
RRh11 [coveredBy disjoint]												2	2	2	2		2	2	2
RRh12 [coveredBy meet]			2	1		1	2				2		2	3	4		1	2	3
RRh13 [coveredBy overlap]			2	2		2	2				2	2		4	4				4
RRh14 [coveredBy covers]			2	3	4	3	2				2	3	4	2	4		1	2	3
RRh15 [coveredBy contains]			2	4	4	4	2				2	4	4	4	2		2	2	2
RRh16 [inside disjoint]																	3	3	3
RRh17 [inside meet]			2			1	2				2	1		1	2	3	2	3	4
RRh18 [inside overlap]			2	2		2	2				2	2		2	3	3			5
RRh19 [inside covers]			2	3	4	3	2				2	3	4	3	2	3	4	5	4
RRh20 [inside contains]			2	4	4	4	2				2	4	4	4	2	3	5	5	5
RRh21 [inside equal]			4	4	4	4	4				4	4	4	4	4	7	7	7	7
RRh22 [inside coveredBy]			4	3		3	4				4	3		3	4	7	6	3	6
RRh23 [inside inide]			4	4		4	4				4	4		4	4	7	7	3	7

Fig. 16. Crispness improvements (in absolute counts) for t_{RR_h} ; t_{R_hR} vs. t_{RR} ; t_{RR} (compositions without improvement left out; darker shading indicates stronger improvement).

Finding 8: From among the 266 compositions with crisper results, 27 (i.e., 10.2%) yield a *complete crispening*, that is a conversion from a universal composition to a unique composition. Complete crispenings occur only for compositions $RR_h a$; $R_h Rb$ with $\pi(RR_h a) = inside$ and $\pi(R_h Rb) = contains$ (Figure 16). Resorting in these cases to the composition of their principal relations would incorrectly imply that these inferences are undetermined.

Finding 9: For all 266 compositions whose results are crisper, on average the crispness of each of these 266 compositions improves by 3.5 counts. Given that the highest possible improvement is seven (for a complete crispening), the average crispness improvement is 50%.

Finding 10: Compositions $\mathrm{RR}_h a$; $\mathrm{R}_h \mathrm{R} b$ are only subject to crispening if $\pi(\mathrm{RR}_h a) \in \{overlap, coveredBy, inside\}$ and $\pi(\mathrm{R}_h \mathrm{R} b) \in \{overlap, covers, contains\}$, yielding nine groups of compositions that feature crispenings (Figure 16). In these groups, the compositions with $\pi(\mathrm{RR}_h a) = inside$ and $\pi(\mathrm{R}_h \mathrm{R} b) = contains$ have the highest crispness improvements, both in absolute counts (319) as well as per composition (5.23, which corresponds to an average crispness improvement of 75%).

8 Conclusions

Most qualitative spatial reasoning has disregarded the inference constraints that cavities of geographic phenomena may impose, because their underlying models either explicitly exclude regions with holes from their domain or assume that the existence of a hole will have no impact on their topological inferences. To overcome these limitations, this paper studied systematically the topological relations of regions with a single hole, offering new insights for spatial reasoning over such regions:

While the 9-intersection captures eight topological relations between two regions, this number increases by 88% to 23 when one of the regions has a hole, yielding refinements of the eight region-region relations. Knowing the relation between a region and the generalized region implies a 63% chance (5 out of 8 relations) of uniquely identifying the complete relation between the two objects without any explicit reference to the relation with the hole.

The 23 relations' compositions over a common region with a hole show that these compositions form subsets—although not necessarily true subsets—of the results obtained from the compositionsBy of regions without a hole. In 36% of the true subsets, the result is unique (i.e., a single relation). Approximately half of the compositions over a region with a hole yield fewer possible relations, with an 8% increase in the average crispness when compared to the results of compositions over a region without a hole. This decrease is due to a general trend of fewer results comprising five or more possibilities, in combination with an increase of the occurrence of results of fewer possibilities (four or less) and by a 10% increase of *complete crispness* (yielding a unique relation) among these improved results. This leads to an average crispness improvement of 50% for those results. These insights relate to people's reasoning performance, because relations that include regions with holes lead to a higher relative number of unique possible results.

These findings provide answers to the questions posed in the motivation: the more constrained composition inferences found for topological relations of a region with a hole are neither anomalies, nor do different inferences occur only in a single case. Since over 50% of the inferences with a hole are more refined than the corresponding inferences over regions without a hole, typically the reasoning over a region with a hole does differ from the well known topological inferences of regions without a hole,

Future work will pursue the derivation of complementary methods for similarity reasoning, such as the 23 relations' conceptual neighborhoods. Initial results indicate that this graph is an asymmetric extension of the graph for the eight region relations. We further intend to pursue the modeling of and inferences from binary topological relations between two regions, each with a hole. Finally, an interesting question for a larger theory of consistent qualitative reasoning across space and time is whether there are analog results to relations over regions with holes in the temporal domain, namely for intervals with gaps.

Acknowledgments. This work for partially supported by the National Geospatial-Intelligence Agency under grant number NMA201-01-1-2003 and a University of Maine Provost Graduate Fellowship.

References

1. Cassati, R., Varzi, A.: Holes and Other Superficialities. The MIT Press, Cambridge (1994)
2. Cohn, A., Gotts, N.: The 'Egg-Yolk' Representation of Regions with Indeterminate Boundaries. In: Burrough, P., Frank, A. (eds.) Geographic Objects with Indeterminate Boundaries, pp. 171–187. Taylor & Francis, Bristol, PA (1996)
3. Clementini, E., Di Felice, P.: An Algebraic Model for Spatial Objects with Indeterminate Boundaries. In: Burrough, P., Frank, A. (eds.) Geographic Objects with Indeterminate Boundaries, pp. 155–170. Taylor & Francis, Bristol, PA (1996)
4. Egenhofer, M.: Spherical Topological Relations. Journal of Data Semantics III, 25–49 (2005)
5. Egenhofer, M.: Query Processing in Spatial-Query-by-Sketch. Journal of Visual Languages and Computing 8(4), 403–424 (1997)
6. Egenhofer, M.: Deriving the Composition of Binary Topological Relations. Journal of Visual Languages and Computing 5(1), 133–149 (1994)
7. Egenhofer, M., Al-Taha, K.: Reasoning about Gradual Changes of Topological Relationships. In: Frank, A.U., Formentini, U., Campari, I. (eds.) Theories and Methods of Spatio-Temporal Reasoning in Geographic Space. LNCS, vol. 639, pp. 196–219. Springer, New York (1992)
8. Egenhofer, M., Clementini, E., Di Felice, P.: Topological Relations Between Regions With Holes. International Journal of Geographical Information Systems 8(2), 129–144 (1994)
9. Egenhofer, M., Franzosa, R.: Point-Set Topological Spatial Relations. International Journal of Geographical Information Systems 5(2), 161–174 (1991)
10. Egenhofer, M., Herring, J.: Categorizing Binary Topological Relations Between Regions, Lines, and Points in Geographic Databases. Technical Report, Department of Surveying Engineering, University of Maine (1994)
11. Frank, A., Kuhn, W.: Cell Graph: A Provable Correct Method for the Storage of Geometry. In: Marble, D. (ed.) Second International Symposium on Spatial Data Handling, Seattle, WA, pp. 411–436 (1986)
12. Freksa, C.: Temporal Reasoning Based on Semi-Intervals. Artificial Intelligence 54(1), 199–227 (1992)
13. Liu, K., Shi, W.: Computing the Fuzzy Topological relations of Spatial Objects Based on Induced Fuzzy Topology. International Journal of Geographical Information Science 20(8), 857–883 (2006)
14. Li, S.: A Complete Classification of Topological Relations Using the 9-Intersection Method. International Journal of Geographical Information Science 20(6), 589–610 (2006)

15. Li, S., Ying, M.: Extensionality of the RCC8 Composition Table. Fundamenta Informticae 55(3-4), 363–385 (2003)
16. Mackworth, A.: Consistency in Networks of Relations. Artificial Intelligence 8(1), 99–118 (1977)
17. OGC: OpenGIS Simple Features Specification for SQL. OpenGIS Project Document 99-049 (1999)
18. Randell, D., Cohn, A., Cui, Z.: Computing Transitivity Tables: A Challenge for Automated Theorem Provers. In: Kapur, D. (ed.) Automated Deduction - CADE-11. LNCS, vol. 607, pp. 165–176. Springer, Heidelberg (1992)
19. Rodríguez, M.A., Egenhofer, M.: A Comparison of Inferences about Containers and Surfaces in Small-Scale and Large-Scale Spaces. Journal of Visual Languages and Computing 11(6), 639–662 (2000)
20. Schneider, M., Behr, T.: Topological Relationships between Complex Spatial Objects. ACM Transactions on Database Systems 31(1), 39–81 (2006)

Geospatial Cluster Tessellation Through the Complete Order-k Voronoi Diagrams

Ickjai Lee[1], Reece Pershouse[1], and Kyungmi Lee[2]

[1] School of Math, Physics & IT,
James Cook University, Douglas Campus,
Townsville, QLD4811, Australia
[2] School of Business and Information Technology,
Charles Sturt University,
P.O. Box 789, Albury, NSW2640, Australia

Abstract. In this paper, we propose a postclustering process that robustly computes cluster regions at different levels of granularity through the complete Order-k Voronoi diagrams. The robustness and flexibility of the proposed method overcome the application-dependency and rigidity of traditional approaches. The proposed cluster tessellation method robustly models monotonic and nonmonotonic cluster growth, and provides fuzzy membership in order to represent indeterminacy of cluster regions. It enables the user to explore cluster structures hidden in a dataset in various scenarios and supports "what-if" and "what-happen" analysis. Tessellated clusters can be effectively used for cluster reasoning and concept learning.

1 Introduction

Geospatial clustering is a series of processes that groups a set $P = \{p_1, p_2, \cdots, p_n\}$ of distinct points into a number of groups exhibiting similar characteristics according to geospatial proximity. Points belonging to the same geospatial cluster are more similar to each other than points belonging to different clusters. Geospatial clustering has been one of the most popular and frequently used approaches for finding undetected or unexpected patterns of concentrations residing in large geospatial-temporal databases [1,2]. It provides hypotheses for "where?" and "when?", and suggests implications into "why?" for postclustering explorations. Clustering can be seen as a starting exploratory analysis tool for a series of geospatial knowledge discovery processes such as cluster reasoning, and concept formation and learning (particularly unsupervised learning). Since the "why?" for geospatial aggregations are of great importance, intelligent postclustering is in high demand [3].

As geospatial clustering becomes mature, several attempts [3,4,5,6] have been made for postclustering. The main idea behind these approaches is to transform a cluster (a set of 0-dimensional points) to a high dimensional object such as 1-dimensional line or 2-dimensional region. In such a way, well developed line matching algorithms or region-based reasoning algorithms (such as RCC (Region Connection Calculus) [7]) can be applied. However, these approaches are application-dependent and too rigid to be general. There still lacks a flexible and consistent way for cluster tessellations.

S. Winter et al. (Eds.): COSIT 2007, LNCS 4736, pp. 321–336, 2007.

In this paper, we propose a new method for defining territories of geospatial clusters. The proposed method provides a set of cluster regions defining different levels of cluster region granularity enabling the user to explore cluster structures hidden in a dataset. Cluster regions are defined by the robust geospatial tessellations modelled by the complete Order-k Voronoi diagrams. The proposed cluster tessellation approach robustly models monotonic and nonmonotonic cluster growth, and also provides fuzzy membership in order to represent indeterminacy of cluster regions. We also propose a unified data structure from which the complete Order-k Voronoi diagrams can be effectively derived. The proposed cluster tessellation method provides a series of cluster tessellations that can be used for postclustering processes such as cluster reasoning, and geospatial concept formation and learning [3,8].

The rest of paper is organized as follows. Section 2 surveys existing methods for extracting cluster shapes and regions that lead to cluster tessellations. Section 3 introduces a new cluster tessellation method through the complete Order-k Voronoi diagram. We provide detailed background and working principle of our approach. This section also discusses how our method models cluster growth, and provides a detailed description of the data structure of our method. Section 4 draws concluding remarks.

2   Geospatial Cluster Tessellation

Determining the region of a set of points (cluster), cluster tessellation, is an inference that has no absolute answer [3] and most approaches are application-dependent [9]. One intuitive way of transforming a cloud of points into a region is to use a base map and the *Point-in-Polygon* algorithm (PiP). Even if this approach has been widely used in the geocomputation community [10], this approach is heavily dependent on the granularity of the base map. Cluster regions are defined by boundaries of the base map. Note that distributions of continuous geographical phenomena are not governed by political or administrative subdivisions of the base map [11].

There exist several other approaches in the literature in order to derive shapes of clusters. They are the *Minimum Bounding Rectangle* (MBR), the *Convex Hull* (CH), and the *Cluster Polygonization Algorithm* (CPA). These shape extraction algorithms can be used for cluster tessellations since they convert a cloud of points into a region. Typically, these algorithms produce fine shapes of clusters (localized regions) and they fail to provide collectively exhaustive tessellated regions of clusters that can be used for concept learning [8]. Since these algorithms produce localized cluster regions (collectively inexhaustive), they have been used for cluster reasoning and association analysis [3].

In order to overcome these localized cluster tessellations, two globalized cluster tessellations (mutually exclusive and collective exhaustive) can be used [12]. These globalized approaches model clusters and represent them with their corresponding influential regions. There have been two dominant approaches in representing clusters: prototype *vs.* exemplar. The former follows the philosophy of learning economy and is widely used in the cognitive science community [12]. Clusters have prototypes and members exhibit central tendency. Therefore, a cloud of points is represented by a representative (prototype). The prototype representation (modelled by the Voronoi diagram of prototypes denoted by VOP) has been the core concept of many learning mechanisms such as k-means [13] and k-medoid [14] clustering approaches. The exemplar (instance) representation has been widely used in the machine learning community [15]. In the instance-based representation (modelled by the Voronoi diagram of individuals denoted by VOI), each member (instance) equally contributes to learning. All members affect a new object's membership and more similar (closer) members have a more (stronger) impact on learning. In this learning, identifying geospatial neighbors (k-nearest neighbors) is an important task.

Figure 1 compares and contrasts these approaches with a study region of 217 urban suburbs around Brisbane, the capital city of Queensland in Australia, and a number of sample incidents of unarmed robbery recorded in 1998. The dataset consists of 3 clusters (circle, rectangle and triangle shaped) and each cluster is composed of 7 crime incidents. Figure 1(b), (c) and (d) show localized cluster tessellations while Fig. 1(e) and (f) depict globalized cluster tessellations. All these tessellations can be used for cluster reasoning using RCC and also can be used for concept learning and formation [3,8,12]. Some of region-based RCC rules are as follows:

- A cluster region of the triangle cluster and a cluster region of the rectangle cluster are *discrete* in PiP, MBR, CH, PTR methods.
- A cluster region of the triangle cluster and a cluster region of the rectangle cluster are *externally connected* in VOI and VOP approaches.

In concept learning and formation, a new crime incident (crossed in Fig. 1 (a)) will be categorized into a new type of crime concept in PiP, MBR, CH and PTR approaches since it does not fall within any of localized cluster regions (shaded regions). However, it will be categorized into the rectangle cluster with VOI whilst it will be into the triangle cluster with VOP. As we can see from this simple example, the localized cluster tessellation approaches generate too compact and tight cluster regions that are not general enough for concept learning, whilst the globalized cluster tessellation approaches generate too exhaustive cluster regions that are not able to learn a new concept. In addition, the globalized approaches generate collectively exhaustive cluster regions that are not suitable for cluster reasoning. A hybrid cluster tessellation of these two extremes is highly required for general purpose postclustering processes.

3 Complete Order-k Voronoi Diagrams Based Cluster Tessellations

3.1 Background

The ordinary Voronoi diagram of a set $P = \{p_1, p_2, \ldots, p_n\}$ of generators tessellates a study region R into mutually exclusive and collectively exhaustive regions. Each region contains a generator closest to it. This geospatial tessellation provides natural neighbor relations that are crucial for many topological queries in geospatial modelling and analysis, whilst its dual graph, the Delaunay triangulation, provides a robust framework for structural arrangements of the Voronoi diagram. This geospatial tessellation has many generalizations and this flexibility provides a robust framework for what-if and what-happens modelling and analysis [16,17].

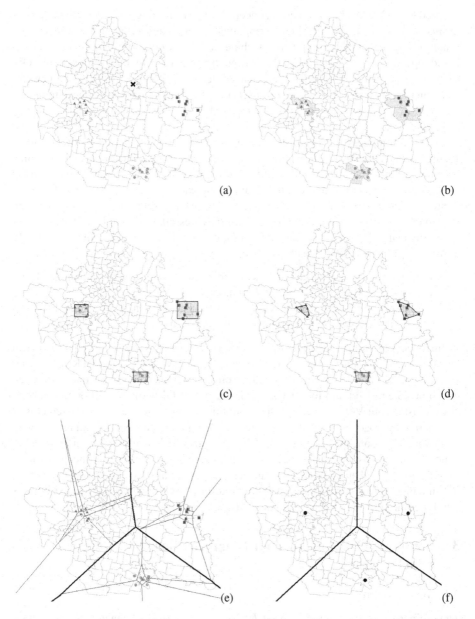

Fig. 1. Cluster tessellations of $|P| = 21$; (a) A dataset in a study region; (b) PiP; (c) MBR; (d) CH and PTR; (e) VOI; (f) VOP

Order-k Voronoi Diagrams (OKVDs) are natural and useful generalizations of the ordinary Voronoi diagram for more than one generator [9]. They provide tessellations where each region has the same k closest sites for a given k. These tessellations are particularly useful for situations where more than one location of interest are not

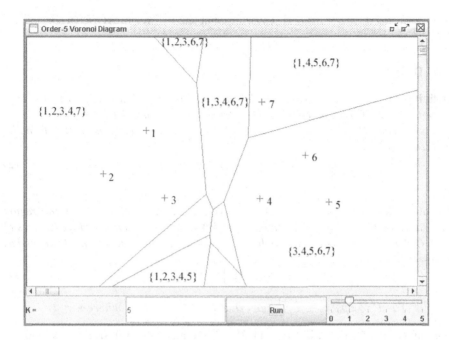

Fig. 2. The Order-5 Voronoi diagram of $P = \{p_1, p_2, p_3, p_4, p_5, p_6, p_7\}$

functioning properly (engaged, busy, closed or fully scheduled) or several locations are required to work together. Each location in the geospace is assigned to a set of k generators in OKVDs. Figure 2 depicts the Order-5 Voronoi diagram of $P = \{p_1, p_2, p_3, p_4, p_5, p_6, p_7\}$. A Voronoi region in the top-right corner has $\{p_1, p_4, p_5, p_6, p_7\}$ as the five closest generators. Since these five generators are the closest points to any location in this region, they are geospatially close to each other. This implies that they are highly likely to belong to the same cluster. If they do, then the region becomes a part of cluster region to which they belong. If they do not, the region is not indicative of any cluster region since the region's generators belong to different clusters. This is the fundamental principle of our proposed cluster tessellation.

Formally, the Order-k Voronoi diagram $\mathcal{V}^{(k)}$ is a set of all Order-k Voronoi regions $\mathcal{V}^{(k)} = \{V(P_1^{(k)}), \ldots, V(P_n^{(k)})\}$, where the Order-$k$ Voronoi region $V(P_i^{(k)})$ for a random subset $P_i^{(k)}$ consisting of k points out of P is defined as follows:

$$V(P_i^{(k)}) = \{p \mid \arg\max_{p_r \in P_i^{(k)}} d(p, p_r) \leq \arg\min_{p_s \in P \setminus P_i^{(k)}} d(p, p_s)\}. \tag{1}$$

Several algorithmic approaches [18,19,20,21] have been proposed to efficiently compute higher order diagrams in the computational geometry community. Dehne [20] proposed an $O(n^4)$ time algorithm that constructs the complete $\mathcal{V}^{(k)}$. Several other attempts [18,19,21] have been made to improve the computational time requirement of Dehne's algorithm. The best known algorithm for the Order-k Voronoi diagram is

$O(k(n-k)\log n + n\log^3 n)$ [22]. Thus, the complete $\mathcal{V}^{(k)}$ requires $O(k^2(n-k)\log n + kn\log^3 n)$ time.

3.2 Definitions

Definition 1. *Given a cloud of points forming a cluster C, the k-th cluster region of C (denoted by $\mathfrak{S}^{(k)}C$) is a union of Order-k Voronoi regions in which generators (member points) of each Order-k Voronoi region all belong to the same cluster. It is formally defined as follows:*

$$\mathfrak{S}^{(k)}C = \bigcup_{\forall P_i^{(k)} \subseteq C} V(P_i^{(k)}). \tag{2}$$

Definition 2. *Given a set of clusters $\mathcal{C} = \{C_1, C_2, \ldots, C_k\}$, the k-th neutral region of $\mathcal{C}$ (denoted by $\wp^{(k)}\mathcal{C}$) is a union of Order-k Voronoi regions in which generators of each Order-k Voronoi region do not belong to the same cluster. It is formally defined as follows:*

$$\wp^{(k)}\mathcal{C} = \bigcup V(P_i^{(k)}), \tag{3}$$

for $P_i^{(k)} \exists p_u, p_v \in P_i^{(k)}$ such that $p_u \in C_s, p_v \in C_t$ where $s \neq t$.

For any C_i and $C_j \in \mathcal{C}$, $\mathfrak{S}^{(k)}C_i$ and $\mathfrak{S}^{(k)}C_j$ are mutually exclusive, and a union of $\bigcup_{\forall C_i} \mathfrak{S}^{(k)}C_i$ and $\wp^{(k)}\mathcal{C}$ is collectively exhaustive. Therefore, a geosaptial tessellation is always obtained.

Let us assume that P in Fig. 2 forms two clusters $\mathcal{C} = \{C_1, C_2\}$ where $C_1 = \{p_1, p_2, p_3\}$ and $C_2 = \{p_4, p_5, p_6, p_7\}$. The Order-2 Voronoi diagram of P, $\mathcal{V}(P^{(2)})$, is shown in Fig. 3. The 2nd cluster region of C_1, $\mathfrak{S}^{(2)}C_1$, is a union of Order-2 Voronoi regions $V(\{p_1, p_2\})$, $V(\{p_2, p_3\})$ and $V(\{p_1, p_3\})$ whilst the 2nd cluster region of C_2, $\mathfrak{S}^{(2)}C_2$, is a union of Order-2 Voronoi regions $V(\{p_4, p_5\})$, $V(\{p_4, p_6\})$, $V(\{p_4, p_7\})$, $V(\{p_5, p_6\})$ and $V(\{p_6, p_7\})$. The 2nd neutral region of $\mathcal{C}$, $\wp^{(k)}\mathcal{C}$, is a union of Order-2 Voronoi regions $V(\{p_1, p_7\})$ and $V(\{p_3, p_4\})$.

3.3 Cluster Tessellations with the Complete OKVDs

The proposed cluster tessellation with the complete OKVDs is a hybrid method of globalized cluster tessellations and localized cluster tessellations. It provides intermediate levels of cluster tessellations that can be adopted for cluster reasoning and concept learning. Figure 4 depicts cluster tessellations based on the complete OKVDs. Neutral regions are shaded in this figure. Since each cluster in P has 7 members, Fig. 4 shows from $\mathcal{V}^{(2)}$ to $\mathcal{V}^{(7)}$. Note that $\mathcal{V}^{(1)}$ is the ordinary Voronoi diagram and it is shown in Fig. 1(e). As can be seen from Fig. 4, neutral regions divide the study region into mutually exclusive cluster regions. As k grows, $\wp^{(k)}\mathcal{C}$ expands. On the other hand, $\mathfrak{S}^{(k)}C$ decreases as k increases. In concept learning, a new object falling into a neutral region is informative of a new concept discovery that leads to forming a new concept. A series of cluster tessellations provided by our approach can be effectively used for this concept learning and formation. Also, a series of cluster tessellations provides a flexible what-if and what-happen framework that is of importance in dynamic decision makings.

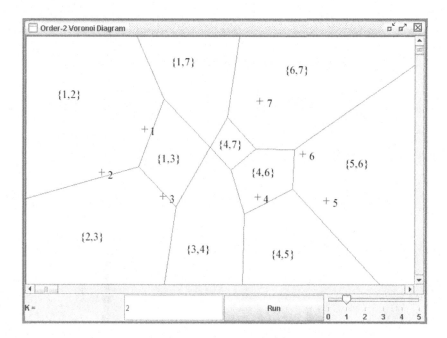

Fig. 3. The Order-2 Voronoi diagram of P (same as in Fig. 2)

3.4   Cluster Growth

Since clusters grow in dynamic situations, it is important to build a model that supports dynamic growth of clusters. Cluster growth is in the same line with unsupervised learning in Machine Learning and concept learning in Artificial Intelligence in the sense that it has monotonic and nonmonotonic learning. Figure 5 shows monotonic cluster growth whilst Fig. 6 exhibits nonmonotonic cluster growth. Monotonic growth occurs when a new object falls within a cluster region to which it actually belongs, while non-monotonic growth happens when a new object does not fall into a cluster region to which it actually belongs.

Note that, localized cluster tessellation approaches (PiP, MBR, CH and PTR) fail to model this monotonic cluster growth. This is because a new object falling into a cluster region to which it actually belongs results in no change. However, it is well modelled by our approach as shown in Fig. 5 where three new objects (appear as crosses) are added in a clockwise direction. Changes in cluster regions are well depicted and demonstrated. When the first object (actually a triangle object) is introduced in Fig. 5(b), it falls within the cluster region of triangles (the shaded region of triangles in Fig. 5(a)). This addition can widen the cluster region of triangles and can lessen cluster regions of its neighbors. Figure 5(b) shows $\mathcal{V}((P + \text{the new triangle object})^{(7)})$ after an introduction of the triangle object. Due to the local update characteristic of the Voronoi diagram [9], only cluster regions of neighbors need to be updated. The monotonic cluster growth is also observed in Fig. 5(c) and (d). When the rectangle cluster monotonically grows as shown in Fig. 5(c), its cluster region could increase whilst its neighboring cluster regions could

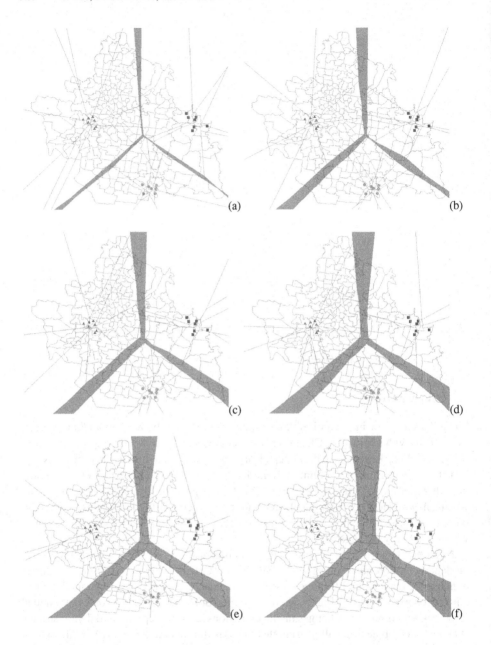

Fig. 4. OKVDs of P (same as in Fig. 1): (a) $\mathcal{V}^{(2)}$; (b) $\mathcal{V}^{(3)}$; (c) $\mathcal{V}^{(4)}$; (d) $\mathcal{V}^{(5)}$; (e) $\mathcal{V}^{(6)}$; (f) $\mathcal{V}^{(7)}$

decrease. In this case, the cluster region of rectangles increases when a new rectangle object (thick cross object shown in Fig. 5(c)) is introduced whilst its neighboring cluster region, the cluster region of circles, decreases. Note that, this monotonic addition does not always decrease its neighboring cluster regions. The cluster region of triangles

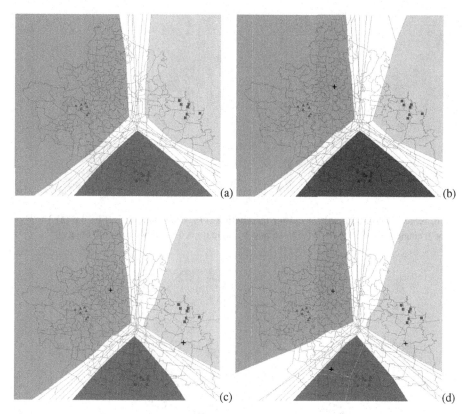

Fig. 5. Monotonic growth of cluster tessellations (P is the same as in Fig. 1): (a) $\mathcal{V}(P^{(7)})$; (b) $\mathcal{V}((P+\text{a new triangle object})^{(7)})$; (c) $\mathcal{V}((P+\text{a new triangle object}+\text{a new rectangle object})^{(7)})$; (d) $\mathcal{V}((P + \text{a new triangle object} + \text{a new rectangle object} + \text{a new circle object})^{(7)})$

is not affected by this addition as shown in Fig. 5(c). A monotonic addition of circle object increases the cluster region of circles whilst it decreases its neighboring cluster region of triangles as shown in Fig. 5(d).

In localized cluster tessellations, nonmonotonic cluster growth never affects cluster regions of other clusters, and only widens the cluster region of a newly introduced object which may result in cluster region overlaps. Unlike localized cluster tessellations, nonmonotonic cluster growth in the OKVD based cluster tessellation not only widens a cluster region of a newly introduced object, but greatly lessens cluster regions of neighbors. Figure 6 shows nonmonotonic cluster growth. Figure 6(a) depicts a scenario when a triangle object is added to the cluster region of rectangles. This addition greatly reduces the cluster region of rectangles and marginally increases the cluster region of triangles as shown in Fig. 5(a) and Fig. 6(a). Figure 6(b) illustrates a situation when a rectangle object is added to the cluster region of triangles. This addition greatly reduces the cluster region of triangles and marginally increases the cluster region of rectangles.

The OKVD based cluster tessellation always produces a new geospatial cluster tessellation regardless of monotonic and nonmonotonic cluster growth. This demonstrates

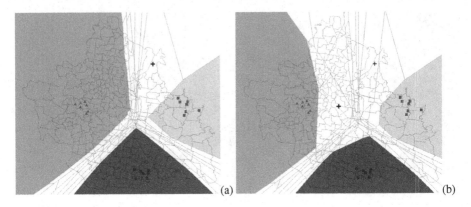

Fig. 6. Nonmonotonic growth of cluster tessellations (P is the same as in Fig. 1): (a) $\mathcal{V}((P +$ a new triangle object$)^{(7)})$; (b) $\mathcal{V}((P + $ a new triangle object $+$ a new rectangle object$)^{(7)})$

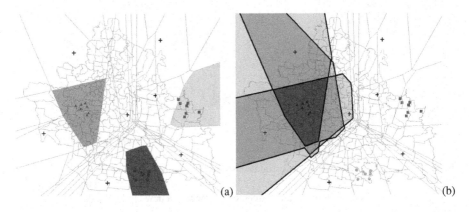

Fig. 7. Cluster tessellations in the presence of noise and fuzzy cluster regions (P is the same as in Fig. 1 + 7 noise points): (a) Cluster regions of $\mathcal{V}((P + 7$ noise points$)^{(7)})$; (b) Fuzzy triangle cluster regions of $\mathcal{V}((P + 7$ noise points$)^{(7)})$

the superior modelling capability of the OKVD based cluster tessellation in highly dynamic situations.

3.5   In the Presence of Noise and Fuzzy Membership

Noise points can be present in many situations. These noise points greatly affect cluster tessellations and must be properly modelled. These noise points are simply ignored and not modelled at all by the localized cluster tessellation approaches. In the globalized cluster tessellation approaches, noise points are somehow modelled by VOI, but not by VOP. In the OKVD based cluster tessellation, noise points significantly affect cluster tessellations and this is well modelled by our proposed approach. It is clearly depicted in Fig. 7(a). Cluster regions become compact around clusters in the presence of noise points.

Another advantage of the OKVD based cluster tessellation approach is that it can express fuzzy (non-crisp) regions of each cluster. It is shown in Fig. 7(b). The darkest region exhibits 100%, its first-order neighbors (Voronoi regions that share an edge with the darkest region) in medium grey exhibit 86% (6/7), second-order neighbors of these first-order neighbors show 71% (5/7) membership. A combination of all these fuzzy regions can represent indeterminate cluster regions.

3.6 A Unified Data Structure Supporting Dynamic Cluster Tessellations

Since cluster tessellations are for cluster reasoning and concept learning in dynamic situations, OKVDs must be dynamically maintained and updated. This can be achieved by developing a cluster tessellation system with a highly flexible data structure. In this subsection, we provide details on a unified data structure from which the complete OKVDs can be derived that supports dynamic cluster tessellations. The unified data structure is similar to the triangle-based data structure as in [16,20]. The data structure is a combination of two tables: generator list and triangle lists. The generator list is a table of points containing the x and y coordinates with a unique identification number. The triangle lists are tables of triangles, each with three vertices, a circumcircle center point and a list of generator IDs lying within the circumcircle.

In the unified data structure, the triangles are assigned to the triangle table based on the number of generators that are lying on the circumcircle for this triangle. These triangles are refer to as Order-k triangles, where k is the number of generators within the circumcircle of this triangle. Table 1 depicts the assignment of triangles shown in Fig. 8. For example, the triangle $\triangle p_1 p_2 p_4$ has a circumcircle that contains p_5 as shown in Fig. 8(b). Therefore since triangle $\triangle p_1 p_2 p_4$ contains one generator, it is classified as an Order-1 triangle and it is assigned to the Order-1 triangle table. In the data structure, with n generators there is at least $(n$ - $3)$ Order-k triangle tables within the data structure.

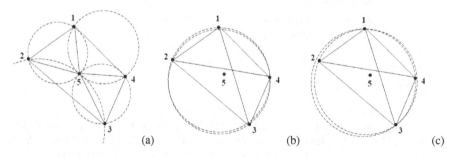

Fig. 8. Circumcircles and data structure for Order-k Voronoi diagrams: (a) Order-0 triangles; (b) Order-1 triangles; (c) Order-2 triangles

3.7 Dynamic Update in the Unified Data Structure

When updates (addition or deletion) occur, the unified data structure has to handle two types of triangles. For p, chosen for deletion or insertion, triangles with p inside their circumcircles and triangles with p as a vertex must be updated. Insertion firstly updates

Table 1. Unified Order-k Delaunay triangle data structure

Order-k Delaunay triangle	Triangle ($\triangle$) ID	Onpoint $\triangle abc$	Inpoint
Order-0 Delaunay triangle	1	$\{p_1, p_2, p_5\}$	ϕ
Order-0 Delaunay triangle	2	$\{p_1, p_4, p_5\}$	ϕ
Order-0 Delaunay triangle	...	...	ϕ
Order-1 Delaunay triangle	5	$\{p_1, p_2, p_4\}$	$\{p_5\}$
Order-1 Delaunay triangle	6	$\{p_2, p_3, p_4\}$	$\{p_5\}$
Order-1 Delaunay triangle	...	...	...
Order-2 Delaunay triangle	9	$\{p_1, p_2, p_3\}$	$\{p_4, p_5\}$
Order-2 Delaunay triangle	10	$\{p_1, p_3, p_4\}$	$\{p_2, p_5\}$

triangles that contain p ($Algorithm$ **Triangle Update**) and generates new triangles having p as a vertex ($Algorithm$ **Generate New Triangles**). Deletion is basically the reverse of insertion, it iterates through all triangles and removes any triangles having p as a vertex and any triangles containing p. The algorithms used are described below.

Insertion. Before new triangles for p are generated, old triangles need to be updated. All the existing triangles in the data structure need to be checked to find whether they contain the new generator p.

$Algorithm$ **Triangle Update**
> Input: A newly added point p and the unified data structure (DS);
> Output: Updated unified data structure;
> 1) begin
> 2) $i \leftarrow 0$;
> 3) do
> 4) $\triangle abc \leftarrow DS[i]$;
> 5)   if (isInCircle($\triangle abc, p$))
> 6) $\triangle abc$.addInPoint(p);
> 7)   end if
> 8) $i \leftarrow i + 1$;
> 9) while($i < |DS|$);
> 10) end

Once the existing triangles are updated new triangles for p can be generated. In a pseudo-code below the algorithm basically iterates through all the possible unordered pairs of $\{p_1, p_2, \ldots, p_n\}$ and creates a triangle including p as the third vertex.

$Algorithm$ **Generate New Triangles**
> Input: A newly added generator p, a set of generators $\{p_1, p_2, \ldots, p_n\}$
> and the unified data structure (DS);
> Output: Updated unified data structure;
> 1) begin
> 2) $i \leftarrow 0$;
> 3) while($i < n$)

```
4)      j ← i + 1;
5)      while(j < n)
6)          numberOfInPoints ← FindInPoints (△ pp_i p_j);
7)          k ← getNumberOfInPoints (△ pp_i p_j);
8)          SaveToDS (△ pp_i p_j, k);
9)          j ← j + 1;
10)     i ← i + 1;
11)end
```

For each of the triangle, it is first evaluated with the FindInPoints() which linearly iterates through generators and looks for those that are within the triangle's circumcircle. These are added to the InPoints in Step 6. Once all the InPoints are found, the triangle is then saved to the corresponding Order-k triangle table based on the number of generators in InPoints list (SaveToDS).

Deletion. The deletion approach finds both triangles that have p as a vertex and whose circumcircles contain p. In the first case, the triangles are simply removed from the data structure. In the second case, the triangles need to be updated as the generator should be removed from the triangle list.

Algorithm **Deletion**

    Input: A deleted point p and the unified data structure (DS);
    Output: Updated unified data structure

```
1) begin
2) i ← 0;
3) do
4)      △abc ← DS[i];
5)      if ( p == a||p == b||p == c )
6)          k ← getNumberOfInPoints (△abc);
7)          RemoveDS(△abc, k);
8)      end if
9)      if (isInCircle(△ abc, p))
10)         k ← getNumberOfInPoints (△abc);
11)         △abc.removeInPoint(p);
12)         RemoveDS(△abc, k);
13)         SaveDS(△abc, k − 1);
14)     end if
15)     i ← i + 1;
16)while(i < |DS|);
17)end
```

Once this data structure is dynamically updated and maintained, then the complete OKVDs can be derived from this unified data structure. Derivations follow the algorithm described in [20]. Figure 9 illustrates the proposed cluster tessellation algorithm interface.

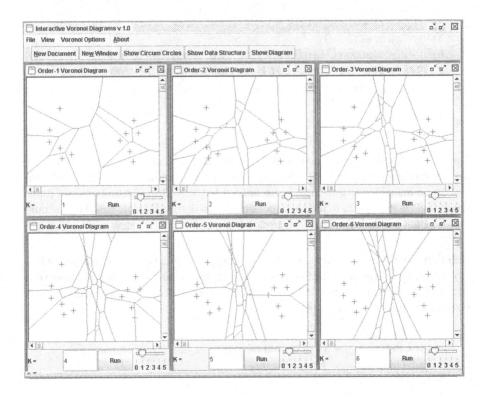

Fig. 9. Interface of cluster tessellation

4   Final Remarks

Geospatial clustering has been one of the popular techniques in geospatial data mining and analysis [23,24]. It produces geospatially and/or temporally aggregated clusters that are indicative of important geospatial/temporal patterns. Understanding the implications of clusters is an important step for many subsequent postclustering analyses.

In this paper, we have proposed a new method for cluster tessellation through the complete Order-k Voronoi diagrams. It overcomes the application-dependency and rigidity of traditional approaches, and bridges the gap between the localized cluster tessellation and the globalized cluster tessellation. It is able to:

- model different levels of cluster tessellations through the complete Order-k Voronoi diagrams;
- model different levels of neutral zones that can be used for geospatial concept formation;
- dynamically model monotonic and nonmonotonic cluster growth;
- effectively model clusters in the presence of noise points;
- provide different levels of non-crisp memberships;
- support dynamic updates for what-if and what-happen analysis.

The Voronoi tessellation is a cognitively economical way of representing information about clusters and concepts [12]. It has been widely used in many GIS-related domains and many generalized variants of the ordinary Voronoi diagrams have been researched [9,25]. OKVDs are one of popular generalizations and the complete OKVDs have the potential for many applications. However, the computational inefficiency of the complete OKVDs is a main bottleneck for dynamic updates. Algorithms for dynamic updates in this paper requires $O(n^3)$ time, and this is computationally intensive. Efficient algorithms for dynamic updates through sophisticated spatial data structures needs to be investigated to support enhanced what-if analysis. Future work also includes investigations into cluster tessellations in complex scenarios such as in the presence of obstacles [26].

References

1. Han, J., Kamber, M., Tung, K.H.: Spatial Clustering Methods in Data Mining. In: Miller, H.J., Han, J. (eds.) Geographic Data Mining and Knowledge Discovery, pp. 188–217. Cambridge University Press, Cambridge, UK (2001)
2. Kolatch, E.: Clustering Algorithms for Spatial Databases: A Survey (2000) Available at http://www.cs.umd.edu/~kolatch/papers/SpatialClustering.pdf
3. Lee, I., Estivill-Castro, V.: Fast Cluster Polygonization and Its Applications in Data-Rich Environments. GeoInformatica 10(4), 399–422 (2006)
4. Galton, A., Duckham, M.: What Is the Region Occupied by a Set of Points? In: Raubal, M., Miller, H.J., Frank, A.U., Goodchild, M.F. (eds.) GIScience 2006. LNCS, vol. 4197, pp. 81–98. Springer, Heidelberg (2006)
5. Knorr, E.M., Ng, R.T., Shilvock, D.L.: Finding Boundary Shape Matching Relationships in Spatial Data. In: Scholl, M.O., Voisard, A. (eds.) SSD 1997. LNCS, vol. 1262, pp. 29–46. Springer, Heidelberg (1997)
6. Son, E., Kang, I., Kim, T., Li, K.: A Spatial Data Mining Method by Clustering Analysis. In: Laurini, R., Makki, K., Pissinou, N. (eds.) Proceedings of the 6th International Symposium on Advances in Geographic Information Systems, pp. 157–158. ACM Press, New York (1998)
7. Cohn, A.G., Bennett, B., Gooday, J., Gotts, N.M.: Qualitative Spatial Representation and Reasoning with the Region Connection Calculus. GeoInformatica 1(3), 275–316 (1997)
8. Lee, I., Lee, K.: Higher Order Voronoi Diagrams for Concept Boundaries and Tessellations. In: Proceedings of the 6th IEEE International Conference on Computer and Information Science, Melbourne, Australia, IEEE Computer Society, Los Alamitos (2007)
9. Okabe, A., Boots, B.N., Sugihara, K., Chiu, S.N.: Spatial Tessellations: Concepts and Applications of Voronoi Diagrams. John Wiley & Sons, West Sussex (2000)
10. Worboys, M.F.: GIS: A Computing Perspective. Taylor & Francis, London (1995)
11. Dent, B.D.: Cartography: Thematic Map Design, 5th edn. WCB McGraw Hill, Boston (1999)
12. Gärdenfors, P.: Conceptual Spaces: The Geometry of Thought. MIT Press, Cambridge (2000)
13. MacQueen, J.: Some Methods for Classification and Analysis of Multivariate Observations. In: Proceedings of the 5th Berkeley Symposium on Maths and Statistics Problems. vol. 1, pp. 281–297 (1967)
14. Kaufman, L., Rousseuw, P.J.: Finding Groups in Data: An Introduction to Cluster Analysis. John Wiley & Sons, New York (1990)
15. Mitchell, T.M.: Machine Learning. WCB/McGraw-Hill, New York (1997)
16. Gahegan, M., Lee, I.: Data Structures and Algorithms to Support Interactive Spatial Analysis Using Dynamic Voronoi Diagrams. Environments and Urban Systems 24(6), 509–537 (2000)

17. Lee, I., Gahegan, M.: Interactive Analysis using Voronoi Diagrams: Algorithms to Support Dynamic Update from a Generic Triangle-Based Data Structure. Transactions in GIS 6(2), 89–114 (2002)
18. Aurenhammer, F., Schwarzkopf, O.: A Simple On-Line Randomized Incremental Algorithm for Computing Higher Order Voronoi Diagrams. In: Proceedings of the Symposium on Computational Geometry, North Conway, NH, pp. 142–151. ACM Press, New York (1991)
19. Chazelle, B., Edelsbrunner, H.: An Improved Algorithm for Constructing kth-Order Voronoi Diagrams. IEEE Transactions on Computers 36(11), 1349–1354 (1987)
20. Dehne, F.K.H.: On $O(N^4)$ Algorithm to Construct all Voronoi Diagrams for K-Nearest Neighbor Searching in the Euclidean plane. In: Díaz, J. (ed.) Proceedings of the International Colloquium on Automata, Languages and Programming, Barcelona, Spain. Lecture Notes in Artificial Intelligence , vol. 154, pp. 160–172. Springer, Heidelberg (1983)
21. Lee, D.T.: On k-Nearest Neighbor Voronoi Diagrams in the Plane. IEEE Transactions on Computers 31(6), 478–487 (1982)
22. Berg, M., Kreveld, M., Overmars, M., Schwarzkoph, O.: Computational Geometry: Algorithms and Applications, 2nd edn. Springer-Verlag, West Sussex (2002)
23. Estivill-Castro, V., Lee, I.: Argument Free Clustering via Boundary Extraction for Massive Point-data Sets. Computers, Environments and Urban Systems 26(4), 315–334 (2002)
24. Miller, H.J., Han, J.: Geographic Data Mining and Knowledge Discovery: An Overview. Cambridge University Press, Cambridge, UK (2001)
25. O'Rourke, J.: Computational Geometry in C. Cambridge University Press, New York (1993)
26. Estivill-Castro, V., Lee, I.: Clustering with Obstacles for Geographical Data Mining. ISPRS Journal of Photogrammetry and Remote Sensing 59(1-2), 21–34 (2004)

Drawing a Figure in a Two-Dimensional Plane for a Qualitative Representation

Shou Kumokawa and Kazuko Takahashi

School of Science & Technology, Kwansei Gakuin University,
2-1, Gakuen, Sanda, 669-1337, Japan
acy85499@ksc.kwansei.ac.jp, ktaka@kwansei.ac.jp

Abstract. This paper describes an algorithm for generating a figure in a two-dimensional plane from a qualitative spatial representation of PLCA. In general, it is difficult to generate a figure from qualitative spatial representations, since they contain positional relationships but do not hold quantitative information such as position and size. Therefore, an algorithm is required to determine the coordinates of the objects while preserving the positional relationships. Moreover, it is more desirable that the resulting figure meets a user's requirement. PLCA is a simple symbolic representation consisting of points, lines, circuits and areas. We have already proposed one algorithm for drawing, but the resulting figures are far from a "good" one. In that algorithm, we generate the graph corresponding to a given PLCA expression, decompose it into connected subgraphs, determine the coordinates in a unit circle for each subgraph independently, and finally determine the position and size of each subgraph by locating the circles in appropriate positions. This paper aims at generating a "good" figure for a PLCA expression. We use a genetic algorithm to determine the locations and the sizes of circles in the last step of the algorithm. We have succeeded in producing a figure in which objects are drawn as large as possible, with complex parts larger than others. This problem is considered to be a type of "circle packing," and the method proposed here is applicable to the other problems in which locating objects in a non-convex polygon.

1 Introduction

Qualitative Spatial Reasoning (QSR) is a method that treats images or figures qualitatively, by extracting the information necessary for a user's purpose [4,14,15]. It also offers methods to handle and reason about unspecific information. Numerous applications use databases of images or figures including Geographic Information Systems (GIS) and navigation systems. In these applications, responding to queries or frequently updating the data requires a large amount of computation. QSR is a promising method that reduces memory and the workspace required for computations that do not involve strict data. In general, it is easy to transform a figure to a symbolic qualitative representation, but it is difficult to generate a figure from a symbolic qualitative representation. The automatic drawing of figures such as Venn diagram from mereological relationships has been studied in the context of diagrammatic reasoning [1]. However, to the best of our knowledge, no system exists for automatically drawing a figure from mereotopological relationships. Symbolic representation in compact form lends itself to calculation

S. Winter et al. (Eds.): COSIT 2007, LNCS 4736, pp. 337–353, 2007.

or storage, but drawings are much beter for visualizing the abstract details and context. It is also interesting to view what shape of a figure is generated from a symbolic representation. Drawing a figure from a symbolic representation is applicable to an important service such as schematic maps.

The difficulty in drawing a figure from a symbolic representation arises from two factors. First, one has to judge whether the expression can be embedded in a two-dimensional plane. Second, one must determine the position of each object in the drawing. The latter is necessary since multiple quantitative representations may exist for a single qualitative expression. Therefore, we cannot determine a unique set of appropriate coordinates. Even if a set of coordinates are determined, the resulting complex figure may not support a user to think or to design.

In this paper, we discuss drawing a figure from a PLCA expression, a qualitative representation method. A PLCA expression represents spatial data by focusing on connected patterns of objects, using four simple elements: $point(P), line(L)$ $circuit(C)$ and $area(A)$ [18,19,20]. In PLCA, no pair of areas has a part in common[1]. The entire space is covered with the areas.

A PLCA expression is considered to be a set of mereotopological relationships between objects including areas. Our goal is to draw a figure for such a representation. This is different from other studies on visualization in which the goal is to generate a figure for a set of conceptual relationships to support human cognition and understanding.

In the previous paper, we proved that realizability for a PLCA expression in a two-dimensional plane is reduced to the planarity of a graph and identified the condition for realizability [21]. We also proposed an algorithm for drawing a figure for a PLCA expression that satisfied the realizability condition in a two-dimensional plane. In that algorithm, we decomposed a PLCA expression into several subexpressions each of which corresponds to a connected graph, determined the position of the objects in a unit circle for each graph independently and combined the related parts of them. However, the resulting figure was far from a "good" one. For example, many objects were drawn in a corner, leaving a large vacant space in the center. The main problem was embedding circles in part of another circle in the last step when related parts were combined. In that last step, the challenge was to determine the location and the size of each circle to produce a "good" figure while preserving the relationships of objects described in the PLCA expression.

The problem of embedding is reduced to one of circle packing, putting n circles of different sizes into a non-convex polygon so that they do not intersect. Circle packing is a well-known optimization problem that is NP-complete in general, and many studies have been undertaken [22,16,17]. However, no algorithm has been proposed that covers the conditions in our problem. In this paper, we address the problem using a genetic algorithm (GA) to determine the location and the size of each circle and produce an approximate solution to optimization, that gives a "good" figure. A "good" figure here means one in which the objects are drawn as large as possible, and a complex parts are drawn larger than the other parts.

---

[1] We use the term $area$ instead of $region$, since $area$ used in this paper is a different entity from the $region$ generally used in qualitative spatial reasoning.

We also discuss another drawing algorithm that does not use a graph-drawing algorithm but generates a figure directly from a PLCA expression. Moreover, we discuss generation of figures from other qualitative representations.

This paper is organized as follows. In section 2, we briefly describe PLCA expressions, and the conditions for their realizability in a two-dimensional plane. In section 3, we introduce the drawing algorithm, and show the results of experiment. In section 4, we present another drawing algorithm for PLCA and discuss drawing for other qualitative representations. And finally, in section 5, we present our conclusions.

2   PLCA Expressions

2.1   Definition of Classes

PLCA has four basic components: $points(P)$, $lines(L)$, $circuits(C)$ and $areas(A)$.

Point is defined as a primitive class.

Line is defined as a class that satisfies the following condition: for an arbitrary instance l of *Line*, $l.points$ is a pair $[p_1, p_2]$ where $p_1, p_2 \in Point$. A line has an inherent orientation. When $l.points = [p_1, p_2]$, l^+ and l^- mean $[p_1, p_2]$ and $[p_2, p_1]$, respectively. l^* denotes either l^+ or l^-. Intuitively, a line is the edge connecting two (not always different) points. No two lines are allowed to cross. Note that multiple lines may have the same pair of points. In Fig. 1(a), the arrows denote the orientation of the lines. All of the lines $l_1.points$, $l_2.points$ and $l_3.points$ are defined to be $[p_1, p_2]$, but they are distinguished by the circuits to which they belong.

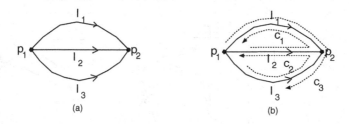

Fig. 1. Multiple lines with the same definition and the associated circuits

Circuit is defined as a class that satisfies the following condition: for an arbitrary instance c of *Circuit*, $c.lines$ is a sequence $[l_1^*, \ldots, l_n^*]$ where $l_1, \ldots, l_n \in Line(n \geq 1)$, $l_i.points = [p_i, p_{i+1}](1 \leq i \leq n)$ and $p_{n+1} = p_1$. $[l_1^*, \ldots, l_n^*]$ and $[l_j^*, \ldots, l_n^*, l_1^*, \ldots, l_{j-1}^*]$ denote the same circuit for any j ($1 \leq j \leq n$). In Fig. 1(b), we have three circuits: $c_1.lines = \{l_1^-, l_2^+\}$, $c_2.lines = \{l_2^-, l_3^+\}$, $c_3.lines = \{l_3^-, l_1^+\}$.

For $c_1, c_2 \in Circuit$, we introduce two new predicates lc and pc to denote that two circuits share line(s) and point(s), respectively. $lc(c_1, c_2)$ is *true* iff there exists $l \in Line$ such that $(l^+ \in c_1.lines) \wedge (l^- \in c_2.lines)$. $pc(c_1, c_2)$ is *true* iff there exists $p \in Point$ such that $(p \in l_1.points) \wedge (p \in l_2.points) \wedge (l_1^* \in c_1.lines) \wedge (l_2^* \in c_2.lines)$. A circuit is the boundary between an area and its adjacent areas viewed from the side of that area.

Area is defined as a class that satisfies the following condition: for an arbitrary instance a of *Area*, $a.circuits$ is a set $\{c_1, \ldots, c_n\}$ where $c_1, \ldots, c_n \in Circuit(n \geq 1)$, and $\forall c_i, c_j \in a.circuits; (i \neq j) \rightarrow (\neg pc(c_i, c_j) \wedge \neg lc(c_i, c_j))$. Intuitively, an area is a connected region which consists of exactly one piece. No two areas are allowed to cross. The final condition means that any pair of circuits that belong to the same area cannot share a point or a line.

The *PLCA expression* e is defined as a five tuple $e = \langle P, L, C, A, outermost \rangle$ where P, L, C and A are a set of points, lines, circuits and areas, respectively, and $outermost \in C$. An element of $P \cup L \cup C \cup A$ is called *a component of e*.

We assume that there exists a circuit in the outermost extremity of the figure called *outermost*. This means that the target figure is drawn in a finite space, and the space can be divided into a number of areas that do not overlap with each other.

In Fig. 2, (a) shows an example of a target figure, and (b) and (c) show the names of the components. Example 1 shows a PLCA expression corresponding to Fig. 2.

Definition 1. *(consistency) A PLCA expression* $e = \langle P, L, C, A, outermost \rangle$, *is said to be consistent iff the following three constraints are satisfied:*

1. constraint on P-L: *For any* $p \in Point$ *there exists at least one line* l *such that* $p \in l.points$.
2. constraint on L-C: *For any* $l \in Line$, *there exist exactly two distinct circuits* c_1, c_2 *such that* $l^+ \in c_1.lines, l^- \in c_2.lines$.
3. constraint on C-A: *For any* $c \in Circuit$ *other than outermost, there exists exactly one area* a *such that* $c \in a.circuits$. *The outermost is not included in any area.*

Due to these constraints, neither isolated lines nor points are allowed.

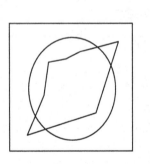 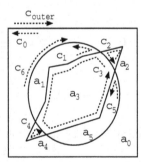 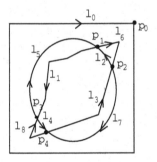

(a) target figure   (b) areas and circuits   (c) lines and points

Fig. 2. Example of a target figure

Example 1

$e.points = \{p_0, p_1, p_2, p_3, p_4\}$ $c_{outer}.lines = [l_0^+]$
$e.lines = \{l_0, l_1, l_2, l_3, l_4, l_5, l_6, l_7, l_8\}$ $c_0.lines = [l_0^-]$
$e.circuits = \{c_{outer}, c_0, c_1, c_2, c_3, c_4, c_5, c_6\}$ $c_1.lines = [l_1^-, l_5^-]$
$e.areas = \{a_0, a_1, a_2, a_3, a_4, a_5\}$ $c_2.lines = [l_2^-, l_6^-]$

$$e.outermost = c_{outer}$$
$$l_0.points = [p_0, p_0]$$
$$l_1.points = [p_4, p_1]$$
$$l_2.points = [p_1, p_2]$$
$$l_3.points = [p_2, p_3]$$
$$l_4.points = [p_3, p_4]$$
$$l_5.points = [p_1, p_4]$$
$$l_6.points = [p_2, p_1]$$
$$l_7.points = [p_3, p_2]$$
$$l_8.points = [p_4, p_3]$$

$$c_3.lines = [l_1^+, l_2^+, l_3^+, l_4^+]$$
$$c_4.lines = [l_4^-, l_8^-]$$
$$c_5.lines = [l_3^-, l_7^-]$$
$$c_6.lines = [l_5^+, l_8^+, l_7^+, l_6^+]$$
$$a_0.circuits = \{c_6, c_0\}$$
$$a_1.circuits = \{c_1\}$$
$$a_2.circuits = \{c_2\}$$
$$a_3.circuits = \{c_3\}$$
$$a_4.circuits = \{c_4\}$$
$$a_5.circuits = \{c_5\}$$

2.2 Two-Dimensional Realizability

We introduce the concept of connectedness for the components of a PLCA expression.

Definition 2. *(d-pcon) Let* $e = \langle P, L, C, A, outermost \rangle$ *be a PLCA expression. For a pair of components of e, the predicate d-pcon is defined as follows.*

1. *d-$pcon(p, l)$ iff $p \in l.points$.*
2. *d-$pcon(l, c)$ iff $l^* \in c.lines$.*
3. *d-$pcon(c, a)$ iff $c \in a.circuits$.*

Definition 3. *(pcon) Let* α, β, γ *be components of a PLCA expression. pcon is the symmetric and transitive closure of d-pcon.*

1. *If d-$pcon(\alpha, \beta)$, then $pcon(\alpha, \beta)$.*
2. *If $pcon(\alpha, \beta)$, then $pcon(\beta, \alpha)$.*
3. *If $pcon(\alpha, \beta)$ and $pcon(\beta, \gamma)$, then $pcon(\alpha, \gamma)$.*

Definition 4. *(PLCA-connected) A PLCA expression e is said to be PLCA-connected iff $pcon(\alpha, \beta)$ holds for any pair α and β of components of e.*

Intuitively, PLCA-connectedness guarantees that all the components including the *outermost* are connected. That is, for any pair of components, there is a trail that can go from one component to the other by tracing components. The PLCA expression in Example 1 is consistent and PLCA-connected. For example, $pcon(c_{outer}, c_6)$ holds since we can move from c_{outer} to c_6 by tracing the components $c_{outer}, l_0, c_0, a_0, c_6$ in that order. On the other hand, the PLCA expression in Example 2 is consistent but not PLCA-connected. It can be divided into two subexpressions: one corresponding to the plane consisting of p_0, l_0, c_{outer}, c_0 and a_0, and the other corresponding to a floating group of other the components. A component of the former is not *pcon* with that of the latter. For example, $pcon(c_{outer}, c_1)$ does not hold.

Example 2
$$e.points = \{p_0, p_1\} \quad c_{outer}.lines = \{l_0^+\}$$
$$e.lines = \{l_0, l_1, l_2\} \quad c_0.lines = \{l_0^-\}$$

$$e.circuits = \{c_{outer}, c_0, c_1, c_2, c_3\} \quad c_1.lines = \{l_1^+, l_2^+\}$$
$$e.areas = \{a_0, a_1, a_2\} \quad c_2.lines = \{l_2^-\}$$
$$l_0.points = [p_0, p_0] \quad c_3.lines = \{l_1^-\}$$
$$l_1.points = [p_1, p_1] \quad a_0.circuits = \{c_0\}$$
$$l_2.points = [p_1, p_1] \quad a_1.circuits = \{c_1\}$$
$$a_2.circuits = \{c_2, c_3\}$$

For a consistent connected PLCA expression, the following theorem holds [21].

Theorem 1. *For a consistent connected PLCA expression $e = \langle P, L, C, A, outermost \rangle$, e can be realized in a two-dimensional plane iff $|P| - |L| - |C| + 2|A| = 0$ holds.*

This theorem shows that the two-dimensional realizability for a PLCA expression is judged only by counting the number of the components.

Definition 5. *(planar PLCA expression) A consistent connected PLCA expression that satisfies $|P| - |L| - |C| + 2|A| = 0$ is said to be planar.*

The PLCA expression shown in Example 1 is planar but the expression in Example 2 is not.

2.3 Orientation of a Circuit

As a preparation, we introduce several concepts from graph theory.

A *(non-directed) graph* is defined to be $G = (V, E)$, where V is a set of vertices and E is a set of edges. An edge of E is defined as a pair of vertices of V. For graphs $G = (V, E)$ and $G' = (V', E')$, if $V' \subset V$ and $E' \subset E$, G' is said to be *a subgraph* of G; if $V \cap V' = \emptyset$ and $E \cap E' = \emptyset$, it is said that G and G' are *disjoint*. Here, when we consider more than one subgraph of G, we assume that they are disjoint. If it is possible to move between any pair of vertices by moving along the edges of the graph, the graph is said to be *connected*; otherwise, it is said to be *disconnected*. A sequence $(v_0, \ldots, v_n)$ where (v_i, v_{i+1}) for each i $(0 \le i \le n - 1)$ is an edge and $v_0 = v_n$, it is said to be *a cycle*. A cycle that is a border of both the graph and the outer infinitely large region is said to be *an outer boundary cycle of g*.

Let $e = \langle P, L, C, A, outermost \rangle$ be a consistent PLCA expression. We can define a non-directed graph $m(e) = (V, E)$ by relating P and L to V and E, respectively. For $p \in P$, $m(p)$ indicates the corresponding vertex, and for $l \in L$, $m(l)$ indicates the corresponding edge. We extend m so that c is mapped to $m(c)$. For each $l_i (i = 0, \ldots, n), l_i^* \in c.lines$, if $m(l_i)$ is contained in a graph g, then we say that $m(c)$ is *contained in g*.

Proposition 1. *Let $e = \langle P, L, C, A, outermost \rangle$ be a consistent connected PLCA expression that satisfies $|a.circuits| = 1$ for any area $a \in A$. Then $m(e)$ is a connected graph [21].*

Each circuit of a planar PLCA expression e has an orientation of *inner* or *outer*. If $m(e)$ is a disconnected graph, then it can be decomposed into connected subgraphs. We determine the orientation of each circuit by considering the relationships among these subgraphs, areas and circuits of e.

[Algorithm: DCO(determine circuit's orientation)]

1. Make a node N *outermost*.
2. $setOuterOrientation(outermost, N outermost)$.

Procedure. $setOuterOrientation(c, N_c)$

1. Set the orientation of c to be *outer*.
2. For the subgraph g such that $m(c)$ is contained in g, make a node N_g and draw an edge from N_c to N_g.
3. For each $m(c')$ contained in g such that $c' \neq c$, do the following:
   (a) Make a node $N_{c'}$ and draw an edge from N_g to $N_{c'}$.
   (b) $setInnerOrientation(c', N_{c'})$.

Procedure. $setInnerOrientation(c', N_{c'})$

1. Set the orientation of c' to be *inner*.
2. For an area a such that $c' \in a.circuits$, make a node N_a and draw an edge from $N_{c'}$ to N_a.
3. For each $c'' \in a.circuits$ such that $c'' \neq c'$, do the following:
   (a) Make a node $N_{c''}$ and draw an edge from N_a to $N_{c''}$.
   (b) $setOuterOrientation(c'', N_{c''})$.

A diagram constructed in this way is called a *DCO diagram*. Each path in the diagram is a sequence of a pattern $N_{c_1} \to N_g \to N_{c_2} \to N_a$ where c_1, c_2 are circuits, a is an area of e, and g is a subgraph of $m(e)$. Fig. 3 is a part of the DCO diagram for Example 1.

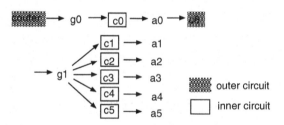

Fig. 3. A part of the DCO diagram for Example 1

Proposition 2. *For a planar PLCA expression e, (i) the orientation of each circuit is decidable, (ii) there exists the unique inner circuit in $a.circuit$ for each area a, and (iii) there exists an outer circuit c such that $m(c)$ is contained in g is an outer boundary cycle of g for each subgraph g [21].*

3   Drawing a PLCA Expression

3.1   Drawing Algorithm

We describe the outline of an algorithm for drawing a figure from a planar PLCA expression in a two-dimensional plane (Fig. 4).

1. Extract the information of points and lines from the planar PLCA to get the corresponding graph expression.
2. Decompose the graph into disjoint connected subgraphs, and determine the coordinates of nodes and edges in a unit circle for each subgraph independently. We utilize an existing graph-drawing algorithm using straight lines in this step [12].
3. Determine the location and the size of these subgraphs using the information on circuits and areas in the PLCA expression.

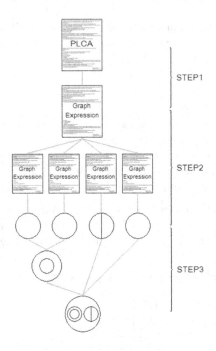

Fig. 4. A drawing process

Definition 6. *(module) Let e be a planar PLCA expression and α be either a point, a line or a circuit of e. If m(e) is decomposed into n disjoint connected subgraphs $g_1, \ldots, g_n$, then we say that e has n modules. If m(α) is contained in g_i, then α is contained in the module corresponding to g_i. For an area a of e, a is contained in the module that contains the inner circuit c in a.circuits.*

Note that each component of the PLCA is contained only in one module.

Definition 7. *(e-circle) A unit circle in which each module of the PLCA is embedded in the second step of the algorithm is called an e-circle.*

Definition 8. *(bridge) An area a such that |a.circuit| ≥ 2 holds is said to be a bridge.*

The third step of the algorithm is the most important since the location and the size of e-circles in a bridge are determined.

For each bridge $a \in A$, do the following: let $a.circuits = \{c_0, c_1, \ldots, c_n\}$, where the orientation of c_0 is *inner* and those of $c_1, \ldots, c_n$ are *outer*. Let $g_1, \ldots, g_n$ be the subgraphs whose e-circles are $ec_1, \ldots, ec_n$, respectively. For each ec_i ($i = 1, \ldots, n$), expand or reduce it and draw it in an appropriate location on the inner part of $m(c_0)$.

Fig. 5(a) shows a part of the DCO diagram that includes bridge a. Fig. 5(b) is a realization of this part. A bridge is actually drawn as a polygon.

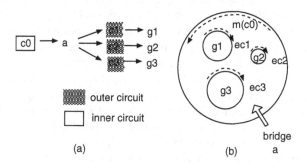

(a)   (b)

Fig. 5. Realization of a bridge

3.2   Circle Packing

Circle packing is an arrangement of circles inside a given boundary such that no two of them overlap and some, or all, of them are mutually tangent [22,16,17]. This is known as an NP-complete problem in general, but optimal solutions have been found in several cases. The studies on circle packing usually treat simple types: the area to be packed is a simple form such as a circle or a rectangle, and few constraints are imposed on the circles to be packed.

The realization of a bridge is considered to be a type of circle packing problem which is formalized as follows:

> **[Problem ∗]** Pack a non-convex polygon with a specified number n of circles with the constraints: (1) all the circles used for packing are as large as possible, and (2) the corresponding circle increases in size with the increasing number of areas in a module.

This problem is a difficult one and none of the existing algorithms can be applied directly. Therefore, we use a Genetic Algorithm, as a more flexible solution.

3.3   Genetic Algorithm

A Genetic Algorithm (GA) is a search technique to find an optimal solution or an approximation to an optimal solution [8]. It is inspired by evolutionary biology concept such as inheritance, mutation, selection and crossover.

In general, after creating the initial populations of chromosomes, each of which is represented by a bit string, the GA involves repeatedly computing the fitness of each chromosome, taking pairs of chromosomes and creating their offspring until a suitable solution is obtained. In creating offspring, crossover (exchanging selected bits between chromosome) and mutation (flipping chosen bits with a certain possibility) are used. Candidate optimal solutions evolve over time.

3.4 Experiment

Gene Encoding. We implemented the above problem [Problem ∗] as follows. Let (x_i, y_i) denote a coordinate of the center of an i-th circle ($1 \leq i \leq n$). In addition, let $M_1, \ldots, M_n$ be the set of modules obtained by decomposing the graph corresponding to a PLCA expression, and n_i be the number of the areas in M_i ($1 \leq i \leq n$).

Each chromosome corresponds to the locations of n circles and it is denoted by an array of the coordinates of their centers $(x_1, y_1), \ldots, (x_n, y_n)$. Each coordinate (x_i, y_i) is encoded as a sequence $a_{i1}, \ldots, a_{iL}, b_{i1}, \ldots, b_{iL}$, where each a_{ij} ($1 \leq j \leq L$) and each b_{ij} ($1 \leq j \leq L$) is a digit either of $0, \ldots, 9^2$, and L is a sufficiently large number. Let $x_{max}, x_{min}, y_{max}$ and y_{min} be the coordinates defined as follows:

x_{max}: the largest x-coordinate of the drawn bridge
x_{min}: the smallest x-coordinate of the drawn bridge
y_{max}: the largest y-coordinate y where (x_i, y) is in the drawn bridge
y_{min}: the smallest y-coordinate y where (x_i, y) is in the drawn bridge

Then, x_i and y_i are calculated as follows so that there is no lethal gene[3](Fig. 6):

$$
\begin{cases}
x_i = (x_{max} - x_{min}) \sum_{j=1}^{n_i} \left(\frac{1}{10} \right)^j a_{ij} + x_{min} \\
y_i = (y_{max} - y_{min}) \sum_{j=1}^{n_i} \left(\frac{1}{10} \right)^j b_{ij} + y_{min}
\end{cases}
$$

The radius of each circle is determined incrementally using distances between the centers of the circles and the boundary of the drawn bridge. We show the pseudo code in the Appendix.

In addition, we use a local search method to compute the fitness for obtaining a better solution.

Parameter Setting. Let S_{ij} ($j = 1, \ldots, n_i$) be the size of an area contained by M_i.

Let S_{total}, N_{total} and S_{av} denote the total size of the areas, the total number of all the areas, and the average size of an area contained in a PLCA expression, respectively. They are defined as follows:

$$
S_{total} = \sum_{i=1}^{n} \sum_{j=1}^{n_i} S_{ij}, \quad N_{total} = \sum_{i=1}^{n} n_i, \quad S_{av} = \frac{S_{total}}{N_{total}}
$$

Fitness is evaluated so that the total size of the circles used for packing an area is as large as possible, and the size of a circle is proportional to the total number of the

---

[2] We used digits here for encoding whereas bits are used in general.
[3] This is the basic definition. The actual calculation of y_i is more complicated.

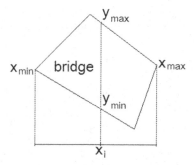

Fig. 6. Calculation of the coordinate

areas that are recursively contained in the corresponding module. Therefore, it can be calculated as:

$$fitness = S_{total} - \sum_{i=1}^{n} \frac{1}{N_i} \cdot \sum_{j=1}^{n_i} |S_{av} - S_{ij}|$$

where N_i is the total number of the areas that are recursively contained in module M_i.

Crossover takes place at n randomly chosen points, and mutation at randomly chosen positions which in our case occurs when the digit is changed to any other digit.

We performed the simulation several times to find appropriate values for the crossover and mutation rates. As a result, the mutation rate is set at the fixed value 1.0, and the crossover rate is set at 0.4 for PLCA expressions with multiple bridges one inside another, and at 0.9 for PLCA expressions with a bridge in which multiple modules are embedded.

Experiment and Evaluation. Experiments are performed using three PLCA expressions corresponding to the figures shown in Fig. 7.

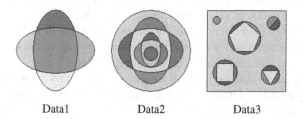

Data1   Data2   Data3

Fig. 7. Figures corresponding to given PLCA expression

1. Data1:This is a simple PLCA expression with one module and no bridge.
2. Data2: This has multiple bridges, one inside another, and is used to check that circles are drawn as large as possible.
3. Data3: This has a bridge in which multiple modules with varying numbers of areas are embedded. This is used to check that the size of a circle is proportional to the total number of areas recursively contained in the corresponding module.

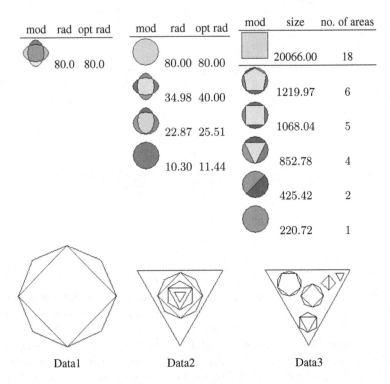

Fig. 8. Results of drawing

We use a $160*160$ rectangle as a drawing field. Fig. 8 shows the result of the drawing. It shows that all the relationships of the PLCA components are preserved. Above all, each module is embedded in the correct bridge.

In the tables for Data1 and Data2, *mod* is the module, *rad* is the radius of an e-circle and *opt rad* is the radius of an inscribed circle of a bridge, that is the biggest size of a module to be located there. In the table for Data3, *mod* is the module, *size* is the total size of all the areas contained in that module, and *no. of areas* shows the total number of areas recursively contained in the corresponding module.

In evaluating the results, we ignore the shape of an *outermost*, which is always an inscribed polygon of an e-circle, and discuss the locations and the sizes of the e-circles.

For Data1, the radius is the biggest radius. It follows that a module is drawn as large as possible.

For Data2, the radii of e-circles are slightly smaller than the optimal ones, since the circle should not be tangent to the boundary of the bridge. It follows that the modules are drawn as large as possible.

For Data3, the size of an e-circle is proportional to the number of areas contained in the corresponding module. It is considered that a module containing the larger number of areas is more complex. Therefore, it follows that complex objects are drawn larger than non-complex objects.

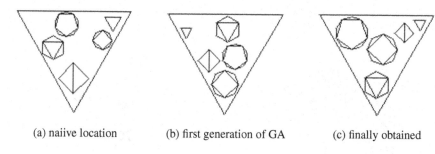

| (a) naiive location | (b) first generation of GA | (c) finally obtained |

Fig. 9. Cognition of "goodness"

We showed the three figures in Fig. 9, equivalent according to PLCA, to twenty test subjects, and asked them to select the "best" figure. Seventeen of them chose (c), the figure produced by our algorithm.

From these results, we conclude that we have obtained the "good figure".

4   Discussion

4.1   Direct Drawing from PLCA

The basic concept of the algorithm used in our experiment was to transform a PLCA expression into a graph, and then draw that graph.

We have proposed another method for drawing a figure directly from a PLCA expression [18]. The basic idea of that algorithm was to draw circuits. Starting from the outermost, draw circuits in the inner area enclosed by the outermost. Draw all the circuits by repeating this procedure recursively.

For that algorithm, we can prove that a figure that preserves the relationship described in the PLCA expression can be drawn, but the method for determining the specific coordinates is not given. This means that we can draw a figure only if we specify proper coordinates. If the circuits share lines or points, a subtle problem exists in determining the size of lines, the location of points, and the shape of circuits. Unfortunately, the actual figure cannot be drawn by automatically using this algorithm.

For example, consider a PLCA expression corresponding to Fig. 10(a). Fig. 10(b) shows three processes for drawing a figure according to this algorithm. If we draw circuits using the process in the right-hand column, then the result is successful. But if we use in the other processes, the drawing fails.

Information from the other circuits to be drawn is not available at the time of drawing the first circuit because this algorithm uses recursive processing. Therefore, appropriate coordinates cannot be determined algorithmically. The algorithm using GA is more flexible in producing a solution.

4.2   Drawing from Other Qualitative Representations

Region Connection Calculus (RCC) [13] and the 9-intersection model [6] are representative frameworks for qualitative spatial reasoning. We examine an algorithm for

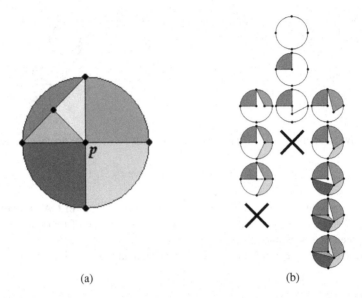

(a)   (b)

Fig. 10. Drawing from a PLCA expression directly

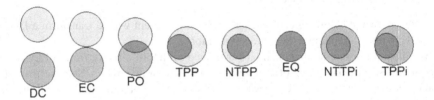

Fig. 11. Fundamental relationships of RCC

drawing a figure in a two-dimensional plane based on these representations. Although the existence of a topological space and planarity for a set of RCC relationships have been discussed in [9,14,23], an algorithm for drawing has not been reported so far.

Fig. 11 shows the eight fundamental relationships of RCC. Consider a set of RCC relationships $S = \{R_1, \ldots, R_n\}$ which is realizable in a two-dimensional plane. If each $R_i(i = 1, \ldots, n)$ is either NTPP, NTPPi, DC or EQ, then S is transformed uniquely into PLCA. In this case, the figure corresponding to S can be drawn using the PLCA drawing algorithm. Otherwise, there exists multiple PLCA expressions. While we can draw one of them, it is an open question as to which one. The 9-intersection model, in which positional relationships between regions are represented in the form of a $3*3$ matrix, results in similar uncertainty.

A major difference between PLCA and the other QSR systems is in the way in which relationships among objects are represented. In the other QSR systems, the entire figure is represented in the form of a set of binary relations, while we do not use binary relations. Moreover, objects in the other QSR systems may share parts with each other which is prohibited in PLCA. PLCA uses a more refined classification for equivalent figures than the other QSR systems. It may be possible to determine a one-to-one mapping to PLCA from the extension of RCC, containing information on the connection patterns of regions [5], or the extension of the 9-intersection model [7,10].

5 Conclusion

In this paper, we have proposed an algorithm for drawing a figure in a two-dimensional plane corresponding to a PLCA expression. In general, it is difficult to draw a figure for a qualitative representation since coordinates are not determined uniquely. We thus have proposed an algorithm using GA, and succeeded in producing drawings. The resulting figures not only preserve the relationships in PLCA expressions, but are also "good" figures in the sense that their objects are drawn as large as possible, and complex parts are drawn larger size than those less complex.

We used the size and the number of areas as parameters in the GA. If a specific requirement such as emphasis on a certain object is integrated, we can automatically draw different figures for the same PLCA expression depending on the application, to meet specific requirement details. In future, we are considering the reduction of time in finding a solution in the GA. We will conduct further study on algorithms to draw directly from PLCA expressions and from other qualitative representations. Furthermore, the solution to the problem of packing multiple circles in a non-convex polygon given in this paper could be useful in other applications of packing.

Acknowledgments. This research was supported by KAKENHI19500134.

References

1. Anderson, M., Meyer, N., Olivier, P. (eds.): Diagrammatic Representation and Reasoning. Springer, Berlin (2002)
2. Chartland, G., Lesniak, L.: Graphs & Digraphs, 3rd edn. Wadsworth & Brooks/Cole (1996)
3. Cohn, A.G.: A hierarchical representation of qualitative shape based on connection and convexity. In: Kuhn, W., Frank, A.U. (eds.) COSIT 1995. LNCS, vol. 988, pp. 311–326. Springer, Heidelberg (1995)
4. Cohn, A.G., Hazarika, S.M.: Qualitative spatial representation and reasoning: an overview. Fundamental Informaticae 46(1), 1–29 (2001)
5. Cohn, A.G., Varzi, A.: Mereotopological connection. Journal of Philosophical Logic 32, 357–390 (2003)
6. Egenhofer, M., Herring, J.: Categorizing binary topological relations between regions, lines and points in geographic databases. Technical Report. Department of Surveying Engineering, University of Maine (1990)
7. Egenhofer, M., Franzosa, R.: On the equivalence of topological relations. International Journal of Geographical Information Systems 9(2), 133–152 (1995)
8. Goldberg, D.E.: Genetic algorithms in search, - Optimization and machine learning. Kluwer Academic Publishers, Dordrecht (1989)
9. Grigni, M., Papadias, D., Papadimitriou, C.: Topological Inference. In: International Joint Conference on Artificial Intelligence, pp. 901–907 (1995)
10. Nedas, K., Egenhofer, M., Wilmsen, D.: Metric Details of Topological Line-Line Relations. International Journal of Geographical Information Science 21(1), 21–48 (2007)

11. Overmars, M., de Berg, M., van Kreveld, M., Schwarzkopf, O.: Computational Geometry. Springer, Berlin (1997)
12. Ochiai, N.: Introduction to Graph Theory: Application for Plane Graph. Nihon-Hyouron-sha (In Japanese) (2004)
13. Randell, D., Cui, Z., Cohn, A.G.: A spatial logic based on regions and connection. In: Proceedings of the Third International Conference on Principles of Knowledge Representation and Reasoning (KR92), pp. 165–176. Morgan Kaufmann, San Francisco (1992)
14. Renz, J.: Qualitative Spatial Reasoning with Topological Information. In: Renz, J. (ed.) Qualitative Spatial Reasoning with Topological Information. LNCS (LNAI), vol. 2293, Springer, Heidelberg (2002)
15. Stock, O. (ed.): Spatial and Temporal Reasoning. Kluwer Academic Publishers, Dordrecht (1997)
16. Stephenson, K.: Circle packings in the approximation of conformal mappings. Bulletin of American Mathematics Society 23, 407–416 (1990)
17. Stephenson, K.: Introduction to circle packing - The theory of discrete analytic functions. Cambridge University Press, Cambridge (2005)
18. Sumitomo, T., Takahashi, K.: DLCS: Qualitative representation for spatial data. In: Twenty-first Annual Meeting of Japan Society for Software Science and Technology (In Japanese) (2004)
19. Sumitomo, T., Takahashi, K.: A qualitative treatment of spatial data. In: The 17th IEEE International Conference on Tools with Artificial Intelligence (ICTAI05), pp. 539–548. IEEE Computer Society Press, Los Alamitos (2005)
20. Takahashi, K., Sumitomo, T.: A framework for qualitative spatial reasoning based on the connection patterns of regions. In: IJCAI-05 Workshop on Spatial and Temporal Reasoning, pp. 57–62 (2005)
21. Takahashi, K., Sumitomo, T., Takeuti, I.: On embedding a qualitative representation in a two-dimensional plane. In: IJCAI-07 Workshop on Spatial and Temporal Reasoning, pp. 101–109 (2007)
22. Williams, R.: Circle packings, plane tessellations, and networks. In: The Geometrical Foundation of Natural Structure: A Source Book of Design. Dover, New York, pp. 34–47 (1979)
23. Wolter, F., Zakharyaschev, M.: Spatio-temporal representation and reasoning based on RCC-8. In: Proc. International Conference on Principles of Knowledge Representation and Reasoning (KR2000), pp. 3–14. Morgan Kaufmann, San Francisco (2000)

Appendix   Determining the Radii of Circles

$C = \{c_1, c_2, \ldots, c_n\}$ where c_i $(1 \leq i \leq n)$ is the coordinate of the center.
$Decided = \{\}$.   % a set of cirlces whose radii are decided
$Undecided = \{\}$.   % a set of cirlces whose radii are undecided

WHILE $(|C| > 0)$
    Take an arbitrary c_i of C.
    Set d_{i1} to the shortest distance between c_i and the boundary of $Area$.
    Set d_{i2} to the half of the shortest distance between c_i and c_j for each j $(i \neq j)$.
    Let r_i be the radius of the circle whose center is c_i.
    IF $(d_{i1} < d_{i2})$
        $r_i = d_{i1}$.
        $Decided = Decided \cup \{c_i\}$.

IF $((d_{i2} < d_{i1}) \wedge (d_{j2} < d_{j1}))$
$\quad r_i = d_{i2}$.
$\quad Decided = Decided \cup \{c_i\}$.
ELSE
$\quad Undecided = Undecided \cup \{c_i\}$.
ENDWHILE
WHILE $(|Undecided| > 0)$
Take an arbitrary c_i of $Undecided$.
Set d_{i2} to the half the shortest distance between c_i and $c_j \in Undecided$ $(i \neq j)$.
Set d_{i3} to the shortest distance between c_i and $c_k \in Decided$.
IF $(d_{i1} = \min(d_{i1}, d_{i2}, d_{i3}))$
$\quad r_i = d_{i1}$.
$\quad Decided = Decided \cup \{c_i\}$
IF $(d_{i3} = \min(d_{i1}, d_{i2}, d_{i3}))$
$\quad r_i = d_{i3}$.
$\quad Decided = Decided \cup \{c_i\}$.
ELSE
$\quad Undecided = Undecided \cup \{c_i\}$.
ENDWHILE

Linguistic and Nonlinguistic Turn Direction Concepts

Alexander Klippel[1] and Daniel R. Montello[2]

[1] GeoVISTA Center, Department of Geography, Pennsylvania State University, PA, USA
klippel@psu.edu
[2] Department of Geography, University of California, Santa Barbara, CA, USA
montello@geog.ucsb.edu

Abstract. This paper discusses the conceptualization of turn directions along traveled routes. Foremost, we are interested in the influence that language has on the conceptualization of turn directions. Two experiments are presented that contrast the way people group turns into similarity classes when they expect to verbally label the turns, as compared to when they do not. We are particularly interested in the role that major axes such as the perpendicular left and right axis play—are they boundaries of sectors or central prototypes, or do they have two functions: boundary and prototype? Our results support a) findings that linguistic and nonlinguistic categorization differ and b) that prototypes in linguistic tasks serve additionally as boundaries in nonlinguistic tasks, i.e. they fulfill a double function. We conclude by discussing implications for cognitive models of learning environmental layouts and for route-instruction systems in different modalities.

1 Introduction

Directions (angles) are basic spatial relations (e.g., [15,17,21]). The processing and representation of directional/angular information is essential for human spatial cognition, especially for wayfinding and the creation of mental spatial representations used in both linguistic and nonlinguistic tasks (e.g., [31,39,44,51]).

One of the first things we can observe about directional knowledge is that we do not conceptualize every potential direction that exists in our environments or to which our bodies could turn. That is, we do not demonstrate infinitely precise directional information. For most situations, qualitative information about directions—in the sense of a fairly small number of equivalence classes—is sufficient. The way this information captures quantitative information about direction, but only imprecisely or approximately, has been referred to as "qualitative metrics" [11,38].

Additionally, environments in which cognitive agents dwell can place supplementary constraints on possible turn directions. In city street networks which constrain possible turn directions, choices of precise angular information are rarely, if ever, necessary. Various studies show that even though humans may perceive angular information precisely, in city street networks—as well as in body and geographic spaces more generally—humans conceptualize and remember it with limited precision (e.g., [5,12,37,43,46]). Verbal route instructions also reflect this qualitativeness; precise, very fine-grained, directional information is exceptional, not typical [2,10].

S. Winter et al. (Eds.): COSIT 2007, LNCS 4736, pp. 354–372, 2007.

If we consider directional categories to be conceptual spatial primitives, questions arise as to how many different categories of directions are necessary and how many categories humans employ in various tasks. Additionally, we can ask whether there are prototypical turning concepts around which less prototypical directions are organized (e.g., [49]). A question of specific interest is the difference between the conceptualization of directional categories in <u>linguistic</u> versus <u>nonlinguistic</u> tasks. In other words, does language influence the conceptualization of directions turns, and if so, how. The central topic of this paper is the question of how language influences the conceptualization of directional turns.

The rest of the paper is structured as follows: We provide a short overview of linguistic and nonlinguistic categorization of directions in spatial tasks. Subsequently we present the results of two experiments. Both use a grouping paradigm that allows us to identify conceptual structures on the basis of participants' similarity classifications. Experiment 1 applied a nonlinguistic task. Experiment 2 applied both a nonlinguistic task and a linguistic task in which participants were made aware that they would verbally label the directional classes. The results are discussed with respect to their implications for cognitive conceptual models of directions—as central parts in mental map theory, cognitive wayfinding assistance systems, and human-machine interaction.

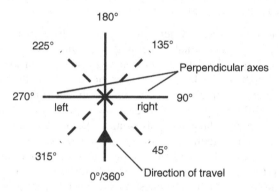

Fig. 1. Depiction of the direction terminology used in this article

We follow the convention of using *italics* for examples of linguistic utterances such as *turn left*; SMALL CAPS are used to refer to a concept, for example, LEFT. To refer to turn directions unambiguously, we apply the following terminology depicted in Fig. 1. We use a full circle (360°) as reference. The counting of the angles starts at 6 o'clock in counter-clockwise direction. 6 o'clock is referred to as 0°/360°, perpendicular right (3 o'clock) is 90°, straight (12 o'clock) is referred to as 180°, and perpendicular left (9 o'clock) is referred to as 270°. Additionally we separate the perpendicular axis into a perpendicular left axis (270°) and a perpendicular right axis (90°). As the setting of the experiments is in the domain of route following, we have the additional perspective of turning as deviation from the main direction of travel (straight = 12 o'clock, 180°). A 90° right turn is equal to 90°, and a 90° left turn is equal to 270°, in our absolute terminology.

2 Linguistic and Nonlinguistic Categorization of Directions

The linguistic and nonlinguistic categorization of direction concepts is a highly active research area (e.g., [8,9,40,48]). Some of the main results, especially for linguistic categorization, can be summarized as follows:

1. Direction changes close to 90° left and right are referred to as (*turn*) *left* and (*turn*) *right* with little variation. Similar findings hold for the concept STRAIGHT, as in *directly straight*. In contrast, direction concepts beyond the main axes of left, right, and straight are not easily associated with a single linguistic term, and a plethora of composite linguistic expressions is used for referring to directions that are between the three main direction concepts—LEFT, RIGHT, STRAIGHT—in two-dimensional space. 'Hedge terms' are used to linguistically indicate gradation effects [6,20,29,50]. We find expressions such as: *turn slightly right, go right 45 degrees, veer right, sharp right bend* and so forth[1].
2. The conceptualization of directions besides the main axes is not straightforward, and the sizes and the boundaries of sectors, or more generally, the semantics and applicability of corresponding spatial prepositions, are ongoing research questions (e.g., [8,12]).
3. The prominence of perpendicular axes as prototypes has been found in nonlinguistic tasks [22,29,43].

While linguistic analysis as a window to cognition allows us to shed light on the question of underlying conceptual structures, the influence of language on the conceptualization of spatial relations (and on cognition in general) is a subject of ongoing debate [3,9,18,19,32,34,41]. There are, of course, extreme positions on the role of language in cognition: That it either completely determines nonverbal thought or that it has no influence at all. However, a majority of researchers probably subscribe to the idea that language has some influence on cognition, including specifically on the conceptualization of spatial relations, even though it is not completely determinative. These theories are often discussed under the term *linguistic relativity* (e.g., [16]) and sometimes further differentiated into weak and strong language-based approaches [14]. The question, however, concerns how profound this influence is and how it is manifested.

One of the major research approaches to addressing this debate is to compare different languages in their expression of spatial relations and how their expressions may in turn influence nonlinguistic thought (see, for example, [4] and a reply by January and Kako [23]). One of the criticisms that January and Kako [23] bring forth is that effects of language on thought should be detectable not only in cross-linguistic studies but also within a single language. The study performed by Crawford et al. [9] found such differences. In three experiments, they analyzed the role of the vertical axis in determining spatial relationships between a reference object and a target. In the linguistic task they used agreement ratings; in the nonlinguistic task, location estimates were employed. Focusing on the vertical axis, they concluded: "[W]e found

---

[1] These example verbalizations are taken from a data set collected at the University of California, Santa Barbara. They were all instructions given for the same turning angle of 140.625°.

an inverse relation between linguistic and nonlinguistic categorization of space. Specifically, the vertical axis, which serves as a category prototype in spatial language, serves as a category boundary in nonlinguistic organization of space." ([9], p. 234). It has to be noted, however, that serving as a prototype and serving as a boundary are not necessarily mutually exclusive. In other words, while in a linguistic task the boundary of a sector is different from a prototypical direction that represents the sector, in a nonlinguistic task prototype and boundary can coincide. This observation motivates our investigation of turn categorization in the present studies, consisting of two experiments in which we explore further the linguistic and nonlinguistic conceptualization of direction concepts in a navigation task. In contrast to the research by Crawford et al. [9], our focus is placed on the horizontal axis.

3 Experiments

We conducted two experiments, both employing a grouping paradigm to assess the similarity—and thereby the underlying conceptual structures—of direction changes at intersections, i.e., turn directions. The first experiment[2] involves a nonlinguistic task in which participants group turns of various angles according to their similarity, followed by a linguistic task in which participants provide verbal labels for their groups. However, participants learn about the linguistic task only <u>after</u> the nonlinguistic task is finished. The second experiment comprises two conditions, administered as a within-subjects factor. The first condition replicates Experiment 1; participants group turns according to similarity, followed by verbal labeling. The second condition requires the same participants to repeat the nonlinguistic grouping task, but this time with the <u>prior</u> knowledge (acquired from participating in the first condition) that they would have to provide verbal labels for the groups. This design allows us to assess whether the linguistic component of the experiments influences similarity groupings and the conceptualization of turn directions.

3.1 Experiment 1

In this experiment, we explored the underlying conceptual structure of directional knowledge. Specifically, we explored how many angular categories should be assumed as a basic set of equivalence relations for knowledge of directions in city street networks, and what the relation between angular sectors and axes is in this domain. We used a grouping task paradigm, which has long been an important method in psychology for investigating conceptual knowledge (e.g., [7]). The motivation behind the grouping task is that people primarily use conceptual knowledge to determine the similarity of given stimuli. Stimuli are placed into the same group of objects if they are regarded as similar, i.e., as instances of the same concepts; they are placed into different groups if they are regarded as dissimilar, i.e., as instances of different concepts. If other aspects of the stimuli and their presentation are controlled, as in our experiment, grouping tasks can provide important insight into the internal structure of conceptual knowledge. To conduct the experiment, we used

---

[2] The first experiment reanalyzes a data set that has been reported in a technical report (Klippel et al. 2004) and places it in the context of linguistic and nonlinguistic conceptualization.

an experimental tool developed by Knauff et al. [27]. The tool implements a procedure that is comparable to card sorting but automatically assists with generating experimental materials, presenting stimuli, and collecting responses as data.

3.1.1 Methods

Participants. Twenty-five students (9 female) from the University of Bremen were paid for their participation. Their average age was 25.6 years.

Materials. We used 112 icons to depict different ways to 'make a turn' at an intersection. We knew that participants would generally not distinguish turns differing by only 1°, and 360 different turn icons would make an unwieldy experiment in any case. So to design the icons, we applied the results of Klippel [24] to the full 360° of two-dimensional space, adding bisection lines incrementally. We started with the prototypical egocentric direction concepts (excluding directly back) of 45°, 90°, 135°, 180°, 225°, 270°, 315°, and bisected the sectors several times until we had turns differing by increments of about 5° (5.625°). Each turn was depicted as a simple intersection with two branches (a path heading into the turn, and a path heading out of it). Turns in the back sector (angles between 337.5° and 22.5°) were excluded for graphical reasons, as very sharp turns were not clearly discriminable on the screen. This produced 56 distinct icons. We doubled the icons as a check on grouping reliability, resulting in 112 icons.

The icons were integrated into the grouping tool. Fig. 1 shows a screenshot of an ongoing experiment. The grouping tool divides the screen in two parts. On the left side, the stimulus material, consisting of all icons depicting possible turns at an intersection, are placed in a different random order for each participant. The large number of icons required scrolling to access all items. (Scrolling is a common procedure in interacting with computer interfaces; no problems were expected nor found during the experiments.) The right side of the screen is empty at the start; participants move icons to this side in order to group them during the experiment. The interface was kept simple so that participants could perform only the following four actions: Create a new group, Delete an existing group, Rearrange, Done.

Procedure. The experiment took place in a laboratory at the University of Bremen. Participants were tested individually. After arriving and obtaining some basic information, we explained to them the general procedure of the experiments and demonstrated the functionality of the grouping tool. Participants were advised to imagine the icons (in German: *Bildchen*) as schematic representations of possible turns at an intersection (in German: *"auf den Bildchen [werden] verschiedene Abbiegemoeglichkeiten an Kreuzungen schematisch dargestellt"*). They were told to place the icons on the left into groups on the right showing what they considered to be similar turning possibilities, in such a way as to maximize similarity within groups and dissimilarity between groups (in German: *"Bitte fasse dir ähnlich erscheinende Abbiegemöglichkeiten in einer Gruppe zusammen, d.h. das Abbiegen soll innerhalb einer Gruppe möglichst ÄHNLICH sein und möglichst verschieden zu den anderen Gruppen"*).

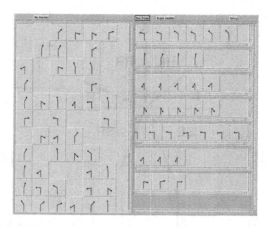

Fig. 2. The grouping tool (snapshot from an ongoing experiment) used in both experiments. On the left side, icons representing turns are presented in random order. On the right side, a participant has started to group icons according to her categories of turns.

After the grouping task, participants were asked to create verbal labels for each of the groups they had created, i.e., a linguistic description for the kind of turn represented by a group (in German: *"Bitte versuche im folgenden sprachliche Beschreibungen für die verschiedenen Möglichkeiten des Abbiegens zu finden, die Du in einer Gruppe zusammengefaßt hast"*). They were additionally advised to keep their description as short as possible.

3.1.2 Results

The groupings of each participant result in a 112 x 112 similarity matrix, the number of the icons used in this experiment on each axis. This matrix allows us to code all possible similarities between two icons simultaneously for all icons in the data set; it is a symmetric similarity matrix with 12,544 cells. Similarity is coded in a binary way; any pair of icons is coded as '0' if its two items are not placed in the same group and '1' if its two items are placed in the same group. The overall similarity of two items is obtained by summing over all the similarity matrices of individual participants. For example, if two icons (called A and B) were placed into the same group by all 25 participants, we add 25 individual '1's to obtain an overall score of 25 in the respective cells for matrix position AB and BA.

To analyze the categorical grouping data, we subjected the overall similarity matrix to a hierarchical cluster analysis with the software CLUSTAN. We applied average linkage (also known as UPGMA), as this method calculates the distance between two clusters as the average distance between all inter-cluster pairs. Average linkage is generally preferred over nearest or furthest neighbor methods (single linkage or complete linkage, respectively), since it is based on information about all inter-cluster pairs, not just the nearest or furthest ones. The cluster structure at the cut-off point for the seven-cluster solution is shown in Fig. 2. The dendrogram is truncated at 56 clusters, as the use of redundant pairs of identical icons had little effect on the similarity ratings—identical icons were placed in the same groups with only two single exceptions.

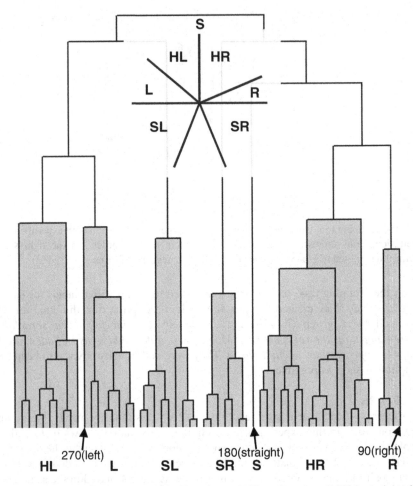

Fig. 3. Results of the cluster analysis in Experiment 1. Turn directions (180° is straight ahead) that are grouped into common clusters are shaded. We truncated the dendrogram at 56 clusters, which led to an under-representation of the width of clusters SL and SR in the dendrogram. The derived angular values are schematically represented in the directions model in the middle of the figure. Most importantly for the analysis here, there is clear perpendicular demarcation of the front and back plane indicated by the bold lines at 90° and 270°.

The following patterns emerge in these data. There are seven distinguishable direction sectors (categories). For convenience, we refer to them as HL (HALF LEFT, translated from the German expression *halb links*), L (LEFT), SL (SHARP LEFT), S (STRAIGHT), HR (HALF RIGHT), R (RIGHT), and SR (SHARP RIGHT). The sectors differ in angular size, with the widest being the sectors SR, HR, HL, and SL, less wide being sectors R and L, and the most narrow being the straight 'sector', which very precisely consists of only the single 'turn' of 180° (±2.8125°). This suggests that direction concepts depend on a combination of axes and sectors. Turns to the left and right are very nearly symmetric. Most importantly, while straight ahead (180°) remains a separate axis, the perpendicular axes (90°/270°) are grouped into the corresponding adjacent (front) sectors. This leads to a clear distinction between the front and the back plane at the perpendicular axes, with less discrimination among back turns than front turns.

From our own experience with verbal and graphical route directions and the conceptualization of turn directions [25,33], we did not expect this clear demarcation at the perpendicular axes into front and back planes. However, this pattern has been noted by researchers interested in differences between linguistic and nonlinguistic conceptualizations of spatial relations (e.g., [9,12]). We therefore extend our analysis to verify that this demarcation is a robust pattern in our data. Following a suggestion by Kos and Psenicka [28], we attempt to validate our interpretation of the cluster analysis by comparing different clustering methods. Fig. 4 shows the outcome of this comparison by juxtaposing three different clustering methods: single linkage, average linkage, and Ward's method. The analysis was performed using CLUSTAN with the default similarity measure of squared Euclidean distance. These three methods provide a range of possible clustering solutions and usually differ with respect to their outcomes (e.g., [1]). Here, we are interested in possible differences with respect to the angle that indicates boundaries between sectors, i.e., the size of the sectors. Fig. 4 clearly shows the proneness of the single linkage method toward chaining as well as the characteristic of Ward's method to create compact clusters (for details, see [1]). The parts of the dendrograms displayed show only the sectors that contain the perpendicular axes (90° and 270°) on the basis of a seven-cluster solution. For our purposes, however, we note two points of major importance:

- First, we confirm that across all methods, the perpendicular left-right axes at 90° and 270° are prominent. In fact, they provide the most prominent direction concepts in the sectors under investigation (the data are not collapsed, so there are two icons each for 90° and 270°).
- Second, all methods reveal the same pattern with respect to the perpendicular left-right axis: It provides the demarcation line between the front and back plane, and both axes are associated with turns in the front plane rather than those in the back plane. Even Ward's method, which is often applied in psychological research (e.g., [36]) because it promotes compact and interpretable clusters, does not change this pattern. (Ward's method did change the emergence of the straight cluster, but this is not our concern here.)

Results of the linguistic labeling task that participants performed subsequent to grouping the turns are not discussed in depth here. As an example, for a participant who created seven individual groups, we find: *gerade aus* (straight), *nach links abbiegen* (turn left), *nach rechts abbiegen* (turn right), *halblinks nach vorne abbiegen* (turn half left to the front), *halbrechts nach vorne abbiegen* (turn half right to the front), *scharf links nach hinten abbiegen* (turn sharp left to the back), *scharf rechts nach hinten abbiegen* (turn sharp right to the back).

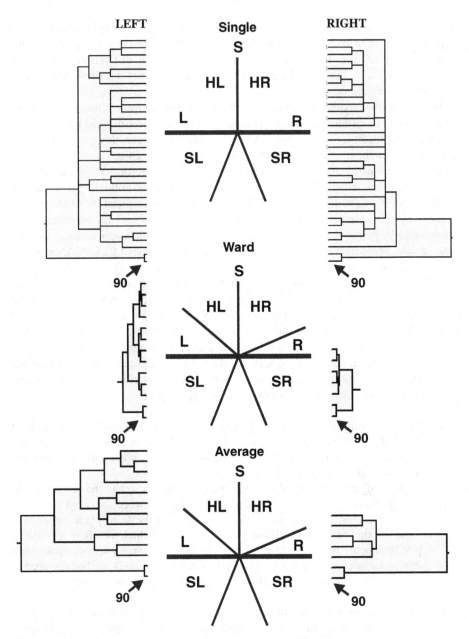

Fig. 4. Results for three methods of cluster analyses in Experiment 1: single linkage (as in the original analysis), Ward's method, and average linkage. The results for all three methods show the same pattern with respect to the perpendicular left- and right axes.

3.1.3 Discussion

The results of Experiment 1 agree with previous research on prototypical, salient directions for the concepts RIGHT (90°), STRAIGHT (180°), and LEFT (270°) (e.g., [24,47]). More importantly, they are consistent (for LEFT and RIGHT) with the interpretation of linguistic and nonlinguistic spatial categorization derived from the work by Crawford et al. [9], namely that axes which serve only as prototypes of linguistic categories serve additionally as the boundaries of nonlinguistic categories. Prototypicality of the perpendicular left and right axes is consistent with findings by Sadalla and Montello [43], i.e., 90°, 180°, and 270° serve as prototype turns in a task that required participants to estimate the size of the turn they had just walked with a circular pointing device; both constant and variable errors were minimized around these axes.

We have to ask, however, what really influenced the similarity groupings that participants applied. Did they really treat this as a nonlinguistic task? Would the results have been different if participants knew in advance that they would have to create verbal labels for the groups that they created? Also, we can ask whether the design of the icons affected the categorization, particularly the fact that no sharp backward turns were provided to categorize? We conducted a second experiment to shed light on these questions.

3.2 Experiment 2

Experiment 2 employed the same setting, task, and materials as in Experiment 1, with a couple of modifications. We included some of the sharp backward turns, the back sector, that we left out in Experiment 1. Also, we did not double the icons this time, as we required participants to perform the turn direction categorization twice in two conditions. The first time, they did it in the same manner as in Experiment 1, not knowing in advance they would have to provide linguistic labels for the groups at the end. The second time, they repeated the categorization task again, but this time, they knew that they would have to verbally label the direction categories (the groups) after creating them. That is, participants were primed by the linguistic component of the first condition of the experiment before they performed the second condition. In this way, we could find out if an expectation to create verbally labeled categories would change the structure of the categories, as compared to when there was no expectation of producing verbal labels.

3.2.1 Methods
Participants. Twenty-four students (12 female) from the University of Bremen were paid for their participation in the experiment. Their average age was 24.9 years.

Materials. We slightly modified the design of the stimulus material from Experiment 1. We used the same angular increments (5.625°) for the turn icons, but we varied their overall appearance. Instead of using arrows we used regular lines without arrowheads (Fig. 5). Additionally, we included icons to represent turn concepts in the back sector. As it is barely possible, given the resolution of our system, to visually represent the sharpest backward directions, we still omitted icons for 0°/360°, 5.625° and 11.25°, and 354.375° and 348.75° (i.e., straight back and the two sharpest turns

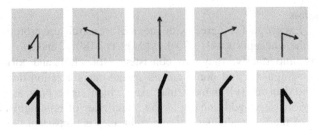

Fig. 5. Upper row shows example icons from Experiment 1; lower row shows example icons from Experiment 2

clockwise and counterclockwise). We also slightly modified the length of the legs to avoid the impression of a clock. Because participants would have to group the turns twice in two different conditions, and because we knew from Experiment 1 that doubling the turns had no effect on grouping, we did not double the icons in Experiment 2. Thus, each condition involved a total of 59 icons. Exactly the same grouping tool was used as in Experiment 1, with the same interface.

Procedure. The procedure for Experiment 2 was the same as for Experiment 1, except that participants repeated the grouping task and the labeling task twice. That means participants performed the same two tasks twice with the only difference being that when they grouped the icons the second time, they would be aware of the linguistic component of the task. This results in the following sequence: Group the icons once, label the groups that were created, group the original set of icons again, label the newly created groups. Participants were not explicitly advised to pay specific attention to anything linguistic in the task. Otherwise, the instructions were the same as in Experiment 1.

3.2.2 Results

Participants created fewer groups in the second condition of the experiment (8.7) than in the first condition (10.3), although this did not quite reach statistical significance, $t(23)=1.99$, $p=.058$. We are particularly interested in whether knowing about the linguistic component of the task influenced the structure of the turn groups in the second condition of the experiment. Specifically, would the role of the perpendicular axis (90° for RIGHT, 270° for LEFT) change, either as a prototype or as a boundary? We therefore compared the results of the two conditions of Experiment 2 with each other and with the results of Experiment 1 with respect to the directional sectors the participants created. To make these comparisons, we assumed a seven-sector model and used the average linkage method in CLUSTAN.

Fig. 6 shows schematic depictions derived from the angular values found for the different sectors, for Experiment 1 and the two conditions of Experiment 2. We label them with the abbreviations for the sectors introduced above. The results for the first condition of Experiment 2 and for Experiment 1, in which participants had no prior knowledge of the verbal labeling task, are very similar. In particular, in both cases, the perpendicular axes of 90° and 270° demarcate the boundary of sectors coinciding with the demarcation of the front and back plane. The differences in the design of the icons between the two experiments apparently did not influence the role of the perpendicular axes; there was no difference in the boundary function of the turning directions 90° and 270° between Experiment 1 and first condition of Experiment 2.

In contrast, in the second condition of Experiment 2, this clear demarcation between front and back plane disappears. That is, the 90° and 270° turn directions are embedded in the sector but do not represent its boundary. For the RIGHT sector, 90° is in fact the bisecting line of the sector; for the LEFT sector, 270° is skewed toward the 'upper' end of the sector.

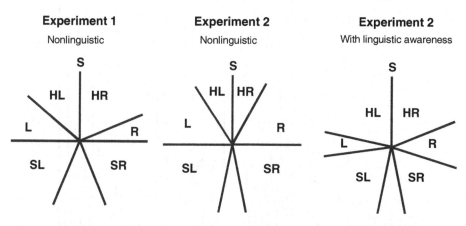

Fig. 6. Analysis of the directional categories using average linkage as a cluster method and assuming a seven-cluster solution. The figure compares the results of Exp. 1 with both conditions of Exp. 2 (without and with advance knowledge of the linguistic labeling task).

The third main axes, the one associated with the concept STRAIGHT, is unaffected by the experimental design. The presence or absence of a linguistic labeling task had no effect on the conceptualization of STRAIGHT as an axis—essentially a very narrow 'sector' of only one turn direction. We also find that the boundaries of the other sectors, particularly the diagonals between the main axes, vary, which agrees with many research results on the difficulty of locating them and on their gradation effects (e.g., [49]).

To provide deeper insight into the underlying conceptualization of the turn directions around the left-right perpendicular axes, we display the results of the hierarchical cluster analysis (average linkage, CLUSTAN) in Fig. 7. This figure is generated in the same way as Fig. 3 for Experiment 1. It reveals that in the second condition of Experiment 2, the 90° and 270° turn directions play different roles in their associated sectors. Here, they are embedded in the corresponding sectors instead of serving as their boundaries. In Experiment 2, one could question whether the seven-cluster solution in this analysis is the best cut-off point, but that is not the focus of this paper and would not affect our conclusion that the RIGHT and LEFT sectors are different when verbal labeling is expected, as compared to when it is not expected.

To delve deeper into the underlying grouping processes in these tasks, Fig. 8 presents a portion of the raw data in order to further reveal the conceptual status of the perpendicular LEFT and RIGHT axes in the nonlinguistic first condition of Experiment 2 (labeled 'without' in the figure) and the linguistic second condition of

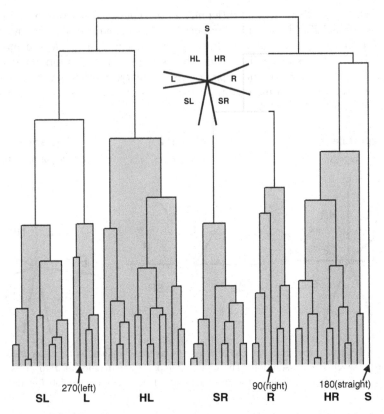

Fig. 7. Results of the cluster analysis for the second condition of Experiment 2, in which participants had advanced knowledge of the linguistic labeling task. Turn directions that are grouped into common clusters are shaded, assuming a seven-cluster solution. Most striking, the clear perpendicular demarcation of front and back planes that we found in Experiment 1 and replicated in the first condition of Experiment 2, has vanished. The 90° and 270° turns of perpendicular left and right are integrated into sectors and are not their boundaries.

Experiment 2 (labeled 'with'). Fig. 8 shows frequency counts for the similarity groupings that were obtained for the 90° (right) turn direction with all other turn directions in the right half-plane (from approximately 16° to 174°). The highest count of being grouped together is obtained for the 90° value, of course, as every icon is placed in the same group with itself (there are no double icons in Experiment 2); hence, the maximum frequency value for this graph is 24. The frequency bars for the different turns illuminate the results presented above (compare to Figs. 6 and 7): In the nonlinguistic task, the similarity rating for 90° is higher toward the front plane (turns from 95° to 174°). In contrast, similarity ratings for the linguistic task yielded an even distribution around the 90° turn direction. Also, we observe a slightly higher tendency toward the back plane in the latter case. This means that in addition to the differences around the 90° turn direction, we also find differences at the boundaries of the half-plane, towards 16° and 174°.

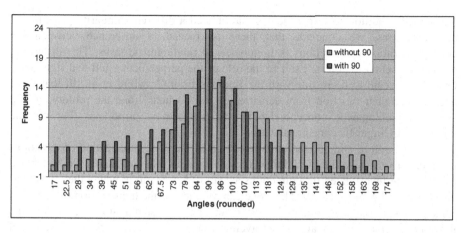

Fig. 8. Frequencies of similarity groupings by participants for the 90° turn (perpendicularly right) for the two conditions of Experiment 2, with and without advance knowledge of the verbal labeling task. As 24 students participated in this experiment, the highest possible similarity is 24 (Y-axis). The X-axis shows the angles displayed by the icons for the right side, ranging from 16° to 174° (rounded). Neither the straight angle (180°) nor any of the turns to the left side were ever placed into the same group as the 90° right angle.

4 General Discussion

We presented two experiments in this paper that examined how directional turns of various sizes are grouped into similarity classes. We are interested in the structure of these groupings that emerges when subjected to cluster analysis, and specifically if there are different underlying conceptual structures for turns depending on whether people expect they will have to verbally label the turns (as in giving route instructions) or not. In our first experiment, in which people were not informed about the verbal task, our results clearly indicate that the perpendicular right-left axes strongly demarcate between the front and back heading planes. Furthermore, these axes are associated with right and left sectors in the front plane, but not in the back plane. Thus, precise right and left turns apparently function as the boundaries of right and left sectors. These conclusions held across several clustering methods.

To validate these findings, we conducted a second experiment in which we repeated the tasks of Experiment 1, with a slightly modified stimulus set. This produced results that largely replicated those from Experiment 1, particularly with respect to the role of the right-left axes. However, we also repeated the grouping and labeling tasks in Experiment 2 with the same set of participants; because they had performed the verbal labeling task already, participants knew they would be generating verbal labels. Thus, their groupings might differ because of an expectation that the groups would have to correspond to verbal classes. In fact, the groupings did differ in this second condition. In particular, the right-left axes no longer emerged as sector boundaries, but as central tendencies of sectors. Our results therefore support the idea that there are differences in the linguistic and nonlinguistic conceptualization of directional turns. Given that it was only the awareness of a linguistic component in the second condition of Experiment 2 that created a different similarity grouping of turn directions, we conclude that there is, indeed, a difference between the conceptualization of directions in linguistic and nonlinguistic tasks. This difference most prominently manifests itself in the role of the perpendicular left and right axes. Our results seem to be in line with the findings from Crawford et al. [9] that these axes change their function from demarcating the boundary (and the prototype) of a sector in nonlinguistic conceptualization to representing prototypes (but not the boundary) in linguistic conceptualization.

While Crawford and collaborators investigated the conceptualization of directions represented vertically and placed their focus on spatial relations that could be labeled as *above*, we investigated turn directions as part of a scenario in which they were part of a horizontal street network. Thus, one can account for the demarcation of front and back planes in our experiments by the nature of the test domain. If we adopt the movement perspective of an agent traveling along a street network, than front and back planes are sensibly demarcated in the way we found. As the name indicates, a turn of more than 90° left or right, i.e., into the sectors <90° or >270° (see Fig. 1), changes the overall direction from forward to backward. For the process of goal directed wayfinding, this is highly pertinent.

That we do not find this distinction in the second condition of Experiment 2, which entailed a linguistic component, might indicate that the conceptualization of directions affording verbalization relies on slightly different principles. It could be related to the linearization that is present in every linguistic description [10,30]. If we think of a route as a sequence of individual decision points, and a linguistic description is provided for every single decision, focus is placed on each decision point as such, without respect for areas beyond the individual decision point. A nonlinguistic task, on the other hand, might focus on larger-scale aspects, in which the position of subsequent decision points is of interest.

Our experiments were not set up around an entire route or environment; their focus was completely on individual intersections. However, the distinction we suggest might contribute to a framework for the underlying cognitive processes for conceptualizing turns that would help explain these differences. Language is sequential, and the focus of verbal route instructions in the case of turn-by-turn instructions is on decision points that are associated with changes in travel direction. When performing the linguistic task of giving route instructions, therefore, this decision-point focus leads to the more classical characterization—in terms of qualitative directions—in which the linguistic concepts of LEFT and RIGHT serve as sectors centered around the orthogonal axes of 90° and 270°.

With respect to modeling directional knowledge generally, we can confirm that a combination of sectors and axes, as, for example, suggested in the double cross calculus by Freksa [13], is a sensible approach. We clearly find that a) the concept STRAIGHT is an axis (independent of linguistic influences) and that b) the perpendicular left and right axes are playing a dominant role. We can also confirm the suggestions by Franklin et al. [12], Montello and Frank [38], and others that sectors have different sizes.

5 Future Work

The similarity classes established in these experiments are symmetric. If a participant places two icons into the same group, the two icons are treated as being equally similar to each other. Early research by Amos Tversky and his colleagues (e.g., [45]) showed that measured similarity relationships are influenced by several factors and are not necessarily symmetric. It would therefore be desirable to assess the similarity of turn directions with methods other than grouping that would allow for establishing asymmetries in similarity.

More important, however, would be research on the effect of contextual factors on turn similarity judgments. The experiments reported in this paper employed a simple, relatively decontextualized setting in order to focus on the differences between linguistic and nonlinguistic categorization. Research on the assignment of linguistic labels to spatial relations (also called spatial terms or projective spatial terms) has provided a great deal of evidence on the relevance of contextual factors. Several theoretical frameworks have been developed that characterize and formally describe these factors (e.g., [8,42]). Our own research on contextual factors in route instructions [26] confirms the importance of these factors for the case of a cognitive agent moving within spatial network structures. We found different conceptualizations depending on the complexity of movements in street networks, and the combination of structural and functional characteristics at an intersection (including the presence of landmarks) that provide a context in which certain actions take place. An example was the shift from a pure direction concept to a combination of coarse direction plus ordering at a complex intersection with turn branches that were themselves equivalent: *take the second left* instead of *veer left*. Another important example was the anchoring of actions by employing landmarks.

The degree to which these results, based on a linguistic analysis of direction concepts, are applicable to nonlinguistic conceptualization is an open question and requires more research. Results like those presented in this paper suggest we cannot assume that a linguistic analysis provides us with sufficient insights into the underlying nonlinguistic conceptual structure of actions taking place in city street networks. We may need a new approach challenging the assumption that a common conceptual structure underlies the linguistic and pictorial externalization of route knowledge [24,47].

An important research issue involves the implications of combining graphic and linguistic representations in navigation systems. For example, should the design criteria for mobile (in the field) assistant systems differ from those for offline systems on the basis of the influence of motion on the conceptualization of environmental information? Does a cognitive agent who is guided through a city street network with instructions for individual decision points build up a better overall mental map of the layout of the city if the instructions—verbal or graphical—are schematized according to the nonlinguistic direction model or to the linguistic direction model?

Finally, questions about designing for individual, cultural, and modality specific conceptualization strategies are ongoing concerns (e.g., [35]). We believe that research on formally characterizing cognitive processes should be diversified in light of evidence on individual and cultural personalization.

Acknowledgements. The experimental research reported here was part of the project MapSpace, Transregional Collaborative Research Center SFB/TR 8 Spatial Cognition, which is funded by the Deutsche Forschungsgemeinschaft (DFG). The grouping tool was developed by Thilo Weigel. We would like to thank Markus Knauff for valuable discussion on the design of Experiment 1, Carsten Dewey for collecting the data and Melissa Bowerman for comments on relevant literature.

References

1. Aldenderfer, M.S., Blashfield, R.K.: Cluster analysis. Sage, Newbury Park, CA (1984)
2. Allen, G.L.: From knowledge to words to wayfinding: Issues in the production and comprehension of route directions. In: Frank, H.S.C. (ed.) Spatial information theory: A theoretical basis for GIS, pp. 363–372. Springer, Berlin (1997)
3. Bloom, P., Peterson, M.A., Nadel, L., Garrett, M.F.: Language and space. The MIT Press, Cambridge (1996)
4. Boroditsky, L.: Does language shape thought? Mandarin and English speakers' conceptions of time. Cognitive Psychology 43, 1–22 (2001)
5. Byrne, R.W.: Memory for urban geography. Quarterly Journal of Experimental Psychology 31, 147–154 (1979)
6. Carlson-Radvansky, L.A., Logan, G.: The influence of reference frame selection on spatial template construction. Journal of Memory and Language 37, 411–437 (1997)
7. Cooke, N.J.: Knowledge elicitation. In: Durso, F.T. (ed.) Handbook of applied cognition, pp. 479–510. John Wiley & Sons, Chichester (1999)
8. Coventry, K.R., Garrod, S.: Saying, seeing, and acting: The psychological semantics of spatial prepositions. Psychology Press, Hove, UK (2004)
9. Crawford, L.E., Regier, T., Huttenlocher, J.: Linguistic and non-linguistic spatial categorization. Cognition 75, 209–235 (2000)
10. Denis, M., Pazzaglia, F., Cornoldi, C., Bertolo, L.: Spatial discourse and navigation: An analysis of route directions in the city of Venice. Applied Cognitive Psychology 13, 145–174 (1999)
11. Frank, A.U.: Qualitative spatial reasoning: Cardinal directions as an example. International Journal of Geographical Information Systems 10, 269–290 (1996)
12. Franklin, N., Henkel, L.A., Zangas, T.: Parsing surrounding space into regions. Memory & Cognition 23, 397–407 (1995)
13. Freksa, C.: Using orientation information for qualitative spatial reasoning. In: Frank, A.U., Campari, I., Formentini, U. (eds.) Theories and methods of spatio-temporal reasoning in geographic space, pp. 162–178. Springer, Berlin (1992)
14. Gennari, S.P., Sloman, S.A., Malt, B.C., Fitch, W.T.: Motion events in language and cognition. Cognition 83, 49–79 (2002)
15. Golledge, R.G.: Primitives of spatial knowledge. In: Nyerges, T.L., Laurini, R., Egenhofer, M.J., Mark, D.M. (eds.) Cognitive aspects of human-computer interaction for geographic information systems, pp. 29–44. Kluwer, London (1995)
16. Gumperz, J., Levinson, S.: Rethinking linguistic relativity. Cambridge University Press, Cambridge (1996)
17. Habel, C., Herweg, M., Pribbenow, S.: Wissen über Raum und Zeit. In: Görtz, G. (ed.) Einführung in die künstliche Intelligenz, 2nd edn. pp. 129–185. Addison-Wesley, Bonn (1995)

18. Hayward, W.G., Tarr, M.J.: Spatial language and spatial representation. Cognition 55, 39–84 (1995)
19. Hermer-Vazquez, L., Spelke, E.S., Katsnelson, A.S.: Sources of flexibility in human cognition: Dual-task studies of space and language. Cognitive Psychology 39, 3–36 (1999)
20. Herskovits, A.: Language and spatial cognition: An interdisciplinary study of the prepositions in English. Cambridge University Press, Cambridge (1986)
21. Hintzman, D.L., O'Dell, C.S., Arndt, D.R.: Orientation in cognitive maps. Cognitive Psychology 13, 149–206 (1981)
22. Huttenlocher, J., Hedges, L.V., Duncan, S.: Categories and particulars: Prototype effects in estimating spatial location. Psychological Review 98, 352–376 (1991)
23. January, D., Kako, E.: Re-evaluating evidence for linguistic relativity: Reply to Boroditsky, Cognition (in press) (2001)
24. Klippel, A.: Wayfinding choremes. In: Kuhn, W., Worboys, M., Timpf, S. (eds.) Spatial information theory: Foundations of geographic information science, pp. 320–334. Springer, Berlin (2003)
25. Klippel, A., Dewey, C., Knauff, M., Richter, K.-F., Montello, D.R., Freksa, C., Loeliger, E.A.: Direction concepts in wayfinding assistance. In: Baus, J., Kray, C., Porzel, R., eds.: Workshop on Artificial Intelligence in Mobile Systems (AIMS'04). Saarbrücken: SFB 378 Memo vol. 84, pp. 1–8 (2004)
26. Klippel, A., Tenbrink, T., Montello, D.R.: The role of structure and function in the conceptualization of directions. To be published in van der Zee, E., Vulchanova, M., eds.: Motion encoding in language and space. Oxford University Press, Oxford (in press)
27. Knauff, M., Rauh, R., Renz, J.: A cognitive assessment of topological spatial relations: Results from an empirical investigation. In: Hirtle, S.C., Frank, A.U. (eds.) Spatial information theory: A theoretical basis for GIS, pp. 193–206. Springer, Berlin (1997)
28. Kos, A.J., Psenicka, C.: Measuring cluster similarity across methods. Psychological Reports 86, 858–862 (2000)
29. Landau, B.: Axes and direction in spatial language and spatial cognition. In: van der Zee, E., Slack, J. (eds.) Representing direction in language and space, pp. 18–38. Oxford University Press, Oxford (2003)
30. Levelt, W.J.M.: Speaking: From intention to articulation. MIT Press, Cambridge, MA (1989)
31. Levinson, S.C.: Frames of reference and Molyneux's question: Crosslinguistic evidence. In: Bloom, P., Peterson, M.A., Nadel, L., Garrett, M.F. (eds.) Language and space, pp. 109–169. The MIT Press, Cambridge, MA (1996)
32. Levinson, S.C., Kita, S., Haun, D.B.M., Rasch, B.H.: Returning the tables: Language affects spatial reasoning. Cognition 84, 155–188 (2002)
33. Lovelace, K.L., Hegarty, M., Montello, D.R.: Elements of good route directions in familiar and unfamiliar environments. In: Freksa, C., Mark, D.M. (eds.) COSIT 1999. LNCS, vol. 1661, pp. 65–82. Springer, Heidelberg (1999)
34. Malt, B.C., Sloman, S.A., Gennari, S.P.: Speaking versus thinking about objects and actions. In: Gentner, D., Goldin-Meadow, S. (eds.) Language in mind, pp. 81–112. The MIT Press, Cambridge, MA (2003)
35. Mark, D.M., Comas, D., Egenhofer, M.J., Freundschuh, S.M., Gould, M.D., Nunes, J.: Evaluating and refining computational models of spatial relations through cross-linguistic human-subjects testing. In: Frank, A.U., Kuhn, W. (eds.) Spatial information theory: A theoretical basis for GIS, pp. 553–568. Springer, Berlin (1995)

36. Meilinger, T., Hölscher, C., Büchner, S.J., Brösamle, M.: How much information do you need? Schematic maps in wayfinding and self localisation. In: Barkowsky, T., Freksa, C., Knauff, M. (eds.) Spatial cognition V, Springer, Berlin (in press)

37. Moar, I., Bower, G.H.: Inconsistency in spatial knowledge. Memory & Cognition 11, 107–113 (1983)

38. Montello, D.R., Frank, A.U.: Modeling directional knowledge and reasoning in environmental space: Testing qualitative metrics. In: Portugali, J. (ed.) The construction of cognitive maps, pp. 321–344. Kluwer Academic, The Netherlands, Dordrecht (1996)

39. Montello, D.R., Richardson, A.E., Hegarty, M., Provenza, M.: A comparison of methods for estimating directions in egocentric space. Perception 28, 981–1000 (1999)

40. Moratz, R., Tenbrink, T.: Spatial reference in linguistic human-robot interaction: Iterative, empirically supported development of a model of projective relations. Spatial Cognition and Computation 6, 63–106 (2006)

41. Munnich, E., Landau, B., Dosher, B.A.: Spatial language and spatial representation: A cross-linguistic comparison. Cognition 81, 171–207 (2001)

42. Regier, T.: The human semantic potential: Spatial language and constraint connectionism. The MIT Press, Cambridge (1996)

43. Sadalla, E.K., Montello, D.R.: Remembering changes in direction. Environment and Behavior 21, 346–363 (1989)

44. Sholl, M.J.: The relationship between sense of direction and mental geographic updating. Intelligence 12, 299–314 (1988)

45. Tversky, A., Gati, I.: Studies of similarity. In: Lloyd, E., Rosch, B.B. (eds.) Cognition and categorization, pp. 79–98. Lawrence Erlbaum, Hillsdale, NJ (1978)

46. Tversky, B.: Distortions in memory for maps. Cognitive Psychology 13, 407–433 (1981)

47. Tversky, B., Lee, P.U.: How space structures language. In: Freksa, C., Habel, C., Wender, K.F. (eds.) Spatial cognition. An interdisciplinary approach to representing and processing spatial knowledge, pp. 157–175. Springer, Berlin (1998)

48. van der Zee, E., Eshuis, R.: Directions from shape: How spatial features determine reference axis categorization. In: van der Zee, E., Slack, J. (eds.) Representing direction in language and space, pp. 209–225. Oxford University, Oxford (2003)

49. Vorwerg, C.: Use of reference directions in spatial encoding. In: Freksa, C., Brauer, W., Habel, W.W., Wender, K.F. (eds.) Spatial cognition III: Routes and navigation, human memory and learning, spatial representation and spatial learning, pp. 321–347. Springer, Berlin (2003)

50. Vorwerg, C., Rickheit, G.: Typicality effects in the categorization of spatial relations. In: Freksa, C., Habel, C., Wender, K.F. (eds.) Spatial cognition. An interdisciplinary approach to representing and processing spatial knowledge. LNCS (LNAI), vol. 1404, pp. 203–222. Springer, Berlin (1998)

51. Waller, D., Loomis, J.M., Haun, D.B.M.: Body-based senses enhance knowledge of directions in large-scale environments. Psychonomic Bulletin & Review 11, 157–163 (2004)

A Uniform Handling of Different Landmark Types in Route Directions

Kai-Florian Richter

Transregional Collaborative Research Center SFB/TR 8 Spatial Cognition
Universität Bremen, Germany
`richter@sfbtr8.uni-bremen.de`

Abstract. Landmarks are crucial for human wayfinding. Their integration in wayfinding assistance systems is essential for generating cognitively ergonomic route directions. I present an approach to automatically determining references to different types of landmarks. This approach exploits the circular order of a decision point's branches. It allows uniformly handling point landmarks as well as linear and areal landmarks; these may be functionally relevant for a single decision point or a sequence of decision points. The approach is simple, yet powerful and can handle different spatial situations. It is an integral part of GUARD, a process generating context-specific route directions that adapts wayfinding instructions to a route's properties and environmental characteristics. GUARD accounts for cognitive principles of good route directions; the resulting route directions reflect people's conceptualization of route information.

1 Introduction

Landmarks are crucial elements in human wayfinding. People use them to identify previously visited places and to orient themselves. Landmarks are a pertinent part of an environment's mental representation. They are also crucial in human direction giving. People refer to them frequently while providing another person with directions to reach a destination; they are mostly used to link actions to be performed to places in the environment. This holds for both verbal route directions and for sketch maps [1].

Integrating references to landmarks in wayfinding assistance systems is essential for automatically generating cognitively ergonomic route directions. We developed a generation process that implements cognitive principles of good route directions and allows for integrating different types of landmarks in instructions. To this end, features of an environment that may serve as a landmark need to be identified and it needs to be checked whether and how they are applicable in communicating the information necessary for route following. In this paper, I present an approach to testing the applicability of different types of landmarks in a uniform way; it is part of the generation process for route directions.

The paper is structured as follows: in the next section, I further illustrate the importance of landmarks for organizing spatial knowledge and their prominence in route directions. Section 3 introduces context-specific route directions, our approach to an automatic generation of route directions. In Section 4, I explain the prerequisites for automatically determining a landmark's functional role in route following; the determination

S. Winter et al. (Eds.): COSIT 2007, LNCS 4736, pp. 373–389, 2007.

is detailed in Section 5. Section 6 then demonstrates how context-specific route directions are generated and how different landmark types are integrated in the directions.

2   Landmarks in Route Directions

Many definitions of the term *landmark* can be found in the literature. Common to all, landmarks are defined as entities that are easily recognizable and memorizable. Presson and Montello provide a general definition stating that everything that stands out of the background may serve as a landmark [2]. In his seminal book 'The Image of the City', Lynch discusses why a feature may serve as a landmark: "Since the use of landmarks involves singling out of one element from a host of possibilities, the key physical characteristic of this class is singularity, some aspect that is unique or memorable in the context" ([3], pp. 78–79). Lynch further elaborates spatial prominence: elements may be established as landmarks either because they are visible from many locations or because of local contrast to nearby elements.

Sorrows and Hirtle pick up these considerations and list features that let a landmark stand out: *singularity, prominence, meaning,* and *prototypicality* [4]. Singularity applies to objects that are in sharp visual contrast with their surroundings. Prominence of entities refers to their spatial location in an environment; they are visible from many other locations or are located at a significant point, such as a major intersection. Some entities may be used as landmarks because they have a meaning common to many people stemming, for example, from their cultural or historical significance. Prototypicality is a characteristics similar to meaning in that such entities are referred to because they are typical representatives of a specific category. According to these characteristics, Sorrows and Hirtle identify three types of landmarks. A *visual landmark* is an entity used as landmark primarily because of contrast with its surrounding, because it has a prominent spatial location, or because its visual characteristics are easily memorizable. A *structural landmark* is defined as having a significant spatial role or location in the structure of space, while a *cognitive landmark* stands out because of its typical or atypical characteristics in the environment.

There is general agreement in the cognitive science literature that landmarks are eminently important for acquiring and organizing knowledge about our surrounding space. For example, there is evidence that spatial knowledge is organized hierarchically [5,6] around landmarks that function as anchor points for this knowledge [7,8]. In acquisition of space, landmarks are learned early on. The model of spatial knowledge acquisition of Siegel and White [9] sets landmarks to be the first kind of spatial knowledge acquired in a new environment. People recognize landmarks as places they have been before; connecting these places is the next step, termed *route learning*. Only in a final step, the different routes are integrated into survey knowledge that allows, for example, calculating shortcuts. Montello [10] questions this framework with regard to the (strict) order of these steps and states that some kind of metric knowledge, i.e. knowledge about distances, is acquired from the very beginning. But he does not question the important role landmarks play in spatial knowledge organization.

Landmarks are not only an important organizing concept for spatial knowledge, they also serve as navigational tool [11]. Landmarks identify decision points, origin and destination of a route, provide verification of route progress, provide orientation cues for homing vectors, and suggest regional differentiating features. This is echoed in research on route directions. People use landmarks to signal crucial actions, locate other landmarks in relation to the referenced landmark, and provide information to confirm that the right track is still followed [12,13,14]. The need to integrate landmarks into computational assistance systems has been confirmed by Tracy Ross and coworkers. In a set of usability studies they show that the use of landmarks in such systems significantly improves users' confidence in correctly executing the instructions and their navigation performance in both car navigation [15] as well as pedestrian wayfinding [16].

Landmarks may have different locations relative to a route. A landmark can either be at a decision point, at a route segment between two decision points [17], or in some distance to, but visible from the route (termed *distant landmark* [13]). Furthermore, there may be *global landmarks* [4]. These are outside the current environmental space, i.e. not immediately reachable by a wayfinder, but their location relative to the current space is known. Landmarks are usually assumed to be point entities. In Lynch's taxonomy of elements that structure spatial knowledge of a city, they are physical objects serving as point references that single out one element from a host of other objects [3]. This assumption is also prominent in research on the mental representation of spatial knowledge. In the model of Siegel and White [9], landmarks are strategic places that a wayfinder travels to and from to keep herself oriented. Hierarchies of spatial knowledge are formed based on point landmarks [5,6]. In their taxonomy of different types of landmarks, Sorrows and Hirtle [4] use the term landmark as it is defined by Lynch. Consequently, computational approaches that integrate landmarks also focus on point landmarks (e.g., [18,19]).

In our work on route directions, we extend the point notion of landmarks and integrate linear and areal entities, as well as structural elements, such as salient intersections [20,21,22]. Examples for linear landmarks include rivers or railway tracks; areal landmarks may be parks or big shopping malls. Typical salient intersections are T-intersections and roundabouts. With this extended view on landmarks, we comply with Presson and Montello's definition of a landmark as everything that sticks out from the background [2] (see also the anchor point theory [7]). Considering different types of landmarks leaves open more options to describe spatial situations, i.e. extends the expressibility of our representation. It also eases adapting instructions to the current situation. And it allows better accounting for human conceptualizations of spatial situations, i.e. it improves communication of the route directions.

3   Context-Specific Route Directions

We developed a computational process for generating route directions that takes different types of landmarks into account [23,20]. The process is termed GUARD, which stands for Generation of Unambiguous, Adapted Route Directions. This reflects that the process generates directions that unambiguously identify each route-segment and that adapt to the current action to be taken in the current surrounding environment. In that, we try to be as precise as possible with as little information as necessary; we take a Gricean perspective [24] in generating references. Our approach is in line with Dey's definition of context: "[...]any information that can be used to characterize the situation of an entity" ([25], p. 5). Accordingly, we coin the route directions generated by our process *context-specific route directions*.

For the adaptation, we need to account for the characteristics (the structure) of the environment in which route following takes place. The structure of an environment strongly influences the kind of instructions that can be given. The embedding of the route in the spatial structure surrounding it, the structure of that route itself, path annotations, and landmarks that are visible along the route all contribute to this influence. Based on an analysis of the basic constituents of routes and route directions, we identified classes of elements that can be used in route directions, as well as the spatial knowledge required to determine and interpret the corresponding references. We group these elements in a systematics on three different levels reflecting their relation to a specific route (see Table 1). They abstract to different degrees from a detailed description of a single decision point, which is captured by different levels of granularity. Global references abstract the most, while egocentric references and landmarks at decision points provide the most detailed description.

Table 1. The systematics of route direction elements. Elements are grouped on three levels according to their relation to the route: *Global References*, *Environmental Structure*, and *Path and Route*.

Global References	Environmental Structure	Path and Route
cardinal directions	edges	egocentric references
global landmarks	districts	landmarks at decision point
	slant	landmarks between decision points
		distant landmarks
		linear and areal landmarks
		path annotations

Based on the systematics, we implemented GUARD. It consists of four steps: in an initial step, every possible description of the action to be performed at each decision point is generated. These are represented in relational statements termed *abstract turn instructions* (ATI). The next step combines ATIs of consecutive decision points into a single instruction, performing so called *spatial chunking* [26]. As the initial chunking is simply syntax based, in the third step the generated chunks are checked against cognitive and representation-theoretic principles defining valid chunks. Finally, the fourth step generates the route directions in an optimization process. These steps are further illustrated in Section 6.

The inclusion of landmarks requires automatically determining their functional role in following a specific route. To this end, first, landmarks need to be assigned to those decision points they might be functionally relevant for. Second, it needs to be checked whether they unambiguously identify the route-segment to take at the decision points they are assigned to. The next section explains the basic representation used in the process and how landmarks are assigned to decision points; Section 5 illustrates how their applicability is checked.

4   Representing Routes and Landmarks

In our approach the environment's representation is coordinate-based. The street-network is represented as a graph, i.e. streets are represented as edges. Nodes that have a degree greater than three represent intersections. Nodes with a degree of two are used to reflect a street's geometry. Each node is associated with a coordinate. In the same coordinate system, features that may serve as landmarks are represented as points, lines, or polygons depending on their geometry.

A route is represented as a directed path in the graph. As already explained, of special interest are decision points, i.e. the nodes with a degree greater two. In the real world, these decision points are not just points. Here, a decision point denotes a certain area around an intersection which contains the point where the streets meet and part of the streets itself. It represents the configuration of streets at an intersection. That is, the branches meeting at a decision point are part of that point. This modeling corresponds to human conceptualization and reflects that humans usually turn in an extended process with deciding on the direction to turn and then changing their direction gradually while keeping their forward movement [1,27]. However, for assigning landmarks to decision points and for determining their applicability, we abstract from this gradual direction change. Decision points are modeled such that all branches meet in a single point (Fig. 1). This divides the area around an intersection into a *region before action* and a *region after action* [28].

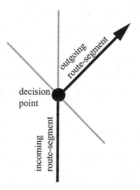

Fig. 1. A decision point with the two functionally relevant branches: the *incoming* and *outgoing* route-segment. The branches' meeting point is the actual decision point.

4.1   Assigning Landmarks to Decision Points

Having set the basic modeling of the environment, the landmarks represented in the annotated graph need to be functionally assigned to the route. As explained above, some landmarks may be functionally relevant for several decision points. Here, we need to distinguish two cases: the landmark spreads along the route which may, for example, hold for a linear landmark such as a river; or the landmark is distant to the route but visible from several decision points, for instance a church spire.

In the first case, the landmark's geometry is represented by a sequence of coordinate points. Even though belonging to the same landmark, these points can be individually assigned to a decision point. Different parts of such a landmark are relevant for different parts of the route, i.e. only that part of a landmark that is in the local surroundings of a decision point is functionally relevant for this decision point. Accordingly, for such landmarks, there is an exclusive assignment of coordinate points to decision points, i.e. each coordinate point is assigned to exactly one decision point. Landmarks distant to the route cannot be assigned exclusively to a specific decision point; this is independent of how they are geometrically represented. That is, next to categorizations of landmarks as, for example, discussed in [22] (see also Section 2), in the context of this work landmarks are distinguished between those that are only locally functionally relevant (i.e. that can be exclusively assigned to decision points) and those that are globally relevant. However, it is important to note that the role of a landmark may change along a route. It may serve as a distant landmark for several decision points, but be located at one of the route's decision points, i.e. serve as a local landmark there. In the following, I will concentrate on landmarks that are locally functionally relevant.

In order to assign these landmarks to decision points, a region around each decision point is defined. The size of this region depends on different parameters including travel mode and visibility[1]. Most importantly, the regions are chosen such that no other decision point falls within the region, i.e. the regions do not overlap. This way, we prevent conflicting assignment of a landmark's coordinate points to decision points; all coordinate points within the decision point's region are exclusively assigned to that decision point. This may lead to different sizes of the regions for different decision points. Decision points that are close together have smaller regions. This reflects that with decision points close together, a wayfinder has to decide on the further way to take more frequently, i.e. has to correctly interpret new spatial situations more often.

4.2   Ordering Information in Route Directions

For determining the functional role of a landmark, we exploit ordering information. Ordering information derives from the linear, planar, or spatial ordering of features [29]. It does not specify any further metric information, such as distances between these entities. This kind of information is a powerful structuring means and it is easy to determine as it only requires knowledge about a neighborhood relation between the relevant entities.

In routes, ordering is closely linked to orientation; a route, being a linear and directed (oriented) entity, induces an order on the entities along that route [30]. That is, the orientation of the route determines in which spatial and temporal order these entities are encountered. Furthermore, the configuration of entities at an intersection, for instance its branches or landmarks located there, can be described using circular ordering information. Figure 2 illustrates both these usages of ordering information.

---

[1] Checking for visibility is kept simple in the system. It is performed on the graph representation using scan-line methods. It just ensures that in the 2D projection an object is in line of sight from the route and in a distance shorter than some threshold.

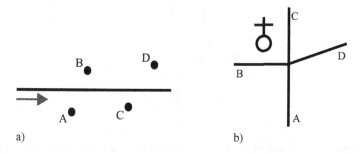

a)   b)

Fig. 2. a) Linear order of entities induced by the orientation of a route. The route is directed from left to right; the order is $A < B < C < D$. b) Circular order of entities at an intersection: starting with A, the order is $A < D < C < Church < B < A$.

5   A Uniform Handling of Landmarks in Route Directions

In the following, I will detail a uniform approach dealing with different types of landmarks, As a basic case, I start with point landmarks at a decision point. This case illustrates the general principles; ongoing from there, extensions needed for handling other types of landmarks will be explained. Generally, the location of a landmark relative to a decision point is required to appropriately reference the landmark in route directions; the relation to incoming and outgoing route-segment and, consequently, to their meeting point (the decision point itself) needs to be determined.

5.1   Point Landmarks at Decision Points

For point landmarks at a decision point, there are three locations that can be distinguished functionally. The turning action may either occur *after* passing a landmark ("turn left after the church"), it may take place *before* the landmark is passed ("turn left before the church"), or the landmark may be *at* the decision point, but not located at a functionally relevant branch ("turn left at the church") (see Fig. 3). In other words, the landmark may be next to the incoming route-segment, next to the outgoing route-segment, or next to any of the other branches of the decision point. In order to

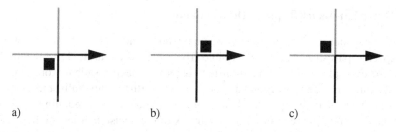

a)   b)   c)

Fig. 3. Three functionally different locations of a landmark at a decision point: a) turning action after passing the landmark; b) turning action before passing the landmark; c) landmark not at a functionally relevant branch

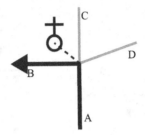

Fig. 4. A virtual branch integrating a landmark at a decision point into the circular order of the decision point's branches

determine the location of the landmark relative to the decision point, we need to determine this *next to* relation. It corresponds to a neighborhood relation of landmark and branch. This neighborhood can be extracted from the circular ordering of the decision point's branches. We introduce a virtual branch ranging from the decision point to the point landmark. This way, the landmark becomes part of the branches' circular ordering (see Fig. 4).

We can now determine whether the virtual branch is neighbored to one of the functionally relevant branches. Two branches are neighbored if one succeeds the other in the ordering. If the virtual branch representing the landmark's location is direct successor or predecessor of the incoming route-segment, the turning action is performed after the landmark is passed. We represent this situation with the relation $lm^<$. The turning action is performed before the landmark if in the ordering the virtual branch is neighbor to the outgoing route-segment, represented by $lm^>$. All situations in which the landmark is not neighbored to a functionally relevant branch are represented by lm^-. To ensure that the landmark is really located before the decision point, two additional virtual branches perpendicular to the incoming route-segment are introduced. These branches demarcate the before-region. Without them, a landmark may be neighbored to the incoming route-segment though it is located after the decision point (see Fig. 5). The same holds for the after-region correspondingly. For relation $lm^>$, further restrictions apply with respect to possible ambiguities in identifying a branch. This is reported in [28], along with further details on this basic case—point landmarks at a decision point.

5.2 Other Landmark Types at Decision Points

Next, we consider features of the environment whose landmark character is dominated by a specific part that may not be visible from everywhere around the feature. This holds, for example, for buildings housing shops that have a salient store-front [18]. These features may be represented as points, as well. But, additionally, the salient side (the façade) needs to be captured; this is done by adding a vector pointing in direction of the salient side. The relation of the landmark to the decision point is determined as illustrated above. Then, the intersection of the vector representing the façade with the functionally relevant branches is calculated. If the landmark is next to the incoming route-segment, the vector needs to intersect with this route-segment for the landmark

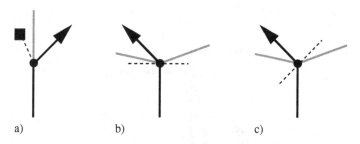

Fig. 5. a) A spatial situation where the landmark is located next to the incoming route-segment, but after the decision point; $lm^<$ is not a valid result here. b) Virtual branches demarcating the before-region... c) ...and the after-region.

to be applicable. The same holds for the outgoing route-segment; the vector needs to intersect it to reference the landmark at the decision point.

The basic principle can also be used for landmarks that are not represented as point features. The three relations $lm^<$, $lm^>$, and lm^- are still sufficient to handle these cases. A landmark's geometry is defined by a sequence of coordinates. For each coordinate point it is possible to determine its relation to a decision point by introducing a virtual branch connecting it to the decision point and then calculating the relation as described above. The functional role of a linear or areal landmark in route following is then determined by the resulting sequence of relations holding for its coordinate points.

For example, the location of a landmark at a decision point that cannot be represented as a single point (see Fig. 6) is addressable in the same way as a point landmark, i.e. there are three functionally different locations (turning action before the landmark, after the landmark, or at the landmark). If the resulting relation is $lm^<$ for all coordinate points, the landmark is passed before the turning action occurs. We can assign $lm^<$ to denote the location of the landmark with respect to the decision point. If all relations are $lm^>$, it is passed after the turning action, and the relation for the landmark is $lm^>$. If the resulting relations differ (especially if some are lm^-) the landmark is not completely located in before- or after-region. Therefore, lm^- is the adequate relation to represent the landmark's location.

5.3 Extended Landmarks Relevant for Several Decision Points

Linear and areal landmarks may stretch along the route, i.e. be applicable for several consecutive decision points. As detailed in Section 4.1, the coordinate points defining their geometry get assigned to different decision points. This way, it is possible to determine for individual decision points the relation to the relevant part of a linear or areal landmark. Using the same approach as before, we can get from the relation of single coordinate points to a decision point to the relation of a landmark to a single decision point. And from there, we can get to the relation of that landmark to a sequence of decision points.

It may, for example, be possible to refer to a river to indicate the direction to take for several decision points in instructions as "follow the river until the gas station". In this case, the river must unambiguously indicate the direction to take at each decision

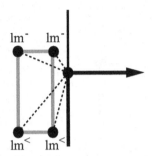

Fig. 6. Example of an extended landmark at a decision point. For each coordinate point, its relation to the decision point is determined. The resulting relation of the landmark to the decision point is lm^- as different relations hold for the points.

point up to the one where the gas station is located. This is the case if for each of these decision points the river is both next to the incoming and the outgoing route-segment. To check for this, for all coordinate points that are assigned to the current decision point the relation to the decision point is determined. This is done in movement direction, i.e. the coordinate points are processed in the order they would be passed along the route. For the linear landmark to be applicable, for the first l coordinate points relation $lm^<$ needs to hold, for the last m coordinate points $lm^>$ needs to hold ($l, m \geq 1$). No coordinate point may be in relation lm^-, and there may be no switch back to $lm^<$ after $lm^>$ has been assigned to a coordinate point. If this condition is fulfilled the linear landmark can be used to indicate the direction to take at the decision point. We denote this with the relation `follow`. With areal landmarks, the same method can be applied to check whether it may be used to guide a wayfinder around that landmark along the route. As long as one part of the coordinate points is next to the incoming route-segment and the other part next to the outgoing route-segment, the route passes along the areal landmark.

In conclusion, the basic principle of determining the location of a point landmark relative to a decision point and the distinction of the three different relations $lm^<$, $lm^>$, and lm^- is sufficient to handle different spatial situations involving different types of landmarks (namely point, linear, and areal landmarks). The situations covered so far all involve passing a landmark, either at a single decision point or walking along it for several consecutive decision points. A simple extension of the basic formalism allows for handling further spatial situations in which the involved landmark's role may be more complex. When crossing a river, for example, the landmark is both to the left and the right of the route. That is, in the corresponding representation of the route, part of the coordinate points representing the landmark's geometry are on the one side of the route-segment, the other part on the other side.

This sideness can be captured using the ordering of branches, as well. The branches' ordering (including the virtual branch connecting a coordinate point) always gets calculated in counterclockwise direction. That is, the order of the branches is ordered, as well. Taking this information into account then allows inferring the sideness of a coordinate point since we know the movement direction along a route-segment; for the incoming

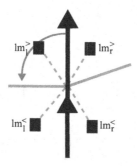

Fig. 7. Distinguishing the sideness of a landmark with respect to a functionally relevant route-segment. The gray arrow shows the direction of determining the branches' order.

route-segment the predecessor in the ordering is left of the branch, the successor right of the branch. For the outgoing route-segment, it is vice versa; the predecessor is right of the branch, the successor left of it (see Fig. 7). We denote this with the indices l and r and end up with five different relations: $\text{lm}_l^<, \text{lm}_r^<, \text{lm}_l^>, \text{lm}_r^>, \text{lm}^-$.

Using these relations, we can capture additional spatial situations. Picking up the example of crossing a linear landmark, a possible sequence of relations that corresponds to this situation is the following: there are l coordinate points for which the relation $\text{lm}_l^<$ holds, followed by m coordinate points with relation $\text{lm}_r^<$. More generally, crossing a linear landmark is represented by an arbitrary number of coordinate points in relation lm^-, followed by l coordinate points in exactly one relation of the set $\text{lm}_l^<, \text{lm}_r^<, \text{lm}_l^>, \text{lm}_r^>$, then m coordinate points being next to the same route-segment but on the other side, for example, switching from $\text{lm}_l^<$ to $\text{lm}_r^<$, followed again by an arbitrary number of coordinate points in relation lm^- (with $l, m \geq 1$).

These examples demonstrate that it is possible to define patterns of relations that capture the functional role of different landmark types in route following. This is based on a simple principle of using ordering information to determine the location of a point landmark at a decision point and by distinguishing five relational terms to capture the functionally relevant possible locations. The presented approach allows identifying spatial situations and generating appropriate references to landmarks. It is important to note that this is not meant to be a model of how humans judge spatial situations and of how they determine adequate descriptions for these situations.

6   Generating Context-Specific Route Directions

The method for determining the applicability of different landmark types is an integral part of GUARD, the process for generating context-specific route directions. As stated in Section 3, this process automatically generates route directions that exploit different features of an environment; it is realized in four steps. In this section, I will provide more details on each step and will give an example for route directions generated by the model. Figure 8 depicts an example route that has been put together using different real-world intersections whose data has been collected as input to GUARD. The following illustration is based on this route.

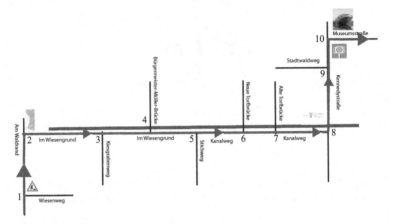

Fig. 8. An example route; the decision points are numbered for future reference

6.1 Generating Abstract Turn Instructions

In the first step of the generation process, for every decision point of the route all possible instructions are generated. To this end, for every element listed in the systematics (see Table 1), it is checked whether it is applicable at all and whether it results in an instruction that unambiguously describes the further direction to take. In the abstract turn instructions, actions to be performed at a decision point are represented as *direction relations*; each kind of action is denoted by a specific relational term. Actions result in a change of heading (which might be 0 if continuing straight on). In the graph representation, this change can be measured as an angular deviation from the previous heading.

For egocentric references, this angle is matched to a category representing a turning direction (e.g., *left*) to determine which relation to use. These categories are represented as angle intervals; we use the direction model of [31] that distinguishes seven different directions. For the first decision point of the example route, for instance, the angular deviation from the current movement direction is in an interval representing going straight, denoted by the relation `straight`. There is also a construction site, modeled as point landmark. Using the mechanism presented in Section 5.1 results in $lm^>$, i.e. the landmark is located after the turning action. Possible abstract turn instructions for the first decision point are (DP_1, `straight`) and (DP_1, `straight construction-site` $lm^>$).

Other kinds of elements, for instance linear landmarks, do not rely on the angular deviation in providing information on the further way to take. The river next to considerable part of the route is a good example. The coordinate points representing the river get exclusively associated with decision points three to eight, respectively. For each of these decision points, one part of the assigned coordinate points are in the before-region and the other in the after-region; checking the landmark's applicability results in the relation `follow`. Accordingly, the landmark is suited to indicate the further direction to take at these decision points. All in all, the first step of the generation process results in a set of possible instructions for each decision point.

6.2   Spatial Chunking

Spatial chunking is a process that combines the actions to be performed at several consecutive decision points into a single action [26,19]. This is a crucial process in conceptualizing routes, i.e. in forming a mental representation of a route. Humans also commonly apply spatial chunking in giving route directions.

GUARD performs spatial chunking in two steps (which cover steps two and three of the generation process). In an initial step, abstract turn instructions are combined based on two simple syntactic rules: first, ATIs that employ the same direction relation can be combined; second, the egocentric direction relation straight can be combined with any other egocentric direction relation. The first rule corresponds to a "do n-times"– or "do until"–instruction, for instance "turn three times right"; the second rule covers situations where going straight at a decision point does not need to be explicitly stated, for instance, in "turn left at the third intersection". Both rules ensure that references to landmarks are sensibly handled, i.e. that each chunk only refers to a single landmark.

Then, it is checked whether the chunks generated using these rules are cognitively and structurally sensible. For example, instructions that refer to following linear landmarks require information on the point along the route upon which the landmark is to be followed. This information, termed *end qualifier*, typically is provided by a point landmark at a decision point. Thus, one of the chunking principles used in this step states that chunks based on linear landmarks need to end with a decision point where another landmark is present. Generally, in this step chunks are pruned until they adhere to the defined cognitive principles.

Performing spatial chunking as a two-step process is advantageous as it allows for more flexibility; different cognitive chunking principles can be implemented and used with the model while always using the same basic chunking mechanism.

6.3   Optimization

The actual generation of the route directions is performed in the fourth step. Here, from all possible abstract turn instructions for each decision point the one that is best gets selected. This is an optimization process; accordingly, 'best' is defined relative to an optimization criterion. Just as with the chunking principles, the model is flexible with respect to the optimization criterion used. In the literature, different principles for good route directions can be found that can be turned into optimization criteria (e.g., [12,13]). A straightforward criterion is to aim for the minimum number of chunks. This reduces the cognitive load of a wayfinder as the communicated route directions get shorter and, hence, the wayfinder needs to remember less information. And the wayfinder's processing of the route directions is also reduced since the instructions are already chunked.

There are two optimization approaches implemented in the model. Local optimization proceeds through the route from origin to destination. For each decision point, it selects the best chunk starting with this decision point. It then checks whether this chunk has to be integrated into the current partial route directions, i.e. whether this chunk improves the directions. If this is the case, the directions are rebuild using this chunk. Local optimization does not necessarily find an optimal solution. Global optimization guarantees this by generating every possible solution (every possible combination of chunks). This is computationally more complex than local optimization. However, computational tests show that in practice both approaches almost always result in the equally optimal route directions.

Table 2 shows the resulting context-specific route directions for the example route of Figure 8 as an XML-specification. Each chunk is grouped between <Direction>-tags. For each chunk, a verbal externalization is listed.[2]

Table 2. The resulting context-specific route directions and their verbal externalization. On the left, the part of the route that is covered by each chunk is depicted (rotated to fit in the table).

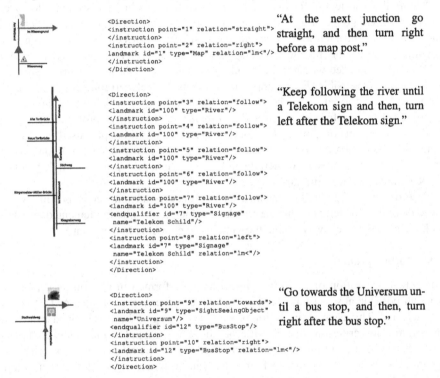

```
<Direction>
<instruction point="1" relation="straight">
</instruction>
<instruction point="2" relation="right">
landmark id="1" type="Map" relation="lm"/>
</instruction>
</Direction>
```

"At the next junction go straight, and then turn right before a map post."

```
<Direction>
<instruction point="3" relation="follow">
<landmark id="100" type="River"/>
</instruction>
<instruction point="4" relation="follow">
<landmark id="100" type="River"/>
</instruction>
<instruction point="5" relation="follow">
<landmark id="100" type="River"/>
</instruction>
<instruction point="6" relation="follow">
<landmark id="100" type="River"/>
</instruction>
<instruction point="7" relation="follow">
<landmark id="100" type="River"/>
<endqualifier id="7" type="Signage"
  name="Telekom Schild"/>
</instruction>
<instruction point="8" relation="left">
<landmark id="7" type="Signage"
  name="Telekom Schild" relation="lm"/>
</instruction>
</Direction>
```

"Keep following the river until a Telekom sign and then, turn left after the Telekom sign."

```
<Direction>
<instruction point="9" relation="towards">
<landmark id="9" type="SightSeeingObject"
  name="Universum"/>
<endqualifier id="12" type="BusStop"/>
</instruction>
<instruction point="10" relation="right">
<landmark id="12" type="BusStop" relation="lm"/>
</instruction>
</Direction>
```

"Go towards the Universum until a bus stop, and then, turn right after the bus stop."

To test the performance of context-specific route directions, we performed an initial user-test as a desktop study. Subjects had to, first, rate context-specific route directions and, second, reach a destination location after memorizing route directions by clicking on the branch to take on intersection photographs. Subjects rated context-specific route directions as useful and well understandable, and even though the setup of the study is somewhat artificial, subjects were well able to reach the destination using context-specific route directions.

---

[2] Verbal externalization of context-specific route directions is not part of the process directly; it is done in collaboration with Prof. John Bateman, Universität Bremen, using their natural language generation system.

7   Conclusions

In this paper I presented an approach to automatically determining the functional role of landmarks in route following. The approach allows integrating different types of landmarks in automatically generated route directions. The circular order that the branches of a decision point form and the order of events in route following that is induced by the directedness of a route are exploited. Distinguishing five different relations, the location of a landmark's coordinate points relative to a decision point are determined. The sequence of these relative locations are compared to patterns representing possible functional roles of the landmark. This way, it is possible to cover different spatial situations based on a single simple principle. The approach is an integral part of GUARD, the generation process for context-specific route directions that automatically generates route directions that adapt to a route's properties and environmental characteristics. Generation of the route directions is a four-step process; optimization is used to decide on which description to choose.

In this process, we integrate different types of landmarks. This is a crucial step in automatically generating cognitively ergonomic route directions. The process also performs spatial chunking which is another basic principle of human direction giving. However, currently it depends on landmarks being known. Except for structural landmarks (e.g., T-intersections), potential landmarks are not identified from a given environmental representation. Integrating approaches that aim for automatically identifying landmarks (e.g., [18,32]) would be another important step for a truly automatic generation of route directions. Another aspect of future work is using the model to find the optimal route applying route selection criteria related to both the difficulty of describing the route and the difficulty of navigating the route (in line with, e.g., [33]), instead of finding the optimal description for a given route as it is done by now. Furthermore, due to its flexibility the model may serve as test-bed for empirical studies testing different principles of generating route directions.

Acknowledgments. This work is supported by the Transregional Collaborative Research Center SFB/TR 8 Spatial Cognition which is funded by the Deutsche Forschungsgemeinschaft (DFG).

References

1. Tversky, B., Lee, P.U.: Pictorial and verbal tools for conveying routes. In: Freksa, C., Mark, D.M. (eds.) Spatial Information Theory - Cognitive and Computational Foundations of Geographic Information Science, pp. 51–64. Springer, Berlin (1999)
2. Presson, C.C., Montello, D.R.: Points of reference in spatial cognition: Stalking the elusive landmark. British Journal of Developmental Psychology 6, 378–381 (1988)
3. Lynch, K.: The Image of the City. MIT Press, Cambridge (1960)
4. Sorrows, M.E., Hirtle, S.C.: The nature of landmarks for real and electronic spaces. In: Freksa, C., Mark, D.M. (eds.) Spatial Information Theory - Cognitive and Computational Foundations of Geographic Information Science, pp. 37–50. Springer, Berlin (1999)
5. Sadalla, E.K., Burroughs, J., Staplin, L.J.: Reference points in spatial cognition. Journal of Experimental Psychology: Human Learning and Memory 6(5), 516–528 (1980)

6. Hirtle, S.C., Jonides, J.: Evidence of hierarchies in cognitive maps. Memory & Cognition 13(3), 208–217 (1985)
7. Couclelis, H., Golledge, R.G., Gale, N., Tobler, W.: Exploring the anchor-point hypothesis of spatial cognition. Journal of Environmental Psychology 7, 99–122 (1987)
8. McNamara, T.P.: Spatial representation. Geoforum 23(2), 139–150 (1992)
9. Siegel, A.W., White, S.H.: The development of spatial representations of large-scale environments. In: Reese, H. (ed.) Advances in Child Development and Behaviour, pp. 9–55. Academic Press, New York (1975)
10. Montello, D.R.: A new framework for understanding the acquistion of spatial knowledge in large-scale environments. In: Egenhofer, M.J., Golledge, R.G. (eds.) Spatial and Temporal Reasoning in Geographic Information Systems, pp. 143–154. Oxford University Press, New York (1998)
11. Golledge, R.G.: Human wayfinding and cognitive maps. In: Golledge, R.G. (ed.) Wayfinding Behavior — Cognitive Mapping and Other Spatial Processes, pp. 5–46. John Hopkins University Press, Baltimore, USA (1999)
12. Denis, M.: The description of routes: A cognitive approach to the production of spatial discourse. Cahiers Psychologie Cognitive 16(4), 409–458 (1997)
13. Lovelace, K.L., Hegarty, M., Montello, D.R.: Elements of good route directions in familiar and unfamiliar environments. In: Freksa, C., Mark, D.M. (eds.) Spatial Information Theory - Cognitive and Computational Foundations of Geographic Information Science, pp. 65–82. Springer, Berlin (1999)
14. Michon, P.E., Denis, M.: When and why are visual landmarks used in giving directions? In: Montello, D.R. (ed.) Spatial Information Theory - Cognitive and Computational Foundations of Geographic Information Science, pp. 400–414. Springer, Berlin (2001)
15. May, A.J., Ross, T., Bayer, S.H., Burnett, G.: Using landmarks to enhance navigation systems: Driver requirements and industrial constraints. In: Proceedings of the 8th World Congress on Intelligent Transport Systems, Sydney, Australia (2001)
16. Ross, T., May, A., Thompson, S.: The use of landmarks in pedestrian navigation instructions and the effects of context. In: Brewster, S., Dunlop, M. (eds.) Mobile Computer Interaction - MobileHCI 2004, pp. 300–304. Springer, Berlin (2004)
17. Herrmann, T., Schweizer, K., Janzen, G., Katz, S.: Routen- und Überblickswissen — konzeptuelle Überlegungen. Kognitionswissenschaft 7(4), 145–159 (1998)
18. Raubal, M., Winter, S.: Enriching wayfinding instructions with local landmarks. In: Egenhofer, M.J., Mark, D.M. (eds.) GIScience 2002. LNCS, vol. 2478, pp. 243–259. Springer, Heidelberg (2002)
19. Dale, R., Geldof, S., Prost, J.P.: Using natural language generation in automatic route description. Journal of Research and Practice in Information Technology 37(1), 89–105 (2005)
20. Richter, K.F., Klippel, A.: A model for context-specific route directions. In: Freksa, C., Knauff, M., Krieg-Brückner, B., Nebel, B., Barkowsky, T. (eds.) Spatial Cognition IV, pp. 58–78. Springer, Berlin (2005)
21. Klippel, A., Richter, K.F., Hansen, S.: Structural salience as a landmark. In: Workshop Mobile Maps 2005, Salzburg, Austria (2005)
22. Hansen, S., Richter, K.F., Klippel, A.: Landmarks in OpenLS - a data structure for cognitive ergonomic route directions. In: Raubal, M., Miller, H.J., Frank, A.U., Goodchild, M.F. (eds.) GIScience 2006. LNCS, vol. 4197, pp. 128–144. Springer, Heidelberg (2006)
23. Richter, K.F., Klippel, A., Freksa, C.: Shortest, fastest, - but what next? a different approach to route directions. In: Raubal, M., Sliwinski, A., Kuhn, W., (eds.) Geoinformation und Mobilität - von der Forschung zur praktischen Anwendung. Beiträge zu den Münsteraner GI-Tagen 2004 IfGIprints, Institut für Geoinformatik, Münster pp. 205–217 (2004)
24. Grice, H.P.: Logic and conversation. In: Cole, P., Morgan, J.L. (eds.) Speech Acts. Syntax and Semantics, vol. 3, pp. 41–58. Academic Press, New York (1975)

25. Dey, A.K.: Understanding and using context. Personal and Ubiquitous Computing 5(1), 4–7 (2001)
26. Klippel, A., Tappe, H., Habel, C.: Pictorial representations of routes: Chunking route segments during comprehension. In: Freksa, C., Brauer, W., Habel, C., Wender, K.F. (eds.) Spatial Cognition III, pp. 11–33. Springer, Berlin (2003)
27. Klippel, A.: Wayfinding Choremes — Conceptualizing Wayfinding and Route Direction Elements. PhD thesis, Universität Bremen (2003)
28. Richter, K.F., Klippel, A.: Before or after: Prepositions in spatially constrained systems. In: Proceedings of Spatial Cognition, Berlin, Springer (to appear)
29. Schlieder, C.: Reasoning about ordering. In: Frank, A.U., Kuhn, W. (eds.) Spatial Information Theory - A Theoretical Basis for GIS, Springer, Berlin (1995)
30. Pederson, E.: How many reference frames? In: Freksa, C., Brauer, W., Habel, C., Wender, K.F. (eds.) Spatial Cognition III, pp. 287–304. Springer, Berlin (2003)
31. Klippel, A., Dewey, C., Knauff, M., Richter, K.F., Montello, D.R., Freksa, C., Loeliger, E.A.: Direction concepts in wayfinding assistance systems. In: Baus, J., Kray, C., Porzel, R., (eds.) Workshop on Artificial Intelligence in Mobile Systems (AIMS'04), SFB 378 Memo 84; Saarbrücken, pp. 1–8 (2004)
32. Elias, B.: Extracting landmarks with data mining methods. In: Kuhn, W., Worboys, M., Timpf, S. (eds.) Spatial Information Theory: Foundations of Geographic Information Science, pp. 375–389. Springer, Berlin (2003)
33. Duckham, M., Kulik, L.: Simplest paths: Automated route selection for navigation. In: Kuhn, W., Worboys, M., Timpf, S. (eds.) Spatial Information Theory: Foundations of Geographic Information Science, pp. 169–185 (2003)

Effects of Geometry, Landmarks and Orientation Strategies in the 'Drop-Off' Orientation Task*

David Peebles[1], Clare Davies[2], and Rodrigo Mora[3]

[1] Department of Behavioural Sciences, University of Huddersfield, Queensgate, Huddersfield, HD1 3DH, UK
D.Peebles@hud.ac.uk
[2] Research Labs C530, Ordnance Survey, Romsey Road, Southampton, SO16 4GU, UK
clare.davies@ordnancesurvey.co.uk
[3] Bartlett School of Graduate Studies, University College London, 1–19 Torrington Place, Gower Street, WC1E 6BT, UK
r.vega@ucl.ac.uk

Abstract. Previous work is reviewed and an experiment described to examine the spatial and strategic cognitive factors impacting on human orientation in the 'drop-off' static orientation scenario, where a person is matching a scene to a map to establish directional correspondence. The relative roles of salient landmarks and scene content and geometry, including space syntax isovist measures, are explored both in terms of general effects, individual differences between participant strategies, and the apparent cognitive processes involved. In general people tend to be distracted by salient 3D landmarks even when they know these will not be detectable on the map, but benefit from a salient 2D landmark whose geometry is present in both images. However, cluster analysis demonstrated clear variations in strategy and in the relative roles of the geometry and content of the scene. Results are discussed in the context of improving future geographic information content.

1 Introduction

Part of the promise of applying a scientific approach to spatial information lies in its potential to enhance the information to improve future geographic data and mapping, together with the systems that manipulate it. If we are to improve geographic visualisations such as maps, whether on paper or screen, it is essential to understand the core tasks that users have to perform when interacting with geographic information. More importantly, as argued by other authors over the years (e.g., [1], [2]) we need to understand the role of the characteristics of the space itself, and of its representation, in influencing human performance on those core tasks.

One such core task is what has been called the "drop-off localisation problem" [3], [4]. In this situation a person is either viewing or is actually immersed within a scene, and has to look for the first time at a map of the area to try to match it to their orientation (and possibly their location). Of course, this problem can arise at a decision point within a navigation task, although the abrupt sense of 'drop off' (i.e. not having used the map

---

* © Crown copyright 2007. Reproduced by permission of Ordnance Survey.

S. Winter et al. (Eds.): COSIT 2007, LNCS 4736, pp. 390–405, 2007.
© Springer-Verlag Berlin Heidelberg 2007

already to progress to this point in space) would only be likely if the wayfinder was suddenly disoriented for some reason, for example if they had just reached an unfamiliar area after traversing a familiar one. There are also many non-wayfinding situations where it may arise. A few examples of this task scenario in an urban setting might include:

1. trying to identify a specific building or object which is not explicitly labelled on the map, e.g., to visit or study it, or in an emergency scenario;
2. trying to match a historic image (e.g., of an old street scene) to a modern-day map, or vice versa;
3. making planning decisions based partly on viewing the current visual landscape (or photographs of it), and partly on a drawn plan or model of a proposed development;
4. trying to judge relative distances and directions to unseen distant locations (whether or not one intends to navigate to them);
5. viewing a 'you-are-here' map signage within a space, where location is indicated but orientation is unclear [5], [6].

For this reason, the task can be seen as an important one to study and model as it may inform the development of geographic information content and visualisation to facilitate orientation across a wide range of uses, from emergency services to tourists and from urban planners to archaeologists. It may also be seen as a precursor to studying the more complex combined task of self-localisation, which inevitably includes simultaneous orientation to some extent.

The few previous studies of this process of orienting with a map in a drop-off scenario have taken an experimental approach, usually using a scene which is limited or simplified in some way. There are important differences, but also important similarities, among these previous studies, which we outline below. In the remaining sections of this paper we will describe the experimental approach we have taken, the results of an initial experiment attempting to focus on the role of the scene geometry in people's strategies to solve the task, and an analysis of those strategies to illustrate the individual differences that occur and the apparent role of different spatial metrics (mostly derived from the field of space syntax research) within each identified strategy.

2   Previous Work

Studies[1] of drop-off orientation that involve physically matching a map to a scene (as opposed to viewing only one of them to infer what the other would look like) have differed in a number of aspects, which makes their results difficult to collate into a single view of this task.

The first source of variation has been the type of space being matched. A common focus (e.g., [7], [8], [4]) has been on matching topographic maps to rural landscapes,

---

[1] For simplicity we have omitted studies that tested orientation processes in the absence of a map, and those where the matching took place from memory, e.g., location/direction judgements made after learning a map. However, it is recognised that these may also involve some common cognitive processes with the task of real-life orientation.

where the primary focus is on the shapes of visible landforms. However there has also been an extensive body of work on orientation within aviation, where the scene view is partly from above rather than immersed within the landscape (e.g., [9], [10], [11], [12]). Meanwhile a different approach [13] required participants to judge their position relative to a single building. Other studies (e.g., [14], [15], [16], [17]) have asked either adults or children to match a map to an even smaller, room-sized space, or to images of it, or to a larger indoor space (e.g., a conference centre).

The second variation lies in the task: some studies (e.g., the field study of Pick et al [4], [18]) have actually immersed participants in a real or a virtual environment and asked them to match it to a map, whereas most other studies have relied on a laboratory simulation where the participant views a static image of a scene. Perhaps more fundamentally, while some of the above studies asked participants to localise (locate) themselves on the map as well as to orientate, others asked only for one or the other—either by marking the participant's position on the map, or by asking for such a mark without asking for direction of view. Arguably, in the latter case, orientation as well as localisation is implied (because the localising task requires one's position relative to nearby objects to be established), but the reverse is not the case. However, since orientation without localisation is often the case in real-world scenarios (e.g., emerging from a subway station, or matching the map to a photograph taken at a named location), this focus has the advantage of narrowing the task and hence the cognitive processes under study (reducing noise in the experimental data) without completely losing ecological validity.

Most of the above studies have focused on response time as the dependent variable indicating task difficulty and performance. However, the average response time is likely to vary greatly with the complexity of the environment and map, from a few seconds in the case of simple displays [19] to (apparently) whole minutes in a field study [4]. Furthermore, the level of accuracy obtained in people's responses also varies greatly, with a typical score of around 50% correct in Pick et al's laboratory study [4], whereas Gunzelmann and Anderson's participants reached near-perfect performance [19].

Despite these many differences of task, focus and outcome, some general conclusions can be drawn about the factors affecting orientation performance. First, there is almost always an effect of the alignment of the map relative to the forward view of the observer. Many studies, both with navigation and with static orientation, have shown that the mental rotation necessitated by map misalignment can have systematic effects on performance (e.g., [19], [13]). Meanwhile, familiarity with the map through its use in previous tasks appears to improve performance if those tasks involved a focus on its geographic content and frames of reference [20].

Another common finding appears to be the role of prominent landmarks and groups of features in people's choice of strategy for solving the task, rather than abstracting the geometry of the scene layout. If a unique landmark exists both in the scene and the map, then matching it between the two can provide an orienting shortcut that saves the observer from having to abstract, rotate and match less salient geometric layout shapes or features—rather like having a 'north' arrow painted on the ground in the scene. This tendency to shortcut the matching task by finding a unique landmark to match instead, has been argued to be a late-developing strategy in human orientation in general [21]. If matching the whole geometry is the default for young children and mature rats, as Hermer and Spelke have famously claimed [21], then it is perhaps surprising to see a role for a landmark (by which in this context we mean any feature whose relative location can be used to aid the matching process) creeping into the drop-off map-matching task, even in extremely sparse scenarios with fairly simple geometric layouts [19]. People's apparent use of such features, often apparently via some kind of propositional description of approximate relative location, may be related to the findings of an apparent tendency to code object location in both an exact and an inexact way in spatial memory [22]. The suggestion that the inexact description of relative location (depending on a landmark or other simplifying cue) may be to some extent a verbal strategy may help explain Hermer and Spelke's failure to find landmark use in their task in very young children or in rats. Although a linguistic explanation has been disputed by some authors [23], it does seem that the use of a landmark is an approximate strategy which functions somewhat as if using a verbal description—relying on an inexact representation of location rather than an exact spatial calculation.

If a landmark is often chosen as a shortcut to matching, what is likely to be chosen? The most systematic study of what makes a landmark more or less suitable, albeit in the context of wayfinding rather than static orientation, is probably that of Winter [24]. This showed that if we assume a landmark is a feature of the scene that is somehow salient to the viewer, this may be a complex mix of visibility (attracting visual attention) and structural salience (relevance to the task). With drop-off static orientation involving a map, however, we may surmise that a key aspect should be the appearance of the landmark on that map.

If a 3D landmark is not labelled on the map such as to make it recognisable and hence easily matchable, and assuming that its most salient feature (e.g., height or unusual roof shape) is not shown in the planimetric 2D view that most maps represent, then it cannot be used effectively for matching. Any attempt to do so is likely to impair performance, either by slowing it or by creating misinterpretations (wrong answers), or both. However, if a salient 2D shape is distinctive and the map is of sufficient detail to reflect that shape, use of the landmark for matching should greatly reduce response latencies and improve accuracy. We therefore tested the relative effects of two- and three-dimensionally salient landmarks in the study we report below.

Previously, apart from observing some role for landmarks such as distinctive clusters of features, no studies have attempted to examine systematically the role of the spatial geometry itself in predicting the ease or difficulty of orientation with a map. In addition, and perhaps surprisingly, few studies have tested orientation (as opposed to navigation) performance within the type of space where the task perhaps most commonly occurs: urban street scenes. Arguably, while the problems of interpreting topography in rural landscapes are well established (e.g., [7]), the opportunity to improve larger-scale urban street mapping to facilitate orientation has been neglected. Furthermore, if landmark matching is indeed an optimal strategy for orientation in any environment, then understanding how this works could help improve all types and scales of geographic spatial representation.

The experiment described below was designed to address these issues by investigating orientation strategies where the scene people viewed contained only the 2D ground layout and the 3D building shapes. The scene images were generated using a 3D model of a UK city (Southampton). All irrelevant details that could distract from the use of an optimal strategy for matching to the map were removed. An example scene and map used in the experiment are shown in Figure 1, together with the corresponding Southampton street location. The map itself contained no name labels or other indicators to distinguish the geographic features. The only remaining salience cues for items within the scene were size (both in terms of ground area and height), shape (again in terms of both roof line and ground layout), and colour (since the same colour scheme was used for both the scene and map, to emphasise the similarity of their 2D geometry and to facilitate its use in matching). In these scenes, therefore, choosing a single 3D item to match to the map was unlikely to be successful, since its 2D geometry was usually ambiguous, although distinctive 2D features and the overall ground layout were not.

3   Method

3.1   Design and Participants

Forty-nine students and members of staff from the University of Huddersfield took part in the experiment. All participants saw the entire set of stimuli in random order. An additional five participants carried out the experiment while having their eye movements and verbal protocols recorded to enable qualitative assessment of their apparent strategies in solving the task. The 49 participants in the main study were encouraged to perform the task as quickly and accurately as possible.

3.2   Materials

The computer based experiment was carried out using PC computers with 17 inch displays and the eye movement and verbal protocol study was conducted using a Tobii 1750 remote desktop eye tracker with a 17 inch display.

Twenty-five scenes from various locations in the city of Southampton, UK were generated using a buildings-only 3D model overlaid on OS MasterMap® Topography Layer and draped on an OS Land-Form PROFILE® terrain model to provide a realistic and accurate representation of height information. The corresponding maps were circular sections of OS MasterMap® Topography Layer at 1:1250 scale. The scenes were selected from photographs of the actual street locations in Southampton (see e.g., Figure 1a) in order to allow subsequent replication of the experiment with the photographs and were chosen to represent a wide range of building shapes, degrees of salience and distinctiveness, together with a range of urban features such as green spaces and road patterns. The colour schemes of the scenes and maps were matched in order to remove any unnecessary distracting information and to facilitate orientation using spatial geometry. Although this procedure reduced the possibility of participants using anything other than the visible geometry, sometimes this still entailed the presence of a landmark (only a completely uniform scene would have no variations in salience).

We hypothesised that a prominent 3D landmark, if its 2D geometry was not especially unique or salient, might distract participants into either slowing their decision or

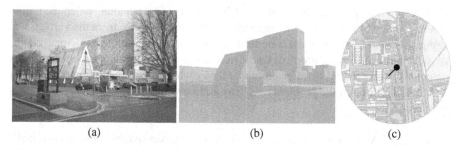

Fig. 1. Street location (a) for example scene (b) and corresponding map (c), used in the experiment. © Crown copyright 2007. Reproduced by permission of Ordnance Survey.

making an incorrect one. An obvious 2D cue in the scene, however, which would be reflected as such on the map, ought to help performance, but in these scenes a 2D shape only tended to be unambiguously salient when it formed part of the foreground layout geometry (e.g., an unusually shaped lawn or pathway). Therefore, we coded the scenes according to occasions when a single object or feature was particularly separable and hence individually salient, either in 3D or in 2D. Thus roughly a quarter of the scenes were deemed to contain each of four possible scene types: an obvious 2D foreground landmark, a prominent 3D landmark, both, or neither. The stimuli were also controlled for alignment by ensuring that the map alignment angle ranged over a roughly even spread from -180 to $+180$ degrees, independently of other aspects of the scenes.

The scene-map pairs were presented sequentially on a 17-inch computer monitor, using specially-programmed software written in tcl/tk which also recorded response locations and times. Each trial presentation, including the five initial practice trials, was separated from the next by a blank screen with a button on it which the participant had to click to move on. This gave all participants a chance to break if needed. It also allowed us to check the eyetracker calibration, where used, by observing the gaze trace on the button between trials.

3.3 Procedure

Participants were introduced to the experiment through the following scenario: "Imagine that you are standing in the street in an unfamiliar town, holding a map. You know where on the map you are standing, but you need to find out which way you are facing". They were then shown an example scene/map pair and told that their task was to work out in which direction they must be facing on the map to see the scene. A black dot in the centre of each map indicated the location of the observer. When the mouse cursor was moved over the map, a short black line of fixed length was drawn from the centre of the dot toward the tip of the cursor (see e.g., Figure 1c). This rotated around the dot in either direction as the mouse was moved around, to follow its position. Participants had to click on the map when they believed they had aligned the pointer towards the centre of the scene on the left of the screen. Participants were asked to respond as rapidly and as accurately as possible. They were told that the maps were all at the same scale, and that they should avoid the natural assumption that the 'upwards' direction on the map indicates 'forward' in the environment [25].

There were five practice trials and twenty experiment trials in total. When the participant responded by clicking on the map the angle of the response from the vertical was recorded, as well as the response time from the onset of the stimulus. Participants in the eye movement and verbal protocol study were asked to talk through each trial as they attempted to solve the problem, in particular to say what they were looking at, how they were thinking through the problem, and why and how they were choosing a particular direction.

Space Syntax Measures. Space syntax measures of urban spaces, although originally focused mainly on understanding paths through it in terms of axial lines, has in recent years also focused on the concept of an *isovist*—the 2D shape that is visible from standing at a particular point in space (and rotating one's body through 360 degrees). Most metrics that can be used to describe an isovist were proposed some years ago [26] although some additional ones have been more recently proposed by other authors (e.g., [27], [28]). A review of the potential of such measures [29] suggested that they may have relevance in helping people to orientate, since the shape and size of the space may make it easier or harder to deduce direction and position. Accordingly, for the twenty scenes used in the main experiment, and looking at both the usual 360° isovist (which of course was visible on the map but not in the scene) as well as on the 60° section of it that was visible within the scene as well, we calculated various metrics as suggested in the space syntax literature. Figure 2 shows (on a simplified, buildings-only version of the map) the 360° and 60° isovists for the scene shown earlier in Figure 1.

Fig. 2. Isovist at ground level for the scene in Figure 1, showing both the full 360° version (darker shading) and the 60° segment (lighter shading) visible in the scene image

The isovist measures that we took for our analyses included all of those that have been identified in the space syntax literature which we felt could have some conceivable role in people's cognition of the scene and map. These included the area and perimeter length of the isovist, its minimum and maximum radius, and these additional measures of its geometry:

1. *Occlusivity*: the extent to which some features of the local environment are hidden by others within the scene.
2. *Compactness*: nearness of the isovist shape to a circle.

3. *Jaggedness*: tending to be inversely related to compactness, this indicates the complexity of the isovist shape (e.g., an isovist from a crossroads in a highly built-up area may be shaped like a long thin cross).
4. *Drift magnitude*: distance of the viewer's location from the centre of gravity of the isovist shape. (Broadly, this and drift angle indicate the level of asymmetry in the isovist; one might expect that an asymmetrical isovist is easier to match unambiguously to a map if the isovist shape is used at all by participants.)
5. *Drift angle*: the angular distance in degrees of the viewer's location from the centre of gravity of the isovist shape. Measured relative to a horizontal line (east).

In addition to these, measures were also calculated that considered the content of the scene. These included the extent to which the isovist perimeter was defined by buildings (as opposed to the edge of the map—we restricted the isovist to the scene that the map depicted, i.e. within the 400m-diameter circle). Other such measures included the proportion of the scene's 2D area that consisted of surface features, since the sidewalks, streets, vegetation and occasional unclassified areas were all distinguished on the map as well as in the scene.

Finally, the above measures were all taken both for the overall 360° isovist, and for the 60° angle subtended by the scene (which is typical of a photograph from a normal camera lens). If a participant was focusing on aspects within the geometry of the scene, either initially or after identifying the broad scene orientation on the map, then it was felt that these versions of the measures might be more relevant than the overall isovist. On the other hand, the overall isovist measures might logically be expected to prove more significant for placing the overall scene geometry within the map's circular area.

4   Results

4.1   General

Responses were scored as correct if the angle of the response line fell within 15 degrees of the true angle in either direction (i.e. within the 30 degree range that it bisected, cf. [13]), at the point when the participant clicked the mouse. Given that the scenes tended to subtend about 60 degrees of visual angle in total, which is also typical of a photograph taken with a normal camera, this meant that the participants had got within 'half a scene' of the exact line.

In general, participants were able to perform the task reasonably accurately, with the proportion of correct responses for each stimulus ranging from .07 to .72 (M = .56, SD = .17). The mean response time for the 20 experiment trials was 40.94 s (SD = 9.76). A Spearman's rho test produced a moderate but non-significant negative correlation between the probability of a correct response and latency, $r_s = -.410$, $p = .072$ indicating that differences do not result from a speed-accuracy tradeoff but suggesting that both measures tended to indicate similarly the relative difficulty of the task for a particular stimulus.

In order to test whether performance was influenced by the presence of salient 3D landmarks, the scenes were coded according to the presence or absence of such a landmark. Ten scenes included at least one. Similarly, scenes were also coded according to the presence of distinctive 2D ground layout information in the foreground of the scene, which would facilitate identification on the grounds of 2D layout. Nine of the 20 scenes included such 2D features, e.g., an extensive and irregularly shaped strip of lawn or pavement in the foreground. For example Figure 1 shows a scene that includes both a salient 3D landmark and distinctive ground layout cues.

The mean response time and percentage of correct responses for the stimuli categorised by the presence or absence of 2D and 3D cues are presented in Figure 3. Separate 2×2 repeated-measures ANOVAs were performed on participants' error rates and response times, with presence or absence of 2D and 3D cues as the two within-subjects factors. For errors, there was a highly significant effect of presence of salient 3D landmarks, $F(1, 48) = 40.35, p < .0001$, and a much smaller but still significant effect of presence of distinctive 2D ground layout, $F(1, 48) = 5.47, p < .05$. There was also a significant interaction between them, $F(1, 48) = 5.26, p < .05$. The directions of these effects showed that while presence of an obvious 2D cue was able to decrease error rates, this was only in the absence of a salient 3D cue which always greatly increased them.

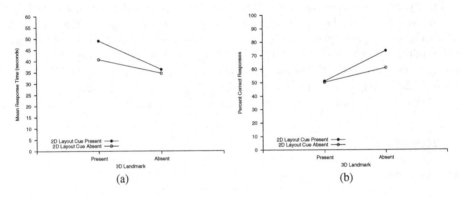

Fig. 3. Mean response time (a) and percentage of correct responses (b) for stimuli categorised by the presence or absence of 2D and 3D cues

The analysis of response times, however, showed that both 3D and 2D cues seemed to slow participants down — again much more so for 3D, $F(1, 48) = 29.7, p < .0001$ than for 2D, $F(1, 48) = 9.28, p < .005$. There was again a mild interaction, $F(1, 48) = 4.37$, $p < .05$, which indicated that the presence of both a 2D and a 3D cue had the most marked effect of all on response times; the presence of a 2D landmark made only a small difference except when a 3D landmark was also present.

Some caution should be expressed with the above analyses since both the response time and error data showed minor deviations from normality; however, the main effects were also checked using non-parametric Wilcoxon signed-rank tests, which showed the same strong significance patterns (but could not, of course, test the interaction effects).

This finding was independently confirmed by qualitative verbal protocol and eye movement analysis of the five additional participants. By far the most commonly reported feature used for solving the problem was 'buildings', and the eye movement patterns in the scenes with the most salient 3D landmarks (e.g., large skyscrapers or church steeples) tended to strongly focus around those landmarks.

4.2 Map Alignment

Previous studies where a map is matched to a scene have tended to find a distinctive 'M' shape pattern in the effect of map alignment with observer position (e.g., [19], [30]). Performance typically is better not only at 0 degrees (where 'up' on the map exactly corresponds to the forward direction within the scene), but also at 90, 180 and 270 (i.e. −90) degrees. It seems that mental rotation to these cardinal directions is easier than with more oblique angles. Although there was a modest effect of map alignment on response time in the current study, the M shape pattern was considerably less well defined than those found in other studies. This is possibly due to the fact that alignment angle was varied semi-randomly rather than at fixed points (such as 0, 45, 90, etc.) in this study and because of the possibility that landmark-based orientation strategies sometimes make mental rotation unnecessary (cf. [19]).

4.3 Space Syntax Measures

A multiple regression analysis incorporating the space syntax measures outlined above was applied to the error data to determine whether they were able to account for the differences in number of correct responses to each scene (described in further detail in [31]). It was found, however, that the space syntax measures appeared to show little clear predictive power for the overall performance across participants. This apparent lack of a clear consistent role for the scene geometry prompted a systematic analysis of individual differences, discussed below.

4.4 Individual Differences

In order to see whether individual differences in strategy could be linked to spatial metrics of the different scenes, the following analysis of the response time data was undertaken. First, for each participant and across the 20 scenes, correlation coefficients were calculated (using non-parametric Spearman correlations due to non-normality of some variables) to indicate the size of effect on performance of the various spatial metrics. These coefficients were then used in a cluster analysis, to examine apparent groupings of participants in terms of which spatial variables appeared to most influence their performance. The cluster method was Ward's linkage [32], since this minimises the variance within groups and thus would most clearly highlight similarities among participant strategies. Squared Euclidean distance (E^2) was used as the similarity measure, as generally recommended in the literature for this clustering method. The Duda-Hart stopping rule [33] indicated that four clusters was the optimum solution in terms of distinguishing clear groups. These are indicated by the horizontal line in the dendogram shown in Figure 4.

Table 1 shows the four identified clusters, the cluster sizes (i.e. number of participants in each group), the key spatial variables whose correlation patterns appeared to strongly distinguish that group's performance (with the mean Spearman r coefficient of

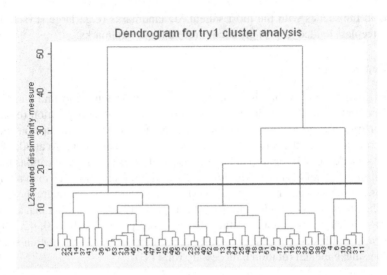

Fig. 4. Dendogram showing the clustering of participants' responses according to the four primary spatial variables described in Table 1. The horizontal line shows where the Duda-Hart stopping rule indicated the optimum number of clusters.

the correlation with response time — so a positive correlation meant slower times), and the apparent broad strategies that were thus implied.

It can be seen that the groups overlap in strategy use, but apparently differ in the most common 'default' strategy (as inferred from the aspects of the space that appear to affect their performance). Groups I and II were affected by the presence of a 3D landmark more than groups III and IV. Group III appeared to use the 'optimum' strategy of focusing on the overall scene geometry and rotating it to match the map; surprisingly, none of the other groups appeared to be slower with greater map misalignment.

The table shows that Group III's response times were facilitated by a strongly jagged scene geometry (which in this environment usually implies a road junction with more than one connecting street), and by a scene where few objects were obscured by others (i.e. a built-up scene whose visible perimeter was largely formed by buildings rather than the obscured spaces behind them). They performed more slowly where the angle of map misalignment was greater, and where there was more vegetation in the scene (probably implying a more open space with more scattered buildings).

One-way ANOVAs to compare the clusters on errors and response times found no overall difference for response times, but did show significant differences in error levels between the groups, $F(3, 45) = 3.00$, p $< .05$). Post hoc contrasts showed that this was due to group III performing significantly better than Groups II or IV (with Group I's performance falling somewhere in between). On average Group III were correct 72% of the time, compared with 63% for Group I, 49% for Group II and 47% for Group IV. This suggests that using the 'correct' strategy of matching overall geometry and performing mental rotation did produce optimum performance on the task, but was only adopted by

Table 1. Identified clusters, contributing correlation coefficients, and the corresponding apparent strategies by participants

Group	N	Key spatial variable effects	Inferred strategies
I	20	Overall isovist area (−0.18); 3D landmark presence (0.29); No mean correlations above 0.3	Little focus on isovist shape. Probably picking a single feature to match: possibly from ground layout, since faster with more open spaces but distracted into trying to use salient 3D landmarks when present.
II	14	Within-scene isovist occlusivity (0.34); Perimeter length (0.36); Drift magnitude (0.28) & area (0.36); 3D landmark presence (0.29).	Abstracting the 2D isovist geometry and then matching it to the map but distracted into trying to use salient 3D landmarks when present.
III	9	Map alignment (i.e. angular bearing of scene centreline from north, 0.26); Overall isovist compactness (0.28) & jaggedness (−0.28); vegetation extent (0.27); proportion of isovist perimeter formed by buildings (−0.36).	Focus on street pattern in built-up areas (enhanced when streets are lined with buildings making their shapes more salient); hence dependence on mental rotation to match to map (worsened by map misalignment)
IV	6	Presence of strong 2D foreground cues (0.51); overall isovist area (0.30), occlusivity (0.37) and perimeter length (0.35); extent of visible area showing as footpaths (sidewalks etc., 0.26), as streets (−0.30), and as undefined ground cover (−0.32); proportion of isovist perimeter formed by buildings (−0.33).	Use of both ground layout patterns and abstracting the overall isovist geometry (easier when isovist smaller and more clearly defined by buildings), but highly distracted by attempt to match foreground 2D cues when available.

a minority of participants. Yet the largest group of participants—around 40% in Group I—did not perform significantly worse in general than this 'optimal' strategy group.

5   Discussion

The results of the experiment are consistent with previous studies in showing that, when possible, people tend to match a single salient landmark between a 2D and 3D representation of a scene, and particularly to pick on a landmark with a distinctive 3D (but not 2D) shape despite the absence of that shape in the 2D map. This is particularly noteworthy given that this strategy was discouraged by the nature of the stimuli. In the scenes used in the present study, as in the studies by Gunzelmann and Anderson [19], the 2D shapes and colours were directly matchable between the scene and the map (though they would not be in real-world scenes or photographs), and all distracting salient cues were removed other than the 3D geometry. Despite this, participants still made errors through attending to the latter rather than the more reliable 2D geometry, most likely due to the particular visual salience of landmarks in the scenes (cf. Winter [24]).

A clue as to a potential reason for people's sometime preference for inappropriate landmark use, rather than sticking to the more reliable 2D geometry, lies in the finding that the presence of a strong 2D cue (whose visual salience would perhaps push participants towards its use) seemed to actually slow people down. It seems reasonable to assume that the process of extracting an overhead 2D geometric configuration from the 3D scene, and then carrying this over to rotate and match to the 2D layout of the map, may sometimes create more cognitive load than finding an alternative such as matching a single feature or taking an approximate, broad account of the approximate layout (e.g., just being aware that one is 'looking down the road').

As well as the obvious implications for understanding human cognition of large-scale spaces, this may also help to explain the public popularity of 'bird's eye' urban maps that show the buildings from an oblique angle rather than from overhead [34]. It also implies that if large-scale maps were to be designed explicitly to aid their use in orientation, it would not be sufficient merely to include orienting landmarks at places where the 2D geometry was an ambiguous cue, since it may not be used efficiently even when unambiguous.

Analysing individual differences via the cluster analysis provides a different perspective however. Here the different aspects of the spatial geometry and features are shown to be relevant to specific strategies for solving the task. Almost half of participants did show a reliance on single salient landmarks, as implied by our overall analysis and by previous studies (e.g., [4]), and a further quarter of the sample would be distracted into this strategy when a salient 3D landmark was offered. However, just over half of participants actually did appear to show some efforts to abstract the 2D isovist geometry, the (simpler) street pattern, or the patterns made by ground layout features. Also, although the overall analysis had suggested that the presence of a 2D foreground landmark generally improved performance, the individual differences analysis showed that it actually slowed down a small minority of participants (possibly because it detracted from their preferred strategy of abstracting a more general sense of the ground layout and/or isovist geometry). Meanwhile, although the minority of participants who adopted the optimal geometry-matching-and-mental-rotation strategy did perform best, it was not significantly better than the largest group of participants who would apparently be helped by being able to reliably reference and match a single feature between the map and the scene—in other words, some kind of landmark. If, say, a church in real life was marked with a church symbol on the correct street corner on the map, then matching would become trivial and highly accurate for this largest group of participants.

It seems, therefore, that the potential role of space syntax measures in interpreting cognitive tasks of this nature is one that is highly dependent on problem-solving strategies, rather than as an overall predictor of task difficulty. It also appears that even when the abstraction of the 2D geometry is the only reliable cue for solving the task, as was the case in the present experiment, a majority of people will still attempt to rely on salient landmark cues within that geometry, whether or not they are discernable from the map.

This confirms the value of landmarks in aiding orientation with maps, but also warns us that the match between the landmark's appearance in the real world and on the map must be unambiguous and rapid if errors are to be avoided. With less congruent representations, e.g., a photograph or actual real-world scene where colours and shapes will usually differ between the scene and the map (and where the map is likely to be smaller-scale and hence subject to greater cartographic generalisation), this is likely to be a greater challenge, although abstraction of the 2D geometry will also be more difficult due to the presence of street furniture, vegetation, cars and other objects. For this and other obvious reasons, increasing congruence by adding 3D realistic landmark representations to the map would not necessarily be the best solution: as decades of cartographic research suggest, along with more recent studies [35], a symbol merely representing the category of object (e.g., church or pub) may be recognised more quickly than an attempt at a photorealistic image of it. In any case, since appearances of real-world objects often change, it would probably be unrealistic to suggest that a mapping agency collect and maintain photorealistic images of the landmarks found on thousands of street corners.

Further experiments will investigate the strategies used under these more realistic circumstances, and the implications for the design of suitable map representations.

Acknowledgements. The authors wish to thank Claire Cannon and Jon Gould for their help in analysing and interpreting the data, Alasdair Turner at UCL for help in extracting the space syntax measures, Isabel Sargent, Jon Horgan and Dave Capstick (creators of the 3D building model), Guy Heathcote and Tim Martin for help in using the model and mapping to create the stimuli, the experiment participants for their time and cooperation, Glenn Gunzelmann and Glen Hart for insights and inspiration, and Ordnance Survey of Great Britain for funding and supporting this work.

This article has been prepared for information purposes only. It is not designed to constitute definitive advice on the topics covered and any reliance placed on the contents of this article is at the sole risk of the reader.

References

1. Robinson, A., Petchenik, B.: The Nature of Maps: Essays towards understanding maps and mapping. University of Chicago Press, Chicago (1976)
2. Montello, D.R., Freundschuh, S.: Cognition of geographic information. In: McMaster, R.B., Usery, E.L. (eds.) A research agenda for geographic information science, pp. 61–91. CRC Press, Boca Raton, FL (2005)
3. Thompson, W.B., Pick, H.L., Bennett, B.H., Heinrichs, M.R., Savitt, S.L., Smith, K.: Map based localization: The 'drop-off' problem. In: Proceedings of the DARPA Image Understanding Workshop, pp. 706–719 (1990)
4. Pick, H.L., Heinrichs, M.R., Montello, D.R., Smith, K., Sullivan, C.N.: Topographic map reading. In: Hancock, P.A., Flach, J., Caird, J., Vicente, K. (eds.) Local applications of the ecological approach to human-machine systems, vol. 2, pp. 255–284. Lawrence Erlbaum, Hillsdale, NJ (1995)
5. Levine, M.: You-are-here maps: Psychological considerations. Environment and Behavior 14, 221–237 (1982)

6. Klippel, A., Freksa, C., Winter, S.: You-are-here maps in emergencies – the danger of getting lost. Journal of Spatial Science 51, 117–131 (2006)
7. Griffin, T.L.C., Lock, B.F.: The perceptual problem of contour interpretation. The Cartographic Journal 16, 61–71 (1979)
8. Eley, M.G.: Determining the shapes of land surfaces from topographical maps. Ergonomics 31, 355–376 (1988)
9. Aretz, A.: Spatial cognition and navigation. In: Perspectives: Proceedings of the 33rd annual meeting of the Human Factors Society, Santa Monica, CA, Human Factors Society, pp. 8–12 (1989)
10. Harwood, K.: Cognitive perspectives on map displays for helicopter flight. In: Perspectives: Proceedings of the 33rd annual meeting of the Human Factors Society, Santa Monica, CA, Human Factors Society, pp. 13–17 (1989)
11. Wickens, C.D., Prevett, T.T.: Exploring the dimensions of egocentricity in aircraft navigation displays. Journal of Experimental Psychology: Applied 1, 110–135 (1995)
12. Gunzelmann, G., Anderson, J.R., Douglass, S.: Orientation tasks with multiple views of space: Strategies and performance. Spatial Cognition and Computation 4, 209–256 (2004)
13. Warren, D.H., Rossano, M.J., Wear, T.D.: Perception of map-environment correspondence: The roles of features and alignment. Ecological Psychology 2, 131–150 (1990)
14. Blades, M., Spencer, C.: The development of 3- to 6-year-olds' map using ability: The relative importance of landmarks and map alignment. Journal of Genetic Psychology 151, 181–194 (1990)
15. Hagen, M.A., Giorgi, R.: Where's the camera? Ecological Psychology 5, 65–84 (1993)
16. Presson, C.C.: The development of map-reading skills. Child Development 53, 196–199 (1982)
17. Meilinger, T., Hölscher, C., Büchner, S.J., Brösamle, M.: How much information do you need? Schematic maps in wayfinding and self localization. In: Spatial Cognition '06, Bremen, Germany, Springer, Berlin (2007)
18. Bryant, K.J.: Geographical spatial orientation ability within real-world and simulated large-scale environments. Multivariate Behavioral Research 26, 109–136 (1991)
19. Gunzelmann, G., Anderson, J.R.: Location matters: Why target location impacts performance in orientation tasks. Memory and Cognition 34, 41–59 (2006)
20. Davies, C.: When is a map not a map? Task and language in spatial interpretations with digital map displays. Applied Cognitive Psychology 16, 273–285 (2002)
21. Hermer, L., Spelke, E.S.: A geometric process for spatial reorientation in young children. Nature 370, 57–59 (1994)
22. Lansdale, M.: Modelling memory for absolute location. Psychological Review 105, 351–378 (1998)
23. Cheng, K., Newcombe, N.S.: Is there a geometric module for spatial orientation? squaring theory and evidence. Psychonomic Bulletin & Review 12, 1–23 (2005)
24. Winter, S.: Route adaptive selection of salient features. In: Kuhn, W., Worboys, M.F., Timpf, S. (eds.) COSIT 2003. LNCS, vol. 2825, pp. 349–361. Springer, Berlin (2003)
25. Shepard, R.N., Hurwitz, S.: Upward direction, mental rotation, and discrimination of left and right turns in maps. Cognition 18, 161–193 (1984)
26. Benedikt, M.L.: To take hold of space: Isovist and isovist fields. Environment And Planning B: Planning & Design 6, 47–65 (1979)
27. Conroy-Dalton, R.: Spatial navigation in immersive virtual environments. PhD thesis, University of London, London (2001)
28. Wiener, J.M., Franz, G.: Isovists as means to predict spatial experience and behaviour. In: Spatial Cognition IV: Reasoning, Action, Interaction, pp. 42–57. Springer, Berlin (2004)

29. Conroy-Dalton, R., Bafna, S.: The syntactical image of the city: A reciprocal definition of spatial elements and spatial syntaxes. In: 4th International Space Syntax Symposium, London, UK (2003)
30. Hintzman, D.L., O'Dell, C.S., Arndt, D.R.: Orientation in cognitive maps. Cognitive Psychology 13, 149–206 (1981)
31. Davies, C., Mora, R., Peebles, D.: Isovists for orientation: Can space syntax help us predict directional confusion? In: Proceedings of the Space Syntax and Spatial Cognition workshop, Spatial Cognition 2006, Bremen, Germany (2006)
32. Ward, J.H.: Hierarchical grouping to optimize an objective function. Journal of the American Statistical Association 58, 236–244 (1963)
33. Duda, R.O., Hart, P.E.: Pattern classification and scene analysis. Wiley, New York (1973)
34. Gombrich, E.H.: The image and the eye: Further studies in the psychology of pictorial representation. Phaidon Press, London (1982)
35. Smallman, H.S., John, M.S., Oonk, H.M., Cowen, M.B.: When beauty is only skin deep: 3D realistic icons are harder to identify than conventional 2D military symbols. In: Proceedings of the 44th Annual Meeting of the Human Factors and Ergonomics Society, Santa Monica, CA, Human Factors and Ergonomics Society, pp. 480–483 (2000)

Data Quality Ontology: An Ontology for Imperfect Knowledge

Andrew U. Frank

Institute for Geoinformation and Cartography, Vienna University of Technology
Gusshausstrasse. 27-29, A-1040 Vienna, Austria
frank@geoinfo.tuwien.ac.at

Abstract. Data quality and ontology are two of the dominating research topics in GIS, influencing many others. Research so far investigated them in isolation. Ontology is concerned with perfect knowledge of the world and ignores so far imperfections in our knowledge. An ontology for imperfect knowledge leads to a consistent classification of imperfections of data (i.e., data quality), and a formalizable description of the influence of data quality on decisions. If we want to deal with data quality with ontological methods, then reality and the information model stored in the GIS must be represented in the same model. This allows to use closed loops semantics to define "fitness for use" as leading to correct, executable decisions. The approach covers knowledge of physical reality as well as personal (subjective) and social constructions. It lists systematically influences leading to imperfections in data in logical succession.

1 Introduction

Data quality, treatment of error in data and how they influence usability of data is an important research topic in GIS. It was research initiative I 1 of the NCGIA research plan [28]. The lack of a formalized treatment of data quality limits automatic discovery of datasets and interoperability of GIS [39]. International efforts to standardize metadata to describe quality of data underline its practical relevance. Despite interesting specific research results, progress has been limited and few of the original goals set forth in the NCGIA program has been achieved. Empirical observations document that users of geographic data do not understand concepts like data quality and related terms [3], which points to a difference in conceptualization of data quality between the research community and practice. An ontological approach to data quality may contribute to bridge this gap.

Ontology is advocated in GI Science as a method to clarify the conceptual foundation of space and time. An ontology for space was the goal of NCGIA research initiative I 2 [28] and the later initiative I 8 extended this to time (to approach space and time separately was most likely a flaw in the research program). Ontology research in GIScience has led to classifications in the formalizations and representation of spatial and temporal data. Ontologists have only rarely considered the imperfections in our knowledge [40,29] and therefore not contributed to data

S. Winter et al. (Eds.): COSIT 2007, LNCS 4736, pp. 406–420, 2007.

quality research. Ontology claims to be useful to lead to consistent conceptualizations and to classify the design of GIS [7,8]. This paper demonstrates this contribution to the data quality discussion.

This paper is structured as follows: Section 2 reviews the ontological commitments of ordinary, perfect knowledge. Section 3 to 7 then detail in turn the ontological commitments with respect to practical causes for imperfection:

- Limitation to partial knowledge,
- observation (measurement) error,
- simplification in processing (object formation),
- classifications, and
- constructions.

Section 8 shows how these commitments for imperfection affect decisions. The application of ontological methods to data quality questions makes it necessary to combine the philosophical approach of an ontology of reality with the information science approach of ontology as a conceptualization of the world. It is necessary to consider both reality and the information model at the same level (an approach I have advocated for agent simulations before [9]).

Closed-loop semantics can then be used to ground semantics—including the semantics of data quality. This gives a novel definition of data quality as "fitness for use" in the sense of leading to correct, executable definitions; this definition is operational and leads to method to assess the influence of imperfections in the data to errors in the decision [14].

2 An Ordinary GIS Ontology

The philosophically oriented ontology research debates different approaches to ontology. Smith has given several useful critiques of philosophical positions [34]. I use here the practical notion of ontology as a "conceptualization of a part of reality for a purpose" [17].

2.1 Different Kinds of "Existence"

Ontology is concerned with what exists or is thought to exist. It appears obvious that the current temperature here, the weight of this apple, democracy in Austria and my loyalty to my employer each exist in a different way and follows different rules in their evolution [24,25]. A tiered ontology separates these forms (Fig. 1); I have suggested five tiers [10,11] and their influence on the conceptualizations and, as will be shown here, imperfections in the data will be discussed in this paper. The tiers explain—with pertinent operations between them—how knowledge of the world is derived from the primitive observations (tier 1), which are the only source of knowledge of the world we have. The tier structure is not unlike the abstraction layers in a communication protocol [19] and represents increasingly abstract forms of knowledge.

Tier O: human-independent reality
Tier 1: observation of physical world
Tier 2: objects with properties
Tier 3: social reality
Tier 4: subjective knowledge

Fig. 1. The five tiers of ontology

2.2 Ontological Commitments

Ontologies should be structured by commitments that state precisely the assumptions used in the construction of the ontology. Consequences from the assumptions become transparent and contradictions between commitments can be avoided. Ontological commitments play in ontology the role of axioms in mathematics; from a different set of commitments different ontologies result (Smith and Grenon give a brief overview over different philosophical –ism resulting from different commitments [34]).

2.3 Commitments to a Physical Reality

The following assumptions represent my understanding of a consensus position and is not more detailed than necessary for current purposes. Several upper ontologies give more details [15,34] and GIS related ontological articles [1,35] give slightly different approaches, but the differences are not consequential for the analysis presented here.

2.3.1 Commitment O 1: A Single World
It is assumed that there is a physical world, and that there is only one physical world. This is a first necessary commitment to speak meaningfully about the world and to represent aspects of it in a GIS. This commitment to a single reality does not exclude descriptions of future planned states of the world or different perceptions of individuals, because these are descriptions and not (yet) part of reality.

2.3.2 Commitment O 2: The World Exists in Space and Evolves in Time
The world exists in space and has states that change in time. This commitment posits both a continuous space and time. The current world states influence future world states [38].

2.3.3 Commitment O 3: Actors Can Observe Some States of the World
The actors, which are part of the physical reality, can observe some of the states of the world (Fig. 2). Observations p give the state v at a point in space x and the current time t (point observation)

$$v = p(\underline{x}, t) \tag{1}$$

2.3.4 Commitment O 4: Actors Can Influence the State of the World
Actors can not only observe the states of the world but also influence them through actions. The effects of actions are changed states of the world and these changed states can again be observed. This gives a closed semantic loop [12], which connects

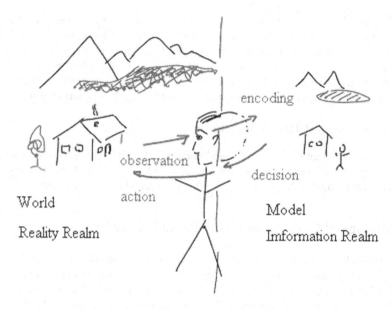

Fig. 2. Reality and Information Realm

the meaning of observations through sensors with the meaning of actions that are reported through proprio-sensors in the body (Fig. 2). The agent with the sensors and actuators in his body give semantic grounding to the observations and actions [4].

2.4 Commitments Regarding Information

To construct a usable ontology for GIS, reality must be separated from the information system. [40] in perhaps the first ontological view on data quality have introduced this separation (see also [9]). This is in contrast to philosophical ontologies that attempt to explain reality and critique the concept-orientation of information system ontologies [33], but also in contrast to Gruber's definition, which considers only the information realm.

2.4.1 Commitment I 1: Information Systems Are Models of Reality
Observations and encoding translate the observable states of reality into symbols is the information system. The information system is constructed as a (computational) model of reality, connected by morphism, such that corresponding actions in model and reality have corresponding results [6,16,30,40].

2.4.2 Commitment I 2: Information Causation Is Different from Physical Causation
The changes in the state of the world are modeled by physical laws, e.g.: "the cause for water flowing downward in the reality realm is gravity." The rules of physics in the reality realm can be modeled in the information realm and allows the computation of expected future states in the information realm, predicting what effects an action has.

A different form of causation is *information causation*. Agents use information to plan actions. The execution of the action causes changes in the physical world (O 4). Actions can be separated into an information process, which I will call decision, and a physical action (Fig. 2). Decisions are in the information realm but they affect—through actions and physical laws—the reality realm. Decisions can have the intended effect only if the action can be carried out and no physical laws contradict it.

2.5 Consequences of These Commitments

These six commitments are realistic, and correspond with our day-to-day experience. They imply other consequences, for example, O 4 with O 1 allows agents to communicate through actions (noise, signs), which are observed by others.

3 Ontological Commitments for Partial Knowledge

Perfect knowledge of the state of the world is not possible. The above "usual" ontological commitments ignore the necessary and non-avoidable imperfections in our knowledge. For many applications it is useful to pretend that we have perfect knowledge, but ignoring the imperfections in our knowledge is hindering the construction of a comprehensive theory of data quality and indirectly the realization of multi-purpose geographic information systems and interoperability between GIS. The following commitments separate individual aspects of data quality—a notion problem documentes by contradiction definitions and use of terms like precision, vagueness, etc. I expect that a computational model for each commitment is possible that then leads to operational definitions for a principled and empirically grounded terminology.

The imperfections in the knowledge are caused by the limitations and imperfections of the processes that construct knowledge in a higher tier from the data in the lower tier. In this section, three commitments regarding the incompleteness of our knowledge are introduced; they state limitations in

- spatial and temporal extent,
- type of observation, and
- level of detail.

The following sections will discuss other causes for imperfections in data.

3.1 Commitment P 1: Only a Part of the World Is Known

Our maps do not show white areas as "terra incognita", but large part of the real world are still unknown (e.g., in microbiology). A complete and detailed model of the world would be at least as big as the world; it is therefore not possible. Our data collections are always restricted to an area of space and a period of time.

3.2 Commitment P 2: Not All States of the World Are Observed

A data collection is limited to the states observable with the sensors used; other states go unobserved and remain unknown. The world may have states for which we know no sensors (e.g., electricity in medieval times).

3.3 Commitment P 3: The Level of Detail Is Limiting

Observation is a physical process and requires finite space and time for observation. The properties of the physical observation process, which is effectively a low-pass filter, limits the level of detail with which the (nearly) infinitely detailed reality can be observed. It is possible to obtain more detailed observations with better instruments, but not everywhere and all the time (P 1).

4 Imperfections in Observations

The observation methods achieve the first step in translating states of the world into symbols in the information system (the second is encoding and not consequential here); they are imperfect physical systems and produce imperfect results.

4.1 Commitment E 1: Observations Are Affected by Errors

The physical observation processes are disturbed by non-avoidable effects and produce results that do not precisely and truly represent the state observed. These effects can be modeled as random disturbation of the true reading and are assumed normally distributed with mean 0 and standard derivation σ.

4.2 Commitment E 2: States of the World Are Spatially and Temporally Autocorrelated

Nearly all world states are strongly autocorrelated, both in space and time. As the value of a state just a bit away or a bit later is most likely very similar to the value just observed. Spatial and temporal autocorrelation are crucial for us to make sense of the world and to act rationally, goal directed. Besides spatial and temporal autocorrelation, correlation between different observation types are also important. One can often replace a difficult to observe state by a strongly correlated, but easier to observe one; for example, color of fruit and sugar content are correlated for many fruits.

5 Imperfections Introduced by the Limitations of Biological Information Processing

Additional imperfections are the result of limitations of biological information processing and how it is adapted to the environment in which biological agents survive.

5.1 Commitment R 1: Biological Agents Have Limited Information Processing Abilities

The structure of our information is not only influenced by our sensors, but also by the systems to process information. The brains of biological agents—including humans—are limited and the biological cost, i.e., energy consumption of information processing, is high. Biological agents have developed methods to reduce the load on their information processing systems to achieve efficient decision making with limited effort and in short time. Humans and higher animal species reorganize the low level *point observations* into information about objects. Karminoff-Smith has shown that humans apply a general cognitive mechanism of re-representation of detailed data into more compact, but equally useful, knowledge [20,21].

5.2   Commitment R 2: Reduction of Input Streams by Concentration of Discontinuities

Most of the world is slowly and continuously changing (E 2); concentrating on discontinuities in this mostly stable environment allows enormous reduction of processing load. This is the same strategy used in technical system (JPEG, run length encoding, etc.) and draws attention to boundaries between larger relatively uniform regions. This is applicable both statically for spatial regions of uniform color, texture, or movement, or dynamically, for uniform periods of light, cold, or economic well-being.

5.3   Commitment R 3: Object Centered Representation

Boundaries separate relatively uniform regions of some property from regions with some other value for the same property (Fig. 3). The relative uniformity in the property value absorbs some of the observation uncertainty (E 1).

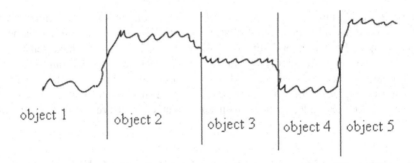

Fig. 3. Discontinuities separate uniform regions (for simplicity one dimensional)

5.4   Commitment R 4: Objects Have Properties Derived from Observable Point Properties

It is possible to aggregate—usually sum—the values of some observable point property over the (2D or 3D) region of the object. This gives ,e.g., the volume, weight, and similar properties of objects.

$$P(o) = \iiint_{V(o)} p(v)\, dV \qquad (2)$$

5.5   Commitment R 5: Object Constructions That Endure in Time Are Preferred

Objects that remain relatively unchanged over time are preferred to reduce the data processing load. Material objects have sharp boundaries where coherence in material is low: the object is what moves together. Such objects have a large number of stable properties (color, weight, shape, etc.).

5.6   Commitment R 6: Object Formation Is Influenced by Interaction with the World

We cut the world in objects that are meaningful for our interactions with the world [22,23]. Our experience in interacting with the world has taught us appropriate strategies to subdivide continuous reality into individual objects. The elements on the tabletop (Figure 4) are divided in objects at the boundaries where cohesion between cells is low and pieces can be moved individually; spoon, cup, saucer each can be picked up individually.

Fig. 4. Typical objects from tabletop space

5.7   Commitment R 7: Multiple Possibilities to Form Objects

For a tabletop the boundaries of objects as revealed when moving them are most salient and it is cognitively difficult to form objects differently (try to see all the food on a table as a single object!). Different objects result from different levels of aggregation: a pen with its cover is first one pen object, once opened, it is two (pen and cover) and occasionally one might take it apart and find more separable pieces.

For non-moveable objects no such salient boundaries are forcing a single, dominating form of object formation. For example, geographic space can be divided into uniform regions (i.e., geographic objects) in many different ways: uniform land-use, land-cover, watersheds, soil quality, etc. A GIS must be prepared to deal with multiple, overlapping, and coexistent geographic objects and must not blindly transfer the exclusive, non-overlapping solid object concept from tabletop space to geographic space [26].

5.8 Commitment R 8: Error Formation Can Be Relative or Absolute

Object boundaries can either be located by a property value passing a threshold, e.g., water content indicating land or lake area (Figure 5):

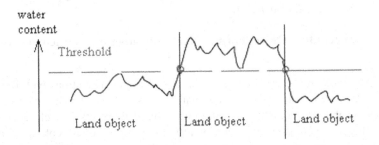

Fig. 5. Boundary located where the property values passes a threshold (for simplicity in one dimension)

Alternatively, the boundary is where the change (gradient) is maximal, e.g., tree density used to delimit a forest area (Figure 6).

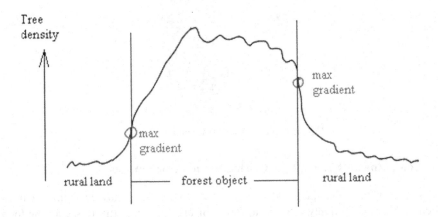

Fig. 6. Boundary located where rate of change is maximal (for simplicity one-dimensional)

Technical and legal systems often use the fixed threshold (e.g.; mountain farming support is available for farms above 1000 m). Human sensors seem to prefer to identify changes relative to the environment and independently of absolute value [5].

5.9 Commitment R 9: Object Boundary Location Is Uncertain

The measurement error (E 1) influence the position of the object boundary. This influences both threshold or maximal gradient used for object formation. Spatial (or temporal) autocorrelation is necessary to assess the effect.

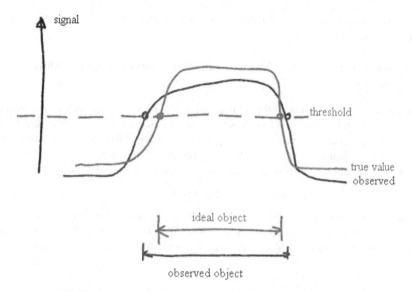

Fig. 7. Error in observation of property results in error of object boundary (for a fixed threshold)

5.10 Commitment R 10: Object Property Values Are Uncertain

The property value obtained for an object is an approximation, influenced firstly by the uncertainty in the object region delimited by the object boundaries (R 9) and secondly by the measurement error of the point property that is integrated (E 1, with Eq 2 from R 4). The aggregation reduces the effect of observation errors, but it is difficult to assess the effects if the property used for delimiting the object and the property aggregated are correlated.

6 Classification

The interactions agents have with the world seem to repeat themselves with small variations in the particular objects involved: I get up every morning, drink tea and eat a toast—it is a different day, tea, and toast every morning, but I can regulate my behavior by using the same patterns every morning. This seems a re-representation like mechanism. The discussion in this section is in terms of physical objects, but the mechanisms are used in the same way for processes.

6.1 Commitment C 1: Objects Are Classified

Physical objects, which result from commitments R 1 to R 9, are classified. Classification selects a property to determine object boundaries (e.g., material coherence) and one or a set of properties of the object to classify it [13]. Prototypical is the classification of animals and plants to species and the hierarchical sub/super class structure created by Linné. Most ontologies currently constructed are taxonomies of object classes.

6.2 Commitment C 2: Class Membership of an Object Is Based on Object Properties

Two mechanisms are used for classification; they can be called set-theoretic (Aristotelean) and radial classes. A set-theoretic category includes all objects that fulfill some property (or properties), e.g., a tree is an individual plant from a tree species with a diameter of more than 10 cm at 1.5 m height above ground. All members of the class tree so defined have this property and no member does not.

$$\{x \in \text{tree} \mid \text{diameter}_{1.5}(x) > 0.10\} \tag{3}$$

A radial category is formed by a number of properties (among them often gestalt) and allows better, more typical and less typical members. Typical *birds* are robin or sparrow, penguin and ostrich are atypical. In contrast, a set-theoretic class does not admit better or worse exemplars.

6.3 Commitment C 3: Object Reasoning Can Use Default Class Values for Properties

In many situations detailed knowledge about the individual object is not available and replaced by the expected usual (default) value for the class. This permits reasoning in the absence of detailed knowledge, for example to analyze future situations, for which detailed knowledge is not available yet. Set-theoretic class are better suited for default reasoning; using default reasoning with a radial category may be wrong. For example. The logic inference:

> Tweacky is a bird & Birds fly
> ⇒ Tweacky flies

is not correct if Tweacky is an atypical bird (e.g., a hen or a penguin).

7 Constructions

Searle has described social constructions with the formula:

$$X \text{ counts as } Y \text{ in context } Z. [32] \tag{4}$$

I suggest here that this formula can be extended to apply equally to personal (subjective), linguistic and not only to social construction.

7.1 Commitment X 1: No Freestanding Y Terms

All constructions are eventually grounded in physical object or process. It is possible that a construction 'B counts as A in context Z' refers to a B, which is a construction and not physically existing, but then B must be part of another construction 'C counts as B in context Z', where C is a physical object or action. Chains of grounding may be longer, but any construction is ultimately grounded in a physical object or action.

7.2 Commitment X 2: A Context Is a Set of Rules

The meaning of a Y term is determined by the context Z; the context Z describes a set of logical rules that give antecedents, consequences potential actions, etc., which are related to Y. Such a context can be modeled as an algebraic structure; Bittner and Navratil have given a formalization of the rules that form the context of landed property ownership and cadastral system [2,27]. The difference between personal (subjective), linguistic, and social constructions seem to be in the applicable context:

- A personal construction is valid in the context of a person, her experience, history, current situation and goals, etc. It is highly variable and not shared.
- A linguistic construction, a word in a language for example, is valid in the current cultural use of this construction; the context gives the rules for entailments; i.e., what another person may conclude from an utterance.
- A social construction in Searle's sense is valid in a shared, often codified set of rules; a national legal system is a prototypical example for such a context.

8 Quality of Data

Data quality can be limited to the data acquisition process, but an analytical connection between data quality and quality of decision becomes feasible in the ontology for imperfect data. The decision process is here considered a black box' and only input data and outcome are of interest, not the possibly complex interactions between multiple persons involved, etc. (for this see Naturalistic Decision Making [41]).

8.1 Quality of Observations and Object Properties

Observations can be assessed by repetition and statistical descriptions of errors and correlations derived (E 1 and E 2). If the rules for object formation are known, the quality of the data describing object properties can be described statistically as well (R 9 and R 10). Such descriptions represent a data acquisition point of view and are not directly relevant to determine if a dataset is useful for a decision [36,37].

8.2 Data Quality Influencing Decision

The important question is whether a dataset can be used in a decision situation, which can be restated as 'How do the imperfections in the data affect the decision?' For decisions about physical actions based on physical properties, as they are typical for engineers, the error propagation law can be used. For a formula $r = f(u,v,w)$ with standard deviations σ_u, σ_v, and σ_w for u, v, w respectively, we obtain σ_r as

$$\sigma_r^2 = \sigma_u^2\left(\frac{df}{du}\right)^2 + \sigma_v^2\left(\frac{df}{dv}\right)^2 + \sigma_w^2\left(\frac{df}{dw}\right)^2 \tag{5}$$

For example, a decision about the dimensions of a load bearing beam reduces to a comparison of resistance R with the expected load L; if resistance is larger than the load then the beam will not break.

$$R > L \; or \; R - L > O \tag{6}$$

In general decision seems either to be of the 'go/no go' type, which reduce to a statistical test for a quantity more than zero (as shown above) or to a selection of the optimal choice from a number of variants (formally the first can be reduced to the second). In practical cases the valuations involved are often unknown, but it seems possible to reconstruct the structure of the decision process after a decision has been taken and then to analyze the influence of data quality on the decision. Decisions usually include also default reasoning with categorized data; a reconstruction of the decision process as a formalization and the identification of the imperfections and their causes through the ontological commitments show what tools can be used: statistics with other than normal distributions, fuzzy set theory, assessment of differences in contexts etc. Even if the results of such assessments are only rough estimates, they will be sufficient to indicate if a dataset can be useful to assist in a decision.

9 Conclusion

An ontological view of data quality is different from a philosophical ontology—which concentrates on the existing world—but also different from the information science ontology—which concentrates on the concepts representing the existing world [18]. It must include both the reality realm and the information realm. Only if both realms can be linked conceptually in the closed semantic loop (Fig. 2) of observation and action, a meaningful discussion of data quality as "fit for use" is possible.

The tiered ontology describes how and through which processes higher level knowledge is derived from lower levels, ultimately grounding all knowledge in primitive observations of properties at a point in psace and time limitations in these processes linking the tiers produce the imperfections in the knowledge. The list of commitments separate individual aspects and are formulated with the construction of computational models in mind, to achieve operational definitions for the plethora of terminologies confusing the user [3]. The list of commitments is a data quality ontology describes properties of the process of data acquisition and processing in very general terms. The commitments indicate what imperfections are present in data. It is then possible to analytically describe these imperfections and the effects they have on decisions—at least on the level of the rough assessment necessary to decide whether a dataset is useful for a decision.

Experience with decision processes has taught us the appropriate levels of data quality necessary to make the decision most of the time correctly. Schneider was using a similar approach reconstructed engineering decisions and found the result satisfactory [31]. In a world of interoperable information systems connected by the web, an analytical and formalizable treatment of data quality is required. The systematic account of imperfections we must cope with leads to prediction of effects data quality has on a decision.

References

1. Bennett, B.: Space, Time, Matter and Things. In: Proceedings of the 2nd international conference on Formal Ontology in Information Systems (FOIS'01), Ogunquit, Maine, USA, ACM, New York (2001)
2. Bittner, S.: An Agent-Based Model of Reality in a Cadastre. Technical University Vienna, Vienna (2001)
3. Boin, A.T., Hunter, G.J.: Facts or Fiction: Consumer Beliefs about Spatial Data Quality. In: Proceedings of the Spatial Sciences Conference (SSC 2007), Hobart, Tasmania (2007)
4. Brooks, R.: New Approaches to Robotics. Science 253, 1227–1232 (1991)
5. Burrough, P.A.: Natural Objects with Indeterminate Boundaries. In: Burrough, P.A., Frank, A.U. (eds.) Geographic Objects with Indeterminate Boundaries, pp. 3–28. Taylor and Francis, London (1996)
6. Ceusters, W.: Towards a Realism-Based Metric for Quality Assurance in Ontology Matching. In: International Conference on Formal Ontology in Information Systems, IOS Press, Baltimore MD (2006)
7. Fonseca, F.T., Egenhofer, M.J.: Ontology-Driven Geographic Information Systems. In: 7th ACM Symposium on Advances in Geographic Information Systems, Kansas City, MO (1999)
8. Fonseca, F.T., Egenhofer, M.J., Agouris, P., Câmara, G.: Using Ontologies for Integrated Geographic Information Systems. Transactions in GIS 6(3), 231–257 (2002)
9. Frank, A.U.: Communication with Maps: A Formalized Model. In: Habel, C., Brauer, W., Freksa, C., Wender, K.F. (eds.) Spatial Cognition II. LNCS (LNAI), vol. 1849, pp. 80–99. Springer, Berlin (2000)
10. Frank, A.U.: Tiers of Ontology and Consistency Constraints in Geographic Information Systems. International Journal of Geographical Information Science 75(5), 667–678 (2001)
11. Frank, A.U.: Ontology for Spatio-Temporal Databases. In: Sellis, T., Koubarakis, M., Frank, A., Grumbach, S., Güting, R.H., Jensen, C., Lorentzos, N.A., Manolopoulos, Y., Nardelli, E., Pernici, B., Theodoulidis, B., Tryfona, N., Schek, H.-J., Scholl, M.O. (eds.) Spatio-Temporal Databases. LNCS, vol. 2520, pp. 9–78. Springer, Heidelberg (2003)
12. Frank, A.U.: Procedure to Select the Best Dataset for a Task. In: Egenhofer, M.J., Freksa, C., Miller, H.J. (eds.) GIScience 2004. LNCS, vol. 3234, pp. 81–93. Springer, Heidelberg (2004)
13. Frank, A.U.: Distinctions Produce a Taxonomic Lattice: Are These the Units of Mentalese? In: International Conference on Formal Ontology in Information Systems, pp. 166–181. IOS Press, Baltimore, MD (2006)
14. Frank, A.U.: Incompleteness, Error, Approximation, and Uncertainty: An Ontological Approach to Data Quality. In: Geographic Uncertainty in Environmental Security. NATO Advanced Research Workshop, Kiev, Ukraine, Springer (to appear)
15. Gangemi, A., Guarino, N., Masolo, C., Oltramari, A., Schneider, L.: Sweetening Ontologies with DOLCE. In: Gómez-Pérez, A., Benjamins, V.R. (eds.) EKAW 2002. LNCS (LNAI), vol. 2473, Springer, Heidelberg (2002)
16. Goguen, J., Harrell, D.F.: Information Visualization and Semiotic Morphisms (2006) Retrieved 01.09.06, 2006 from http://www.cs.ucsd.edu/users/goguen/papers/sm/vzln.html
17. Gruber, T.R.: TagOntology - a way to agree on the semantics of tagging data (2005) Retrieved October 29, 2005 from http://tomgruber.org/writing/tagontology-tagcapm-talk.pdf
18. Gruber, T.R.: A Translation Approach to Portable Ontologies. Knowledge Acquisition 5(2), 199–220 (1993)

19. ISO: The EXPRESS language reference manual, ISO TC 184 (1992)
20. Karmiloff-Smith, A.: Reprint: Precis of: Beyond Modularity A Developmental Perspective on Cognitive Science. Behavioral and Brain Sciences 17(4), 693–745 (1994)
21. Karmiloff-Smith, A.: Beyond Modularity A Developmental Perspective on Cognitive Science. MIT Press, Cambridge, MA (1995)
22. Lifschitz, V. (ed.): Formalizing Common Sense - Papers by John McCarthy. Ablex Publishing, Norwood, NJ (1990)
23. McCarthy, J., Hayes, P.J.: Some Philosophical Problems from the Standpoint of Artificial Intelligence. In: Meltzer, B., Michie, D. (eds.) Machine Intelligence 4, pp. 463–502. Edinburgh University Press, Edinburgh, UK (1969)
24. Medak, D.: Lifestyles - A Formal Model. Chorochronos Intensive Workshop '97, Petronell-Carnuntum, Austria, Dept. of Geoinformation, Technical University Vienna (1997)
25. Medak, D.: Lifestyles - A Paradigm for the Description of Spatiotemporal Databases. Vienna, Technical University Vienna (1999)
26. Montello, D.R.: Scale and Multiple Psychologies of Space. In: Campari, I., Frank, A.U. (eds.) COSIT 1993. LNCS, vol. 716, pp. 312–321. Springer, Heidelberg (1993)
27. Navratil, G.: Formalisierung von Gesetzen. Vienna, Technical University Vienna (2002)
28. NCGIA,: The U.S. National Center for Geographic Information and Analysis: An Overview of the Agenda for Research and Education. International Journal of Geographical Information Systems 2(3), 117–136 (1989)
29. O'Hara, K., Shadbolt, N.: Issues for an Ontology for Knowledge Valuation. In: Proceedings of the IJCAI-01 Workshop on E-Business and the Intelligent Web (2001)
30. Riedemann, C., Kuhn, W.: What are Sports Grounds? Or: Why Semantics Requires Interoperability. In: Včkovski, A., Brassel, K.E., Schek, H.-J. (eds.) Interoperating Geographic Information Systems. LNCS, vol. 1580, pp. 217–230. Springer, Heidelberg (1999)
31. Schneider, M.: Vague Spatial Data Types. In: ISSDQ '04. Bruck a.d. Leitha, Department of Geoinformation and Cartography pp. 83–98 (2004)
32. Searle, J.R. (ed.): The Construction of Social Reality. The Free Press, New York (1995)
33. Smith, B.: Beyond Concepts: Ontology as Reality Representation. In: Proceedings of FOIS 2004. International Conference on Formal Ontology and Information Systems, pp. 73–84. IOS Press, Turin (2004)
34. Smith, B., Grenon, P.: SNAP and SPAN: Towards Dynamic Spatial Ontology. Spatial Cognition and Computing 4(1), 69–103 (2004)
35. Tecknowledge: (2004) Overview of the SUMO From http://ontology.teknowledge.com/arch.html
36. Timpf, S., Frank, A.U.: Metadaten - vom Datenfriedhof zur multimedialen Datenbank. Nachrichten aus dem Karten- und Vermessungswesen Reihe I(117), 115–123 (1997)
37. Timpf, S., Raubal, M., Kuhn, W.: Experiences with Metadata. In: 7th Int. Symposium on Spatial Data Handling, SDH'96, Delft, The Netherlands (August 12-16, 1996), IGU pp. 12B.31–12B.43 (1996)
38. Tobler, W.R.: A Computer Model Simulation of Urban Growth in the Detroit Region. Economic Geography 46(2), 234–240 (1970)
39. Vckovski, A.: Interoperability and spatial information theory. In: International Conference and Workshop on Interoperating Geographic Systems, Santa Barbara, CA (1997)
40. Wand, Y., Wang, R.Y.: Anchoring Data Quality Dimensions in Ontological Foundations. Communications of the ACM 39(11), 86–95 (1996)
41. Zsambok, C.E., Klein, G. (eds.): Naturalistic Decision Making (Expertise: Research and Applications Series). Lawrence Erlbaum Associates, Mahwah (1996)

Triangulation of Gradient Polygons: A Spatial Data Model for Categorical Fields

Barry J. Kronenfeld

Department of Geography, George Mason University,
4400 University Drive, MS 1E2, Fairfax, VA, USA
bkronenf@gmu.edu

Abstract. The concept of the categorical gradient field is introduced to encompass spatially continuous fields of probabilities or membership values in a fixed number of categories. Three models for implementing categorical gradient fields are examined: raster grids, epsilon bands and gradient polygons. Of these, the gradient polygon model shows promise but has not been fully specified. A specification of the model is developed via a four-step process: 1) the constrained Delaunay triangulation of the polygon is created, 2) vertices are added to the polygon edge to ensure consistency, 3) a skeleton of the medial axis is produced and flat spurs are identified, and 4) additional vertices are added along each flat spur. The method is illustrated on a hypothetical transition zone between four adjacent regions, and evaluated according to five general criteria. The model is efficient in terms of data storage, moderately flexible and robust, and intuitive to build and visualize.

1 Introduction

Area-class maps have proven useful for representing the geographic pattern of various types of phenomena, including vegetation, soils, climate and generalized ecological regions [2,30]. For many such phenomena, the exact locations of boundaries between regions are difficult to determine, posing problems in interpretation of the spatial data model. The problem of indeterminate boundaries has led to numerous research volumes [7], specialist meetings [4,23], research priorities [16] and special journal issues [4,14]. Despite this body of research, methods for characterizing boundary indeterminacy are still limited.

The problem may be conceptualized in at least three ways, depending on one's view of ontological crispness and the source(s) of uncertainty. According to some authors, geographic regions are crisply bounded by definition; the alternative is nonsensical [29]. Under this view, uncertainty in the location of boundaries can derive from one of two sources. First, there may be vagueness or disagreement in the definition, and therefore the spatial extent, of the region. In this case, one may use rough set theory to identify the region of indeterminacy, as well as regions where presence or absence of the region is clear [5,9]. Any attempt to further quantify indeterminacy will entail precisification of something that is inherently vague. We restrict our attention to the remaining cases.

S. Winter et al. (Eds.): COSIT 2007, LNCS 4736, pp. 421–437, 2007.

Uncertainty in the location of crisp boundaries may also derive from inaccurate, imprecise or incomplete observation. A given location may not have been observed directly, may have been observed only at a coarse granularity, or the relevant observation(s) may simply be incorrect. Under this conceptualization, boundary indeterminacy is considered a problem of accuracy, and it is often desirable to quantify potential errors in boundary location and/or probabilities of membership in candidate classes for specific locations (e.g. [32]).

Some authors accept the notion of regions with boundaries that are not crisp (e.g., [6,28]). Adherents to this view allow that border locations can belong simultaneously to adjacent regions, and that regions can have extended boundaries. Indeed, if transitions between regions occur over broad zones, it is argued that any attempt to isolate crisp boundaries will result in a Sorites-like paradox [11].

Although conceptually distinct, the latter two concepts of indeterminacy lead to similar models. Let $X = \{x\}$ represent a domain of spatial locations. A conceptual model of indeterminacy for a k-category system can be defined as a set of categorical affinity functions $a_1, a_2, ..., a_k$ on X, such that the following two constraints hold for all x:

$$0 \le a_i(x) \le 1, \forall i \tag{1}$$

$$\sum_{i=1}^{k} a_i(x) = 1 \tag{2}$$

The former constraint defines the range of the affinity function, while the latter *unity sum constraint* enforces mutual exclusivity. Depending on whether or not boundaries are held to be ontologically crisp, the results of the affinity functions may be conceptualized as either probabilities or degrees of membership. In the latter case, researchers often claim to be applying fuzzy set theory [31], but reviews of fuzzy set theory in geographical domains [11,22,25] fail to explain the rationale behind the unity sum constraint, which is enforced in nearly all domain-specific applications involving more than one region (e.g., [6,10,12,21,22,33]; but see [34] for a counter-example).

The concepts of ontological crispness and gradation need not be incompatible with each other. I have argued [18] that, in many situations, classes, regions and boundaries should be viewed as components of a descriptive model. The concept of gradual transitions between regions is useful in describing patterns of vegetation or soils in the same way that the concept of an elevation profile is useful in describing the shape of a hill. The unity sum constraint is explained cognitively by the need for categories to contrast with each other [26]. Indeed, this can be viewed as a mechanism for enforcing the law of the excluded middle. For descriptive purposes, it is not necessary that the conceptualized 'fuzzy' regions represent mereological subsets of reality; when such ontological entities are required, practitioners will 'defuzzify' their maps to produce them (e.g. [6]).

Further insight is gained when one recognizes that, although the concepts of boundary uncertainty and gradation are philosophically distinct, both tend to apply to similar situations in the real world. Namely, whenever underlying characteristics such as soil moisture or species composition change continuously across a gradient, ontologically crisp boundaries will be difficult to identify with certainty, and at the same time, gradient-based descriptions will be useful and informative. The correspondence is not universal: when underlying characteristics vary randomly across a spatial domain, boundary uncertainty will remain but gradient-based descriptions will fail to provide useful information. Nevertheless, they apply frequently enough to the same phenomenon that confusion between the two concepts is not surprising.

Given that the same formalization may be used to characterize both boundary uncertainty and gradation, it is not strictly necessary to distinguish between these two concepts for the purposes of developing spatial data models. In what follows, the term *categorical gradient* will be used broadly to encompass both boundary uncertainty and gradation. Similarly, the term *affinity function* should be interpreted as either a categorical probability or a descriptive membership value, depending on the context of application. Proper attention to these distinctions should be applied when interpreting any spatial data model.

1.1 Ideal Characteristics of a Spatial Data Model

Categorical gradients may be modeled conceptually as a collection of field layers specifying affinity values in a set of categories at all locations in a spatial domain, subject to the constraints in eqs. 1 & 2. A spatial data model that satisfies these conditions will be referred to as a *categorical gradient field*. Before examining candidate spatial data models for categorical gradient fields, it will be useful to discuss the criteria of an ideal model. Five such criteria are proposed:

1. The model should be *efficient*, representing instantiations of the conceptual model with a relatively small number of data elements.
2. The model should be *flexible*, with the ability to represent all (or at least most) possible instantiations of the conceptual model.
3. The model should be *robust* to unusual instantiations. There should be *no* ambiguity in specification; categorical affinities at each location should be uniquely defined in all cases.
4. An instantiation of the model should be *easy to produce* using both manual and automated methods.
5. The model should *lend itself to a variety of visualization techniques*.

In practice different models are likely to be advantageous on some of these criteria and disadvantageous on other. Their purpose is to allow characterization of the strengths and weaknesses of alternative data models, not to judge an absolute winner.

1.2 Raster Grids, Epsilon Bands and Transition Zones

At least three spatial data models have been proposed to represent categorical gradient fields. Figure 1 depicts these three models for a hypothetical 2-category system on a rectangular domain: (a) raster grid with affinity values, (b) polygon tessellation with epsilon bands, and (c) polygon tessellation with transition zones.

424   B.J. Kronenfeld

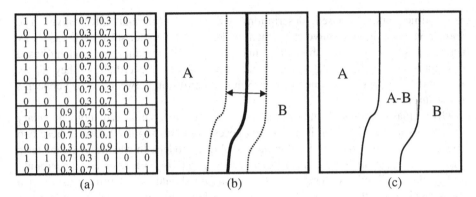

(a)   (b)   (c)

Fig. 1. Spatial data models for representing categorical gradients. (a) raster grid, (b) epsilon band, (c) transition zone.

1.3 Raster Grids

The most commonly implemented model has been the square raster grid, in which affinity values (often termed fuzzy membership values) in the full set of k categories is stored for each grid cell [6,10,12,21,22,33]. Indeed, this may be the only spatial data model that has been paired with affinity functions in a fully functional implementation (but see epsilon band model below). The model allows for precise and sophisticated mapping, especially when detailed source data is available in raster format.

Despite ample research in the application of fuzzy set theory on raster grids, two drawbacks of their practical use in modeling affinity gradients are evident. First, the model requires a large volume of data storage to achieve acceptable results. The large storage requirements of raster grids are well known, but in representing categorical gradients the required volume increases linearly with k. A second and perhaps more serious problem is that manual creation and editing is impractical, due in part to the large data volume and inflexibility of the spatial units. This is significant because soil maps and vegetation maps are often drawn by experts out in the field. The fact that raster-based visualizations of categorical gradients have been found difficult to interpret [13] is related, since interpretive ease seems to be correlated with the ability to identify regions [15,19]. A raster-based model does not lend itself easily to the types of object-based cartographic manipulations that might enhance readability.

1.4 Epsilon Bands

The epsilon-band model was originally proposed as a means of generalizing cartographic lines, but its applicability to the problem of uncertainty in area-class maps has not gone unnoticed. In this model, an error (epsilon) value is associated with each boundary edge in a polygon tessellation. The epsilon values can be interpreted as signifying the width of the zone of uncertainty or gradient. Affinity values at a specific location within this zone can be determined based on distances from the boundary between regions or edge of the epsilon band relative to the epsilon width.

The affinity function can be linear or based on a sine or other curve. Greater flexibility can be obtained by allowing epsilon values to differ on either side of an edge, or be assigned individually to each edge segment or vertex.

Despite its intuitive appeal, it is not clear whether a practical and generalizable implementation of the epsilon-band model has ever been developed. Several authors have cited a conference presentation or unpublished manuscript by Honeycutt in 1987 (cited in [8]) in which such an implementation was presented. However, a broad search has failed to turn up either a description or images of the working implementation. Indeed, it has been commented without objection that "the epsilon band model is largely a conceptual undertaking" ([20] p. 608).

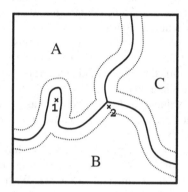

Fig. 2. Ambiguities of the epsilon-band model

The apparent simplicity of the epsilon band model masks specification challenges in certain atypical (but not uncommon) configurations. In these configurations, ambiguities exist concerning the assignment of affinity values to specific locations within the epsilon zone. Two such configurations are illustrated in Figure 2, which depicts an epsilon band model applied to a tessellation of three regions (A, B and C). One type of ambiguity results from boundary curvature. The border between A and B is sinuous and includes a deep indentation, the width of which is less twice that of the epsilon band. Calculation of affinity values for point #1, located within this indentation, will differ depending on whether distances are measured to the region boundary, epsilon boundary, or both.

A second ambiguity occurs in the vicinity of 3-way junctions between regions. At such locations one would expect to find non-zero affinity values to all three neighboring regions. If this were not the case, sharp discontinuities would occur between gradients involving each pair of adjacent regions/classes. For example, in Figure 2 a sharp boundary would occur between the B-A and B-C gradient zones. Since the location of this boundary is not readily discernable, there would exist ambiguity regarding the affinity values of point #2. Clearly, a smooth 3-category transition is desirable at such junctions, but it is not immediately obvious how this would be accomplished using epsilon bands.

A final problem with the epsilon band model is that it appears somewhat difficult to implement using manual cartographic methods. The specification of varying epsilon widths at each vertex along a polygon boundary is likely to be a tedious and time-consuming process. Although none of these problems are insurmountable, they demonstrate idiosyncrasies in the model that have not been previously noted.

1.5 Transition Zones

One common cartographic method of representing gradients is to designate certain polygons in a tessellation as transitional between other polygons. The transition zone or "ecotone" is considered an important theoretical concept in ecology [24], and such zones are often depicted on ecological maps at various scales (e.g., [30]). In describing his principles used to map ecoregions, Bailey ([3] p. S18) states explicitly that "the mosaic of ecosystems found in major transitional zones (ecotones) should be delineated as separate ecoregions." Thus, the designation of transitional polygons has a long tradition in environmental mapping.

The implicit assumption behind the concept of the transition zone is that polygons representing transitional gradients are fundamentally different from other polygons in terms of their function within the data model. Different locations within a gradient polygon do *not* to have the same characteristics, and therefore would not be expected to have equivalent affinity values. No formal method has been developed, however, to identify affinity values at specific locations within a gradient polygon. Thus, the conceptual model has never been fully specified.

Their historical use in environmental cartography, however, suggests that the conceptual model of transition zones may be lead to an intuitive and therefore useful data model for representing environmental gradients. With this in mind, a suitable method was sought to specify affinity values at specific locations within a gradient polygon. After some trial and error, a TIN-based representation built upon the medial axis (skeleton) of the gradient polygon was found to provide a reasonably intuitive and computationally efficient means for accomplishing this task. The representation involves triangulation of affinity values within gradient polygons. The purpose of the present paper is to develop and evaluate the triangulation procedure as a way of modeling categorical gradient fields.

2 Gradient Polygon Triangulation

An obvious alternative to the raster grid representation of categorical gradients is a triangulated irregular network (TIN). The TIN model for topographic representation is easily extended to multiple values recorded at each vertex. It is easy to show that if affinity values at each vertex conform to the unity sum constraint, then a weighted average of these values based on triangular for points within a triangle will also conform to the unity sum constraint.

To model categorical gradient fields within a gradient polygon, a logical method is therefore to create a Delaunay triangulation of the polygon. Affinity values for each vertex along the edge of the polygon are determined by association with adjacent polygons, and additional vertices are created based on geometric rules.

Consider the polygon tessellation shown in Figure 3. Polygons 001-004 represent homogenous regions, whereas polygon 005 is marked as a transition zone. A simple data table records category membership of every polygon, with a unique code signifying a gradient polygon (Table 1). Affinity values for non-gradient polygons are defined as 1 in the recorded category and 0 in all other categories (here affinity values are restricted to 0 and 1 for the sake of simplicity, but the model can easily be extended to allow non-gradient polygons with non-zero affinity values in more than one category). The triangulation within the gradient polygon is not stored in the database but created on the fly. For the time being, it is assumed that there are no null polygons adjacent to the gradient polygon.

Table 1. Attribute table for polygon tessellation in Figure 3

ID	Category
001	A
002	B
003	C
004	D
005	transition

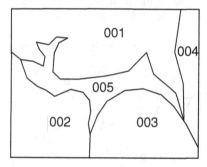

Fig. 3. Polygon tessellation with transition zone

The problem of identifying affinity values at a user-specified point within a transition polygon is analogous to that of interpolating elevations at points in between contours on a topographic map [27]. The differences are that 1) multiple values must be interpolated on the categorical gradient map, as opposed to a single elevation value on the contour map, and 2) the bounding edges of a gradient polygon may be numerous, whereas on a contour map there will always be only two. The first difference is inconsequential as it simply involves applying the same interpolation technique to multiple field values. The second difference is more significant and will be discussed further in section three. The interpolation problem is solved via a four-step process:

1. a constrained Delaunay triangulation is created from the gradient polygon
2. vertices are added to the edge of the polygon to promote consistency
3. the skeletal approximation of the medial axis is created and flat spurs are identified
4. vertices are added along each spur

The resulting TIN, consisting of the original polygon vertices and those added to the polygon edge and along the medial axis, is used to define affinity values at any point within the gradient polygon. The four steps to creating the triangulation are discussed in detail below.

2.1 Step 1: Delaunay Triangulation

Creation of the constrained Delaunay triangulation of a polygon from its vertices has been treated abundantly elsewhere in the literature. The key points to note here relate to the assignment of affinity values to each TIN vertex. The following rules correspond to an intuitive interpretation:

1. Each vertex that is contained in the edge of a single adjacent polygon is assigned the affinity values of that polygon
2. If a vertex is the intersection node of exactly two adjacent polygons, it is assigned the average of the affinity values of these two polygons
3. If a vertex is the intersection node of more than two adjacent polygons, it is assigned the average of the affinity values of the two outer polygons that share an edge with the gradient polygon.

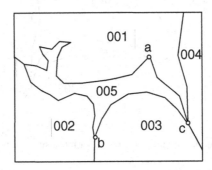

Fig. 4. Selected vertices on edge of transition polygon

The three cases are illustrated in Figure 4. Vertex *a* will be assigned the affinity values associated with polygon 001 since it is the only adjacent polygon. Vertex *b* will be assigned the average of the affinity values of polygons 002 and 003 to ensure a smooth gradient. Similarly, vertex *c* will be assigned the average of the affinity values of polygons 001 and 003; polygon 004 is considered to have no influence because it does not share an edge with the gradient polygon.

2.2 Step 2: Edge Filling

The constrained Delaunay triangulation of a polygon will differ depending on the number of recorded vertices even if the boundary remains the same. As the number of vertices increases, the mid-points of interior edges of the Delaunay triangulation will fall closer and closer to the medial axis, until in the extreme case they define the medial axis [1]. When vertices are sparse, however, the triangulation will depend

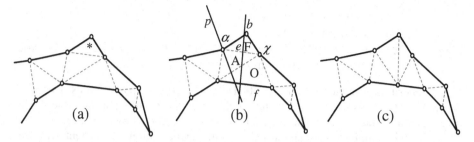

Fig. 5. Addition of vertex to eliminate flat triangle

substantially on the placement of vertices. These differences will be most pronounce near flat triangles, where all three vertices have exactly the same affinity values, located along small spurs that enter into the adjacent polygons. For example, consider the triangle marked by the asterisk in Figure 5a. The triangle is flat and represents a small spur in the medial axis. The skeleton produced from the triangulation will also have a spur, but this spur will be lopsided due to the asymmetrical nature of the two triangles located directly below the asterisk. Addition of vertices to the polygon edge is necessary to mitigate the irregularity of the skeleton.

The number of vertices that should be added will depend on the importance of consistency and the needs of computational efficiency. One strategy is to add vertices at a relatively small, constant spacing, but this may result in an unnecessarily high number of vertices. For sampling points along curves, Amenta et al. [1] suggest minimum spacing between vertices be computed locally as a constant factor r (≤ 1) of the distance between the polygon edge and medial axis. However, implementation for non-smooth curves is impossible since the medial axis extends to the polygon boundary. Furthermore, such a strategy will likely produce more vertices than strictly necessary for the present application, since it is only necessary to approximate the medial axis, not the polygon boundary.

If computational efficiency is important, more efficient strategies may be devised by selectively searching for flat triangles, and determining for each such triangle if a vertex should be added to its opposing edge. Analysis of the properties of the medial axis led to the following algorithm for situations such as that depicted in Figure 5a:

Algorithm for Eliminating Small, Lopsided Spurs

```
For each flat triangle F in gradient polygon TIN
    For each interior edge e of F
        Create the perpendicular bisector b
        For each endpoint α of e
            Create the perpendicular p of the adjacent
                triangle edge through v
            Let β denote the intersection of b and p
            Let A denote the triangle adjacent to F that
                contains v as a vertex
            Let O denote the triangle adjacent to A that does
                not contain α
```

```
Let χ denote the endpoint of e opposite α
Let f denote the edge of O opposite χ
If  side(β,f)  <>  side(F,f)  then  create  a  new
    vertex at the intersection of b and f
```

The algorithm is demonstrated Fig. 5b. The intersection of p and b falls outside of the triangle O, suggesting that a vertex could be added to the opposite edge f. The triangulation resulting from the addition of this vertex is shown in Fig. 5c. The algorithm will eliminate real spurs and so will not be entirely consistent with the medial axis representation. However, these spurs will be small, and by avoiding lopsidedness, the resulting triangulation will closely approximate the full triangulation based on an infinitely small vertex interval.

2.3   Step 3: Skeletal Approximation of Medial Axis and Identification of Flat Spurs

The skeleton is produced from the TIN by connecting midpoints of internal triangles and edges with end vertices of each triangle with two exterior edges. In contrast to the true medial axis, the midpoints of internal triangles are defined as the center of gravity to assure that these midpoints fall within the gradient polygon. The skeleton is stored in a tree structure to facilitate identification of spurs. Each node in the tree represents a linear segment of the skeleton that is homogenous with respect to adjacency to neighboring polygons. The tree can be built iteratively beginning with any triangle in the TIN. Figure 6 shows the skeleton for the gradient polygon used in Figures 1-4.

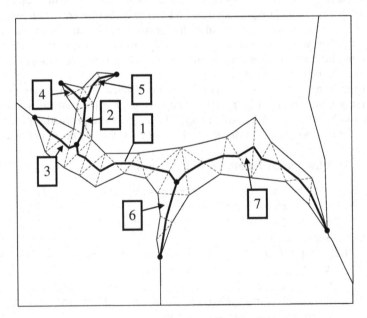

Fig. 6. Skeleton approximation of medial axis

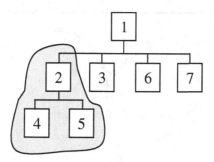

Fig. 7. Hierarchical tree structure for skeleton in Fig. 6

Table 2. Identification of flat spur segments

Seg	left_cat	right_cat
1	B	A
2	A	A
3	B	A
4	A	A
5	A	A
6	B	C
7	C	A

The associated tree structure is shown in Figure 7. To identify flat spurs, categories of polygons on the left and right sides of each segment of the skeleton are recorded (Table 2). Connected chains of segments for which both sides are the same are identified within the skeleton tree (shaded region in Fig. 7).

2.4 Step 4: Addition of Vertices Along Skeleton Within Flat Spurs

The final step is to add nodes to the TIN at every point where the skeleton intersects the TIN edges within flat spurs (Figure 8). Nodes are placed at the midpoints of TIN edges and interior triangles, as described above. Thibault and Gold [27] suggest that values for added nodes can be determined by comparing the radius of the inscribed circumcircle at each added node with that of a reference point taken at the origin of the spur. Let r_{ref} denote the radius of the circumcircle defined by the skeleton junction point where the spur originates, a_{ref} denote the affinity values at this junction point, r_i denote the radius at an added node i, and a_{end} denote the affinity values of a point at the end of the spur. The affinity values a_i assigned to node i are calculated as:

$$a_i = \frac{r_{ref} - r_i}{r_{ref}} a_{ref} + \frac{r_i}{r_{ref}} a_{end} \tag{3}$$

Note that a hierarchical network of spurs is easily handled by this method. Problems will occur if the circumcircles of spur nodes are larger than that of the reference node, but this may easily be corrected by setting an upper limit on the relative weight of the reference node.

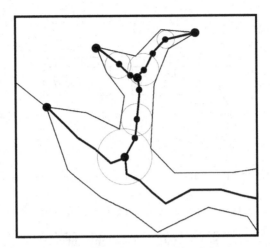

Fig. 8. Circumcircles to determine affinity values for added spur nodes. Bottom circle is the reference circle.

The result of the use of inscribing circumcircles to define affinity values at added nodes is that the shape of the transition gradient is defined by the width of the spur. This seems intuitive but other legitimate solutions exist. One natural alternative is to define spur node affinity values based on the distance along the longest path from the reference point to the spur end. A weighted combination of the two approaches is also possible.

3   Evaluation of the Proposed Data Model

The previous section defined a method for triangulating a transition zone to yield categorical gradients between polygonal regions. Five criteria were proposed for evaluating such a model: *efficiency, flexibility, robustness, ease of production* and *ease of visualization*. This section provides an informal discussion of the characteristics of the triangulate transition zone model in relation to each of these criteria.

3.1   Efficiency

In terms of data storage, the proposed model requires only an ordinary polygon tessellation with associated data table. Thus, storage efficiency of the data model will be similar to that of a polygon tessellation. Specifically, the proposed model will be efficient in terms of data storage when homogenous regions are large relative to the size of transition zones, and when boundaries between regions are not very complex. For complex landscapes, the advantages over a raster grid or TIN will be smaller.

Implementation of the model requires on-the-fly creation of the gradient TIN, so computational efficiency is a consideration that needs to be addressed. Computational efficiency will be a function of the number of polygon vertices and, to a lesser degree, the complexity of the skeleton segmentation hierarchy. Thus, visualization and analysis will be slow for extremely complex landscapes. Further work is needed to characterize the degree of complexity that can be handled in various working environments.

3.2 Flexibility

Similar to the epsilon band model, gradient polygons are only able to produce non-zero affinity values in three categories at any single location. However, the gradient polygon model exhibits greater flexibility in the shape of these gradients, allowing for variable width gradients and spurs into neighboring regions. Numerous variations on the basic data model can be envisioned to deal with 4-way gradients, null boundaries, etc. The viability and justifiability of such variations is left for future work.

3.3 Robustness

It is assumed that the data input is a topologically correct polygon tessellation. The robustness of the data model will depend upon the ability to handle various topological configurations between the gradient and neighboring polygons. A sample of such configurations is shown in Figure 9. The top row illustrates configurations involving two adjacent regions without null polygons. Spurs and compound spurs (a & b) have already been discussed in section two. The configurations in (c) and (d) involve "tunnel spurs" in which a flat section of the gradient polygon is sandwiched between two non-flat sections; (d) differs from (c) in that the tunnel spur contains secondary spur within it.

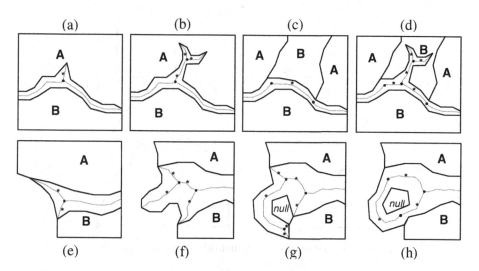

Fig. 9. Irregular gradient polygon configurations

The bottom row of Figure 9 depicts possible configurations involving a null polygon. Configuration (e) involves a single null edge. In (f), the edge has a spur intruding into the null polygon. Configurations (g) and (h) exhibit null holes in the gradient polygon. In (h), the gradient polygon completely wraps around the null polygon. The author has sought methods for identifying and handling each of these configurations. Although plausible solutions have been found, these solutions are at present *ad hoc* and have not been theoretically evaluated. In addition, the set of configurations examined is not exhaustive. Thus, further synthesis is required to develop a general framework for robustness under a variety of configurations.

It is easy, however, to confirm whether or not a gradient polygon is normally configured and therefore robust to the triangulation technique described in section two. This simply requires checking for the absence of tunnel spurs and adjacent null polygons. Thus, the present model is robust but limited in flexibility.

3.4 Production Ease

The ease by which gradient polygons can be created via manual cartographic methods is perhaps the biggest advantage of the proposed model. As noted in section one, transition zones have long been presented in ecological maps. Drawing gradient polygons is simpler than specifying epsilon widths, and requires virtually no technical expertise on the part of the cartographer.

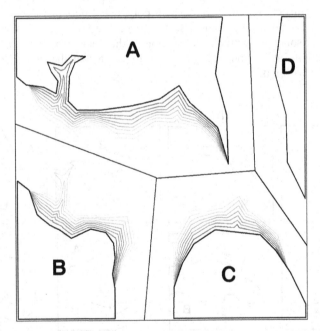

Fig. 10. Contour visualization of individual categorical gradients

Automated production from quantitative data is also not difficult, although it requires generalization of the source data. If a TIN is derived from the source data, then a tessellation of gradient and non-gradient polygons can be created by aggregating TIN triangles on the bases of slope measures. Triangles with steep slopes are aggregated into gradient polygons, while those with flat slopes are aggregated into regular (non-gradient) polygons. The process is simple and efficient, and although some loss of information is inevitable, the generalization process will often result in a more aesthetically pleasing map.

3.5 Ease of Visualization

Because the categorical gradients within a transition zone are defined by a TIN, the results may be visualized using a variety of methods. Figure 10 shows the distribution of affinity values in each category for the categorical gradient field from section two using 0.125 interval contours extracted from the TIN. The results are intuitive and visually appealing. Other visualization methods may be developed to take advantage of the triangulation, including color gradients and hill-shading effects for a 3-dimensional appearance.

4 Discussion

A model was developed to specify values of categorical affinity functions for points within a designated gradient polygon. The model is a specification of a cartographic technique that has long been used in environmental mapping, and requires as input only an ordinary polygon tessellation with a coded value field to designate the gradient polygon(s). Through a process of skeletonization and triangulation, affinity values at any location within the polygon can be determined. These affinity values may be interpreted as either categorical probabilities or membership values, depending on the application. The concept of a *categorical gradient field* was introduced to encompass both of these interpretations.

Triangulation of gradient polygons appears to be a useful and powerful method for implementing categorical gradient fields. It has advantages over the raster grid model in terms of data storage, manual cartographic production and visualization. In all of these aspects it is similar to the epsilon band model. Indeed, the two models can be seen as duals of each other in the simplest case. That is, a simple gradient polygon of uniform width without spurs can be transformed into an epsilon band and vice versa. The skeleton of the gradient polygon will correspond to the polygonal boundary of the dual epsilon band, while the edge of the epsilon band will correspond to the gradient polygon boundary.

The gradient polygon easily handles simple and compound spurs; further work is required for robust handling of tunnel spurs, adjacent null polygons, and topological combinations of these elements. The epsilon band model, on the other hand, may provide more flexibility in the specification of non-linear and asymmetrical gradients but is limited in its ability to model spurs. It might be fruitful to explore the possibility of a hybrid model that would combine the advantages of both approaches.

While both the gradient polygon and epsilon band models may be extended in the ways mentioned above, the extensions to the gradient mentioned above do not require additional user input beyond delineation of the gradient polygon. There are of course limits to this flexibility, and the precision modeler may want to adjust affinity value contours individually. A fully flexible model would require explicit storage of the internal triangulation, which could be modified by an end user.

In sum, this paper has presented a fully specified data model for a basic range of categorical gradient fields using gradient polygons. With the possible exception of a single unpublished manuscript, no other alternative to the raster grid model has been fully specified. The proposed model is simple to instantiate and intuitive to interpret, and should provide a means for explicit representation of categorical gradient fields in situations where raster grids are not feasible or appropriate.

References

1. Amenta, N., Bern, M., Eppstein, D.: The crust and the (-skeleton: Combinatorial curve reconstruction. Graphical Models and Image Processing 60(2), 125–135 (1998)
2. Bailey, R.G.: Ecoregions: the ecosystem geography of the oceans and continents. Text volume (176 pp.) and two folded maps. Springer, New York (1998)
3. Bailey, R.G.: Identifying ecoregion boundaries. Environmental Management 34(1), S14–S26 (2005)
4. Bennett, B.: Introduction to special issue on spatial vagueness, uncertainty and granularity. Spatial Cognition and Computation 3(2/3), 93–96 (2003)
5. Bittner, T., Smith, B.: Vague reference and approximating judgements. Spatial Cognition and Computation 3(2/3), 137–156 (2003)
6. Burrough, P.A.: Natural objects with indeterminate boundaries. In: Burrough, P.A., Frank, A.U. (eds.) Geographic Objects with Indeterminate Boundaries, Taylor & Francis, London (1996)
7. Burrough, P.A., Frank, A.U.: Geographic Objects with Indeterminate Boundaries. Taylor & Francis, London (1996)
8. Castilla, G., Hay, G.J.: Uncertainties in land use data. Hydrology and Earth System Science Discussions 3, 3439–3472 (2006)
9. Cohn, A.G., Gotts, N.M.: The 'egg-yolk' representation of regions with indeterminate boundaries. In: Burrough, P.A., Frank, A.U. (eds.) Geographic Objects with Indeterminate Boundaries, pp. 171–187. Taylor & Francis, London (1996)
10. Equihua, M.: Fuzzy clustering of ecological data. Journal of Ecology 78, 519–534 (1990)
11. Fisher, P.F.: Sorites paradox and vague geographies. Fuzzy Sets and Systems 113(1), 7–18 (2000)
12. Foody, G.M.: A fuzzy sets approach to the representation of vegetation continua from remotely sensed data: An example from lowland heath. Photogrammetric Engineering and Remote Sensing. 58, 221–225 (1992)
13. Goodchild, M.F., Chih-chang, L., Leung, Y.: Visualizing fuzzy maps. In: Hearnshaw, H., Unwin, D. (eds.) Visualization in Geographical Information Systems, John Wiley and Sons, New York (1994)
14. Goodchild, M.F.: Introduction: Special issue on Uncertainty in geographic information systems. Fuzzy Sets and Systems 113, 3–5 (2000)
15. Hengl, T., Walvoort, D.J.J., Brown, A., Rossiter, D.G.: A double continuous approach to visualization and analysis of categorical maps. International Journal of Geographical Information Science 18, 183–202 (2004)
16. Kronenfeld, B.J., Mark, D.M., Smith, B.: Gradation and Objects with Indeterminate Boundaries. Short-term research priority, University Consortium for Geographic Information Science (UCGIS) (2002)

17. Kronenfeld, B.J.: Implications of a data reduction framework for assignment of fuzzy membership values in continuous class maps. Spatial Cognition and Computation 3(2/3), 221–238 (2003)
18. Kronenfeld, B.J.: Incorporating gradation as a communication device in area-class maps. Cartography and Geographic Information Science 32(4), 231–241 (2005)
19. Kronenfeld, B.J.: Gradation and map analysis in area-class maps. In: Mark, D.M., Cohn, A.G. (eds.) Proceedings of the Conference on Spatial Information Theory (COSIT), Ellicottville, NY, USA, Springer, Berlin (2005)
20. Leung, Y., Yan, Y.: A Locational Error Model for Spatial Features. International Journal of Geographical Information Science 12(6), 607–620 (1998)
21. McBratney, A.B., Moore, A.W.: Application of fuzzy sets to climatic classification. Agricultural and Forest Meteorology 35, 165–185 (1985)
22. McBratney, A.B., Odeh, I.O.A.: Applications of fuzzy sets in soil science: fuzzy logic, fuzzy measurements and fuzzy decisions. Geoderma 77, 85–113 (1997)
23. Peuquet, D.J., Smith, B., Brogaard, B.: The ontology of fields: Report of a specialist meeting held under the auspices of the Varenius Project. Santa Barbara: National Center for Geographic Information and Analysis (1998)
24. Risser, P.G.: The status of the science examining ecotones. Bioscience 45(5), 318–325 (1994)
25. Robinson, V.B.: A perspective on the fundamentals of fuzzy sets and their use in geographic information systems. Transactions in GIS 7(1), 3–30 (2003)
26. Rosch, E.: Principles of categorization. In: Rosch, E., Lloyd, B.B. (eds.) Cognition and Categorization, Lawrence Erlbaum Associates, Hillsdale, NJ (1978)
27. Thibault, D., Gold, C.M.: Terrain reconstruction from contours by skeleton reconstruction. GeoInformatica 4, 349–373 (2000)
28. Usery, E.L.: A conceptual framework and fuzzy set implementation for geographic features. In: Burrough, P.A., Frank, A.U. (eds.) Geographic Objects with Indeterminate Boundaries, pp. 71–85. Taylor & Francis, London (1996)
29. Varzi, A.C.: Vagueness in geography. Philosophy & Geography 4(1), 49–65 (2001)
30. Walter, H., Harnickell, E., Mueller-Dombois, D.: Climate-diagram maps of the individual continents and the ecological climatic regions of the earth. Text volume (Vegetation Monographs; 36 pp.) and 9 folded maps issued together in a case. Springer, Berlin (1975)
31. Zadeh, I.: Fuzzy Sets. Information and Control 8, 338–353 (1965)
32. Zhang, J., Stuart, N.: Fuzzy methods for categorical mapping with image-based land cover data. International Journal of Geographical Information Science 15(2), 175–195 (2001)
33. Zhang, L., Liu, C., Davis, C.J., Solomon, D.S., Brann, T.B., Caldwell, L.E.: Fuzzy classification of ecological habitats from FIA data. Forest Science 50(1), 117–127 (2004)
34. Zhu, A.: A similarity model for representing soil spatial information. Geoderma 77, 217–242 (1997)

Relations in Mathematical Morphology with Applications to Graphs and Rough Sets

John G. Stell

School of Computing, University of Leeds, Leeds, LS2 9JT, UK
jgs@comp.leeds.ac.uk

Abstract. Rough sets have been applied in spatial information theory to construct theories of granularity – presenting information at different levels of detail. Mathematical morphology can also be seen as a framework for granularity, and the question of how rough sets relate to mathematical morphology has been raised by Bloch. This paper shows how by developing mathematical morphology in terms of relations we obtain a framework which includes the basic constructions of rough set theory as a special case. The extension of the relational framework to mathematical morphology on graphs rather than sets is explored and new operations of dilations and erosions on graphs are obtained.

1 Introduction

The need to present, acquire, and transform data at multiple levels of detail is found across many different disciplines, and in the context of spatial data this need is particularly evident. Theories of multi-resolution spatial data have been developed from several viewpoints and have practical application especially in the model-oriented aspects of generalization [10]. Two approaches which can be seen as multi-resolution are rough set theory [11] and mathematical morphology [13,6]. This paper shows how rough sets relate to mathematical morphology in a way that significantly extends the previous work on this topic. The framework developed for this purpose also provides the basis for a theory of morphology on graphs.

Mathematical morphology is generally studied as an aspect of image processing. As digital images are usually two-dimensional arrangements of pixels, where spatial relationships between elements of the image are essential features, it is perhaps surprising that there has been little interaction between the spatial information theory community and the world of image processing. A notable exception to this has been the work of Bloch [1,2]. In [1], Bloch pointed out the close similarities between rough set theory and mathematical morphology, and provided some results relating these two theories. She showed how two morphological operations, dilation and erosion, can be obtained from a relation. In the case that the relation is an equivalence relation, the same construction yields the upper and lower approximations for a rough set. Bloch did not, however, go so far as to provide a single framework which included the principal constructions both of mathematical morphology and of rough set theory. This is made clear when she writes [1, p1491] "Even if it is not as obvious as for dilation and erosion to find an expression for opening and closing based on a relation, these operators have interesting properties that deserve to consider them as good operators for constructing rough sets".

S. Winter et al. (Eds.): COSIT 2007, LNCS 4736, pp. 438–454, 2007.

In the present paper, the issue raised by Bloch about the relationship of rough set theory to mathematical morphology is settled. In particular, it is shown how all the basic constructions of mathematical morphology (opening and closing as well as dilation and erosion) can be obtained from a single framework which also includes the principal aspects of rough set theory. Essentially a reformulation of mathematical morphology based on relations is provided, and this is then used to begin the development of morphology for graphs.

The structure of the paper is as follows. I begin in Section 2 with a brief review of the essentials of mathematical morphology, and Section 3 gives a similar outline of the most basic aspects of rough sets. The relationship between these two areas is developed in Section 5 and in order to do this we need some algebraic ideas which are summarized in Section 4. The possibility of extending the framework from relations on sets to relations on graphs is explored in Section 6 and some conclusions and directions for further work are given in the final section.

2   Mathematical Morphology

Mathematical morphology is usually presented as a collection of techniques for image processing [5, Chap. 9]. Although physical images are in practice bounded, the basic theory is most easily understood by taking a binary (black-white) image as specified by a subset of $\mathbb{Z}^2$, the set of all two-dimensional vectors with integer components.

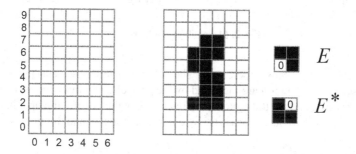

Fig. 1. Part of $\mathbb{Z}^2$, an image as a subset, and a structuring element: E

We can visualize the subset by marking the selected pixels black as shown in Figure 1, although the opposite convention (that the background is black and the selected pixels are white) is also commonplace in image processing. The basic idea of mathematical morphology, as described by Heijmans [6, p1], is

"to examine the geometrical structure of an image by matching it with small patterns at various locations in the image. By varying the size and shape of the matching patterns, called *structuring elements*, one can extract useful information about the shape of the different parts of the image and their interrelations."

The structuring element (see E in Figure 1) is an arrangement of pixels with a specified origin or centre (marked by 0 in the figure). We can take E to be some subset of

$\mathbb{Z}^2$ and consider translates of the form $x + E = \{x + e \mid e \in E\}$ for $x \in \mathbb{Z}^2$. Each $x + E$ places the structuring element with its origin at x and the pixels in E (which need not include the origin) then lie over pixels in $\mathbb{Z}^2$ which may or may not be part of the image.

The two primary operations are called dilation and erosion. The dilation of $A \subseteq \mathbb{Z}^2$ by E is denoted $A \oplus E$, and the erosion of A by E is denoted $A \ominus E$. These operations are defined as follows

DILATION: $A \oplus E = \{x \in \mathbb{Z}^2 \mid \exists y \in (x + E^*) \cdot y \in A\}$,

EROSION: $A \ominus E = \{x \in \mathbb{Z}^2 \mid \forall y \in (x + E) \cdot y \in A\}$,

where $E^* = \{-e \mid e \in E\}$. The set E^* is usually called a reflection of E, but is effectively a half-turn rotation about the origin.

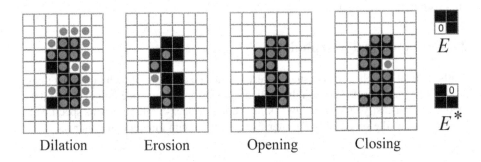

Dilation   Erosion   Opening   Closing

Fig. 2. Morphological Operations

Figure 2 illustrates these operations. The set A is shown as black pixels and the pixels present in the dilation are indicated by the grey discs on the leftmost image. The remaining three images in the figure show respectively the erosion, the opening and the closing. These last two operations arise by combining erosion and dilation as follows:

OPENING: $A \circ E = (A \ominus E) \oplus E$,

CLOSING: $A \bullet E = (A \oplus E) \ominus E$.

From a granularity perspective, opening and closing provide coarser descriptions of the set A. The opening describes A as closely as possible using not the individual pixels but by fitting (possibly overlapping) copies of E within A. The closing describes the complement of A by fitting copies of E^* outside A. The actual set is always contained within these two extremes: $A \circ E \subseteq A \subseteq A \bullet E$, and the informal notion of fitting copies of E, or of E^*, within a set is made precise in these equations:

$$A \circ E = \bigcup \{x + E \mid x + E \subseteq A\},$$

$$A \bullet E = X - \bigcup \{x + E^* \mid x + E^* \subseteq (X - A)\}.$$

The literature contains alternative definitions of these operations. For example, it is possible to have a different dilation by using E instead of E^*, which initially seems simpler but requires modified definitions of opening and closing. This approach is found in Bloch [1, p1488] which I discuss later.

Mathematical morphology has been situated in a more general context than pixels arranged in a rectangular grid. The use of lattice theory is well established [6], and much of the theory is known to be expressible by defining dilations and erosions to be arbitrary join preserving and meet preserving functions between complete lattices. Dilations are associated to erosions via the concept of adjunction or Galois connection. However, the possibility of developing a comprehensive framework based on relations, as in Section 5 below, does not seem to have appeared in the literature.

3   Rough Sets

A set consists of distinct individuals. Taking a coarser or less detailed view of these individuals, we can (for certain applications) group then into clusters where each individual belongs to exactly one cluster.

If we have a selection of some of the individuals, we can attempt to describe this selection in a coarser way. If we limit ourselves to saying what is in or not in our selection at the level of clusters, there are two approximate descriptions of the selection available. That is, we can select

- those clusters containing one or more of the selected individuals, or
- those clusters containing only selected individuals and no others.

In more formal language, we start with a set X and then impose an equivalence relation on it. This can be described as a function $r : X \to X/R$, where X/R denotes the set of equivalence classes. The function r takes each individual to the set of individuals in the cluster to which it belongs. Then for each $A \subseteq X$ two subsets provide approximate descriptions of A:

UPPER APPROXIMATION:	$A \boxplus r = \{x \in X \mid \exists y \in r(x) \cdot y \in A\},$
LOWER APPROXIMATION:	$A \boxminus r = \{x \in X \mid \forall y \in r(x) \cdot y \in A\}.$

The actual set A is contained between these approximations: $A \boxminus r \subseteq A \subseteq A \boxplus r$.

A rough set is defined as a pair $\langle A \boxminus r, A \boxplus r \rangle$ and can be seen as a generalization of the usual notion of a subset to one where there is some imprecision as to exactly what the elements of the subset are. That is, for some elements of the larger set we do not know whether they are in the subset or not, while for other elements this is known. One interpretation of these structures is as fuzzy sets for a 3-valued logic: elements in $A \boxminus r$ are definitely in, elements in $(A \boxplus r) - (A \boxminus r)$ are possibly in, and elements in $X - (A \boxplus r)$ are definitely out.

The idea of rough sets can be generalized in various ways. One possibility is to replace the equivalence relation by an arbitrary binary relation, or equivalently an arbitrary function $r : X \to \mathcal{P}X$. This is discussed, for example, in [17, p296] under the name 'generalized approximation space'. We shall see later that mathematical morphology can be developed for binary relations on sets, and that by doing so rough sets arise as a special case.

4 Algebraic Notions

In this section we gather together various algebraic concepts we need to relate rough sets to mathematical morphology.

4.1 Relations

A relation R on a set X can be defined formally as a subset of the cartesian product $X \times X$. These structures can be visualized by drawing the elements of X as dots and each $(x, y) \in R$ as an arrow from x to y. The infix notation $x \, R \, y$ is usually adopted in place of $(x, y) \in R$.

Two operations on relations are important for our purposes. Given relations R and S on the same set X their composite $R \, ; S$ is defined

$$R \, ; S = \{(x, z) \in X \times X \mid \exists y \in X \cdot x \, R \, y \text{ and } y \, S \, z\}.$$

Visually, the arrows of $R \, ; S$ correspond to pairs of arrows, one in R followed by one in S. Note that composition is written here in the 'diagrammatic' order which is opposite to the notation $S \circ R$ which is often seen. The other operation we need is the converse, denoted R^*, and defined by

$$R^* = \{(x, y) \in X \times X \mid y \, R \, x\}.$$

Visually, taking the converse corresponds to reversing all the arrows. Note the interaction between converse and composition: $(R \, ; S)^* = S^* \, ; R^*$.

4.2 Monoid Actions

A monoid is a set equipped with an associative binary operation, and an identity for that operation. If M is a monoid, with identity 1, and the operation is written as juxtaposition, then for all $m, m_1, m_2, m_3 \in M$ we have

$$m1 = m = 1m$$
$$(m_1 m_2)m_3 = m_1(m_2 m_3)$$

For any set X, the set $\mathcal{R}X$ of all relations on X is a monoid with ; (composition) as the binary operation and the relation $x \sim y$ iff $x = y$ as the identity element, 1. Another example of a monoid is the set $\mathbb{N}$ of natural numbers under addition, with the number 0 as the identity. For mathematical morphology, the monoid $\mathbb{Z}^2$ of two-dimensional vectors under addition with the zero vector as identity is particularly important. Of course this example is not only a monoid, but also a group, since there is an inverse operation, but this additional structure is not essential in order to define dilations and erosions.

Given a monoid M and a set X we say that M acts on X when each $m \in M$ gives rise to a function from X to itself and the collection of these operations works coherently with the operation in M. Formally, a left action of M on X is a function $\cdot : M \times X \to X$ such that

1. $m_1 \cdot (m_2 \cdot x) = (m_1 m_2) \cdot x$ for all $m_1, m_2 \in M$ and all $x \in X$.
2. $1 \cdot x = x$ for all $x \in X$.

and a right action of M on X is a function $\cdot : X \times M \to X$ such that

1. $(x \cdot m_1) \cdot m_2 = x \cdot (m_1 m_2)$ for all $m_1, m_2 \in M$ and all $x \in X$.
2. $x \cdot 1 = x$ for all $x \in X$.

Any monoid acts on itself by taking the action to be the binary operation. For another example, consider a clock registering hours and minutes. From any given initial setting (i.e. time showing on the clock) any number of minutes later will produce a certain configuration of the hands. Here the monoid $(\mathbb{N}, 0)$ acts on the set of positions on the clock. In this example different elements of the monoid may have the same effect.

In many cases of interest, a monoid acts not just on a set X, but on a set equipped with some additional structure. In particular X may be a lattice, where for each $A \subseteq X$ we can form the join, or least upper bound, $\bigvee A$, and the meet, or greatest lower bound, $\bigwedge A$. In this case, we say that a right action of M on X is join preserving if $(\bigvee A) \cdot m = \bigvee_{a \in A}(a \cdot m)$. The notion of meet preserving is defined analogously.

4.3 Adjunctions

Imagine two sets, M and T, where M is a set of machines partially ordered by cost of use, and T is a set of sets of tasks partially ordered by inclusion. There is a function $g : M \to T$, taking each machine to the set of tasks it performs. This function is assumed to be order preserving, that is $m_1 \leqslant m_2 \Rightarrow gm_1 \leqslant gm_2$ for all $m_1, m_2 \in M$. Given a set, t, of tasks in T, we want to find the cheapest machine that does at least t. There may be no machine m such that $gm = t$, so we want the best m such that $t \leqslant gm$. If such a best machine exists for every set of tasks, we have a function $f : T \to M$ such that for every $t \in T$ and every $m \in M$, $t \leqslant gft$, and $t \leqslant gm \Rightarrow ft \leqslant m$. This function, f, something weaker than an inverse to g, and is called a left adjoint.

Definition 1. *Let $g : Y \to X$ be an order preserving function between posets. An order preserving function $f : X \to Y$ is a **left adjoint** to g if for all $x \in X$ and all $y \in Y$, we have (1) $x \leqslant gfx$, and (2) $x \leqslant gy \Rightarrow fx \leqslant y$.*

The notation $f \dashv g$ is common for f is left adjoint to g. Instead of saying f is left adjoint to g we can say g is right adjoint to f. It can be shown that, when they exist, right and left adjoints are unique, so we can refer to 'the right adjoint' etc. The whole situation, f, g and X and Y is called an adjunction. There are many equivalent formulations of the definition of adjunction, the next result contains two useful ones.

Lemma 1. *Let $f : X \to Y$ and $g : Y \to X$ be order preserving functions between posets. Each of the following two conditions (quantified over all $x \in X$ and all $y \in Y$) is equivalent to f being left adjoint to g.*

> *a) $x \leqslant gfx$, and $fgy \leqslant y$,*   *b) $x \leqslant gy \Leftrightarrow fx \leqslant y$.*

The next result provides a practical means of calculating adjoints to a given function.

Lemma 2. *Suppose the posets X and Y have joins and meets for all subsets. A function $f : X \to Y$ has a right adjoint iff f is join preserving. In this case the right adjoint, g, is given by $gy = \bigvee\{x \in X \mid fx \leqslant y\}$. Dually, a function $g : Y \to X$ has a left adjoint* *iff g is meet preserving. In this case the left adjoint, f, is given by $fx = \bigwedge\{y \in Y \mid x \leqslant gy\}$.*

5   A Relational Account of Mathematical Morphology

5.1   Structuring Elements Describe Relations

Conceptually, a spatial domain consists of primitive entities, which are in some sense places, together with a way of saying how these places relate spatially to each other. One way is by the imposition of a topology on the set of primitive entities; another example would be a metric, assigning to each pair of places the distance between them. Systems of qualitative spatial reasoning such as the RCC use a binary relation of connection between the primitive places, which are thought of as spatial regions. One approach to discrete space [4] uses a relation of adjacency between primitive cells.

In the case of mathematical morphology for image processing we have already noted that it is conventional to identify the set of pixels with $\mathbb{Z}^2$. It is advantageous, however, to separate the underlying set of pixels from the group of translations in the plane. This type of separation leads naturally to the algebraic context of a group action on a set, as in [12]. The set provides the primitive places and the action relates them spatially. However, it is not necessary to have the full group structure in order to be able to treat structuring elements and the erosions and dilations to which they give rise. The weaker structure of a monoid action is sufficient for the basic features of mathematical morphology. In fact, although the details will not be given here, the still weaker notion of a partial monoid action [7] can be used, and is necessary for bounded images where an element of the monoid might 'move pixels beyond the image boundary'.

Given a set X and a monoid M with action $\cdot : X \times M \rightarrow X$, a subset $E \subseteq M$ gives rise to a relation, R, on X by $x\, R\, y$ iff $y = x \cdot e$ for some $e \in E$. When X is a set of pixels in the plane equipped with the usual action of $\mathbb{Z}^2$, and a structuring element provides $E \subseteq \mathbb{Z}^2$, the relation can be interpreted as follows. For $x, y \in X$ we have $x\, R\, y$ if when the origin of E is positioned at x then y is 'under one of the pixels in E'. It is this relation, rather than the monoid action itself, that provides the key structure which enables the morphological operations to be defined. Working just with a relation is more general, as not every relation arises from a monoid action, and the following parts of this section show how the basic features of mathematical morphology can all be derived in this setting.

5.2   Dilations and Erosions from Relations

Let R be a relation on a set X. For each $x \in X$ we have the set of places we can get to by following the arrows. More generally, for any subset $A \subseteq X$ we have the set of places we can get to from at least one of the elements of A. Denoting this subset by $A \oplus R$ we obtain the definition of the dilation operation:

$$A \oplus R = \{x \in X \mid \exists a \in A \cdot a\, R\, x\}.$$

Given two relations on X, say R and S, the places you can get to following an arrow in R then following an arrow in S are exactly the places you can get to by following an arrow in $R\,;S$. Formally, we have a right monoid action:

Lemma 3. *The operation just defined,* $\oplus$, *provides a right action for the monoid of relations on* X *on the power set. Specifically, for all* $A \in \mathcal{P}X$ *and all* $R, S \in \mathcal{R}X$,

$$A \oplus 1 = A,$$
$$A \oplus (R \,;\, S) = (A \oplus R) \oplus S.$$

The action also preserves joins (unions of subsets): $(\bigcup A_i) \oplus R = \bigcup (A_i \oplus R)$.

Because the operation $- \oplus R$ preserves joins, it has a right adjoint which we shall write as $R \ominus -$, and which will be called erosion. The erosion can be constructed, using Lemma 2, as

$$R \ominus A = \bigcup \{B \in \mathcal{P}X \mid B \oplus R \subseteq A\}.$$

Thinking of R as representing an accessibility relation, $R \ominus A$ consists of those elements of X from which nowhere outside A is accessible, i.e. $X - A$ is inaccessible. Note that elements from where nowhere is accessible lie in $R \ominus A$ for every A. The two functions are illustrated in the following diagram for a set A with three elements and the relation as indicated.

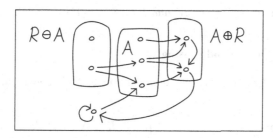

The next result follows readily from the properties of adjunctions.

Lemma 4. *The operation* $\ominus$ *is a meet preserving (the meets in this case being intersections of subsets) left action of the monoid* $\mathcal{R}X$ *on the lattice* $\mathcal{P}X$. *Thus the following equations are satisfied*

$$1 \ominus A = A,$$
$$(R \,;\, S) \ominus A = R \ominus (S \ominus A),$$

and also $R \ominus (\bigcap A_i) = \bigcap (R \ominus A_i)$.

The importance of adjunctions is already well-known in mathematical morphology [6], but relations figure rarely in the literature (two exceptions are [8] and [9]), and their fundamental importance does not seem to have been appreciated. The approach of defining erosions and dilations to be adjoint pairs is (when working with the powerset lattice) exactly equivalent to working with relations. This is because relations are equivalent to union preserving endofunctions on $\mathcal{P}X$, and every adjoint pair between $\mathcal{P}X$ and $\mathcal{P}X$ has the form $- \oplus R \dashv R \ominus -$. For given any union preserving $f : \mathcal{P}X \to \mathcal{P}X$ we can define a relation R on X by $x \, R \, y$ iff $y \in f\{x\}$. We can then recapture f from R since $f = - \oplus R$. One advantage of relations is that they provide a practical means of describing each adjunction on $\mathcal{P}X$ as a simple structure on X itself.

5.3 Anti-dilations and Anti-erosions

Dilations and erosions using not R but its converse provide what are called anti-dilations and anti-erosions. These can also be obtained by composing the adjoint pair $- \oplus R \dashv R \ominus -$ with two other adjunctions formed from the complement on the powerset as in the following diagram, where $(\mathcal{P}X)^{\mathrm{op}}$ denotes the opposite lattice (the same elements with the reversed ordering). These two adjunctions are in fact isomorphisms since $\neg\neg A = A$ for any $A \in \mathcal{P}X$.

$$(\mathcal{P}X)^{\mathrm{op}} \quad \underset{\neg}{\overset{\neg}{\underset{\cong}{\rightleftarrows}}} \quad \mathcal{P}X \quad \underset{R\ominus -}{\overset{-\oplus R}{\underset{\dashv}{\rightleftarrows}}} \quad \mathcal{P}X \quad \underset{\neg}{\overset{\neg}{\underset{\cong}{\rightleftarrows}}} \quad (\mathcal{P}X)^{\mathrm{op}}$$

This gives us a join preserving left action and its adjoint, a meet preserving right action, defined by $R \oplus^* A = A \oplus R^*$ and $A \ominus^* R = R^* \ominus A$. The definitions of $\oplus$ and $\ominus$ easily lead to the following Lemma.

Lemma 5. $A \oplus R = \neg((\neg A) \ominus^* R)$ *and* $R \ominus A = \neg(R \oplus^* (\neg A))$.

5.4 Relational Opening and Closing

The adjoint pair gives rise to two closure operators on $\mathcal{P}X$. These can be described as follows:

Lemma 6 (opening). $(R \ominus A) \oplus R = \bigcup\{B \oplus R \subseteq X \mid B \oplus R \subseteq A\}$

Proof.   Use the description that $R \ominus A = \bigcup\{B \subseteq X \mid B \oplus R \subseteq A\}$ and the fact that the operation $- \oplus R$ is union preserving. $\square$

Lemma 7 (closing). $R \ominus (A \oplus R) = \neg \bigcup\{R \oplus^* B \subseteq X \mid R \oplus^* B \subseteq \neg A\}$

Proof.   From $A \oplus R = \bigcap\{B \subseteq X \mid A \subseteq R \ominus B\}$ and $R \ominus B = \neg(R \oplus^* \neg B)$, we get $\neg(A \oplus R) = \bigcup\{\neg B \subseteq X \mid R \oplus^* \neg B \subseteq \neg A\} = \bigcup\{B \subseteq X \mid R \oplus^* B \subseteq \neg A\}$. Hence $\neg(R \oplus^* \neg(A \oplus R)) = \neg \bigcup\{R \oplus^* B \subseteq X \mid R \oplus^* B \subseteq \neg A\}$, and the result follows by Lemma 5. $\square$

These results show that the idea of opening as fitting copies of the structuring element within A and the idea of closing as fitting copies of its rotated form outside A holds in the relational context where there is no spatial structure on the underlying set. Conceptually, the imposition of the relation R on the underlying set X moves us from describing subsets of X in terms of the individual elements of X to a coarser or filtered view in which we can no longer see the individual elements $x \in X$, but where we do have access to sets of the form $\{x\} \oplus R$ and of the form $R \ominus \{x\}$. We can think of sets of the form $\{x\} \oplus R$ as neighbourhoods induced by R, and any $A \oplus R$ can be obtained as a union of these neighbourhoods. In the special case of R being the identity relation on X, the neighbourhoods do give us the singleton subsets, or elements, of X.

5.5   Rough Sets in the Framework

We get the basic features of rough set theory by taking the special case that R is an equivalence relation.

Lemma 8. *For any set X, and subset $A \subseteq X$, and R any equivalence relation on X, $A \oplus R$ is the upper approximation of A with respect to R, and $R \ominus A$ is the lower approximation of A.*

While the upper and lower approximations can be obtained in this way, and this is essentially the approach taken by Bloch [1], it is more useful to see them as openings and closings.

Theorem 9. *For an equivalence relation, R, we have $(R \ominus A) \oplus R = R \ominus A$ and $R \ominus (A \oplus R) = A \oplus R$, so the upper approximation of A by R is the closing of A by R and the lower approximation is the opening.*

This makes it clear that the morphological interpretation of opening as the union of all the sets of the form $\{x\} \oplus R$ fitting inside the set A includes the lower approximation of rough set theory because when R is an equivalence relation these sets are just the equivalence classes. The upper approximation of A is generally seen as the union of those equivalence classes intersecting A, but we can take the equivalent formulation as the complement of the union of the equivalence classes disjoint from A. This reveals the upper approximation as just a special case of the morphological interpretation of closing, since for an equivalence relation $R^* = R$.

This essentially makes rough set theory the special case of mathematical morphology in which the relation is an equivalence relation. The account in [1] gets close to this connection, but by using dilations paired with anti-erosions and anti-dilations paired with erosions, rather than the adjoint pairs, does not reveal the full picture.

6   Graphs

We now see how the operations of dilation and erosion can be extended from sets to graphs. It should be noted that the term 'graph morphology' has already been used by Heijmans and Vincent [9], but this is something different from our concern here. Essentially Heijmans and Vincent use 'graph' to mean a symmetric relation but their approach is not the same as in Section 5 above. The term 'graph' will be used here to mean a directed graph where multiple edges and loops are permitted. It is convenient to formalize these structures in the way found in [3], where they are called reflexive digraphs.

Definition 2. *A **graph** consists of a set G with source and target functions $s, t : G \to G$ such that for all $g \in G$ we have $s(t(g)) = t(g)$ and $t(s(g)) = s(g)$.*

The elements of G are the nodes and the edges of the graph, and the nodes can be distinguished as those elements g for which $s(g) = g = t(g)$. The subgraphs of G in this setting are those subsets of G which are closed under the source and target functions.

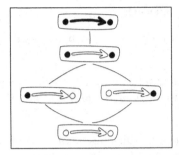

Fig. 3. The lattice of subgraphs of a graph with one edge and two nodes

6.1   Subgraphs

The set of all subsets of a set forms a boolean algebra, but the set of all subgraphs of a graph will not necessarily have this form. The structure is a bi-Heyting algebra, as discussed by Stell and Worboys [15], and illustrated by an example in Figure 3. An important feature is that there are two distinct operators similar to the complement in a boolean algebra. One is the pseudocomplement characterized by $A \wedge B = 0$ iff $B \leqslant \neg A$, so that $\neg A$ is the largest subgraph disjoint from A. The other is the dual pseudo-complement or supplement $\sim A$ characterized by $A \vee B = 1$ iff $\sim A \leqslant B$, so that $\sim A$ is the smallest subgraph it is necessary to add to A to get the whole graph. The fact that neither $\sim\sim A$ nor $\neg\neg A$ need equal A makes some aspects of the morphological operations on graphs more complex than in the case of sets.

6.2   Relations on Graphs

The appropriate notion of a relation on a graph for mathematical morphology is just an ordinary relation on the underlying set which interacts in the right way with the source and target.

Definition 3. *A relation, R, on the set of elements of a graph G is **graphical** if for all $g, h \in G$, it satisfies*

1. *if $g\,R\,h$ then $g\,R\,s(h)$ and $g\,R\,t(h)$, and*
2. *if $s(g)\,R\,h$ or $t(g)\,R\,h$ then $g\,R\,h$.*

That this is the right definition is justified in Theorem 10 by the fact that such relations correspond to the join preserving functions on the lattice, $\mathcal{P}G$, of subgraphs of G. This means that they provide adjoint pairs just as in the case of sets. It is not hard to see that the set of all graphical relations on a graph is closed under composition of relations and that the identity graphical relation, I, is given by $x\,I\,y \Leftrightarrow y \in \{s(x), x, t(x)\}$.

Given a graphical relation R on G and a subgraph $H \leqslant G$ we immediately obtain a subset $H \oplus R \subseteq G$ by using the dilation for sets. However, since R is graphical it follows that $H \oplus R$ is a subgraph. The dilation operation on subsets of a set using arbitrary relations can thus be restricted to one on subgraphs using graphical relations, and it is straightforward to check that this gives a right action of the monoid of graphical relations on the set of all subgraphs of G. The dilation operation on subgraphs is join preserving because joins of subgraphs are just unions of the underlying sets of elements and dilation for sets preserves these unions.

There is a one-to-one correspondence between graphical relations on G and join preserving functions on the lattice $\mathcal{P}G$. Given a join preserving function, $\delta : \mathcal{P}G \rightarrow \mathcal{P}G$, we can define a graphical relation R by $g\,R\,h$ iff $h \in \delta\{s(g), g, t(g)\}$. It then follows that $- \oplus R = \delta$. In the other direction, starting with a graphical relation R, and obtaining the join preserving function $- \oplus R$, we can define a relation R' by $g\,R'\,h$ iff $h \in \{s(g), g, t(g)\} \oplus R$. The relations R and R' are equal, for if $g\,R\,h$ then $h \in \{g\} \oplus R$ so $g\,R'\,h$, and if $g\,R'\,h$ then $g\,R\,h$ as $s(g)\,R\,h$ implies $g\,R\,h$ and similarly for $t(g)$.

The above discussion establishes the main points of the following result.

Theorem 10. *On any graph G the graphical relations form a monoid, with composition as for arbitrary relations and identity $x\,I\,y \Leftrightarrow y \in \{s(x), x, t(x)\}$. For any subgraph $H \leqslant G$ and any graphical relation R we have a subgraph $H \oplus R$ and this assignment is a right action of the monoid of graphical relations on $\mathcal{P}G$. The assignment $R \mapsto - \oplus R$ is an isomorphism of the monoid of graphical relations with the monoid of join preserving functions on $\mathcal{P}G$ under function composition.*

6.3   Presentations for Graphical Relations

The relations identified as graphical by Definition 3 can be drawn with arrows in the usual way. As the elements of the graphs are the nodes together with the edges we can have arrows in the relation which go from edges to nodes or vice versa as well as between elements of the same kind. The conditions on a graphical relation would require that whenever we draw an arrow from one edge to another we also draw arrows from the first edge to the source and target of the second edge. The second condition for a relation to be graphical requires further arrows to be present. When all the required arrows are drawn, diagrams can quickly become rather complex. It is more convenient to use a simplified representation in which some arrows are omitted – that is we draw a relation on the elements of the graph from which a unique graphical relation can be generated. From an arbitrary relation, R, on G we can generate the least graphical relation containing R by adding the appropriate arrows. Thus in the examples which follow, the relations drawn are not actually graphical but they are presentations for graphical relations and the presented graphical relation can easily be derived by adding the required arrows.

Two examples are provided in Figure 4 where the edges of the graph are single arrows and the arrows of the relation are double shafted arrows. In the example on the left, the corresponding graphical relation would have four additional arrows: each of the uppermost edges being related to each of the nodes D and E. The advantage of using a simplified representation is evident even in this straightforward example.

The same example shows the flexibility of the relations we work with. The subgraph containing B alone is not related to anything (or alternatively is mapped to the empty subgraph). In terms of applications this allows us to relate two stages in a network (perhaps train routes between stations) to a single stage in another network without being forced to relate the intermediate node to anything.

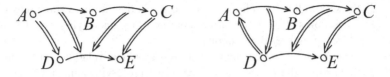

Fig. 4. Presentations of graphical relations

6.4   Morphological Operations on Graphs

Given a graphical relation R on a graph G, the corresponding join preserving function from Theorem 10 provides the dilation operation for graphs. This operation takes a subgraph H to its dilation, the subgraph $H \oplus R$. Because this is a join preserving function we automatically obtain, via Lemma 2, the erosion operation on graphs as

$$R \ominus H = \bigvee \{K \in \mathcal{P}G \mid K \oplus R \leqslant H\}.$$

where $\mathcal{P}G$ denotes the lattice of all subgraphs of G. We now examine some examples which illustrate these two operations, and the opening and closing on graphs which are obtained from them.

In Figure 5 we have a graph G with the edges shown by single shafted arrows, and a graphical relation R on G presented by an ordinary relation drawn with the double shafted arrows. The right hand side of Figure 5 shows a subgraph H indicated by the heavy edges and nodes.

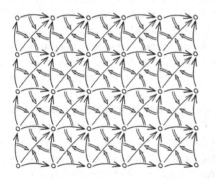
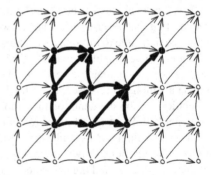

Fig. 5. A presentation of a graphical relation R on a graph G (left) and a subgraph H of G (right)

In Figure 6 the dilation and erosion of H by R are shown as the heavy edges and nodes. In this example the dilation has one of the diagonal edges in the graph whenever at least one of its attendant horizontal or vertical edges lies in H. The erosion has one of the horizontal or vertical edges if and only if both the diagonal edges on either side of it are present in H, but all the diagonal edges and all the nodes are included in the erosion.

In relating this example to classical mathematical morphology, it is the nodes of the underlying graph G that should be identified with the pixels. The subgraph H is not

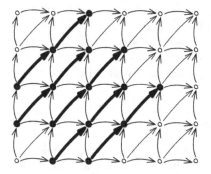 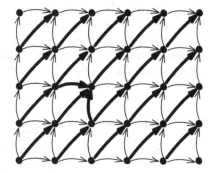

Fig. 6. The dilation $H \oplus R$ (left) and the erosion $R \ominus H$ (right)

given by just taking a subset of the pixels, it also involves a choice of which edges in the graph are selected too. Every subgraph consisting entirely of nodes, with no edges, will be mapped by the dilation to the empty subgraph. We can interpret this as saying that regarded purely as a set of pixels, the morphological operations for this particular graphical relation are unable to distinguish one set from another – only when the additional spatial structure is provided can we distinguish between subgraphs.

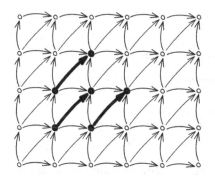 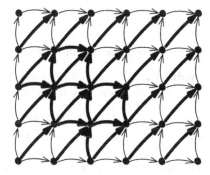

Fig. 7. The opening $(R \ominus H) \oplus R$ (left) and the closing $R \ominus (H \oplus R)$ (right)

The opening and closing are shown in Figure 7. It follows from basic properties of adjunctions in Lemma 1(a) that H lies between its opening and its closing. The opening of H by R is still expressible as the join of all the subgraphs of the form $K \oplus R$ contained within H,

$$(R \ominus H) \oplus R = \bigvee \{K \oplus R \leqslant G \mid K \oplus R \leqslant H\},$$

but it is not immediately obvious how to obtain an analogue of the description of closing for the case of sets as in Lemma 7. The difficulty in doing this lies in the use of anti-dilation and the negation both of which have significantly weaker properties in the context of graphs.

Lemma 7 is only one possible granular interpretation of the closing operation, and an alternative does provide one that remains valid when considering graphs rather than merely sets. Whether we are working with sets or with graphs, it can be shown that the closing of H under R is the largest K such that $K \oplus R = H \oplus R$. We can think of the relation R as coarsening our view so that entities with the same dilation are indistinguishable, and the closing as picking out a particular one from this collection of indistinguishables. Dually, opening H provides the least entity with the same erosion as H.

6.5 Anti-dilations and Anti-erosions for Graphs

We have seen how in providing a relational account of mathematical morphology on sets, the converse of a relation plays a significant role. In the set case the anti-erosion and anti-dilation operations admit two descriptions: one in terms of the erosion or dilation for the converse and the other by composing the ordinary dilation or erosion with the boolean negation. In the case of graphs, the presence of two kinds of negation, $\neg$ and $\sim$, makes anti-dilation and anti-dilation more subtle. These two kinds of negation are no longer isomorphisms of $\mathcal{P}G$, the lattice of subgraphs of a graph, with its opposite $(\mathcal{P}G)^{op}$. However they do provide adjunctions as in the following diagram.

$$
(\mathcal{P}G)^{op} \xrightleftharpoons[\sim]{\sim} \mathcal{P}G \xrightleftharpoons[R\ominus-]{-\oplus R} \mathcal{P}G \xrightleftharpoons[\neg]{\neg} (\mathcal{P}G)^{op}
$$

Hence we can define anti-dilation and anti-erosion for graphs.

Definition 4. *For a graph G carrying a graphical relation, R, the anti-dilation and anti-erosion of a subgraph H are given by*

$$R \oplus^* H = \sim(R \ominus \neg H) \quad \text{for the anti-dilation, and}$$
$$H \ominus^* R = \neg((\sim H) \oplus R) \quad \text{for the anti-erosion.}$$

These do provide a pair of adjoint operations, but properties familiar from classical morphology may only hold in weaker versions. The following result shows that the anti-opening has the expected form, but we cannot replace $\sim K$ by K in the anti-closing, since not every subgraph is of the form $\sim K$.

Theorem 11. *For a graphical relation R on a graph G, the anti-opening and anti-closing of any subgraph H can be expressed as follows*

$$R \oplus^* (H \ominus^* R) = \bigvee \{R \oplus^* K \mid R \oplus^* K \leqslant H\},$$
$$(R \oplus^* H) \ominus^* R = \neg \bigvee \{(\sim K) \oplus R \mid (\sim K) \oplus R \leqslant \neg H\}.$$

The anti-dilation provides a means of defining a notion of converse for graphical relations. We can define the converse, R^*, of R to be the graphical relation which corresponds (via Theorem 10) to the function $H \mapsto \sim(R \ominus \neg H)$, but the details of how this appears as a relation on the underlying set of the graph are beyond the scope of the present paper. It can, however, be observed that this converse is much weaker than that found in the case of relations on sets, and, in particular,the property $(R^*)^* = R$ no longer holds. In fact, neither $((R^*)^*)^* = R^*$ nor $(R\,;S)^* = S^*\,;R^*$ holds in general, and the algebraic structure is evidently weaker than a monoid action for the anti-dilation and anti-erosion.

7   Conclusions and Further Work

This paper has shown how rough sets relate to mathematical morphology thus settling the issue first raised by Bloch [1]. In doing this, mathematical morphology has been developed in the context of a relation on a set. It has been shown that this structure is sufficient to define all the basic operations: dilation, erosion, opening and closing, and also to establish their most basic properties. Of course there are many aspects of mathematical morphology which do depend essentially on a richer geometric structure. Many of these require a group action rather than a monoid action, and in some cases properties of the action (e.g. transitivity) will be significant. Further work is needed to establish how much of the very extensive body of results in mathematical morphology can benefit from a relational approach. One area for extension is to images which are grey scale rather than just black and white; these would be expected to use relations taking values in a lattice of fuzzy truth values.

The notions of dilation, erosion, opening and closing for graphical relations described here appear to be novel, and there is further work needed in order to determine the algebraic properties of these operations. In terms of practical applications, a theory of rough graphs, perhaps in connection with other forms of granularity for graphs [16,14], might be used to model networks (for example in gas, water or electricity distribution) at multiple levels of detail.

Acknowledgements. John Stell is partially supported by EPSRC grant EP/C014707/1 Mapping the Underworld: Knowledge and Data Integration.

References

1. Bloch, I.: On links between mathematical morphology and rough sets. Pattern Recognition 33, 1487–1496 (2000)
2. Bloch, I.: Spatial reasoning under imprecision using fuzzy set theory, formal logics and mathematical morphology. International Journal of Approximate Reasoning 41, 77–95 (2006)
3. Brown, R., Morris, I., Shrimpton, J., Wensley, C.D.: Graphs of Morphisms of Graphs. Bangor Mathematics Preprint 06.04, Mathematics Department, University of Wales, Bangor (2006)
4. Galton, A.: The mereotopology of discrete space. In: Freksa, C., Mark, D.M. (eds.) COSIT 1999. LNCS, vol. 1661, pp. 251–266. Springer, Heidelberg (1999)
5. Gonzalez, R.C., Woods, R.E.: Digital Image Processing. Prentice-Hall, Englewood Cliffs (2001)
6. Heijmans, H.: Mathematical morphology: A modern approach in image processing based on algebra and geometry. SIAM Review 37, 1–36 (1995)
7. Hollings, C.: Partial actions of monoids. Semigroup Forum (to appear)
8. Hsueh, Y.C.: Relation-based variations of the discrete Radon transform. Computers and Mathematics with Applications 31, 119–131 (1996)

9. Heijmans, H., Vincent, L.: Graph Morphology in Image Analysis. In: Dougherty, E.R. (ed.) Mathematical Morphology in Image Processing, ch. 6. Marcel Dekker, pp. 171–203 (1993)

10. Müller, J.C., Lagrange, J.P., Weibel, R. (eds.): GIS and Generalisation: Methodology and Practice. Taylor and Francis, London (1995)

11. Orłowska, E. (ed.): Incomplete Information – Rough Set Analysis. Studies in Fuzziness and Soft Computing, vol. 13. Physica-Verlag, Heidelberg (1998)

12. Roerdink, J.B.T.M.: Mathematical Morphology with Non-Commutative Symmetry Groups. In: Dougherty, E.R. ed.: Mathematical Morphology in Image Processing, ch. 7. Marcel Dekker, pp. 205–254 (1993)

13. Serra, J.: Image Analysis and Mathematical Morphology. Academic Press, London (1983)

14. Stell, J.G.: Granulation for graphs. In: Freksa, C., Mark, D.M. (eds.) COSIT 1999. LNCS, vol. 1661, pp. 417–432. Springer, Heidelberg (1999)

15. Stell, J.G., Worboys, M.F.: The algebraic structure of sets of regions. In: Frank, A.U. (ed.) COSIT 1997. LNCS, vol. 1329, pp. 163–174. Springer, Heidelberg (1997)

16. Stell, J.G., Worboys, M.F.: Generalizing graphs using amalgamation and selection. In: Güting, R.H., Papadias, D., Lochovsky, F.H. (eds.) SSD 1999. LNCS, vol. 1651, pp. 19–32. Springer, Heidelberg (1999)

17. Yao, Y.Y.: Two views of the theory of rough sets in finite universes. International Journal of Approximate Reasoning 15, 291–317 (1996)

Author Index

Lecture Notes in Computer Science

For information about Vols. 1–4416

please contact your bookseller or Springer

Vol. 4580: B. Ma, K. Zhang (Eds.), Combinatorial Pattern Matching. XII, 366 pages. 2007.

Vol. 4576: D. Leivant, R. de Queiroz (Eds.), Logic, Language, Information and Computation. X, 363 pages. 2007.

Vol. 4547: C. Carlet, B. Sunar (Eds.), Arithmetic of Finite Fields. XI, 355 pages. 2007.

Vol. 4546: J. Kleijn, A. Yakovlev (Eds.), Petri Nets and Other Models of Concurrency – ICATPN 2007. XI, 515 pages. 2007.

Vol. 4545: H. Anai, K. Horimoto, T. Kutsia (Eds.), Algebraic Biology. XIII, 379 pages. 2007.

Vol. 4533: F. Baader (Ed.), Term Rewriting and Applications. XII, 419 pages. 2007.

Vol. 4528: J. Mira, J.R. Álvarez (Eds.), Nature Inspired Problem-Solving Methods in Knowledge Engineering, Part II. XXII, 650 pages. 2007.

Vol. 4527: J. Mira, J.R. Álvarez (Eds.), Bio-inspired Modeling of Cognitive Tasks, Part I. XXII, 630 pages. 2007.

Vol. 4525: C. Demetrescu (Ed.), Experimental Algorithms. XIII, 448 pages. 2007.

Vol. 4514: S.N. Artemov, A. Nerode (Eds.), Logical Foundations of Computer Science. XI, 513 pages. 2007.

Vol. 4513: M. Fischetti, D.P. Williamson (Eds.), Integer Programming and Combinatorial Optimization. IX, 500 pages. 2007.

Vol. 4510: P. Van Hentenryck, L.A. Wolsey (Eds.), Integration of AI and OR Techniques in Constraint Programming for Combinatorial Optimization Problems. X, 391 pages. 2007.

Vol. 4507: F. Sandoval, A. Prieto, J. Cabestany, M. Graña (Eds.), Computational and Ambient Intelligence. XXVI, 1167 pages. 2007.

Vol. 4501: J. Marques-Silva, K.A. Sakallah (Eds.), Theory and Applications of Satisfiability Testing – SAT 2007. XI, 384 pages. 2007.

Vol. 4497: S.B. Cooper, B. Löwe, A. Sorbi (Eds.), Computation and Logic in the Real World. XVIII, 826 pages. 2007.

Vol. 4494: H. Jin, O.F. Rana, Y. Pan, V.K. Prasanna (Eds.), Algorithms and Architectures for Parallel Processing. XIV, 508 pages. 2007.

Vol. 4493: D. Liu, S. Fei, Z. Hou, H. Zhang, C. Sun (Eds.), Advances in Neural Networks – ISNN 2007, Part III. XXVI, 1215 pages. 2007.

Vol. 4492: D. Liu, S. Fei, Z. Hou, H. Zhang, C. Sun (Eds.), Advances in Neural Networks – ISNN 2007, Part II. XXVII, 1321 pages. 2007.

Vol. 4491: D. Liu, S. Fei, Z.-G. Hou, H. Zhang, C. Sun (Eds.), Advances in Neural Networks – ISNN 2007, Part I. LIV, 1365 pages. 2007.

Vol. 4490: Y. Shi, G.D. van Albada, J. Dongarra, P.M.A. Sloot (Eds.), Computational Science – ICCS 2007, Part IV. XXXVII, 1211 pages. 2007.

Vol. 4489: Y. Shi, G.D. van Albada, J. Dongarra, P.M.A. Sloot (Eds.), Computational Science – ICCS 2007, Part III. XXXVII, 1257 pages. 2007.

Vol. 4488: Y. Shi, G.D. van Albada, J. Dongarra, P.M.A. Sloot (Eds.), Computational Science – ICCS 2007, Part II. XXXV, 1251 pages. 2007.

Vol. 4487: Y. Shi, G.D. van Albada, J. Dongarra, P.M.A. Sloot (Eds.), Computational Science – ICCS 2007, Part I. LXXXI, 1275 pages. 2007.

Vol. 4484: J.-Y. Cai, S.B. Cooper, H. Zhu (Eds.), Theory and Applications of Models of Computation. XIII, 772 pages. 2007.

Vol. 4475: P. Crescenzi, G. Prencipe, G. Pucci (Eds.), Fun with Algorithms. X, 273 pages. 2007.

Vol. 4474: G. Prencipe, S. Zaks (Eds.), Structural Information and Communication Complexity. XI, 342 pages. 2007.

Vol. 4459: C. Cérin, K.-C. Li (Eds.), Advances in Grid and Pervasive Computing. XVI, 759 pages. 2007.

Vol. 4449: Z. Horváth, V. Zsók, A. Butterfield (Eds.), Implementation and Application of Functional Languages. X, 271 pages. 2007.

Vol. 4448: M. Giacobini (Ed.), Applications of Evolutionary Computing. XXIII, 755 pages. 2007.

Vol. 4447: E. Marchiori, J.H. Moore, J.C. Rajapakse (Eds.), Evolutionary Computation, Machine Learning and Data Mining in Bioinformatics. XI, 302 pages. 2007.

Vol. 4446: C. Cotta, J. van Hemert (Eds.), Evolutionary Computation in Combinatorial Optimization. XII, 241 pages. 2007.

Vol. 4445: M. Ebner, M. O'Neill, A. Ekárt, L. Vanneschi, A.I. Esparcia-Alcázar (Eds.), Genetic Programming. XI, 382 pages. 2007.

Vol. 4436: C.R. Stephens, M. Toussaint, D. Whitley, P.F. Stadler (Eds.), Foundations of Genetic Algorithms. IX, 213 pages. 2007.

Vol. 4433: E. Şahin, W.M. Spears, A.F.T. Winfield (Eds.), Swarm Robotics. XII, 221 pages. 2007.

Vol. 4432: B. Beliczynski, A. Dzielinski, M. Iwanowski, B. Ribeiro (Eds.), Adaptive and Natural Computing Algorithms, Part II. XXVI, 761 pages. 2007.

Vol. 4431: B. Beliczynski, A. Dzielinski, M. Iwanowski, B. Ribeiro (Eds.), Adaptive and Natural Computing Algorithms, Part I. XXV, 851 pages. 2007.

Vol. 4424: O. Grumberg, M. Huth (Eds.), Tools and Algorithms for the Construction and Analysis of Systems. XX, 738 pages. 2007.

Vol. 4423: H. Seidl (Ed.), Foundations of Software Science and Computational Structures. XVI, 379 pages. 2007.

Vol. 4422: M.B. Dwyer, A. Lopes (Eds.), Fundamental Approaches to Software Engineering. XV, 440 pages. 2007.

Vol. 4421: R. De Nicola (Ed.), Programming Languages and Systems. XVII, 538 pages. 2007.

Vol. 4420: S. Krishnamurthi, M. Odersky (Eds.), Compiler Construction. XIV, 233 pages. 2007.

Vol. 4419: P.C. Diniz, E. Marques, K. Bertels, M.M. Fernandes, J.M.P. Cardoso (Eds.), Reconfigurable Computing: Architectures, Tools and Applications. XIV, 391 pages. 2007.